HIGH-THROUGHPUT SYNTHESIS

HIGH-THROUGHPUT SYNTHESIS

Principles and Practices

edited by
Irving Sucholeiki

Solid Phase Sciences Corporation
Medford, Massachusetts

CRC Press
Taylor & Francis Group
Boca Raton London New York

CRC Press is an imprint of the
Taylor & Francis Group, an **informa** business

CRC Press
Taylor & Francis Group
6000 Broken Sound Parkway NW, Suite 300
Boca Raton, FL 33487-2742

First issued in paperback 2019

ISBN-13: 978-0-8247-0256-4 (hbk)
ISBN-13: 978-0-367-39754-8 (pbk)

Library of Congress Cataloging-in-Publication Data

High-throughput synthesis: principles and practices / edited by Irving Sucholeiki.
    p. cm.
  Includes bibliographical references and index.
  ISBN 0-8247-0256-5 (alk. paper)
  1. Combinatorial chemistry. I. Sucholeiki, Irving
QD262 .H54 2000
547′.2—dc21   00-050928

Foreword

Preparing new compounds and finding their practical application have always been among the most important goals of chemical science. These compounds have supplied the materials for new pharmaceuticals, pesticides, dyes, plastics, catalysts, components of fragrances, etc. When it first became known that large series of compounds were needed to produce new pharmaceuticals, these compounds were prepared and tested one by one. This conventional method of research was in obvious contrast to the technical possibilities available even two decades ago, since fabricating large series of items through the use of production lines and automation was extensively applied in industry. The advent of combinatorial chemistry radically changed this situation. It brought about a revolution in pharmaceutical research that is gradually expanding to other fields in which new compounds or materials are produced. The title of this book, *High-Throughput Synthesis: Principles and Practices*, indicates the current importance of combinatorial methods.

There are two main approaches for high-throughput synthesis: one of them comprises the *real combinatorial methods*; the other includes the *parallel procedures*. Real combinatorial methods apply tricks to substantially reduce the number of operations/compounds needed. The reduced labor requirement makes it possible to prepare in less than a week more compounds than were made in the whole previous history of chemistry. These methods were responsible for the outbreak of the combinatorial revolution. In parallel procedures the number of operations needed to carry out a synthesis is practically the same as in the conventional approach, and for this reason productivity cannot be nearly as high as when the previously mentioned methods are used. The parallel execution of the reactions, however, as well as extensively applied automation and other advantages made this type of library synthesis very popular. The parallel method also proved indispensable in realization of one-pot reactions, in solution-phase syntheses, and in searching for new materials.

The capability to synthesize a large number of new compounds or to make new materials by mixing compounds is very important in the present era, since new useful compounds and materials can be found experimentally by making and testing a large number of candidates. The practical application of new compounds or composite materials depends largely on molecular interactions, and the fact that the combinatorial methods are so extensively used nowadays just reflects how incomplete our knowledge is of the rules that govern these interactions. One hopes, however, that the analysis of the huge amount of data accumulating from testing combinatorial libraries sooner or later will help us to better understand the relationship between molecular interactions and structure. It will be very important to make these data publicly available and not store them in the safes of companies. In the long run, understanding molecular interactions will eliminate the need for combinatorial chemistry and open the way for safe application of molecular design and molecular engineering. This won't happen very soon, and until it does we must stick to combinatorial chemistry. This book will certainly contribute to its more successful application. It was an excellent idea to present it in the cookbook format. Listing the chemicals and equipment used in the syntheses and description of the procedures and even the postsynthetic analyses in sufficient detail will be very useful for readers and will help them to design their own combinatorial experiments.

Árpád Furka
Department of Organic Chemistry
Eötvös Loránd University
Budapest, Hungary

Preface

This book grew out of discussions that I had in the early months of 1998 with Anita Lekhwani of Marcel Dekker, Inc. She was interested in creating a follow-up to an earlier book they had published—*High Throughput Screening,* edited by John P. Devlin—which was a review of state-of-the-art biological screening. She asked me to develop a chemistry version of that book, tentatively entitled *High Throughput Synthesis,* that would be an edited review of the field of combinatorial chemistry. Because my experience with a couple of edited review-type books in the past had been that they tended to get out of date even before they went to print, my first reaction was not very positive. Then I realized that what the field of combinatorial chemistry needed was a practical laboratory manual that would describe the mechanics of how to make large numbers of compounds. Too often, published reviews and papers leave out detailed information that, although not crucial to the overall understanding of the author's work, is important in reproducing the methods used. This book would target not only chemists in industry but also graduate students and faculty members in academia who may want to incorporate combinatorial or solid-phase chemistry methods into their graduate research program.

I had envisioned a recipe-type cookbook modeled after one of my all-time favorite undergraduate chemistry textbooks, *Vogel's Textbook of Practical Organic Chemistry* (4th ed., Longman Group, 1978). For those not having had the good fortune of using the book, it was an all-encompassing manual presenting the gamut of "how-to" methods in organic chemistry from preparing and running samples for infrared analysis to running multigram scale nitration reactions. The combinatorial cookbook would contain recipes or case studies sorted by category. For example, there would be a section called "high-throughput synthesis for drug discovery'' that would contain case studies illustrating the mechanics of making a large number of compounds for biological study. Other sections would have case studies on other aspects of producing large numbers of compounds, such as purification strategies and automation methods. Each case study would be written by a contributing chemist(s) with an emphasis on procedure and with very little theory. However, for a cookbook in combinatorial chemistry to succeed it must assume some knowledge on the part of the reader, such as familiarity with the theory behind various aspects of combinatorial chemistry and/or multiple parallel synthesis. Because most universities do not offer a course in combinatorial or solid-phase chemistry, I felt that for the book to be of benefit to graduate students and scientists not familiar with the field of combinatorial chemistry, it would need to include introductory sections or reviews covering the topics brought out by the case studies.

Therefore, this book begins with an introduction to solid-phase organic chemistry, which is followed in Chapter 2 by a series of practical case studies written by J. Manuel Perez of Solid Phase Sciences Corporation covering various qualitative and quantitative methods for determining the loading capacity on solid support. Chapter 3 provides a perspective on how combinatorial chemistry came about, by Michael Pavia of Millennium Pharmaceuticals, Inc., who was one of the industry's pioneers in the use of combinatorial chemistry for small-molecule, nonpeptide drug discovery. Chapter 4 covers the field of high-throughput synthesis for drug discovery. It begins with a review by Stephen Wilson and Kathryn Reinhard of New York University and is immediately followed by several case studies contributed by various chemists from both industry and academia.

Chapter 5, on purification strategies, begins with an overview by Mark Suto of DuPont Pharmaceuticals, followed by a series of case studies that highlight various methods for purifying combinatorial libraries. Chapter 6 covers the use of automation and robotics in the production of compound libraries. Its author is Paul D. Hoeprich, Jr., of Agilent Technologies (currently with Genicon Sciences), who has been involved from the beginning in the development of automated synthesizers for general organic chemistry. Several case studies follow this review that, although not all based on the use of automated systems, are highly innovative techniques that readers may find useful in enhancing the efficiency of their compound library production.

The book then takes a dramatic shift away from the theme of producing compounds for biological study to producing molecules for other uses. There is a chapter on new materials (Chapter 7) by Xiao-Dong Xiang of Lawrence Berkeley National Laboratory and one (Chapter 8) on new catalyst development by Richard Willson, David Hill, and Phillip Gibbs of the University of Houston. Both chapters contain case studies that apply high-throughput synthetic methods to either the development of new materials or the discovery of new catalysts.

In Part IV, the book turns from currently practiced methods to possible future applications or methods in the area of high-throughput synthesis. Chapter 9, by Nicholas Hodge, Luc Bousse, and Michael Knapp of Caliper Technologies, covers the area of microchemistry and further expounds on the "lab-on-a-chip" concept. Chapter 10, by David Walt of Tufts University, delves into the uses of fiberoptic sensors as a way to screen for compounds. Finally, Chapter 11 covers the application of some new solid-phase technologies for improving the automation of solid-phase organic synthesis. Chapters 9–11 do not contain case studies, since they discuss methods or concepts that are still relatively new and under development.

Alas, the reader may find that not every aspect of high-throughput synthesis has been covered in this book. For instance, there is no chapter on the planning of a combinatorial library through diversity analysis, nor on cataloging the library once it has been made. There are several reasons for this, not the least of which was the need to keep the book to a reasonable length. Another reason was my interest in focusing the book solely on the "hands-on" mechanics of production. Since there are already some good books out there that focus on those other subjects, I did not feel that there was a dire need to include them in this book. Even Vogel's textbook had its limits! For those who find other omissions or flaws in either the content or the format of the book, I suggest that they contact Marcel Dekker, Inc., with their comments and/or suggestions so that they may be incorporated into future editions of this book.

There are many whom I would like to thank, most notably the many contributors of case studies without whose participation this book could never have been created. I would also like to express my gratitude to Marcel Dekker, Inc., for having kept the faith over these past two years. Lastly, I thank my wife for her patience and understanding during all those work-at-home weekends.

Irving Sucholeiki

Contents

Contents

Contributors

Fariba Aria   Kimia Corporation, Santa Clara, California

Carmen M. Baldino   ArQule, Inc., Woburn, Massachusetts

Sylvie Baudart*   Argonaut Technologies, San Carlos, California

B. John Bergot**   Protein Products, Applied Biosystems, Foster City, California

Luc Bousse   Caliper Technologies Corporation, Mountain View, California

Christine Burger   Trega Biosciences, Inc., San Diego, California

James B. Campbell   International Lead Discovery Department, Zeneca Pharmaceuticals, Wilmington, Delaware

Kathy Cao   Cobalt Unit Chemistry, Affymax Research Institute, Palo Alto, California

Hauyee Chang   Lawrence Berkeley National Laboratory, Berkeley, California

Surendrakumar Chaturvedi†   R&D Laboratory, Applied Biosystems, Foster City, California

---

Current affiliations:
* Consultant, New York, New York.
** Retired.
† Orchid Biosciences, Princeton, New Jersey.

Yuewu Chen Trega Biosciences, Inc., San Diego, California

Andrew P. Combs Chemical and Physical Science Department, DuPont Pharmaceuticals Company, Wilmington, Delaware

Dennis P. Curran Department of Chemistry, University of Pittsburgh, Pittsburgh, Pennsylvania

Thomas A. Dice Parallel Medicinal and Combinatorial Chemistry, Searle Discovery Research, Pharmacia Corporation, St. Louis, Missouri

Lun-Cong Dong Cobalt Unit Chemistry, Affymax Research Institute, Palo Alto, California

Yongjun Duan* Department of Chemistry, Boston University, Boston, Massachusetts

Peter V. Fisher R&D Laboratory, Applied Biosystems, Foster City, California

Phillip R. Gibbs Department of Chemical Engineering, University of Houston, Houston, Texas

Brian M. Glass Chemical and Physical Science Department, DuPont Pharmaceuticals Company, Wilmington, Delaware

David R. Hill Department of Chemical Engineering, University of Houston, Houston, Texas

C. Nicholas Hodge Caliper Technologies Corporation, Mountain View, California

Paul D. Hoeprich, Jr.** Hewlett Packard/Agilent, Wilmington, Delaware

Amir H. Hoveyda Merkert Chemistry Center, Boston College, Chestnut Hill, Massachusetts

Yonghan Hu† Argonaut Technologies, San Carlos, California

Alan P. Kaplan ArQule, Inc., Woburn, Massachusetts

Michael R. Knapp Caliper Technologies Corporation, Mountain View, California

Current affiliations:
* Symbollon Corporation, Framingham, Massachusetts.
** Genicon Sciences, San Diego, California.
† Genetics Institute, Cambridge, Massachusetts.

Petr Kocis International Lead Discovery Department, Zeneca Pharmaceuticals, Wilmington, Delaware

Kevin W. Kuntz Merkert Chemistry Center, Boston College, Chestnut Hill, Massachusetts

Richard A. Laursen Department of Chemistry, Boston University, Boston, Massachusetts

Lanchi Le Cobalt Unit Chemistry, Affymax Research Institute, Palo Alto, California

Michal Lebl Spyder Instruments, Inc., San Diego, California

Rongshi Li ChemRx/IRORI, Discovery Partners International, San Diego, California

Bruno Linclau Department of Chemistry, University of Southampton, Southampton, England

James P. Morken Department of Chemistry, University of North Carolina at Chapel Hill, Chapel Hill, North Carolina

Michael C. Needels Cobalt Unit Chemistry, Affymax Research Institute, Palo Alto, California

Yidong Ni Trega Biosciences, Inc., San Diego, California

K. C. Nicolaou Department of Chemistry, Skaggs Institute for Chemical Biology, Scripps Research Institute, and Department of Chemistry and Biochemistry, University of California, San Diego, La Jolla, California

Kenneth M. Otteson Solid-Phase Organic Chemistry, Applied Biosystems, Foster City, California

Patrick D. Owens Solid Phase Sciences Corporation, Medford, Massachusetts

John J. Parlow Parallel Medicinal and Combinatorial Chemistry, Searle Discovery Research, Pharmacia Corporation, St. Louis, Missouri

Michael R. Pavia Millennium Pharmaceuticals, Inc., Cambridge, Massachusetts

J. Manuel Perez Solid Phase Sciences Corporation, Medford, Massachusetts

Jaylynn C. Pires Spyder Instruments, Inc., San Diego, California

John A. Porco, Jr.*   Argonaut Technologies, San Carlos, California

Kathryn Reinhard   Department of Chemistry, New York University, New York, New York

Lynn F. Schneemeyer   Applied Materials Research, Bell Laboratories, Lucent Technologies, Murray Hill, New Jersey

Marc L. Snapper   Merkert Chemistry Center, Boston College, Chestnut Hill, Massachusetts

Michael S. South   Parallel Medicinal and Combinatorial Chemistry, Searle Discovery Research, Pharmacia Corporation, St. Louis, Missouri

Richard B. Sparks   International Lead Discovery Department, Zeneca Pharmaceuticals, Wilmington, Delaware

Irving Sucholeiki   Solid Phase Sciences Corporation, Medford, Massachusetts

Mark J. Suto   Chemical and Physical Sciences Department, DuPont Pharmaceuticals Research Laboratories, San Diego, California

Steven J. Taylor   Department of Chemistry, University of North Carolina at Chapel Hill, Chapel Hill, North Carolina

David Tumelty   Cobalt Unit Chemistry, Affymax Research Institute, Palo Alto, California

R. Bruce van Dover   Physics of Electronic and Photonic Materials, Bell Laboratories, Lucent Technologies, Murray Hill, New Jersey

David R. Walt   Department of Chemistry, Tufts University, Medford, Massachusetts

Zhengxin Wang†   Department of Chemistry, Boston University, Boston, Massachusetts

Richard Wildonger   International Lead Discovery Department, Zeneca Pharmaceuticals, Wilmington, Delaware

Richard C. Willson   Department of Chemical Engineering, University of Houston, Houston, Texas

---

Current affiliations:
* Boston University, Boston, Massachusetts.
† Rockefeller University, New York, New York.

Stephen R. Wilson   Department of Chemistry, New York University, New York, New York

Xiao-Dong Xiang   Lawrence Berkeley National Laboratory, Berkeley, California

Xiao-Yi Xiao   ChemRx/IRORI, Discovery Partners International, San Diego, California

Caizhi Zhu*   Department of Chemistry, Boston University, Boston, Massachusetts

* *Current affiliation*: Joslin Diabetes Center, Harvard Medical School, Boston, Massachusetts.

Introduction

Combinatorial chemistry represents the confluence of a global approach to problem solving with the mature discipline of chemistry. It is not the first example of such an approach when we consider the 1997 defeat of Kasparov by the IBM chess-playing computer Deep Blue, wherein Deep Blue used a brute force combinatorial-move generator coupled to a scoring function to outperform a human intuitive pattern recognition approach. At the other extreme of usefulness, computer hacking to gain unauthorized access can be considered a very simple combinatorial exercise, mindlessly trying all possible password combinations until the correct one is found. In this context, a combinatorial strategy equates to the two activities of enumerating large numbers of candidate solutions to a particular problem and deciding the merits or success of each of these possibilities.

Combinatorial chemistry can be broken down into the implementation of an algorithm (chemical synthesis) to enumerate chemical entities followed by a measurement (assay) of the degree to which each of the entities meet the desired outcome. As, for example, with computer hacking, the number of allowed characters and the length of the required password describe the landscape of possible solutions, and the number of issued passwords then defines the successes within this landscape. Similarly, in 1984, the library of candidate epitopes we in principle synthesized was defined by the set of genetically coded amino acids (20) and the notion that linear epitopes comprised seven or fewer of those amino acids. In contrast, the application of general organic chemistry to the synthesis of libraries of chemical entities is more open-ended or unbounded, and the achievable library size is usually limited by the availability of suitable monomers.

In practice, most of the libraries synthesized today are numerically small compared to what would be an exhaustive enumeration of the chemistry used, and are produced as discrete entities using automated parallel synthesizers. This obviates the need for a method of encoding the use of monomers during library synthesis, as well as the code-reading step after actives are identified by assay. One can speculate that smaller libraries of discrete and better characterized compounds are being targeted because of the widespread experience of not being able to confirm a positive assay outcome by resynthesis of the expected compound. Alternatively, a view that is often expressed is that today's computational methods can better define molecules having a greater likelihood of activity against a given target, and therefore a smaller library of candidates is a more cost-effective solution.

I believe that there is still a real opportunity to improve methods of producing numerically larger, better characterized libraries of chemical entities to provide a more complete exploration of the chemical landscape defined by a given chemistry. This will require more effective encoding methodologies, preferably nondestructive, compatible with a split-and-mix synthetic strategy, coupled to better analytical procedures that can progressively remove failed or low-yielding chemical outcomes. Economics dictates that quantities of each compound of numerically larger libraries will need to be small, sufficient only to achieve a successful outcome in a designated number of assays. A smaller scale of synthesis matched to the number of active assays is an attractive option, without the attendant problems of providing suitable storage, the long-term stability of the compounds in storage, solubility during storage, and retrieval from storage.

In the last 10 years combinatorial methods applied to organic synthesis have become pervasive in the chemical and pharmaceutical industry, as well as resulting in a significant number of start-up enterprises. Many conferences now focus on this and related topics, and a wealth of information is available in the journal literature as well as in a smaller number of dedicated books. Chemists today can draw on this collective experience to help make decisions about, for example, the numerical scale of the library, solid-phase versus solution-phase synthesis, choice of solid-phase and linker, synthesis scale, synthesis equipment, quantitation of synthesis outcome, and purification of products.

It is of more significance that a combinatorial approach is finding applications in more technically challenging problems such as materials science, catalysis, polymer science, and molecular biology with the search for gene products with novel or improved properties. Another application of a combinatorial approach is to identify all possible chemical reactions systematically, a problem that has a very large solution space defined by the pairwise interaction of all functional groups under all possible reaction conditions, solvents, and various temperatures, to name some of the parameters. It is difficult to envisage a time when a first-principles computational methodology will provide the optimal—or at least an acceptable—solution to a given problem in any of the above fields.

From a personal point of view, I rather enjoy the excitement of the gambling aspects associated with "creating" a set of possible solutions to a problem, determining the success or otherwise of each, hoping that I have a "winner." Given our demonstrated ability to find creative solutions to even the most technically challenging problem, it is certain that a combinatorial approach will contribute to valuable products or outcomes in an ever-increasing number of disciplines in the next decade and on.

More important, even than, for example, the discovery of a truly high-temperature superconductor, which would have widespread impact, is that researchers of today and tomorrow can think bigger and better and enjoy the "high" that comes with the sense of a difficult job well done. The opportunity to use increasingly sophisticated technology in all fields of endeavor is limited only by our imagination, and I hope to continue to be a participant.

H. Mario Geysen
Diversity Sciences Department
GlaxoWellcome
Research Triangle Park, North Carolina

HIGH-THROUGHPUT SYNTHESIS

Part I
Theory and Methods in Solid
Phase Organic Synthesis

Chapter 1
An Introduction to Solid Phase Organic Synthesis

Irving Sucholeiki
Solid Phase Sciences Corporation, Medford, Massachusetts

I. INTRODUCTION

Throughout this book are references to the use of a technique termed "solid phase" or "solid-supported" organic synthesis. This is a method whereby a reaction is run on an insoluble support. The fact that the support is insoluble to the reacting solvent does not negate the fact that reactions can still occur at the support's surface. The key steps in solid phase organic synthesis begin with the covalent attachment of an organic molecule or "scaffold" (**2**) to a linker molecule (**1**) that is already attached to the surface of the insoluble support (Scheme 1). This linker molecule should be cleavable under very selective conditions but also stable to the reaction conditions used to modify the scaffold. The scaffold is then exposed to reaction conditions that chemically modify it in some way, such as the covalent attachment of some molecule (**4**). After the reaction is complete, the support is washed and filtered to remove any unreacted starting material and/ or excess reagents. The process can then be repeated to further modify the scaffold (**5**). After a series of reactions followed by washes and filtrations, the now modified scaffold can then be isolated by chemical cleavage from the linker. A simple filtration step removes the chemically cleaved scaffold from the insoluble support (**6**) (Scheme 1). From this process one can see that one advantage of using solid phase organic synthesis is the ability to rapidly separate out products from the soluble components of the reaction mixture by filtration. One can also use this rapid ability to separate the soluble components from the resin-bound components to drive reactions to completion. To fully understand how this can be done, let us first look at the hypothetical bimolecular solution phase reaction

3

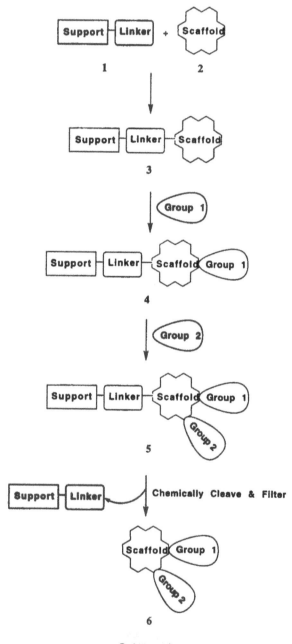

Scheme 1

A + B = C. The initial rate of formation of C can be described by the rate Eq. (1).

$$\text{Initial rate} = d[C]/dt = k[A][B] \tag{1}$$

where the formation of C is dependent on both the concentration of A and B times some constant k (termed the reaction rate constant). If B is present in a large excess, its concentration of B will remain virtually unchanged throughout the initial stages of the reaction and therefore the initial reaction becomes pseudo–first order and can be represented as Eq. (2).

$$d[C]/dt = k'[A] \qquad \text{where } k' = k[B] \tag{2}$$

Of course, if one were to use a large excess of B, there would be the problem of separating B from C at the end of the reaction. If one were to attach A on a support, one could use a very large excess of B and very easily remove this excess by simple filtration. If at the end of the reaction one still has some unreacted A present, one can then reexpose the support to a large excess of B and further diminish the amount of residual support-bound A.

Whether one actually observes an increase in the reaction rate by using a large excess of B depends on several factors, most notably on the solid phase diffusion rate of component B. If support-bound A is easily accessible to the molecules in solution, then one should see the expected rate enhancement. If, however, support-bound A is poorly accessible, then the rate of diffusion of B will be the rate-determining step. Following Le Chatelier's principle, which states that when a stress is applied to a system that is in equilibrium, the system will adapt in a way to offset that stress, increasing the concentration of B in solution drives more of B to accumulate throughout the support's matrix, thereby enhancing the solid phase diffusion rate of B and possibly the observed reaction rate as well. We will see later in this chapter how other factors can come into play that also can dramactically affect the observed reaction rate. For now, let us continue to focus on the main advantages for utilizing solid phase organic chemistry. In many cases these advantages can be very subtle and best conveyed through a brief historical overview of the development of solid phase organic chemistry.

II. A HISTORICAL OVERVIEW OF EARLY SOLID PHASE ORGANIC CHEMISTRY

No review can truly do justice to all of the people involved in the development of a particular field. In fact, most often one finds that work done in a tangential field can contribute significantly to the evolution of a new field but inadvertently may be forgotten as the new field matures. This is no exception when discussing solid phase organic chemistry. Early work by scientists in the field of heteroge-

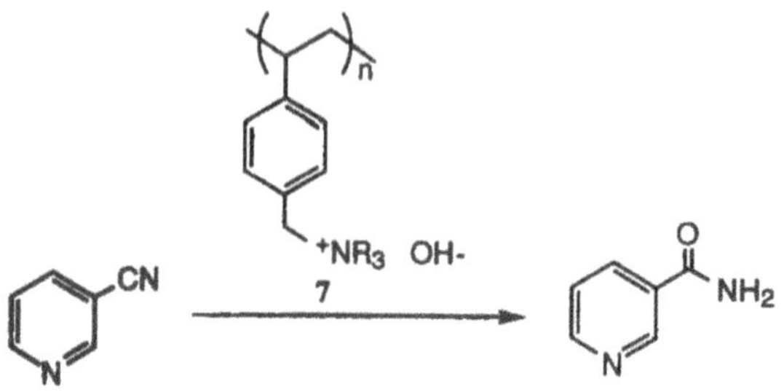

Scheme 2

neous catalysis and, more specifically, the development and use of polymer-supported catalysts ultimately has to be credited for giving credence to the idea that
one can perform organic chemistry at the interface of a solid and a liquid [1].
Although it is not the intention of this review to cover the field of heterogeneous
catalysis some early notable exceptions are worthy of mentioning simply for the
fact that they use supports that are almost identical to many of the supports used
today for solid phase synthesis. For example, in 1948, Alexander Galat was one
of the first to demonstrate the use of a newly, commercially available anion exchange resin to catalyze the hydrolysis of a nitrile to an amide [2].

The support consisted of cross-linked polystyrene functionalized with a
quaternary ammonium hydroxide group (7) (Scheme 2). It had been known that
one could use basic conditions to achieve the hydrolysis of a nitrile (8) to an
amide (9) in solution, but there were substantial problems with the further hydrolysis of the amide to the corresponding acid. Galat's purpose in using the anion
exchange support (7) was to limit this side reaction as well as to facilitate an
easy way to separate the product from the base. This resulted in elimination of
the acid byproduct and easy isolation of the desired amide product (9) in high
yield (Scheme 2). Astle and Abbot used a similar commercially available polystyrene based anion exchange support to catalyze the aldol condensation of nitro-
alkanes with various aldehydes and ketones [3].

Solid phase organic synthesis as we know it today was first described in
1963 by Professor R. B. Merrifield of Rockefeller University in his seminal paper
entitled "Solid phase peptide synthesis. I. The synthesis of a tetrapeptide" [4].
At that time, peptides were synthesized in solution and purified by chromatography and/or recrystallization. Although this strategy worked fine for short peptides, Merrifield saw that problems with both the solubility and purification of
the growing peptide made it impractical for the synthesis of longer polypeptides.
Merrifield reasoned that if one could covalently attach the first amino acid to an

insoluble support one could then couple succeeding amino acids in a stepwise fashion and literally grow the peptide on the support which, when completed, could be removed chemically and filtered from the support. Merrifield tried various polymeric materials (cellulose, polyvinyl alcohol, polymethacrylate among others) but ultimately found that derivatized polystyrene cross-linked with 2% divinylbenzene (10) worked the best (Scheme 3). Merrifield came up with solid phase peptide synthesis not only for the obvious reason that one could very quickly and easily remove the reagents and byproducts of the reaction mixture by manual filtration of the solid support but also as a way to facilitate the introduction of automation into peptide synthesis. As quoted in his 1963 paper, "It is hoped that such a method will lend itself to automation and provide a route to the synthesis of some of the higher molecular weight polypeptides." Merrifield quickly followed this work a year later with the solid phase synthesis of the natural peptide bradykinin as way to prove that peptides synthesized by a solid phase approach exhibited the same biological activity as those obtained naturally [5]. Two years later, Merrifield reported the details of the first fully automated synthesizer for the stepwise synthesis of peptides following a solid phase synthesis strategy [6].

As was mentioned earlier, Merrifield found that a support made up of 2% cross-linked polystyrene exhibited the best physical properties for solid phase organic synthesis. He found 1% cross-linked supports to be too fragile, which caused the polymer beads to fracture. At higher levels of cross-linking (8%), Merrifield observed slower reaction rates due to poor penetration of reagents into the interior of the support as a result of poor bead swelling. Because Merrifield for the most part relied on commercially obtained resins, he was not able to further optimize his support. Lestsinger et al. were not only quick to embrace Merrifield's solid phase approach to peptide synthesis, but because they made their own polymer resins they were able to better optimize the support's overall swelling properties [7,8]. They also found that one could incorporate functionality directly into the polymer by simply copolymerizing styrene and divinylbenzene with functionalized vinyl monomers such as p-vinylbenzylalcohol, p-vinylbenzoic acid, and p-vinylbenzoate. Lestsinger et al. later extended Merrifield's technique to the synthesis of polynucleotides [9].

Fridkin et al. used solid phase methods in a slightly different way. They attached an activated amino acid on a support (11), which they then reacted with an amino acid (14) or peptide in solution to give the coupled peptide back in solution (15) (Scheme 4). The result was the stepwise solution synthesis of a peptide using polymer-bound activated amino acids (Scheme 4) [10]. One interesting aspect of this work was the use of centrifugation rather than filtration to separate the support (11 and 16) from the soluble components of the reaction mixture.

Non-peptide chemists seeing Merrifield's work saw an opportunity to apply

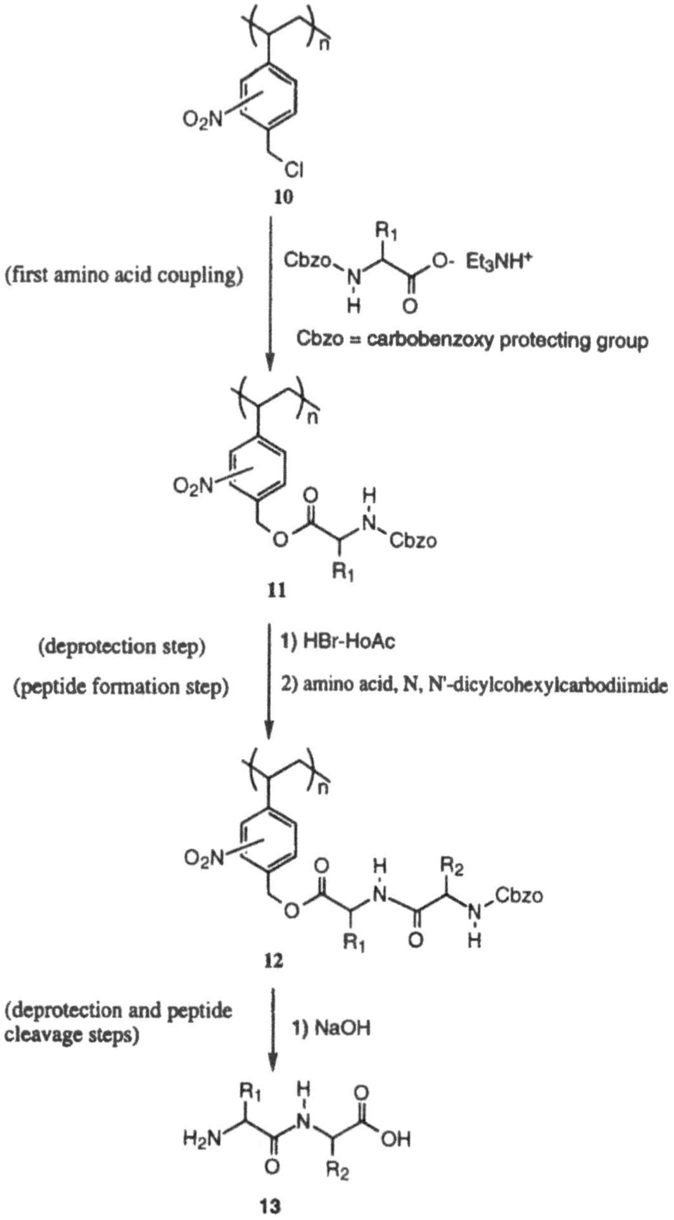
O2N
Cl
10
(first amino acid coupling)
Cbzo
N
H
R1
O- Et3NH+
O
Cbzo = carbobenzoxy protecting group
O2N
O
H
N
Cbzo
O
R1
11
(deprotection step)
(peptide formation step)
1) HBr-HoAc
2) amino acid, N, N'-dicylcohexylcarbodiimide
O2N
O
H
N
R2
N
Cbzo
O
R1
O
H
12
(deprotection and peptide cleavage steps)
1) NaOH
R1
H
O
H2N
N
OH
O
R2
13

Scheme 3

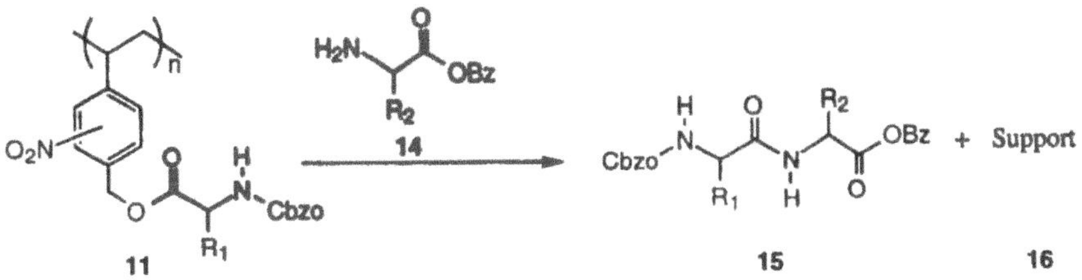

Scheme 4

solid phase methods in the synthesis of other synthetic targets in organic chemistry. Fréchet and Schuerch, for example, applied solid phase methods to the synthesis of oligosaccharides [11]. They first attached a protected glycosyl bromide (**18**) to an allyl alcohol functionalized polystyrene support (**17**) to give the resulting resin-bound glucose (**19**) (Scheme 5). Selective deprotection at the C-6 position gave the resulting resin-bound alcohol (**20**), which was then treated with more glycosyl bromide (**18**) to complete the cycle. Oxidative cleavage gave the resulting disaccharide (**22**). Fréchet also worked extensively on developing efficient methods for derivatizing various cross-linked polystyrene supports. Unlike Lestsinger, who incorporated various functionality within the monomers used in the polymerization process itself, Fréchet used the lithium salt of preformed cross-linked polystyrene to form various hydroxyl, thiol, carboxyl, and silyl derivatives of the support [12,13].

For those synthetic chemists whose targets were not made up of chemically similar repetitive units, solid phase synthesis was a way of modifying polyfunctionalized molecules. The solid support acted as a form of protecting group, allowing one to tie up one end of a polyfunctionalized molecule while leaving the other end available for future modification. Leznoff et al. promoted this type of solid phase synthetic approach as a way of rapidly accessing various natural and unnatural products [14–16]. Solid phase methods have also been exploited in other areas of classical organic chemistry. For example, Kawana and Emoto were the first to perform asymmetrical synthesis on a solid support [17]. Solid phase methods have also been utilized as a way of performing organic chemistry under high-dilution conditions [18].

III. CHEMISTRY AT THE POLYMER–LIQUID INTERFACE

Early researchers using polymer-based supports in solid phase organic synthesis quickly learned that they could get better results when they used certain solvents with their solid support than with others. For example, many of the solid phase

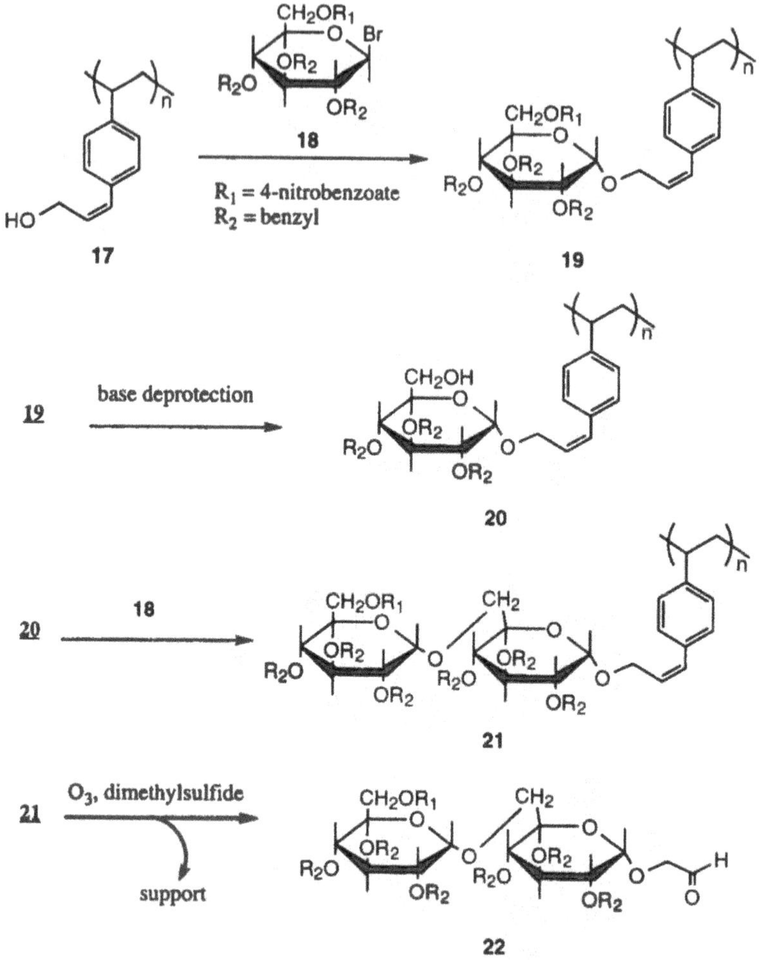

Scheme 5

reactions previously mentioned were run using dimethylformamide. To understand why this is the case one must first understand the physical properties of many polymer-based supports. Low cross-linked polystyrene (with 1–2% divinyl benzene) can be seen as a sponge with the majority of the active sites located in the cavities. These types of supports are sometimes termed *microporous* or gel-type supports (Fig. 1B). Solvents such as dimethylformamide can cause the support to expand up to 3 times its dry volume, while solvents such as methanol cause the support to dramatically contract. Higher levels of cross-linking reduce

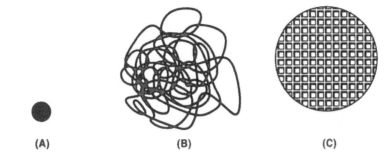

Figure 1 (A) Polystyrene-based polymer bead. (B) Magnification of a microporous-type polymer bead in its solvated state. (C) Magnification of a macroporous-type polymer bead.

the extent to which the support expands and contracts and also affect the level of reactivity. Supports that have a high degree of cross-linking yet have large internal pores are termed *macroporous* supports (Fig. 1C). Work by Regen using electron spin resonance (ESR) spectroscopy of nitroxide radical probes showed that a bound substrate was more restricted than a substrate dissolved in the swollen particle [19,20]. Regen also showed that a higher degree of swelling was associated with a decrease in the rotational correlation time. A decrease in the rotational correlation time can be interpreted to mean that the bound nitroxide probe exhibits a greater mobility in the expanded state of the cross-linked polystyrene. In addition, Regen provided evidence that with greater cross-linking,which translates to less expansion of the support, the internal viscosity of the solvent increased. The use of gel phase ^{13}C-NMR has also been used to evaluate the extent of mobility of bound substrates [21–26]. Giralt et al. successfully used gel phase ^{13}C-NMR to correlate the line widths of amino acids bound to various supports to their peptide coupling yields [23–25]. They found that a broadened ^{13}C line spectrum was related to restricted mobility and that narrower line widths correlated with a higher coupling yield for the bound amino acid. One can dramatically sharpen the line widths by attaching a spacer arm between the bound molecule and the support. Bayer and Rapp have found that attaching a long polyethylene glycol (PEG) spacer (MW = 2000–3000 Da) can dramatically sharpen the line widths and increase the relaxation time of the bound component [27,28]. It can be seen then that the extent of mobility of the bound substrate can be affected, not only by the internal viscosity of the medium surrounding the bound substrate but by the length and solubility properties of the spacer arm connecting the substrate to the support. When the spacer arm is PEG the viscosity properties of the support itself can be affected. Rapp has shown that PEG-grafted polystyrene particles display a greater sorption for the dye Oracett blue 2 as compared to

Table 1   Comparison of Swelling Properties Between Polystyrene (Merrifield, **23**)–
and PEG–modified (Tentagel S RAM, **24**) Supports (volume in mL/g) [31]

Solvent	23	24
Tetrahydrofuran	7.7	4.0
Dimethylsulfoxide	1.8	3.8
Acetonitrile	1.8	4.0
Methanol	1.8	3.6

standard 1% cross-linked polystyrene [29]. This greater sorption is also observed
with very large molecules, such as proteins. For example, Quarrell et al. looked
at the penetration of the enzyme trypsin (23.5-kDa) on the PEG-grafted support,
Tentagel, bound with a strong trypsin peptide antagonist [30]. Using confocal
microscopy, the researchers showed that fluorescein-labeled pancreatic trypsin
was capable of penetrating to the core of the Tentagel bead. The researchers
observed that trypsin was able to diffuse preferentially between contacting Tentagel beads at their interface. The researchers concluded that the Tentagel beads
were acting as a separate gel phase enhancing the diffusion rate of the large
trypsin molecules, preferentially, between contacting beads. Table 1 shows a
comparison of the swelling properties between polystyrene cross-linked with 2%
divinylbenzene (**23**) and a PEG-grafted support (**24**) in different solvents [31].

The rate of a solid phase reaction depends on the diffusion rate of reagents
and reaction products that enter and leave the pores of the solid matrix. Variables
such as particle size, pore size, solvation effects, as well as the polarity of the
solvent can have a large effect on the diffusion rate of solute molecules in a
support's matrix and hence on the overall reaction rate. For example, the rates of
9-fluorenylmethoxycarbonyl (Fmoc) deprotection using 25-μm Tentagel particles
were found to be more than 3 times faster than with 90-μm Tentagel particles
[29]. This indicates that simply reducing the particle size of the support can dramatically enhance the reaction rate. What this also means is that under diffusion
control, if one were to run a solid phase reaction using a support having a broad
particle size range, the rate of the reaction would ultimately be limited by the
diffusion rate of the largest particles used in the mixture.

What type of mixing is best for a solid phase reaction? This is a question
that many have asked, but unfortunately data comparing various methods of agitation have been few and sometimes contradictory. Rapp has found that shaking

a resin or perturbing it through gas bubbling had no affect on the rate of amide formation in solid phase peptide synthesis [29]. However, Rapp found that vortexing and continuous flow systems exhibited a pronounced affect on the solid phase peptide coupling rates. On the other hand Li and Yan, reported very different results when they looked at the yields of hydrazone formation on solid support under varying agitation conditions. Using single-bead infrared (IR) and fluorescence spectroscopy, they found that agitation through gas bubbling or rotating the reaction mixture gave slightly higher yields as compared to orbital shaking, magnetic stirring, or wrist action shaking [32].

IV. OTHER TYPES OF SOLID PHASE SYNTHESIS SUPPORTS

Although it is apparent that polystyrene based supports have played a large part in the early development of solid phase organic synthesis, they do have some shortcomings that have caused some researchers to develop other types of supports. One of the main shortcomings that has already been alluded to earlier is the large variability in swelling volume exhibited by polystyrene-type supports in the presence of different solvents. For example, Li and Yan showed that when comparing the PEG-grafted polystyrene support (Tentagel) to a non-PEG-grafted polystyrene support (Wang Resin) they found that neither support was universally better than the other and that some reactions worked better using the PEG-grafted support and vice versa. They concluded that there was no single universally superior support and that one needed to look closely at the solvent that was going to be used to swell the support [33]. If a particular chemistry requires a solvent that cannot expand a gel-type support then it most likely will exhibit a poor reaction rate profile. For example, many of the solvents and reagents used in solid phase DNA synthesis do not expand low cross-linked polystyrene very well and is one of the reasons other materials are used in solid phase DNA synthesis. Even PEG-grafted supports have been found to give poorer yields and purities as compared to non-gel-like supports. For instance, in the synthesis of oligonucleotides having more than 20 nucleotides, controlled pore glass (CPG) and Teflon exhibited higher coupling yields per cycle (>98%) as compared to the PEG-grafted, Tentagel, support (90–91%) [34]. In another example, it was found that higher purity and yields could be obtained when synthesizing oligonucleotides on macroporous polyethylene than could be obtained on microporous polystyrene or any of a number of PEG-based supports [35]. Another area that has researchers develop other types of materials as supports is the area of continuous flow solid phase synthesis. In continuous flow synthesis the reagents are continuously run through a sealed column containing the support-bound scaffold. Under these conditions, large column pressures can be produced if a support is used that exhibits a large variability in its expanding and contracting properties. The solution to this problem is the development of various macroporous supports such as Kieselguhr (dia-tonaceous earth) and highly cross-linked macroporous polystyrene.

V. LISTING OF VARIOUS SOLID PHASE SUPPORTS

The organic chemist interested in utilizing these and other supports must consider two factors: the level of substitution on the resin and the extent to which the bound component's mobility may be affected by the support's environment. Many solid phase organic chemists making combinatorial libraries tend to use commercially available supports that exhibit very high loading capacities. This is due to the fact that many of the targets are low molecular weight compounds (<300 MW), and to isolate a few milligrams of the desired product (within a small reaction volume), a support with a sufficiently high loading capacity (0.2–3 mmol reactive groups/g of support) is required. On the other hand, DNA and some peptide chemists tend to use supports that exhibit lower loading capacities (<0.2 mmol reactive groups/g of support). Below is a nonexhaustive listing in alphabetical order of various microporous and macroporous supports used today in various solid phase organic synthesis applications.

A. Cellulose-Based Supports

Cellulose supports in the form of paper sheets have been used for solid phase peptide synthesis. An amino acid is attached to the surface of the cellulose support through the formation of an ester bond at the C-terminal end of the amino acid. Typical loading levels for this support are in the range of 0.5–0.6 μmol reactive groups/cm^2 [36]. Cotton has also been used as a support for the synthesis of peptide libraries exhibiting loading levels of around 0.1–0.4 mmol reactive groups/g (1–3 μmol/cm^2) [37].

B. Controlled Pore Glass (CPG)

This is a highly porous form of silica beads (96% SiO$_2$) that are available in many pore diameters in the range of 40–2500 Å and exhibiting a very narrow pore diameter. This support has been primarily used in solid phase DNA synthesis, with typical loadings running in the range of 0.06–0.17 mmol amine/g [38].

C. Cross-Linked Ethoxylate Acrylate Resins (CLEAR)

These are a series of supports that are composed of trimethylolpropane ethoxylate triacrylate cross-linked to a series of different methacrylate monomers, such as 2-aminoethylmethacrylate and poly(ethylene glycol-400)dimethylmethacrylate [39]. Although the supports are highly cross-linked, they exhibit swelling characteristics similar to hydrophilic gelatinous resins both in polar organic solvents and in water. Loading capacities for these supports are reported to be in the range of 0.13–0.29 mmol amine/g.

D. Grafted Polyethylene-Based Supports

Geysen synthesized a grafted polymethacrylate-polyethylene support which was molded to the shape of a pin for peptide synthesis. The pins were later composed of a polypropylene-polymethacrylate-dimethylacrylamide composite and has found wide use for not only peptide but also for nonpeptide combinatorial library production. Several types of grafted pin crowns and "lanterns" are available (Mimotopes) having a wide range of loading capacities [40,41].

E. Kieselguhr–Polyacrylamide Composite

This is a composite consisting of polyacrylamide gel trapped in the porous structure of kieselguhr (diatomaceous earth or clay). The loading is typically 0.1–0.2 mmol amine/g resin [42].

F. Merrifield Resin

The support is made of polystyrene that is 1–2% cross-linked with divinylbenzene and is functionalized with a chloromethyl group. The loading capacity ranges from 0.4 to 3.0 mmol chloromethyl/g (**23**, Table 1).

G. Polyethylene Glycol–Polyacrylamide Composite (PEGA)

This is a polyacrylamide-PEG composite produced through the copolymerization of *N,N*-dimethylacrylamide, bis-2-acrylamidoprop-1-yl-PEG-1900, and sarcosine ethyl ester (**25**, Fig. 2). The support has a reported loading capacity in the range of 0.4–0.8 mmol amine/g [43,44].

H. Polyethylene Glycol–Polystyrene (POEPS) and Polyethylene Glycol–Polyoxypropylene (POEPOP)

The POEPS (**26**, Fig. 2) and POEPOP (**27**, Fig. 2) supports consist of polystyrene or polypropylene that have been crosslinked with PEG. They are made by polymerizing PEG-containing styrene or propylene monomers. These supports have loading capacities in the range of 0.1–0.6 mmol hydroxyl groups/g [45,46].

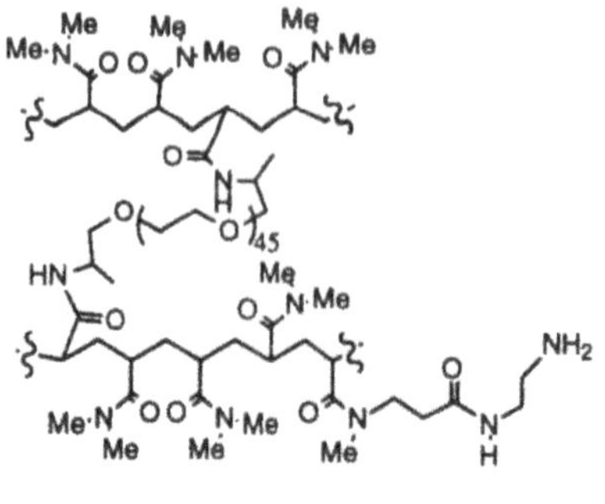

25

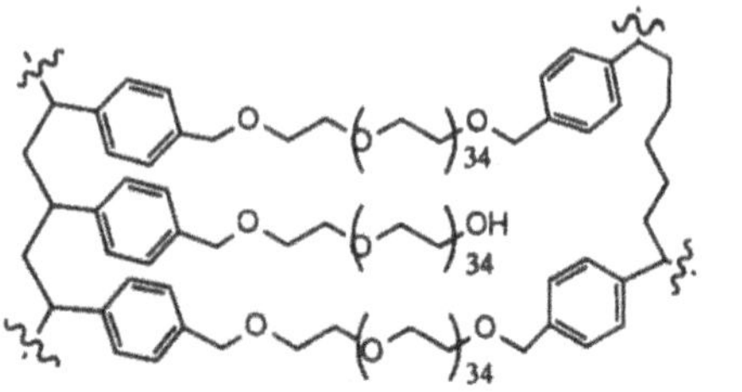

26

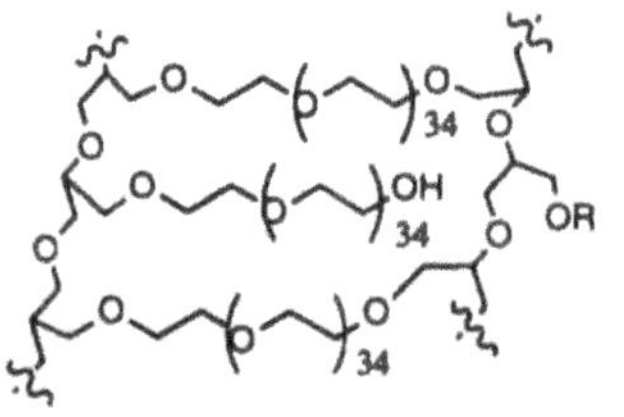

27

Figure 2   The PEGA **25**, POEPS **26**, and POEPOP **27** polyoxyethylene-based resins.

I.   Polyethylene Glycol (PEG) Grafted to Polystyrene

Unlike the POEPOP support, these PEG-based supports are synthesized by graft-
ing various sized PEGs (500–300 Da) onto existing 1–2% cross-linked polystyrene. One of these types of supports that has been previously mentioned is sold
under the name Tentagel (**28** and **24**) and is made by polymerizing ethylene oxide
onto a primary alcohol located on the cross-linked polystyrene (Table 1 and Fig.
3) [47–49]. A second form, sold under the name PEG-PS (**29**), is made by attaching, through the formation of an amide bond, an already formed amino terminal polyethylene glycol chain (Jeffamine) to 1% cross-linked polystyrene (Fig.
3) [50,51]. The loading capacities for both types of supports are typically in the
range of 0.2–0.3 mmol amine/g resin. A third support, which is sold under the
name ArgoGel (**30A**), has two PEG chains (30–40 units long) branching from
a single carbon attachment point, which gives it a reported loading capacity of
0.5 mmol amine/g of support (Fig. 3) [52,53]. Both the Tentagel and PEG-PS
supports have incorporated lysine branching to substantially raise their available
amine substitution level to as high as 1.0 mmol amine/g. An example of a highly
branched PEG-grafted support is Dendrogel (**30B**), which has 61% PEG incorpo-

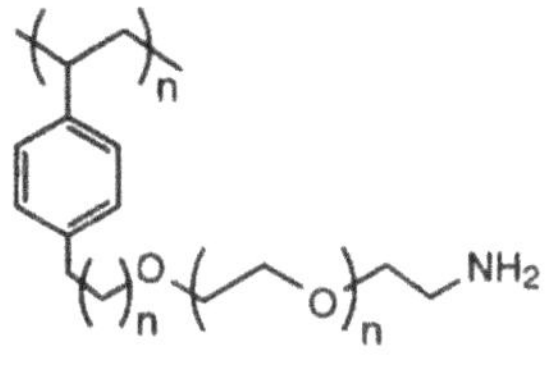

28

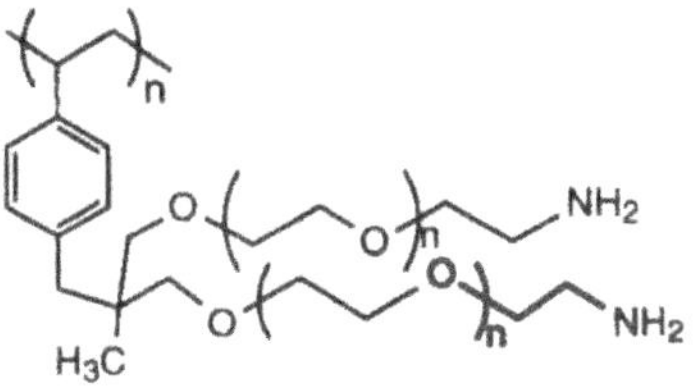

30A

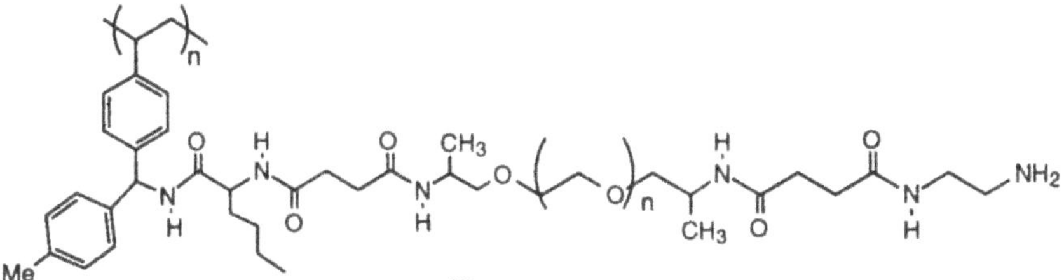

29

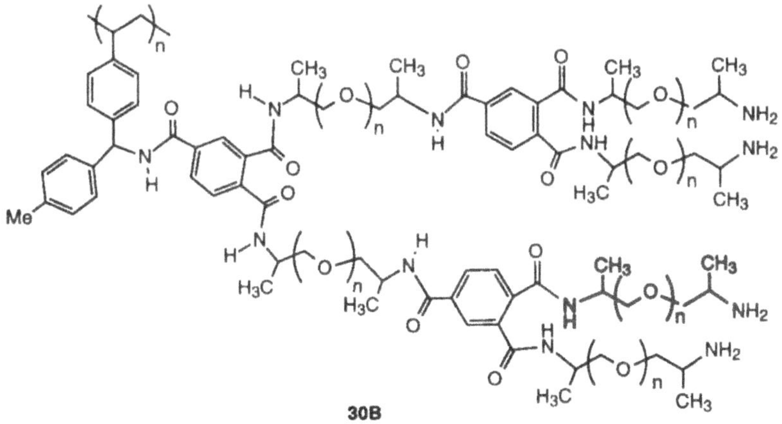

30B

Figure 3 Tentagel resin **28**, PEG-PS resin **29**, ArgoGel resin **30A**, and Dendrogel resin **30B**.

ration and a reported loading capacity of around 0.23 mmol amine/g of support (Fig. 3) [54].

J. Polystyrene-Based Paramagnetic Supports

These supports consist of magnetite (Fe_3O_4) encapsulated in polystyrene beads. These types of supports are attracted to a magnetic field but do not become perma-

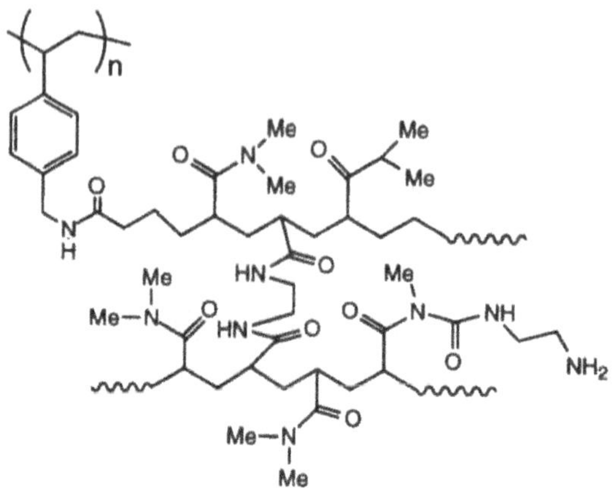

Figure 4 The Polyhype support **31**.

nently magnetic once the external magnetic field has been removed. The main advantage of these types of supports is that one can separate the support from the surrounding liquid using a magnet rather than by filtration. Loading capacities can reach as high as 1 mmol reactive groups/g of support [55–57].

K. Polystyrene–Polyacrylamide Composite (Polyhype-Based Composite)

This is a composite of cross-linked polystyrene containing a very high pore volume (around 90%) containing polyacrylamide within the cavities. Both covalently attached polystyrene–polyacrylamide (**31**) (Fig. 4) and noncovalently attached polystyrene–polyacrylamide composites are available with substitutions ranging from 0.1 to 2.0 mmol amine/g resin [58,59].

L. Silica Gels and Other Silica-Based Supports

Silica gels have been successfully used as supports in solid phase organic syntheses. The silica gels are usually derivatized through treatment with some amino alkyl trialkoxysilanes to incorporate amino functionality to the inner and outer surface of the support. Loading capacities in the range of 0.8–1.2 mmol amine/g of support can be achieved with this method. A major advantage of using these types of supports is their very high temperature stability as compared to polymer-based supports [60].

VI. LINKERS FOR SOLID PHASE ORGANIC CHEMISTRY

The area of solid phase linkers is vast and will be only briefly reviewed here. There are a few good sources to which the reader can go to get a more thorough exposure of not only other types of linkers available but also the subtleties involved in their identification [61–63]. For the most part, the majority of commercially available linkers owe their existence to the field of solid phase peptide synthesis. Therefore, it is not surprising that many of the methods employed to attach molecules to a support rely heavily on the formation of an ester or an amide bond. Table 2 lists some of the more common solid phase peptide synthesis linkers that have also been applied to the production of nonpeptide compounds. In every case shown, the type of bond formed on the support is either an amide- or an ester-type linkage. Cleavage of the linker usually involves the use of a strong acid or base to produce an appendage on the scaffold having an amide, ester, or acid functionality.

The use of light to cleave resin bound molecules selectively has been an active area of research for years. Since the early 1960s there has been a great deal of research on photoactive protecting groups that upon irradiation in solution release the active group [70–72]. One of the first to adapt these types of photoactive protecting groups to a heterogeneous system for the purposes of cleaving product from a support was Rich and Gurwara [73]. They synthesized a 3-nitro-4-bromomethylbenzoic acid linker, which they then attached to an aminomethylated cross-linked polystyrene support to give the resulting amide-coupled linker **38** (Fig. 5). The carboxylic acid end of a protected amino acid was then attached followed by deprotection and further solid phase couplings with other protected amino acids to give a resin-bound polypeptide **39**. Upon irradiation with 350-nm light, the benzyl carbon-oxygen bond was observed to cleave releasing the peptide from the support. One year later, Wang attached an α-methylphenacyl ketone directly onto 1% cross-linked polystyrene followed by attachment of the carboxylic acid end of an amino acid to give the resulting resin-bound amino acid ester **40**, which he then cleaved photochemically [74]. Later, Tjoeng and Heavner incorporated the α-methyl ketone as a separate linker (**41**) (Fig. 5) [75]. Pillai synthesized a peptide on a support consisting of 1% cross-linked polystyrene containing a 3-nitro-4-*N*-methylaminomethyl moiety (**42**). Pillai showed that one can successfully cleave the polymer-bound carbon–nitrogen bond upon irradiation with 350-nm light to give C-terminal N-methylated peptide amides [76].

Only a handful of research groups have exploited the use of photocleavable linkers in the production of combinatorial libraries. One photosensitive linker that has found broad use in both peptide and nonpeptide solid phase synthesis is the 6-nitroveratryl-based linker (**43**), which was based on an earlier photosensitive protecting group (Fig. 5) [77,78]. Like some of the photolinkers mentioned previously, photocleavage of this linker leaves an amide appendage on the scaf-

Table 2   Free and Resin-Bound Linkers Used in Solid Phase Organic Chemistry

Resin-bound linker or free linker	Type of bond formed between scaffold and linker	Cleavage conditions (ref.)
 2-Chlorotrityl chloride resin (32)	Ester	Acetic acid–trifluoro-ethanol–methylene chloride (64)
 4-Hydroxymethylbenzoic acid (33) (HMBA)	Ester	Sodium methoxide (60)
 Hydroxymethylphenoxyacetic acid (34) (HMPA)	Ester	Trifluoroacetic acid (65)
 Knorr Linker (35)	Amide	Trifluoroacetic acid–methylene chloride (66)
—CH_2Cl Merrifield (23)	Ester	Anhydrous HF (67)
 Rink amide linker (36)	Amide	Trifluoroacetic acid (68)
 Wang resin (37)	Ester	Trifluoroacetic acid (69)

= 1-2% cross-linked polystyrene

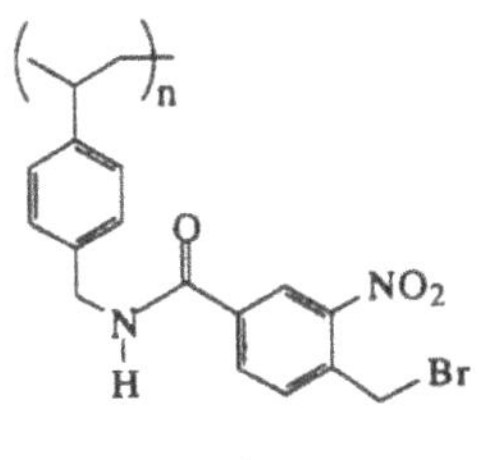

Figure 5   Various light-sensitive linkers used in solid phase organic synthesis.

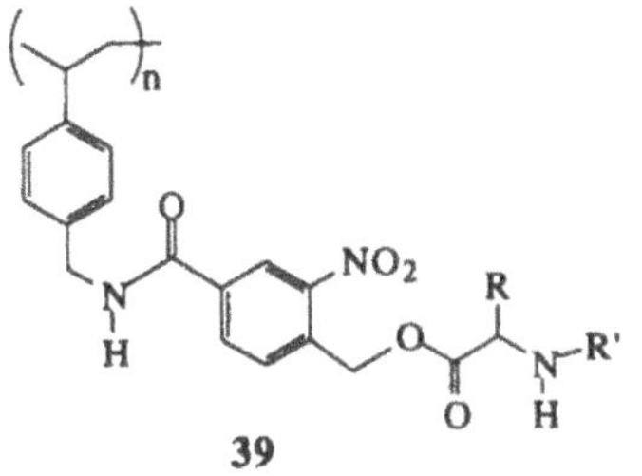

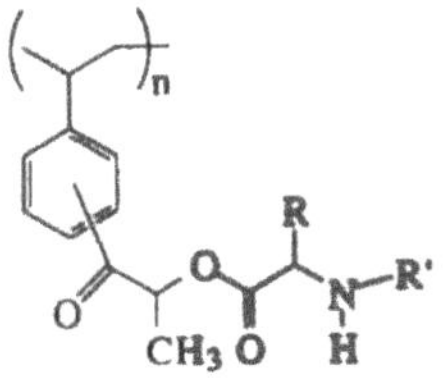

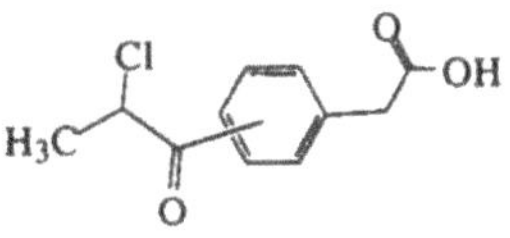

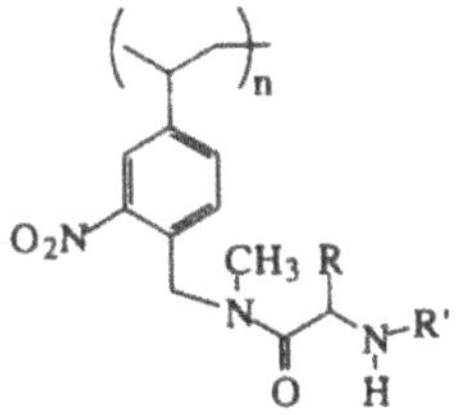

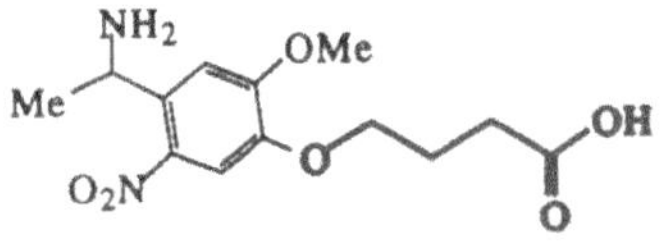

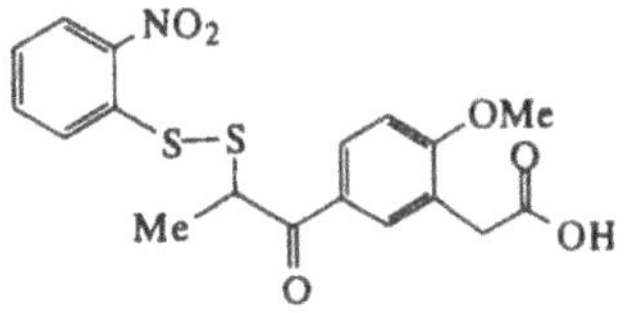

fold. Many in the combinatorial chemistry community who were interested in producing molecules via a solid phase synthesis format did not want molecules having an amide, ester, or acid appendage as a result of the chemistry used to cleave the linker. In 1994, a photosensitive linker was reported that for the first time cleaved under a mechanism that did not produce these types of artifacts on the cleaved scaffold [79,80]. This photosensitive linker was an α-mercapto-substituted phenyl ketone (**44**), which is protected as the disulfide (Fig. 5). Instead of an ester, acid, or amide appendage, only a methyl group was left on the cleaved scaffold. Later, this concept of a general linker that would leave no trace, or that would be ''traceless'' on the cleaved scaffold, took hold with the goal being that hydrogen would be the only acceptable group allowed on the cleaved scaffold. One of the first of these types of linkers employed an arylsilane linkage **45** and

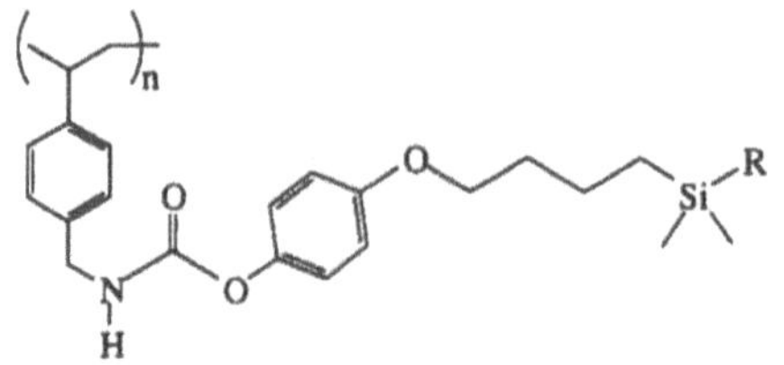

45 (R = phenyl containing scaffold) 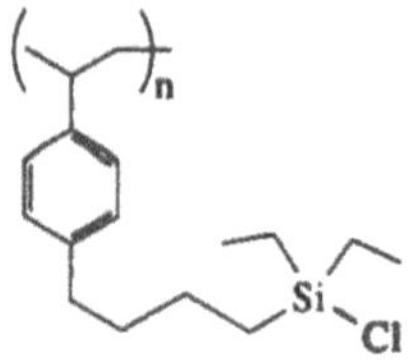   **46**

Figure 6   Traceless linkers used in solid phase organic synthesis.

used hydrogen fluoride to cleave the scaffold off the support (Fig. 6) [81]. Later this type of linker was modified by other researchers to incorporate a silyl halide handle to make it easier to attach and cleave molecules from the support (**46**, Fig. 6) [82].

VII. A BRIEF SUMMARY OF ANALYTICAL METHODS AVAILABLE FOR MONITORING THE SOLID SUPPORT

No matter what support and linker one uses in a solid phase synthesis approach, some analytical method must be used to evaluate the success or failure of that approach [83]. When the chemistry is run in solution, there is not much of a problem, but when the chemistry occurs on a support new methods and techniques must be developed to handle the insoluble nature of the support itself. In addition, certain types of supports may not be compatible with all of the available methods of analysis. The most common methods of measuring the loading capacity of the support is measuring the absorbance of a UV-active group cleaved off the support and elemental analysis of the support itself. Both methods can give very good quantitative data; however, the uncertainty in the measurement will vary depending on the group that is being measured. For example, halogen and nitrogen elemental analysis of a solid support typically produces data with very small levels of uncertainties in the measurement, whereas sulfur elemental analysis has intrinsic in the technique greater levels of uncertainty. Other techniques that are more qualitative yet give more structural information about the bound molecule include gel phase carbon nuclear magnetic resonance (NMR), gel phase proton (NMR, Fourier transform infrared spectrometry FTIR), and laser desorption mass spectrometry.

In general, inorganic supports, such as those based on silica, rely on elemental analysis, FTIR, and spectrophotometric measurements of a cleaved UV-active molecule to gauge the efficiency of a solid phase reaction. Polymeric supports, in addition to those techniques, can also be monitored using gel phase carbon and proton NMR. The most conservative approach and still the most informative one is the cleavage of a portion of the support after each synthetic step and the subsequent analysis of the crude cleaved product by HPLC coupled with mass spectrometry and/or NMR. Unfortunately, such an approach can become costly when applied to the analysis of large compound arrays, and some form of statistical sampling may have to be used. Additionally, multiple types of detection systems (UV, evaporative light scattering, total ion mass spectrometry) may be required to get a more objective view of the quantity of material present. With time, major strides in miniaturization will dramatically lower the cost of using such techniques in a high-throughput mode and allow individual researchers outside of the well-funded academic institution or company an opportunity to incorporate these methods into their solid phase synthesis strategy.

VIII. SUMMARY AND CONCLUSION

Solid phase organic synthesis is a very useful tool for the synthesis of organic compounds in that it allows one a very quick and easy method for separating the synthesized product from the soluble components of the reaction mixture. Because of the ease in separating the product from the reaction mixture, one is able to use a large excess of starting materials and reagents to drive a reaction to completion. Of course, with this capability come several issues that one must address, such as the supports loading capacity, its stability under the planned reactions conditions, as well as its overall solute diffusion properties. After the support has been chosen, one must look at the type of linker that one will use, making certain that the conditions necessary to cleave the linker will be compatible with the final product. Finally, once the compound has been synthesized, a series of analytical methods must be deployed to evaluate the extent to which the reaction has gone to completion. Following these steps will not guarantee success but it will help greatly reduce the time required in determining whether or not success can be achieved. Although this review has not gone into great depth on the various analytical methods available to the solid phase organic chemist in determining the extent to which a reaction has gone to completion, it has touched on the various approaches. The next chapter in this volume will have a more in-depth discussion about some of the various chemical methods one can adopt on a solid support that can give one a better idea as to how a solid phase reaction has proceeded. These techniques are fairly quick and inexpensive to perform and should be learned by anyone seriously interested in performing solid phase organic synthesis.

REFERENCES

1. For an excellent book covering the field, see *Synthesis and Separation Using Functional Polymers*, 1988 (Sherrington, D.C. and Hodge, P.; eds.), John Wiley and Sons, New York.
2. Galat, A. *J. Am Chem. Soc.* 1948, *70*, 3945.
3. Astle, M.J.; Abbot, F.P. *J. Org. Chem.* 1956, *21*, 1228–1232.
4. Merrifield, R.B. *J. Am. Chem. Soc.* 1963, *85*, 2149–2154.
5. Merrifield, R.B. *J. Am. Chem. Soc.* 1964, *86*, 304–305.
6. Merrifield, R.B.; Steward, J.M.; Jernberg, N. *Anal. Chem.* 1966, *38*, 1905–1910.
7. Letsinger, R.L.; Kornet, M. *J. Am. Chem. Soc.* 1963, *85*, 3045–3046.
8. Letsinger, R.L.; Kornet, M.J.; Mahadevan, V. Jerina, D.M. *J. Am. Chem. Soc.* 1964, *86*, 5163–5165.
9. Letsinger, R.L.; Mahadevan, V. *J. Am. Chem. Soc.* 1965, *87*, 3526–3527.
10. Fridkin, M.; Patchornik, A.; Katchalski, E. *J. Am. Chem. Soc.* 1966, *88*, 3164–3165.
11. Fréchet, J.M.; Schuerch, C. *J. Am. Chem. Soc.* 1971, *93*, 492–494.
12. Farrall, M.J.; Fréchet, J.M. *J. Org. Chem.* 1976, *41*, 3877–3882.
13. Fréchet, J.M; de Smet, M.D.; Farrall, M.J. *Polymer* 1979, *20*, 675–680.
14. Leznoff, C.C.; Wong, J.Y. *Can J. Chem.* 1973, *51*, 3756–3758.
15. Wong, J.Y.; Manning, C.; Leznoff, C.C. *Angew. Chem., Int. Ed. Engl.* 1974, *13*, 666–667.
16. Fyles, T.M., Leznoff, C.C.; Weatherston, J. *J. Chem. Ecol.* 1978, *4*, 109–116.
17. Kawana, M.; Emoto, S. *Tetrahedron Lett.* 1972, 4855–4858.
18. Fridkin, M.; Patchornik, A.; Katchalski, E. *J. Am. Chem. Soc.* 1965, *87*, 4646–4647.
19. Regen, S.L. *J. Am. Chem. Soc.* 1974, *96*, 5275–5276.
20. Regen, S.L. *Macromolecules* 1975, *8*, 689–690.
21. Epton, R.; Goddard, P.; Irvin, K.*J. Polymer* 1980, *21*, 1367–1368.
22. Fréchet, J.M.J. *Tetrahedron* 1981, *37*, 663–683.
23. Giralt, E.; Albericio, F.; Bardella, F.; Eritja, R.; Feliz, M.; Pedroso, E.; Pons, M.; Rizo, J. In: *Innovations and Perspectives in Solid Phase Synthesis* (Epton, R., ed.), 1990, Hartnolls Ltd, Bodmin, Cornwall, pp. 111–120.
24. Albericio, F.; Pons, M.; Pedroso, E.; Giralt, E. *J. Org. Chem.* 1989, *54*, 360–366.
25. Giralt, E.; Rizo, J.; Pedroso, E. *Tetrahedron*, 1984, *40*, 4141–4152.
26. Look, G.C.; Holmes, C.P.; Chin, J.P.; Gallop, M.A. *J. Org. Chem.* 1994, *59*, 7588–7590.
27. Bayer, E.; Rapp, W. In *Poly(ethylene Glycol) Chemistry: Biotechnology and Biomedical Applications* (Harris, J.M., ed.), 1992, Plenum Press, New York, 325–345.
28. Bayer, E.; Albert, K.; Willisch, H.; Rapp, W.; Hemmasi, B.*Macromolecules* 1990, *23*, 1937–1947.
29. Rapp, W.E., In Wilson, S.R. and Czarnik, A.W. (Eds.) *Combinatorial Chemistry: Synthesis and Application*, 1997, John Wiley and Sons, New York, pp. 65–93.
30. Quarrell, R.; Claridge, T.D.W.; Weaver, G.W.; Lowe, G., *Molecular Diversity*, 1996, *1*, 223–232.
31. Santini, R.; Griffith, M.C.; Qi, M. *Tetrahedron Lett.* 1998, *39*, 8951–8954.
32. Li, W.; Yan, B. *Tetrahedron Lett.* 1997, *38*, 6485–6488.
33. Li, W.; Yan, B. *J. Org. Chem.* 1998, *63*, 4092–4097.

34. Zhao, B.P.; Panigrahi, G.B.; Sadowski, P.D.; Krepinsky, J.J. *Tetrahedron Lett.* 1996, 3093–3097.

35. Devivar, R.V.; Koontz, S.L.; Peltier, W.J.; Pearson, J.E.; Buillory, T.A.; Fabricant, J.D. *Bioorgan. Medic. Chem. Lett.* 1999, *9*, 1239–1242.

36. Frank, R. *Tetrahedron* 1992, *48*, 9217–9232.

37. Kramer, A.; Reineke, U.; Malin, R.; Schleuning, W.; Vakalopoulou, E.; Scholz, P.; Mergener, J.S. *Pept. Res.* 1993, *6*, 314–318.

38. Truffert, J.C.; Lorthioir, O.; Asseline, U., Thuong, T.N.; Brack, A. *Tetrahedron Lett.* 1994, *35*, 2353–2356.

39. Kemp, M.; Barany, G. *J. Am. Chem. Soc.*, 1996, *118*, 7083–7093.

40. Geysen, H.M.; Meloen, R.H.; Bareling, S.J. *Proc. Natl. Acad. Sci. USA.* 1984, *81*, 3998–4002.

41. Bray, A.M.; Maeji, N.J.; Geysen, H.M. *Tetrahedron Lett.* 1990, *31*, 5811–5814.

42. Minganti, C.; Ganesh, K.N.; Sproat, B.S.; Gait, M.J. *Anal. Biochem.* 1985, *147*, 63–74.

43. Renil, M.; Meldal, M., *Tetrahedron Lett.* 1995, *33*, 4647–4659.

44. Meldal, M.; Auzanneau, F.I., Hindsgaul, O.; Palcic, M.M. *J. Chem. Soc., Chem. Commun.* 1994, 1849–1850.

45. Renil, M.; Meldal, M., *Tetrahedron Lett.* 1996, *37*, 6185–6188.

46. Buchardt, J.; Meldal, M. *Tetrahedron Lett.* 1998, *39*, 8695–8698.

47. Bayer, E.; Rapp, W. *Chem. Pept. Prot.* 1986, *3*, 3–8.

48. Bayer, E.; US 4,908,405, 1990, 5 pages.

49. Bayer, E.; Rapp, W. In *Poly(Ethylene Glycol) Chemistry: Biotechnical and Biomedical Applications*, 1992 (Harris, J.M., ed.), Plenum Press, New York, pp. 325–345.

50. Zalipsky, S.; Albericio, F.; Barany, G. In *Peptides: Structure and Function. Proceedings of the Ninth American Peptide Symposium*, 1985 (Hruby, V.J.; Deber, C.N.; Kopple, K.D., eds.), Pierce Chemical Co., Rockford, IL, pp. 257–260.

51. McGuiness, B.F.; Kates, S.A.; Griffin, G.W.; Herman, L.W.; Solé, N.A.; Vágner, J.; Albericio, F.; Barany, G. In *Peptides: Chemistry and Biology: Proceedings of the 14th American Peptide Symposium* (Kaumaya, P.T.P.; Hodges, R.S., eds.), 1996, Mayflower Worldwide, Birmingham, U.K., pp. 125–126.

52. Labadie, J.W.; Deegan, T.L., Jr.; Tran, T.W.; van Eikeren, P. *Polym. Mater. Sci. Eng.* 1996, *75*, 389–390.

53. Gooding, O.W.; Baudart, S.; Deegan, T.L.; Heisler, K.; Labadie, J.W.; Newcomb, W.S.; Porco, J.A., Jr.; van Eikeren, P. *J. Comb. Chem.* 1999, *1*, 113–122.

54. Adams, J.H.; Cook, R.M.; Hudson, D.; Jammalamadaka, V.; Lyttle, M.H.; Songster, M.F. *J. Org. Chem.* 1998, *63*, 3706–3716.

55. Szymonifka, M.J.; Chapman, K.T. *Tetrahedron Lett.* 1995, *36*, 1597–1601.

56. Sucholeiki, I.; Perez, J.M. *Tetrahedron Lett.* 1999, *40*, 3531–3534.

57. Rana, S.; White, P.; Bradley, M. *Tetrahedron Lett.* 1999, *40*, 8137–8140.

58. Small, P.W.; Sherrington, D.C. *J. Chem. Soc. Chem. Commun.* 1989, 1589–1591.

59. Bhaskar, N.K.; King, B.W.; Meyers, P.; Westlake, J.P. in *Innovation and Perspectives in Solid Phase Synthesis, Peptides, Proteins and Nucleic Acids* (Epton, R. ed.), 1994, Mayflower Worldwide Ltd, Birmingham, UK, pp. 451–452.

60. Sucholeiki, I.; Pavia, M.R.; Kresge, C.T.; McCullen, S.B.; Malek, A.; Schramm, S. *Mole. Divers.* 1998, *3*, 161–171.

61. Fields, G.B.; Noble, R.L. *Int. J. Pept. Prot. Res.* 1990, *35*, 161–214.

62. Balkenhouhl, F.; von dem Busshe-Hünnefeld, C.; Lansky, A.; Zechel, C. *Angew. Chem., Int. Ed. Engl.* 1996, *35*, 2288–2337.

63. A good source that reviews the various linkers commercially available for solid phase chemistry including their corresponding cleavage conditions is the yearly updated "synthesis notes" produced by Calbiochem-Novabiochem.

64. Barlos, K.; Gatos, D.; Karpolos, S., Phaphotiu, G.; Schafer, W.; Wenquing, Y.; *Tetrahedron Lett.* 1989, *30*, 3947–3950.

65. Sheppard, R.C.; Williams, B.J. *Int. J. Pept. Prot. Res.* 1982, *20*, 451–454.

66. Bernatowics, M.S.; Daniels, S.B.; Köster, H. *Tetrahedron Lett.* 1989, *30*, 4645–4648.

67. Barany, G.; Merrifield, R.B., in *The Peptides*, Vol. 2 (Gross E, Meienhofer, J; Eds.), Academic Press, New York, 1980, pp. 1–284.

68. Rink, H. *Tetrahedron Lett.* 1987, *28*, 3787–3790.

69. Wang, S.-S. *J. Am. Chem. Soc.* 1973, *95*, 1328–1333.

70. Barltrop, J.A.; Schofield, P. *Tetrahedron Lett.*, 1962, *16.*, 697–699.

71. Barltrop, J.A.; Plant, P.J.; Schofield, P., *Chem. Commun.* 1966, *22*, 822–823.

72. Chamberlin, J.W. *J. Org. Chem.* 1966, *31*, 1658–1659.

73. Rich, D.H.; Gurwara, S.K. *J. Am. Chem. Soc.*, 1975, *97*, 1575–1579.

74. Wang, S.-S. *J. Org. Chem.* 1976, *41*, 3258–3261.

75. Tjoeng, F.S.; Heavner, G.A. *Tetrahedron Lett.* 1982, *23*, 4439–4442.

76. Pillai, V.N.R.; Ajayaghosh, A. *Ind. J. Chem.* 1988, *27B*, 1004–1008.

77. Fodor, S.P.A.; Read, J.L.; Pirrung, M.S.; Stryer, L.; Lu, a.T.; Solas, D. *Science* 1991, *251*, 767–773.

78. Holmes, C.P.; Jones, D.G. *J. Org. Chem.* 1995, *60*, 2318–2319.

79. Sucholeiki, I.; *Tetrahedron Lett.* 1994, *35*, 7307–7310.

80. Forman, F.W., Sucholeiki, I. *J. Org. Chem.*, 1995, *60*, 523–528.

81. Plunkett, M.J.; Ellman, J.A. *J. Org. Chem.* 1995, *60*, 6006–6007.

82. Hu, Y.; Porco, Jr., J.A.; Labadie, J.W.; Gooding, O.W. *J. Org. Chem.* 1998, *63*, 4518–4521.

83. For a good review see: Fitch, W.L. in *Annual Reports in Combinatorial Chemistry and Molecular Diversity*, 1997, Moos, W.H.; Pavia, M.R.; Kay, B.K.; Ellington, A.D. (editors), ESCOM, Leiden, 59–68.

Chapter 2
Methods of Analysis to Determine the Progress of a Chemical Reaction on Solid Support

J. Manuel Perez
Solid Phase Sciences Corporation, Medford, Massachusetts

I. INTRODUCTION

The quantification of functional groups on a solid support is crucial during solid phase organic synthesis, not only for the determination of the loading capacity on the solid support but for monitoring the progress of the solid phase chemical reaction. The loading capacity, or number of active functional groups available for reaction on a solid support, is typically expressed in millimoles of functional group per gram of resin. As a chemical reaction proceeds on a solid support, the molecular weight of the resin-bound compound increases, causing a reduction in the millimolar amount of such compound per gram of support, resulting in a decrease in the relative substitution (loading) per gram of resin. For example, if a molecule with a molecular weight of 200 g/mol is coupled to a resin-bound amine with a loading of 1 mmol/g, a decrease in loading to 0.83 mmol of molecule per gram of resin will be observed, assuming a 100% coupling efficiency for the reaction. If instead, a molecule with a molecular weight of 1000 g/mol is coupled, the loading will decrease to 0.5 mmol/g. This theoretical loading capacity can be calculated using the following equation:

$$\text{Loading capacity (mmol/g of resin)} = \frac{I_{LC}}{1 + [(I_{LC} \times FW_{Add})/1000]}$$

This chapter includes contributions in the form of case studies by J. Manuel Perez, Irving Sucholeiki, and Richard A. Laursen.

where I_{LC} is the initial loading in mmol/g of resin and FW_{Add} is the formula weight of the molecule coupled to the solid support.

It is important to assess how far a particular reaction has proceeded in order to optimize the reaction yields on the solid support. Quantifying the loading capacity—or the amount of functional groups attached to a solid support after a solid phase reaction—is important in estimating the completeness of the reaction. For example, if one calculates the theoretical loading capacity of a solid support to be 0.83 mmol/g and finds experimentally after running the reaction that the loading capacity is only 0.4 mmol/g, then one can conclude that the reaction has only gone to 50% completion and will require either an additional coupling step or further optimization to guarantee a higher yield.

Experimentally, the actual loading of a support has been typically obtained by cleaving a portion of the resin and either weighing the released product ("cleave-and-weigh" method) or analyzing it by high-performance liquid chromatography (HPLC) [1,2]. A drawback with these methods is that determination of the loading capacity might be affected by complications from the cleavage reaction. Elemental analysis is a better way to determine the loading level on a solid support, and it is most often used when coupling a particular molecule that results in the addition of a particular element (C, N, F, Cl, Br, I) on the support [2,3]. The increase in elemental mass of that element is then related to the loading on the solid support. This method has the convenience that no cleavage reaction is necessary to obtain the loading value; however, it does have some disadvantages. For example, it is expensive, it takes time to get the results if the analysis is subcontracted out, and the accuracy depends on the element measured.

All methods mentioned so far may not be practical for monitoring the progress of multiple solid phase reactions in a timely manner. Therefore, other simpler and faster methods have been developed to monitor the progress of a solid phase reaction. Most if not all of these methods are limited to the determination of the amine loading on a solid support since they were originally developed to monitor the progress of solid phase peptide synthesis. These methods are based on the release of an ultraviolet (UV)–active molecule from the resin. Methods to quantify the presence of other functional groups on a resin have been limited to thiols [4,5], aldehydes and ketones [6], and secondary amines [7], among others.

Here we will only discuss the most widely used chemical methods of analysis to determine the amine loading capacity on a solid support. Among these methods, the ninhydrin and the picric acid tests are based on the reaction of free amino groups on the solid support with a reagent (either ninhydrin or picric acid) to form a chromophore that can be monitored spectrophotometrically after being cleaved off the resin. In contrast, in the Fmoc test, a molecule containing a fluorenylmethoxycarbonyl (Fmoc) group (usually an Fmoc–amino acid) is coupled to the resin, cleaved off, and measured spectrophotometrically. Here we will present

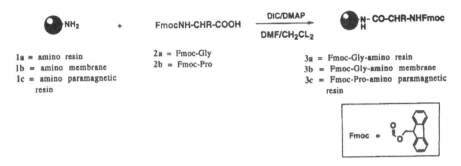

Scheme 1

examples using a variety of solid supports, such as aminomethypolystyrene resin (**1a**), aminoethylmethacrylamide-coated polypropylene membrane (**1b**) [8], and amino-methylated paramagnetic resin (**1c**) [9] (Scheme 1).

CASE STUDY 2-1: COUPLING OF AN FMOC–AMINO ACID TO A SOLID SUPPORT

J. Manuel Perez and Irving Sucholeiki   *Solid Phase Sciences Corporation, Medford, Massachusetts*

The coupling of an Fmoc–amino acid to a solid support can be done by a variety of methods. Coupling techniques currently used include the use of activated esters (HOBt or OPfp), activating agents (PyBOP, HBTU, etc.), and preformed symmetrical anhydride using diisopropylcarbodiimide (DIC). Here we used preformed symmetrical anhydrides to couple either Fmoc–glycine or Fmoc–proline to the amino supports (**1a**), (**1b**), and (**1c**) in Scheme 1. The formation of the symmetrical anhydride of the corresponding Fmoc–amino acid is done using DIC and 4-(dimethylamino)pyridine (DMAP). The preformed anhydride is then added to the solid support and the reaction continued overnight.

Chemicals

Aminomethylpolystyrene resin (Aldrich)
Aminoethylmethacrylamide-coated polypropylene membrane (Millipore)
Aminomethylated paramagnetic resin (Solid Phase Sciences)
Diisopropylcarbodiimide (DIC) (Aldrich)
4-(Dimethylamino)pyridine (DMAP) (Aldrich)
N-Fmoc-L-glycine-OH (Fmoc-Gly) (Novabiochem)

N-Fmoc-L-proline-OH (Fmoc-Pro) (Novabiochem)
Dimethylformamide (DMF) (EM Science)
Methylene chloride (CH_2CL_2) (J.T. Baker)
Methanol (J.T. Baker)
Piperidine (Aldrich)

Equipment and Supplies

Stirring bars
One round-bottom flask (25 mL)
Eppendorf brand pipetor (200 µL and 1000 µL)
Three plastic bottles (25 mL)
Fritted funnel

Procedure

1. Six equivalents (based on the solid support amino loading) of Fmoc–amino acid (**2a** or **2b**) was added to a 25-mL round-bottom flask and dissolved in a minimal amount of DMF (1–2 mL)

2. Two milliliters of methylene chloride was added to the Fmoc–amino acid solution and stirred at 0°C.

3. Three equivalents of DIC was dissolved in 1–2 mL of methylene chloride and added to the Fmoc–amino acid solution while stirring at 0°C.

4. The reaction mixture was stirred for 20 min at 0°C under a nitrogen atmosphere to keep the reaction free of moisture.

5. Additional methylene chloride (2–3 mL) was added because the suspension became viscous and therefore difficult to stir.

6. The reaction mixture was then transferred to a plastic or glass reaction vessel containing 1 equivalent of solid support (**1a, 1b,** or **1c**) and 0.1 equivalent of DMAP at room temperature.

7. The suspension was then agitated gently overnight.

8. After 24 h, the resulting Fmoc–amino acid–coupled supports **3a** and **3b** were filtered using a fritted funnel and manually washed with DMF, methanol, and methylene chloride, in that order.

9. The coupled paramagnetic support **3c** was not recovered by filtration; instead, the support was separated from the reaction solution using a magnet and the solution aspirated off. The DMF, methanol, and methylene chloride washes were done again using a magnet to attract the support and aspiration to take out the washes (see Chapter 11 for more details).

10. Finally, the resulting solid support (**3a**, **3b**, or **3c**) was dried under vacuum.

CASE STUDY 2-2: FMOC TEST

J. Manuel Perez and Irving Sucholeiki *Solid Phase Sciences Corporation, Medford, Massachusetts*

This version of the Fmoc method, developed by Green and Bradley [10], measures the release of the fluorenylmethoxycarbonyl (Fmoc) after treatment with piperidine (Scheme 2). This method has been used extensively in monitoring the progress of a solid phase peptide synthesis reaction using Fmoc–amino acids. After coupling the Fmoc–amino acid to the solid support, the resin is washed and dried, and a sample is treated with 20% piperidine in DMF to release the Fmoc group in the form of a dibenzofulvene–piperidine adduct (**5**) (Scheme 2). The concentration of this adduct is determined spectrophotometrically at 301 nm using an extinction coefficient of 7200 mL mmol^{-1} cm^{-1} and related to the amount of amino groups present on the support. This method can be used to determine the loading of any Fmoc-containing molecule on a solid support. Here we will use this method to estimate the loading on the Fmoc–Gly–amino membrane **3b** and the Fmoc–Pro–amino paramagnetic resin **3c** (Scheme 2).

Chemicals

Dimethylformamide (DMF) (EM Science)
Methanol (Aldrich)
Piperidine (Aldrich)

Equipment and Supplies

Fritted funnel
Eppendorf brand pipetor (200 µL and 1000 µL)

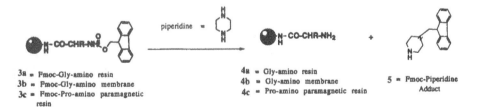

3a = Fmoc-Gly-amino resin
3b = Fmoc-Gly-amino membrane
3c = Fmoc-Pro-amino paramagnetic resin

4a = Gly-amino resin
4b = Gly-amino membrane
4c = Pro-amino paramagnetic resin

5 = Fmoc-Piperidine Adduct

Scheme 2

Table 1 Amine Loading on Fmoc–Glycine Membrane **3b** Determined by the Fmoc Method

Abs(301)	Wt (g)	Vol (mL)	Loading
0.280	0.0067	10	0.058 mmol/g
			3.9×10^{-4} mmol/cm^2
0.267	0.0060	10	0.062 mmol/g
			3.7×10^{-4} mmol/cm^2
0.285	0.0067	10	0.059 mmol/g
			3.9×10^{-4} mmol/cm^2

Volumetric flask (10 mL)
UV-Vis spectrophotometer

Procedure

1. The derivatized solid support **3b** or **3c** was weighted out (5–10 mg) and treated with 1 mL of 20% piperidine in DMF for 20–30 min under mild stirring.
2. The piperidine solution was then transferred to a 10-mL volumetric flask and diluted to 10 mL with methanol.
3. The absorbance of the solution was then measured at 301 nm.
4. The loading in mmol/g of support was calculated with the following equation, using an extinction coefficient (e) value of 7200 mL mmol^{-1} cm^{-1}:

$$\text{Loading (mmol/g of support)} = \frac{\text{Absorbance (301 nm)}}{\text{Wt of sample (g)}} \times \frac{\text{Vol (mL)}}{e} \qquad (1)$$

Results

The Fmoc assay test was first used to estimate the loading on samples of amino membrane after they had been derivatized with Fmoc–glycine (**3b**). Membrane samples (1 cm^2 each) were cut, weighed, and treated with 1 mL of 20% piperidine. The absorbance readings at 301 nm and the corresponding Fmoc loading values for these samples are shown in Table 1. The amine loading on a membrane support can be expressed either in mmol/g of membrane or mmol/cm^2 of membrane by only substituting the weight (in g) by area (in cm^2) in Eq. (1).

Table 2 Amine Loading on Fmoc–Proline Resin
3c Determined by the Fmoc Method

Abs(301)	Wt (g)	Vol (mL)	Loading
0.168	0.0022	50	0.53 mmol/g
0.351	0.0047	50	0.52 mmol/g
0.131	0.0017	50	0.54 mmol/g

$$\text{Loading (mmol/g of support)} = \frac{\text{Absorbance (301 nm)}}{\text{Area (cm}^2)} \times \frac{\text{Vol (mL)}}{e} \tag{2}$$

The average amine loading for membrane **3b** using the Fmoc method was found to be 0.060 mmol/g (60 µmol/g) or 3.8×10^{-4} mmol/cm^2 (380 nmol/cm^2). The reported amine loading for this membrane is 300–400 nmol/cm^2 [8].

The Fmoc test was also used to estimate the Fmoc loading on samples of amino paramagnetic resin after they had been derivitized with Fmoc-Pro (**3c**) (Table 2). The average Fmoc loading for this Pro-derivatized paramagnetic support is 0.53 mmol/g.

CASE STUDY 2-3: NINHYDRIN TEST

J. Manuel Perez and Irving Sucholeiki *Solid Phase Sciences Corporation, Medford, Massachusetts*

The ninhydrin test is based on the reaction of ninhydrin (**6**) with free amino groups on the solid support resulting in the formation of a chromophore (**7**) that absorbs at 570 nm (Scheme 3). This test was originally developed by Kaiser [11] to qualitatively monitor the progress of a solid phase peptide coupling reaction

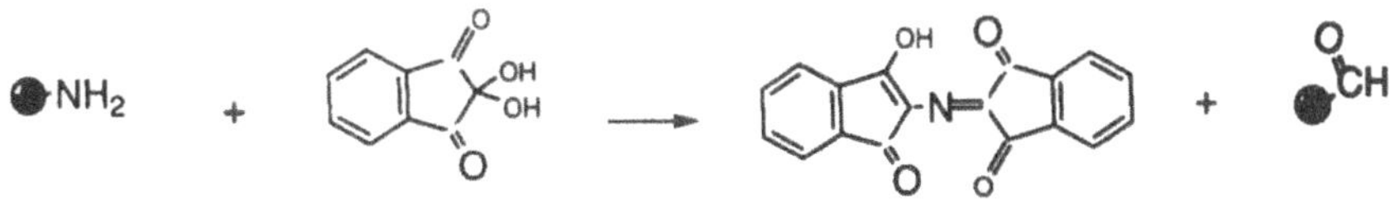

4a = Gly-amino resin   6 = Ninhydrin   7 = One of several possible
4b = Gly-amino membrane   Ninhydrin Adducts

Scheme 3

by visually monitoring the intensity of the blue color found both in solution and on the resin. A quantitative method for the determination of primary amino groups during solid phase synthesis was later developed by Sarin and Merrifield [12]. This method has been widely used in the determination of the amino loading levels of most amino acids and peptides attached to a solid support. However, the authors found that the extinction coefficient used to quantify the level of amino groups present varies with the nature of the amino group and peptide under study. For example, the extinction coefficient at 570 nm for a Leu resin was found to be 16,500 mL mmol^{-1} cm^{-1}, as compared to nonderivatized aminomethyl resin, which was 14,900 mL mmol^{-1} cm^{-1}. For this reason, extending this method to determine the presence of other types of primary amine on a solid support requires the determination of the extinction coefficient for that particular amine if an accurate amine loading is expected. In most cases, however, an average extinction coefficient value of 15,000 mL mmol^{-1} cm^{-1} is used. Here we describe the application of the ninhydrin test to estimate the amine loading on two different supports that has been derivatized with glycine, Gly–amino resin (**4a**), and Gly–amino membrane (**4b**) (Scheme 3).

Chemicals

Phenol (Aldrich)
Ninhydrin (Aldrich)
Potassium cyanide (KCN)
Ethanol (Aldrich)
Pyridine (Aldrich)
Tetraethylammonium chloride (Aldrich)
Methylene chloride

Equipment and Supplies

Fritted funnel
Eppendorf brand pipetor (200 μL and 1000 μL)
Beaker
Volumetric flask
UV-VIS spectrophotometer
Test tubes
Hot plate

Procedure

1. The following two solutions were prepared:

     A) **Phenol/KCN solution**: Composed of solution 1 and 2 mixed together:

     Solution 1: 40 g of phenol dissolved in 10 mL ethanol

     Solution 2: 2 ml of a 0.65 mg/mL KCN diluted up to 100 mL with pyridine

     B) **Ninhydrin solution**: 2.5 g ninhydrin dissolved in 50 mL ethanol

2. Two to five milligrams of solid support (**4a** or **4b**) was added to a test tube.

3. As a control, 2–5 mg of the precursor solid support (**3a** or **3b**) was used.

4. To a test tube containing 2–5 mg of solid support, 100 µL of the phenol/KCN solution and 25 µL of the ninhydrin solution were added. In addition, a test tube containing both solutions and no resin was used as a blank.

5. The test tubes were heated at 100°C for 10 min and then cooled to room temperature.

6. To each test tube 1 mL of 60% ethanol in water was added.

7. Next, the mixture was filtered through a Pasteur pipet containing a filter plug and rinsed twice with 200 µL of a 0.5 M tetraethylammonium chloride solution in methylene chloride. To the reagent blank was added 400 µl of 0.5 M tetraethylammonium chloride solution.

8. The filtrate was transferred to a 10-mL volumetric flask. At this point, the filtrate might be composed of two layers that will come in solution when diluted to 10 mL with 60% ethanol in water.

9. The absorbance of this solution was measured at 570 nm for each of the amino support samples (**4a** and **4b**), including the precursor Fmoc supports (**3a** and **3b**).

10. The loading of the support was calculated with Eq. (3) using an extinction coefficient (e) of 15,000 mL mmol^{-1} cm^{-1}:

$$\text{Loading (mmol/g of support)} = \frac{\text{Absorbance (570 nm)}}{\text{Wt of sample (g)}} \times \frac{\text{Vol (mL)}}{e} \tag{3}$$

11. The loading can also be expressed in mmol/cm^2 by simply substituting weight by cm^2 in Eq. (3).

Results

The ninhydrin test was used to determine the amine loading on samples of Gly–amino resin (**4a**) and Gly–amino membrane (**4b**) (Scheme 3). As a control, the

Table 3 Amine Loading on Membrane 4b Determined by the Ninhydrin Method

Abs(570)	Wt (g)	Vol (mL)	Amine loading
0.538	0.0059	10	0.061 mmol/g 3.60×10^{-4} mmol/cm^2
0.523	0.0053	10	0.066 mmol/g 3.5×10^{-4} mmol/cm^2
0.570	0.0064	10	0.060 mmol/g 3.80×10^{-4} mmol/cm^2

amine loading value of the precursor Fmoc supports **3a** and **3b** (Scheme 1) was determined, but they showed no significant absorbance readings at 570 nm. The absorbance readings at 570 nm were obtained for three different samples of the amino membrane **4b** (0.5 cm^2 each), and the corresponding amine loadings were calculated in terms of mmol/g and mmol/cm^2 (Table 3). The average amine loading for membrane **4b** using the ninhydrin method was found to be 0.062 mmol/ g (62 μmol/g) or 3.6×10^{-4} mmol/cm^2 (363 nmol/cm^2). The reported amine loading for this membrane is 300–400 nmol/cm^2 [8]. For three samples of amino resin **4a** the following readings at 570 nm and their corresponding amine loading values were obtained (Table 4). The average amine loading for this resin using this method was found to be 0.85 mmol/g (84 μmol/g). The reported amine loading for this resin is 1 mmol/g.

CASE STUDY 2-4: PICRIC ACID TEST

J. Manuel Perez   *Solid Phase Sciences Corporation, Medford, Massachusetts* **and Richard A. Laursen**   *Boston University, Boston, Massachusetts*

The picric acid test is based on the reaction of picric acid (**8**) with free amino groups on a solid support as described originally by Gisin [13]. As a result, a

Table 4 Amine Loading on Support **4a** Determined by the Ninhydrin Method

Abs(570)	Wt (g)	Vol (mL)	Amine loading
0.605	0.0047	100	0.86 mmol/g
0.543	0.0043	100	0.84 mmol/g
0.566	0.0044	100	0.85 mmol/g

picrate salt (**9**) is formed between the amino group on the polymer and the picric acid (Scheme 4). The support is then washed thoroughly to remove unbound picric acid. Base treatment quantitatively releases the picrate from the support into the solution. The concentration of picrate in this solution is determined spectrophotometrically at 358 nm using an extinction coefficient of 14,500 mL mmol^{-1} cm^{-1} and reflects the amine loading on the support. This method is a convenient and easy way to quantitatively determine the amine loading on a support. The method is highly reproducible and seems to be independent of the nature of the amino group. In addition, unlike the ninhydrin test, the loading of proline and other secondary amines on a support can be determined by this method. We will use this method to determine the amine loading on the nonderivatized amino membrane **1b**. [*Caution: Picric acid is explosive and should be handled with care. Solid picric acid should be stored in a moist (with water) condition.*]

Chemicals

Picric acid (35% water) (Aldrich)
Methylene chloride (EM Science)
Diisopropylethylamine (Aldrich)
Ethanol (Aldrich)

Equipment and Supplies

Fritted funnel
Eppendorf brand pipetor (200 μL and 1000 μL)
Volumetric flask (10 mL)
UV-Vis spectrophotometer
Fritted funnel

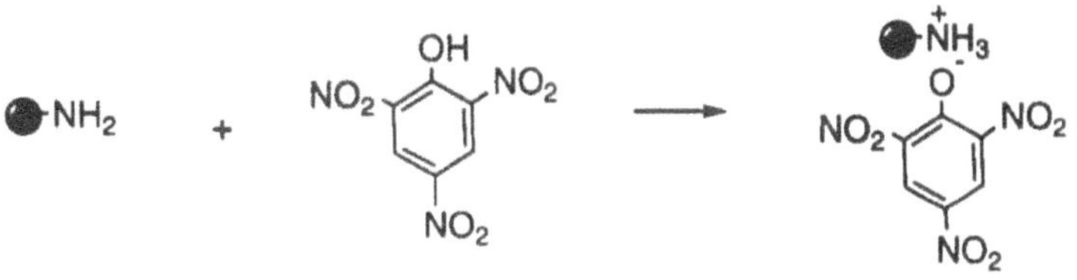

1b = amino membrane   8 = Picric Acid   9 = Picric Acid Adduct

Scheme 4

Procedure

1. The amino membrane **1b** was weighed out (2–10 mg) and washed 5 times with a 5% diisopropylethylamine in methylene chloride in a fritted funnel.
2. Next the support was washed 5 times with 1-mL portions of methylene chloride and then 5 times with 1-mL portions of a 0.1 M picric acid solution in ethanol.
3. The support was washed 5 times with methylene chloride until all non-bound picric acid was washed off (judged visually by the disappearance of yellow coloration coming from the support).
4. The support was then washed with 1 mL of 5% diisopropylethylamine in methylene chloride and the filtrate collected. The support was repeatedly washed with diisopropylethylamine until no yellow coloration is observed (usually 5 times) and all the washes combined.
5. All washes were transfered to a 10-mL volumetric flask (Vol) and diluted with ethanol.
6. The absorbance of the solution was measured at 358 nm using ethanol as a blank.
7. The loading in mmol/g of support was calculated with Eq. (4) using an extinction coefficient (e) of 14,500 mL mmol^{-1} cm^{-1}:

$$\text{Loading (mmol/g of support)} = \frac{\text{Absorbance (358 nm)}}{\text{Wt of sample (g)}} \times \frac{\text{Vol (mL)}}{e} \tag{4}$$

Results

The picric acid test was used to estimate the loading on four 0.5-cm^2 samples of nonderivatized amino membrane (**1b**). These pieces were weighed out and washed with methylene chloride before being treated with a 0.1 M picric acid solution as previously described. The corresponding absorbance readings at 358 nm and calculated amine loadings on these samples are shown in Table 5. The average amine loading for this membrane (**1b**) using the picric acid method was found to be 0.076 mmol/g (76 µmol/g) or 4.6×10^{-4} mmol/cm^2 (460 nmol/cm^2), a value that is slightly higher than that reported (300–400 nmol/cm^2) and also higher than the values determined by the Fmoc and ninhydrin methods.

II. CONCLUDING REMARKS

The use of chemical methods of analysis to determine the amine loading capacity on a solid support is a quick and easy way to monitor the progress of a solid phase

Table 5 Amine Loading on Membrane **1b** Determined by the Picric Acid Test

Abs(358)	Wt (g)	Vol (mL)	Amine loading
0.343	0.0031	10	0.076 mmol/g
			4.73×10^{-4} mmol/cm^2
0.340	0.0030	10	0.078 mmol/g
			4.69×10^{-4} mmol/cm^2
0.345	0.0033	10	0.072 mmol/g
			4.76×10^{-4} mmol/cm^2
0.320	0.0028	10	0.079 mmol/g
			4.41×10^{-4} mmol/cm^2

chemical reaction if the molecule being coupled on the solid support contains an amino group. No particular method is superior over the others, and it should be advised that one should determine the amine loading by at least two independent methods. In our hands, the picric acid test gave slightly higher values than the ninhydrin and Fmoc tests. This most probably is due to nonspecific binding of the picric acid to the solid support.

REFERENCES

1. Green, J. *J. Org. Chem. 60* (1995) 4287.
2. Forman, F.W. and Sucholeiki, I. *J. Org. Chem. 60* (1995) 1053.
3. Tietze, L.F. and Steinmetz, A. *Angew. Chem., Int. Ed. Engl. 35* (1986) 651.
4. Ellman, G.L.; *Arch. Biochem. Biophys. 82* (1959) 70.
5. Sucholeiky, I.; Lansbury, P.T. *J. Org. Chem. 58* (1993) 1318.
6. Yan B., Li, W. *J. Org. Chem. 62* (1997) 9354.
7. Chu, S.S.; Reich, S.H. *Biorg. Med. Chem. Lett. 5* (1995) 1053–1058.
8. Daniels, S.B., Bernatowicz, M.S., Coull, J.D. and Koster, H. *Tetrahedron Lett. 30* (1989) 4345.
9. Sucholeiki, I.; Perez, J.M. *Tetrahedron Lett. 40* (1999) 3531.
10. Green, J., Bradley K. *Tetrahedron 49* (1993) 4141.
11. Kaiser, E., Colescott, R.L., Bossinger, C.D., Cook, P.I. *Anal. Biochem. 34* (1970) 595.
12. Sarin, V.K., Kent, S.B.H., Tam, J.P., Merrifield, R.B. *Anal. Biochem. 117* (1981) 147.
13. Gisin B., *Anal. Chim. Acta 58* (1972) 24.

Part II
High-Throughput Synthesis for Drug Discovery

Chapter 3
High-Throughput Organic Synthesis: A Perspective Looking Back on 10 Years of Combinatorial Chemistry

Michael R. Pavia
Millennium Pharmaceuticals, Inc., Cambridge, Massachusetts

High-throughput organic synthesis is quickly becoming a standard technique practiced by many, if not most, medicinal chemists in the pharmaceutical and biotechnology industries. The technique has fundamentally changed the way medicinal chemists practice their art. We used to learn through an iterative one-at-a-time design, synthesize, and evaluate paradigm. Today we can generate large numbers of compounds and huge quantities of information very rapidly, which fundamentally changes the way we approach medicinal chemistry problems.

It has been only about 10 years since the concepts of combinatorial chemistry, molecular diversity, and parallel synthesis came to attract significant attention from a small group of scientists. How did we get to where we are today? What was unique about the environment 10 years ago that allowed this area of research to blossom? Where did those early efforts focus? How did they affect subsequent direction? Where are we today and in what directions are we headed? Has this technology been embraced and fully incorporated into the discovery process? These are the questions I will address in this introductory chapter.

I. WHAT WAS UNIQUE ABOUT THE ENVIRONMENT 10 YEARS AGO?

Until about 10 years ago there was no real high-throughput screening, there was no combinatorial chemistry, and there was no genomics. Biological targets were selected largely through suggestive but time-consuming pharmacological studies, mostly in whole animals. Lead chemical series were selected mostly from similarity to other known series or through a very low-volume screening approach involving hand-selected natural products and synthetic chemicals from the company's historical collection. And most of the screening was done in whole animals or isolated tissues. The information the chemist received from screening these compounds was used to design a very small number of compounds (sometimes one!) which were then synthesized and tested. This information would then be used to design the next molecule or series, and so on. In retrospect, the pharmaceutical industry's progress over the hundred years or so of its existence is amazing.

About 10 years ago, two major events took place that I believe led to high-throughput organic synthesis. The first was the introduction of the first high-throughput screening systems employing single molecular target proteins. We quickly were able to evaluate hundreds or even thousands of compounds. The historical company compound collections at the time usually consisted of tens or hundreds of thousands of compounds. While this probably seemed sufficient to many in the industry, a few scientists became aware of the need to obtain larger numbers of compounds.

I believe the second major event was the publication (and maybe more importantly bringing these publications to the attention of the medicinal chemistry community) of several papers describing the synthesis of large numbers of peptides. These papers included Mario Geysen's work [1] detailing the synthesis of peptides in multiple spatially addressable format, Richard Houghten's paper [2] describing the synthesis of peptides in "tea-bags," and George Smith's paper [3] describing the generation of huge quantities of peptides via phage display. Undoubtedly, there were other papers that I am not including. The purpose of this chapter is not to assign credit for the foundations of this field; many others have tried this over the years [4].

Interestingly, these papers didn't have a major influential effect from day one; they were seeds that when planted took a number of years before many scientists involved in new-drug discovery took serious notice. Speaking from my own experience as a medicinal chemist, I did not generally read literature focusing on peptides, and in turn the majority of peptide scientists had little involvement in the pharmaceutical discovery arena. I don't recall becoming aware of this literature until 1989.

II. WHERE DID THOSE EARLY EFFORTS FOCUS? HOW DID THEY AFFECT SUBSEQUENT DIRECTION?

As already mentioned, much of the initial work in the field focused on peptide library methods, primarily due to the ready availability of the natural and unnatural amino acids and a well-established coupling methodology. These methods were used to rapidly generate hundreds to millions of small to medium-size peptides for identifying novel leads or to help elucidate the chemical basis of known ligand–ligate interactions.

Key issues differentiating the various approaches were the size of the libraries, whether they were generated as single compounds or mixtures, and whether the final libraries were tested attached to a solid support or in solution. In the methods affording single compounds, the active member was identified by its physical location. In the case of mixtures, the active compound was identified either by its tag for encoded libraries or through deconvolution, where an active compound is identified by iterative synthesis and screening of mixtures. Since these early methods have previously been reviewed extensively [5,6], I will comment only briefly on a few key methodologies to illustrate the early thinking.

Two key methods illustrate the early work in spatially differentiated single-compound synthesis. Using the pin technology, Geysen [1] demonstrated that peptides could be synthesized in numbers several orders of magnitude greater than by conventional one-at-a-time methods. The peptides were synthesized on polyethylene rods (or pins) arranged in a microtiter plate format allowing 96 separate peptides to be simultaneously synthesized and tested for activity at the tips of the rods. Another classic example of individual-compound synthesis was reported by workers at Affymax [7] and was given the name VLSIPS (very large scale immobilized polymer synthesis). This method combined solid phase synthesis with photolithography and miniaturization on a silicon surface to achieve light-directed, spatially addressable, parallel chemical synthesis. It should be noted that both of these methods as originally used called for the peptides to be screened while attached to the solid surface, which in the opinion of many of us limited their utility.

The alternative approach was to rapidly prepare large mixtures of compounds. This approach was largely made possible by the introduction of the split-pool method [2]. An important consideration in synthesizing large mixtures of compounds is to ensure that each final mixture component is present in approximately equimolar concentrations. This issue was effectively addressed in the split-pool approach whereby the solid support material is physically segregated into equal portions for coupling to each individual initial reactant. This affords uniform coupling because competition between reactants is eliminated. The individual polymers are combined in a single vessel for washing and deprotection and then divided again into individual portions for the next coupling. The resulting synthetic products exhibit a statistical distribution of sequences. Based on this approach, a complete set of possible molecular combinations is rapidly prepared in approximately equimolar amounts.

As mentioned earlier, when a large mixture of compounds is prepared, one must then determine the structure of the compound of interest. This process is referred to as *deconvolution* and a number of clever methods were introduced to accomplish this task. An iterative deconvolution procedure was used with the tea-bag method. Physical isolation of beads containing active compounds was used by investigators at Selectide [8]. In this method the library is prepared on polymeric beads by the split-pool method and incubated with a tagged ligate. Those ligates with bound peptides are identified by visual inspection, physically removed, and microsequenced. In an alternative method, encoded libraries are introduced whereby an ''identifier'' tag is attached to the solid support material coincident with each monomer, again using a split-pool synthesis procedure. The structure of the molecule on any bead is obtained by decoding the identifier tags. Numerous methods of tagging the beads have been reported [9,10]. Again, at this point many of us had concerns about whether screening of mixtures was an effective method of identifying active compounds. In addition, several of these methods still evaluated compounds while attached to the solid-support material.

Finally, a biological method of generating huge numbers of peptides was reported using phage libraries [3,11]. These libraries contain tens of millions of filamentous phage clones, each displaying a unique peptide sequence on the bacteriophage surface. The phage genome contains the DNA sequence encoding for the peptide. The ligate of interest is used to affinity purify phage that displays binding peptides, the phage propagated in *E. coli*, and the amino acid sequences of the peptides displayed on the phage are identified by sequencing the corresponding coding region of the viral DNA. Tens of millions of peptides can be rapidly surveyed for binding. Limitations of this method are that only naturally occurring amino acids can be used and little is known about the effect of the phage environment.

Clearly, early work in the field of high-throughput synthesis was pursued from a number of different directions. Fundamental differences existed in regard to how many compounds were prepared, if single compounds or mixtures were produced, if the compounds were evaluated on solid support or tested in solution, etc. One thing all of the methods had in common was that they focused on the synthesis of peptides. (In most cases, small peptides are not suited as drugs due to in vivo instability and lack of oral absorption. Furthermore, the conversion of a peptide chemical lead into a pharmaceutically useful, orally active, nonpeptide drug candidate is more difficult than identifying the original peptide lead, and no general solution yet exists for designing effective peptide mimics.)

Although each of these methods demonstrated great scientific creativity, their usefulness for pharmaceutical drug discovery was limited.

In 1989, when I first became interested in this field, the first question we asked was, ''What do we ultimately want to achieve by this method?'' The answer seemed clear to us: to prepare large numbers of drug-like molecules (defined in greater detail below) and test them in solution that would represent what the target would see in the natural situation. Therefore, the approach we pursued was based on two major beliefs: (1) the technology needed to be applied to the synthesis of low molecular weight nonpeptide compounds and (2) single compounds needed to be tested in solution. The methods described earlier in this chapter sparked our imaginations to what was possible. Our challenge was to develop a derivative that possessed maximal usefulness for drug discovery.

My first foray into this field was during my time at Parke-Davis and resulted in an approach we termed "diversomers" [12]. At the time we believed that target molecules would best be prepared by solid phase chemistry. However, minimal work had been done on performing organic synthesis on solid support materials, with the exception of peptide and oligonucleotide chemistry. The little work that had been done suggested that the concept was feasible but raised some concerns [13,14].

We wanted to demonstrate that solid-supported chemistry could be used to prepare a pharmacologically useful class of nonpolymeric molecules. A search of the solid phase literature revealed that the solid phase synthesis of the benzodiazepine nucleus had been accomplished [15] in 1974! Although this particular synthetic route did not meet our requirements, we believed there was enough precedent in this paper to warrant a full-blown project. My co-workers Sheila Hobbs DeWitt, John Kiely, Mel Schroeder, and Chuck Stankovic did a great job in developing an elegant route that allowed us to generate a large number of structurally diverse benzodiazepines. One of the key attributes of this route was the fact that the site of attachment to the solid support was not present in the final compound. In fact, the final step of the synthesis formed the benzodiazepine structure while removing the molecule from the solid support. We very efficiently synthesized the desired molecule and met our need of generating compounds in solution for testing. During the course of this work the group (which now also included Donna Reynolds Cody) also prepared another pharmacologically important class of molecules, the hydantoins. At the same time that we were working out the solid phase chemistry, Sheila and Donna were busy developing an apparatus that would allow for simultaneous parallel synthesis. The reactions were performed on a polystyrene support placed inside of standard gas dispersion tubes. These tubes were originally arrayed into a block containing eight tubes allowing for the preparation of eight different molecules. The tubes (or "pins") were dipped in individual reaction wells into which different reagents were placed. The resin-containing tubes could be removed from the reaction wells after reaction for washing. The advantages of this "resin-in-a-pin" approach were that larger quantities of product was formed than with the original Geysen pin approach, and the chemistry we employed allowed for the compound to be removed from the resin. The apparatus was also designed to allow for heating, cooling, and operation under inert atmospheres. This work allowed us to demonstrate the overall feasibility of rapidly performing the synthesis of a wide range of structural analogs for biological testing. These were very exciting (and frequently frustrating) times for all of us in the group.

While our manuscript was in preparation, another group published a solid phase route to benzodiazepines [16]. However, the method was limited by the need to have an attachment group on the benzodiazepine (in the form of a hydroxyl group or carboxylic acid).

These two papers [12,16] set the stage for using the early peptide techniques to synthesize nonpeptide molecules by a large number of groups, and these activities continue to this day.

In 1992, I left Parke-Davis and moved to the Boston area to join a start-up biotech company named Genesis. This company was soon sold to Sphinx Pharmaceuticals. It was at Sphinx that I had the opportunity to continue my efforts in high-throughput synthesis. At Sphinx we initiated programs in two major areas: the first was to design a scaffold specific for a particular class of biological targets (kinases in this case) that was amenable to a wide range of modifications. The other major project was to build a versatile nonpeptide scaffold that would allow us to display a wide range of important functional groups in a vast number of orientations. The goal was to generate a "universal library" that would have a high probability of containing hits for a range of biological targets.

An additional goal of our efforts was to construct a simple apparatus that allowed for solid phase organic synthesis in 96-well format, the format commonly used at the time for high-throughput screening. We were ultimately successful in meeting all of these objectives. The credit goes to a fantastic team of co-workers, including Harold Meyers, Tim Powers, Irving Sucholeiki, Michael Cohen, Mark Hediger, Hong Zhu, Tracy Durgin, Oscar Almeida, and others.

After meeting our initial objectives we turned to expanding the scope of our successes, as well as continuing to improve our equipment. Each of these efforts are briefly described in more detail below.

Once again, we decided to prepare individual molecules of known structural identity using a parallel-synthesis approach. The individual chemical reactions were carried out via multistep organic synthesis in a spatially addressable and parallel format, so that one single and well-defined compound was prepared at each synthesis site and multiple syntheses were carried out simultaneously in multimilligram quantities. This approach allowed us to avoid the issues inherent in synthesizing and screening mixtures of compounds and allowed us to prepare a sufficient quantity of each compound to allow for evaluation in multiple assays over a period of years. Compounds were selected to be molecules with low molecular weights (generally <600 amu) to afford the best chance of oral bioavailability and to be "drug-like" (nonpeptide or oligomeric).

First, a few comments on the design of our apparatus. We decided to pursue parallel array synthesis using 96-well commercial microtiter plates and to apply automation only to select operations. As a result, we were successful in developing a low-cost, simple approach that could be easily utilized by combinatorial chemists and could also be easily transferred to traditional medicinal chemistry labs. Our basic reaction vessel consisted of a commercially available polypropylene 96-deepwell plate that was modified for filtration by the drilling of a small hole in the bottom of each well, followed by placement of a porous polyethylene fritted filter at the bottom of each well [17]. An aluminum plate clamp was made as a two-piece assembly, consisting of a solid base clamp fitted with four removable corner stainless steel studs, and a frame clamp that fits atop the plate and is secured with wing nuts. An inert gasket was utilized on the base clamp to prevent leakage of well contents.

As stated earlier, we decided to apply automation only to select tasks that would free the chemists for more productive endeavors and ensure consistency in repetitive procedures. Among the procedures that were automated were liquid transfer (solvents and reagents), solid transfer (reagents and resins), filtration, solid phase extraction, and several analytical procedures.

Our initial efforts in preparing libraries were once again focused on solid phase organic chemistry because this method has many advantages over traditional solution-based methods (at least at that point in time). For example, large excesses of starting materials can be employed to drive reactions to completion without fear of complicating the workup procedure; simple filtration and resin washing are typically sufficient for removal of unreacted starting materials and reaction byproducts, and resin-bound materials are easily manipulated. Nevertheless, solid phase synthesis also has many disadvantages. Solid phase reactions are more difficult to monitor by conventional techniques. Reactions are frequently significantly retarded on solid phase, and heterogeneous reagents cannot be employed. Many resins, particularly cross-linked polystyrene, display significant swelling and shrinking properties that can severely affect reaction rates and site accessibility; frequently the optimal solvent for resin swelling is not the optimal solvent for the desired reaction. However, we addressed many of these disadvantages with promising results at both Parke-Davis and Sphinx.

With a simple and economical procedure for carrying out library synthesis we set out to make libraries. Our initial target molecules were designed as potential protein kinase inhibitors. A general structural feature shared among known protein kinase (e.g., serine/threonine and tyrosine) inhibitors is phenolic or polyhydroxylated aromatic functionalities. As such, we designed a series of differentially functionalized bisbenzamide phenols (as well as the sulfonamide and urea analogs) [17].

Our synthesis strategy was to batch-prepare key resin-bound free aniline intermediates, which could then be functionalized at nitrogen in a divergent-synthesis strategy for constructing numerous libraries. For example, it was demonstrated that acylation, sulfonylation, and isocyanation are all feasible chemistries to provide libraries of bisbenzamide phenols, bisbenzosulfonamide phenols, and bisbenzourea phenols. For synthesis in a 96-well plate (an 8 × 12 array), 8 or 12 unique resins were distributed across 8 rows or down 12 columns in a plate, and cross-reacted with 12 or 8 unique electrophiles, respectively. For convenience, we chose to cross-react 8 unique resins in a plate with 12 electrophiles.

For the resin chemistry, we sought to employ a solid support linker for our phenolic substrates that would maintain its integrity across the range of anticipated reaction conditions. It was also desirable to employ a linker that could be readily cleaved under mild conditions to maintain the integrity of the scaffold and its attached functional groups. This strategy would enable us to monitor reactions and characterize products in solution at each synthetic step if desired. Hence, the benzoate ester proved to be a satisfactory linker group and could be cleaved readily under basic conditions.

In this initial library project we demonstrated a facile, multistep array synthesis of hundreds of small, nonpeptidyl compounds with moderate to excellent purity levels and in sufficient quantities for direct screening in various biological assays. The limited human resources required for this effort illustrated the economic and productivity advantages of this technology for small-molecule synthesis and readily allowed for the construction of thousands of molecules in a 1-year time frame. This was a key turning point, now knowing that it was possible to achieve efficient and rapid high-throughput organic synthesis. Of course, an expansion of the breadth of chemistries, additional applications, analytical improvements, and so forth were still needed.

Success with the "kinase" library led our colleagues at Sphinx in Durham, NC, under the direction of Steve Hall to continue making a collection of diverse, drug-like scaffolds that could be readily functionalized with a wide range of chemical functionalities. By early 1997 the team had prepared over 40 unique scaffolds. For example, the synthesis of two additional scaffolds was described in a recent paper [18]. In the first example, a thiophene library, the scaffold was prepared first in solution. The library was then prepared via nucleophilic substitution and acylation reactions. In the second example involving the preparation of highly functionalized pyrrolidines, the scaffold was synthesized from various diversity element–containing building blocks via sequential carbon–carbon bond–forming reactions. By this time the group had made significant automation and equipment improvements, achieved highly improved analytical methodologies, and developed more than 100 different organic transformations that could be routinely used in library construction.

It is worth mentioning one additional undertaking, an effort we called the "universal library approach." In this method, we designed a versatile class of molecules that allows for the rapid display of multiple functional groups in large numbers of spatial arrangements using one general class of scaffolds. The relative geometry of the functional groups is altered simply by changing the substitution patterns of the groups around the scaffold. The biphenyl scaffold was selected as our initial class of target molecules [19]. This drug-like scaffold allows for facile introduction of three or four functional groups in a large number of spatial arrangements simply by altering the substitution pattern on each aromatic ring. Furthermore, we had built into our design a simple method of changing the biphenyl scaffold to easily change the size, shape, and physical properties of the final products. It is important to note that the final products contain a pendant methyl group since we desired not to have an invariant hydroxy or carboxy group in our final product.

The biphenyl molecules described were synthesized by a combination of solution and solid phase chemistries. The scaffold was functionalized with appropriate side chains on the solid support using the Mitsunobu reaction and the final products are then cleaved from the solid support for subsequent modification and testing. This approach resulted in the preparation of tens of thousands of analogs for testing.

While the chemistry groups at Sphinx initially focused on solid phase organic chemistry, our colleagues at Lilly (who had acquired Sphinx in September 1994) began to take advantage of a complementary approach to solid phase chemistry that largely circumvents its disadvantages discussed but maintains many of its advantages. This involves carrying out library construction in solution but using either polymer-supported reagents to effect reaction or using polymer-bound scavengers to remove impurities and/or unwanted reactants or products [20]. This method, which is discussed in more detail in Chapter 5, has proven to be the preferred method when it can be applied.

III. HAS THIS TECHNOLOGY BEEN EMBRACED BY THE COMMUNITY? HAS IT BEEN FULLY INCORPORATED INTO THE DISCOVERY PROCESS AND THE EDUCATIONAL PROCESS?

I can speak from personal experience that as recently as 5 years ago many in the pharmaceutical, academic, and financial communities firmly believed that these technologies would simply not work and would have little application in the industry. Of course, there were some forward-looking organizations who saw the potential value and had either initiated collaborations with small companies or had initiated their own in-house efforts. However, I believe this group was in the minority. Today the picture has changed dramatically. Most companies have now set up separate groups or even departments of combinatorial chemistry. Even with this progress the acceptance of the techniques by mainstream medicinal chemists was quite slow and in many cases a high level of resistance was felt. But the evolution is continuing. High-throughput synthesis is becoming increasingly accepted, and these specialized groups are disappearing as the chemistry community gets accustomed to these techniques. The groups are being integrated into the traditional lead discovery and optimization groups, and high-throughput organic synthesis is becoming just another, albeit powerful, technique available to medicinal chemists.

It is also interesting to note the initial resistance of the academic world to this technology. In the early days it was very difficult to even get a publication accepted. One leading organic chemistry journal even expressed the opinion that they would not publish work in solid phase or parallel-synthesis techniques. One result was the introduction of the journal *Molecular Diversity* [21] to supply a publishing forum for workers in this field. Clearly this situation has changed. Today each week a significant number of manuscripts are published in the field.

I even recall a number of leading academic chemists making comments such as "this isn't science" or "this is garbage chemistry." Hopefully, they now realize the importance of these techniques to many industries where chemistry leads to new products.

IV. WHERE ARE WE TODAY AND IN WHAT DIRECTIONS ARE WE HEADED?

Clearly, over the past 10 years we have come a long way. Many scientists at small companies, big pharma, and academic labs have contributed to this technology.

Recent advances in genomics and molecular biology are supplying the pharmaceutical industry with large numbers of novel biological targets. The industry is responding by making use of high-throughput screening technologies coupled with compound libraries to rapidly identify and optimize ligands for these targets. Clearly, chemically generated screening and optimization libraries are being widely embraced by researchers in both industry and academia. There has been steady development of new chemistries and equipment applied to library generation, so that it is now possible to synthesize almost any desired class of compound using these methodologies. Additional methods and chemistries are being reported on a weekly basis [22].

Finally, it is worth mentioning the "pendulum phenomena" in this field. It is evident from the early work reported above that a major focus was to prepare very large libraries of compounds that "maximally covered diversity space." This was the complete opposite approach of many of our scientific colleagues who for the past 20 years have been pursuing the rational design approach to drug discovery. These two techniques represent the two extremes of the pendulum's swing. Happily, the pendulum is settling to its natural position in the middle. Very few successful groups today rely on one technique or the other. Groups with a rational design preference are routinely using chemical libraries to gather rapid information for their design process, and groups with a preference for library approaches are taking a greater interest in the rational design of smaller, smarter libraries. The techniques are converging to a result greater than the sum of the individual parts. We are using the best of both techniques to better achieve our ultimate goal: the discovery and development of novel therapeutic agents to improve human health.

REFERENCES

1. HM Geysen, RH Meloen, SJ Barteling SJ. *Proc Natl Acad Sci USA 81*: 3998–4002, 1984.
2. RA Houghten. *Proc Natl Acad Sci USA 82*: 5131–5135, 1985.
3. GP Smith. *Science 228*: 1315–1318, 1985.
4. MJ Lebl. *J Comb Chem 1*: 3–24, 1999.
5. WH Moos, GD Green, MR Pavia. In: JA Bristol, ed. *Annu Rep Med Chem* 315–324, 1993.
6. EM Gordon, RW Barrett, WJ Dower, SPA Fodor, MA Gallop. *J Med Chem 37*: 1385–1401, 1994.
7. SPA Fodor, JL Read, MC Pirrung, L Stryer, AT Lu, D Solas. *Science 251*: 767–773, 1991.
8. KS Lam, SE Salmon, EM Hersh, VJ Hruby, WM Kazmierski, RJ Knapp. *Nature 354*: 82–84, 1991.
9. S Brenner, RA Lerner. *Proc Natl Acad Sci USA 89*: 5381–5383, 1992.
10. MHJ Ohlmeyer, RN Swanson, LW Dillard, JC Reader, G Asouline, R Kobayashi, M Wigler, WC Still. *Proc Natl Acad Sci USA 90*: 10922–10926, 1993.
11. JK Scott, GP Smith. *Science 249*: 386–90, 1990.
12. SH DeWitt, JS Kiely, CJ Stankovic, MC Schroeder, DMR Cody, MR Pavia. *Proc Natl Acad Sci USA 90*: 6909–6913, 1993.
13. CC Leznoff, JY Wong. *Can J Chem 50*: 2892–2893, 1972.
14. JI Crowley, H Rapoport. *J Am Chem Soc 92*: 6363–6365, 1970.
15. F Camps, J Castells, J Pi. *An Quim 70*: 848–849, 1974.
16. BA Bunin, JA Ellman. *J Am Chem Soc 114*: 10997–10998, 1992.
17. HV Meyers, GJ Dilley, TL Durgin, TS Powers, NA Winssinger, H Zhu, MR Pavia. *Mol Divers 1*: 13–20, 1995.
18. MR Pavia, SP Hollinshead, HM Meyers, SP Hall. *Chimia 51*: 826–831, 1997.
19. MR Pavia, MP Cohen, GJ Dilley, GR Dubuc, TL Durgin, FW Forman, ME Hediger, G Milot, TS Powers, I Sucholeiki, S Zhou, DG Hangauer. *Bioorg Med Chem 4*: 659–666, 1996.
20. SW Kaldor, MG Siegel, JE Fritz, BA Dressman, PJ Hahn. *Tetrahedron Lett 37*: 7193–7196, 1996.
21. *Molecular Diversity* (ISSN 1381–1991). Kluwer Academic, Dordrecht, The Netherlands.
22. RE Dolle, KH Nelson. *J Comb Chem 1*: 235–282, 1999.

Chapter 4
Techniques and Strategies for Producing Compound Libraries for Biological Screening

Stephen R. Wilson and Kathryn Reinhard
New York University, New York, New York

Combinatorial chemistry has recently become a critical tool for most pharmaceutical companies, and it is rapidly spreading and being adapted to other discovery areas [1,2]. Along with the acceptance of combinatorial chemistry in basic research have been advancements in instrumentation and robotics. This short review is intended (1) to survey some of the now classic combinatorial chemistry concepts and (2) to introduce a selection of "case studies" that detail actual examples of the application of combinatorial chemistry and high-throughput synthesis to real problems. Six such studies follow this chapter and include the preparation of a 125-member library γ-turn mimics using robotics and solid phase synthesis; the preparation of a 1600-member library of α-ketoamides using parallel solution phase synthesis; manual preparation by solid phase synthesis of an encoded 46,656-member benzimidazole library; peptide synthesis and microassay on polypropylene membranes; and, finally, synthesis using microwave-assisted solid phase chemistry.

We begin by drawing a distinction between combinatorial chemistry and high-speed synthesis. High-speed synthesis and related methods do not necessarily have combinatorial aspects. Points of structural variation—the R groups that lead to molecular diversity—are called *inputs*. Strictly speaking, combinatorial chemistry involves preparing a multidimensional array of compounds wherein more than one "input" is varied during the synthesis (Fig. 1). For example, note the "Rubik's cube" shown in Fig. 1. This $3 \times 3 \times 4$ array is made of 36 blocks, each representing one discrete compound. The compounds can be prepared from

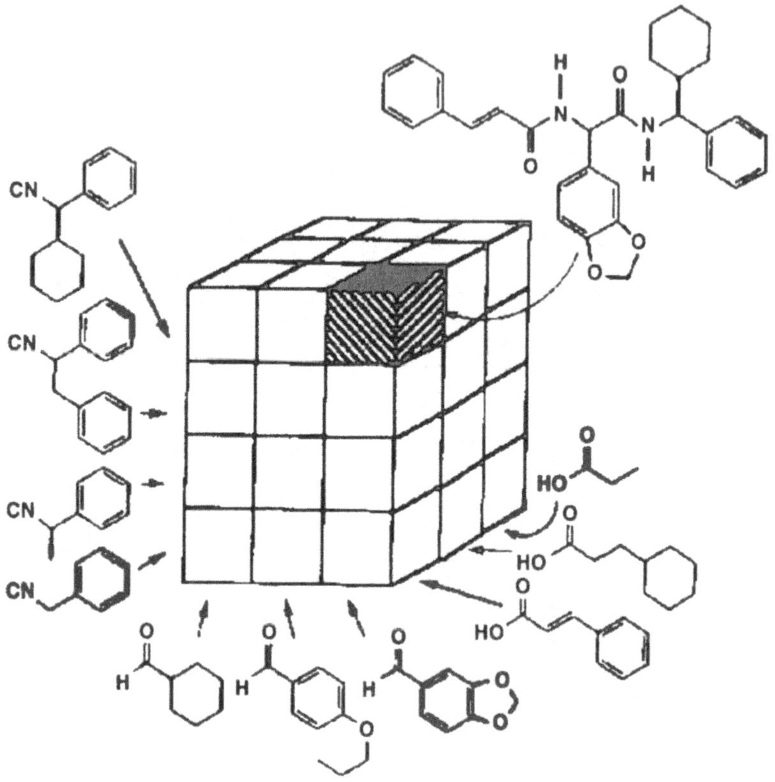

Figure 1 "Rubik's cube" representing 3 × 3 × 4 array of discrete compounds.

three types of inputs, illustrated on each of the three axes. This situation is combinatorial, i.e., the number of products produced is the arithmetic product of the number of inputs (3 × 3 × 4 = 36).

On the other hand, one can imagine high-speed synthesis being carried out in a linear fashion that could also yield the same large number of compounds by varying one R group at a time. Both methods employ similar strategies and processes for the rapid synthesis of large collections of compounds, called "libraries." In general, both combinatorial and high-speed synthesis produce collections of molecularly diverse compounds that are used for screening for biological activity. Whether or not a library has in some way been designed or made more or less at random depends on the reasons for preparing the compounds. Both high-speed synthesis and combinatorial chemistry also require the integration of some different concepts into the synthetic chemistry process. These concepts derived from the need to increase the speed and parallelism of chemistry and to automate some or all of the steps. Often this also means miniaturization of the chemistry, which leads to less solvent and reagent use, removal of the need to purify compounds, and only partial (or even no) characterization of individual library components.

A combinatorial chemistry discovery project typically has four phases: (1) identification of the "cost function" (the bioassay result or other property) that one wishes to optimize; (2) planning a library synthesis; (3) making the collection of compounds; and (4) finding the best compound. Thus, a critical element from the outset is the assay for finding the best compound. In fact, this issue must be addressed first, since the format of the test and its sample requirements dictate the types of high-speed synthesis or combinatorial chemistry approach that can be employed. Screening for biological activity, as practiced in the pharmaceutical industry, can sometimes be compared to finding a needle in a haystack. This reduces the problem to (1) haystack construction and (2) needle searching.

Case Study 4-4 illustrates a novel method for synthesis of 400 dipeptides attached to polypropylene/polyaminoethylacrylamide membrane disks [3]. The preparation also uses the concept of split synthesis described in more detail below. Most importantly, the synthesis experiments were planned with the final assay in mind, i.e., the peptides were tagged with a dye so that when the enzyme cleavage assay is carried out the cleaved peptides leave the membrane support and can be detected and measured in solution. In Case Study 4-5, the peptides were synthesized as distinct spots on the membrane matrix. The membrane-bound peptides were then subjected to an enzyme-linked immunosorbant assay (ELISA) using a method that directly detected the spots on the support.

The description of these experiments and the novel techniques employed reveal the close connection between the "haystack" construction phase—the array of membrane-supported peptides—and the "needle searching phase"—the bioassay. The method itself also shows how structural characterization of the sample is avoided because one knows from the method of synthesis what peptide structure must be attached at each spot on the membrane.

The origin of combinatorial chemistry has its roots in the pioneering work of Bruce Merrifield [4]. Merrifield's solid phase peptide synthesis laid a foundation for one of the key elements of combinatorial chemistry—techniques for preparation of individual compounds attached to a solid support, such as a polystyrene bead. This advance means that the individual product molecules can be separated from other materials (side products, excess reagents, etc.) by physical means—filtration or even (if you have good eyesight!) with tweezers. The fact that the compounds are attached to beads allows mixture libraries to be prepared, as discussed below.

Merrifield reported in 1963 the first examples of solid phase synthesis of peptides using chloromethylated polystyrene containing immobilized N-pro-

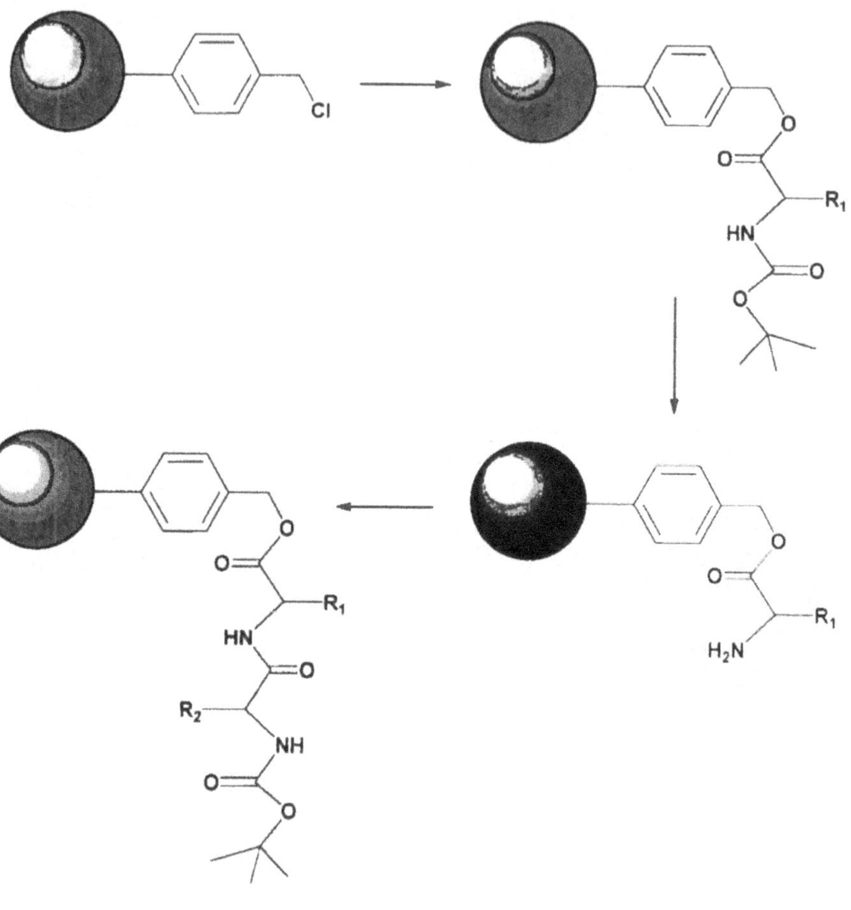

Figure 2 Merrifield's peptide synthesis. Solid phase synthesis on polystyrene beads.

tected amino acid building blocks. Figure 2 shows the Merrifield approach. An insoluble polymer bead containing $-CH_2Cl$ groups is prepared by chloromethylation of the copolymer of styrene and p-divinylbenzene. These chloromethyl groups can be esterified with N-protected amino acid building blocks. The amino group is iteratively deprotected and coupled with new N-protected amino acids to build the growing chain. Merrifield used the tert-butyloxycarbonyl protecting group (BOC group) to protect the free amino terminus during the coupling step. This allows deprotection with acid (CF_3COOH) before coupling with the next N-protected residue. The cycle can be continued until the desired sequence is obtained. By products and excess reagents are not attached to the resin and can be removed from the resin beads during washing cycles.

The first report describing a solid phase combinatorial chemistry synthesis was in a 1984 paper by Mario Geysen [5] titled the "Use of peptide synthesis to probe viral antigens for epitopes to a resolution of a single amino acid." Another early pioneer was Dr. Arpad Furkas (University of Budapest) [6] who introduced the commonly used pool-and-split (split/mix) method. This technique involves carrying out several separate synthetic reactions, combining (pooling) the resulting beads, mixing them up, and then portioning them out again into individual reactors. Repetition of this process allows rapid preparation of mixture libraries. An example of this process may be found in Case Study 4-1, which describes the development of a 125-member model library based on 4-*cis*-aminoproline as a molecular scaffold using the pool-and-split protocol. Both manual and automated methods were applied in the synthesis of the library, the model library being used to explore techniques needed to create a larger combinatorial library containing several tens of thousands of compounds.

Since the amount of compound on each polystyrene bead is small (a typical 200-mesh, 106-μm bead weighs about 1μg and only contains 0.1 pmol of compound per bead), another important advance involved introduction of the so-called tea bag method for rapid synthesis [7]. This technique allows a much larger quantity of polystyrene beads to be retained in separate polypropylene mesh bags. The bags can then be mixed for simultaneous chemistry.

Here is how this pool-and-split tea bag method works. Instead of making one peptide at a time, imagine preparing 8000 small tea bags. In each tea bag 10 g of polystyrene resin is placed that contains a derivatized resin with 0.1 mmol/g, which means that each bag contains 1 mmol of starting compound. Next, the 8000 tea bags are separated into 20 different reaction flasks, with 400 bags in each flask. A suitable solvent is added to each flask and 20 amino acids are coupled to the bags in the flask, a different amino acid for each flask. When the reactions are finished, the bags are fished out (remember to label them!) Next, 20 of each kind of bag (remember each bag contains polymer beads containing only one amino acid) is placed in another 20 different reaction flasks, so that, while each flask contains 400 tea bags, there actually are 20 of each type of bag in each flask. Again, all 20 amino acids are coupled to the tea bags, and after the reaction is complete the bags are retrieved and labeled. One now has 8000 bags containing 20 duplicate sets of 400 dipeptides. The final reaction series involves placing the 20 sets of 400 into another 20 reaction flasks and again separately coupling 20 amino acids. We now have in hand, in a total of three reaction steps, a set of 8000 tea bags. Each bag contains 10 g of resin and 1 mmol of a single tripeptide. This is an 8000-member combinatorial library. An alternative process for preparing this library would be to take one tea bag and carry out the same three steps on it. In that case, the resin in the bag would also be a 8000-member combinatorial library. However, the 8000 compounds would all be mixed together.

Two problems might now become obvious. The first is "how do you label all those individual tea bags?" especially if the are thousands of them. Several labeling strategies have been developed including a technique called encoding, that introduces labels directly on individual beads. Case Study 4-3 reports the preparation of a 46,656-member library of substituted benzimidazoles (36 × 36 × 36 inputs) synthesized by solid phase combinatorial chemistry using a manual split-and-mix protocol. A combination of 71 different primary amines and 35 different aldehydes was used to diversify the three substituted positions on the final structure, and the structures were labeled using an encoding system in which unique alkylamine tags were incorporated onto each of the beads prior to the two reactions with the various amines, so that the identity of the substituents at those two positions could be determined once the library was constructed.

The second obvious problem is, if you elected to adopt the mixture approach, how do you find a single active peptide in a mixture with 7999 inactive ones? The solution to the problem is called *deconvolution*. One type of deconvolution is as follows: imagine constructing the tea bag library described above, starting not with all 20 amino acids but with only 19, leaving out one specific (known) residue. Subsequent coupling with 20 × 20 amino acids leads to a library of only 7600 tripeptide compounds. If this library is now inactive, we have learned that position 1 of the tripeptide has the residue that was omitted from the synthesis. If you continue this process, methodically omitting one amino acid after another, a single-deletion library still has the active compound. Thus, a maximum of 20 experiments are needed to define position 1. Now, with the active first residue fixed, 20 more experiments are carried out to define the best residue in position 2. There are $20 + 20 + 20 = 60$ deletion libraries required to define the complete tripeptide sequence of the active component in 8000 compounds. This principle is the mathematical advantage: $20 + 20 + 20 = 60$ (linear) vs. $20 \times 20 \times 20 = 8000$ (exponential) that underlies the principle of combinatorial chemistry [8,9]. This iterative synthesis and screening strategy has been discussed in detail by Dooley and Houghten [9]. A careful analysis of methods for deconvolution of libraries and their intrinsic problems are described by Freier. [10]

As you can see from this example, strategies for combinatorial chemistry rely on two important points. First, the synthesis of the mixture must be fast and easy because you need to prepare many variations of compounds to find the best. Second, the testing must also be fast and easy because you need to test many compound mixtures rapidly. Fast and easy synthesis and testing suggests automation and robotics. Combinatorial chemistry has grown quickly over the last few years, in part because good automation exists in both areas. Many labor-intensive aspects of combinatorial library synthesis can be automated. Case Study 4-1 de-

Table 1   Typical Data for Polystyrene–Bead Solid Support [11,12]

Styrene-divinylbenzene copolymer	(1–2% cross-linked) (1 g of 200 to 400-mesh resin has ~4–10 × 10^6 beads)
Thermal stability	105–130°C
Bead sizes	100 mesh (212 μm), 140 mesh (150 μm), 200 mesh (106 μm), 325 mesh (75 μm), 400 mesh (45 μm)
Swelling	DMF (3.5 mL/g), THF (5.1 mL/g), CH$_3$OH (1.5 mL/g), H$_2$O (1.5 mL/g), CH$_2$Cl$_2$ (5.3 mL/g), CHCl$_3$ (5.9 mL/g).
Loading	0.1–0.4 mmol/g, ~100–400 pmol/g

scribes in detail instrument prototyping, i.e., developing the chemistry needed to prepare a library and validating on a small (in this case 125 members) library.

Besides the reaction chemistry, solid phase and combinatorial chemistry methods also require consideration of several new technical issues. For example, the following table reports some characteristics of the most common form of resin bead: cross-linked polystyrene. In planning a synthesis, one must also consider loading, the level of substitution on the beads, and swelling characteristics of the resin. As can be seen in Table 1, cross-linked polystyrene beads swell quite a bit in volume with different solvents. Of course, the physical size of the beads used in a solid phase synthesis also effects the weight of each bead and therefore the amount of product on each bead. While cross-linked polystyrene resin beads (Merrifield-type resin) are the most common for historical reasons, a popular new material called Tentagel has come into wide use [13]. This material contains a polystyrene core (PS) with polyethylene (PEG) spacer arms (PS-PEG), so that the attached reacting groups are held out in solution rather than close to the polymer backbone, providing better reactivity and more closely paralleling solution phase chemistry.

The combinatorial method for producing mixtures of thousands or millions of compounds has always presented an awkward deconvolution process. Sometimes a mixture confuses the bioassay or even fails to achieve results. In principle, a better strategy would be to prepare the compounds in pure form, in separate, labeled bottles. It was recognized that it is possible to prepare one compound per solid phase bead. The mixture of beads then represents a physically separable collection of compounds. This approach has been widely used and only requires a means to determine what is attached to a single bead. Recall from the table that the amount of compound on each bead is very small. While it might be possible to determine the structure on a bead by mass spectrometry, an alternative approach has been developed called "bead encoding"—recording the synthesis history of each bead on the bead. Instead of direct structural determination of the compound attached to a single bead, the synthesis history of the bead is recorded on the bead. Each bead can be derivatized with a "tag" that contains information on the structure attached to the bead. Therefore, instead of identifying the compound, one identifies the code.

This concept was first described for beads encoded with a genetic DNA tag [14]. Another coding method used peptide sequences for a reaction code [15]. Two orthogonally protected points of extension allow the primary peptide (or organic) molecule to be formed at one point and a coding sequence built at the second. The code can be "read" by selective cleavage of the coding peptide, followed by sequencing.

Another approach, especially applicable to organic compounds, is the method of Pharmacopia and Clark Still [16]. This method involves attachment (at the level of a few percentage points of the real compound) of a binary code of chlorophenyl groups at each synthesis step. Using electron capture GC analysis, 0.1 pmol of code attached to a single 50- to 80-μm bead is more than enough for detection [17]. The presence or absence of one of the four GC separable chlorophenyl derivatives gives a binary code of 0 to 1. Thus the binary codes such as 0000, 0001, 0010, 0100, etc., can be used to define the synthetic steps applied to the bead.

Tea bags also provide an option for split- and-mix but this time with much larger quantities of materials. Since almost any size polypropylene mesh bag can be used, one compound per bag is the rule. The coding of each bag could involve direct labeling of each bag, but an interesting new coding method uses Rf transponder chips, similar to those used for tracing animals. The memory chip can be placed in the tea bag (or other reaction vessel) and coded with radiofrequency pulses [18]. These memory chips are similar to EPROM chips (erasable, programmable, read-only memory) and are available from several sources.

A final approach to structure coding involves spatial separation of combinatorial library collections based on preparing each compound immobilized on a separate spot on a surface. This method then allows the x/y coordinates of the spot to be related to the structure. This process was reported in the classic *Science* paper "Light-directed, spatially addressable parallel chemical synthesis" [19]. In this approach, solid phase peptide synthesis is carried out on surface-derivatized glass substrate using photolabile protecting groups and photolithographic techniques.

A more accessible version of this method uses simultaneous parallel synthesis on cellulose disks (filter paper) [20]. Peptide synthesis on a 50-nmol scale can be achieved in an array of 100 spots on a 1-cm² area of filter paper. A modification of this process is described in Case Studies 4-4 and 4-5, which report two applications of the technique using polypropylene membrane disks. Case Studies 4-2 and 4-6 also demonstrate methods of spatially separate parallel synthesis in solution and on the solid phase, respectively, using 96-well plates.

Finally, several other novel tricks of the trade have been used in combinatorial chemistry, These involve such techniques as supporting the chemistry on string [21], using nonorganic miscible "fluorous phase" reagents for liquid-liquid extractions [22], using soluble polymeric phases that can be organic phase-soluble or alternatively can be precipitated and filtered [23]. Solid phase reagents [24] can also be used, as can a technique called resin capture [25,26]. Using resin capture, one can carry out solution phase chemistry followed by chemical trapping of the desired product with an activated solid phase resin. After washing the product free from unwanted side products, the desired product is freed from the resin. Finally, the sometimes sluggish reactions on solid phase is shown to be accelerated using microwave irradiation (Case Study 4-6).

We have tried to survey many of the concepts and techniques that originated during the early days of "the combinatorial explosion." Following this section are the case studies mentioned, that illustrate the practical application of high throughput synthesis methods.

REFERENCES

1. Dolle, R.E.; Nelson, K.H. *J. Comb. Chem.* 1999, *1*, 233–282.
2. Dolle, R.E. *Mol. Divers.* 1998, *3*, 199–233.
3. Duan, Y.; Laursen, R.A. *Anal Biochem.* 1994, *216*, 431–438.
4. Merrifield, R.B. *J. Amer. Chem. Soc.* 1963, *85*, 2149–54.
5. Geysen, H.M.; Meloen, R.H.; Barteling, S.J. *Proc. Natl. Acad. Sci. USA* 1984, *81*, 3998–4002.
6. Furka, A.; Sebestyen, M.; Dibo, G. *Abstr., 14th Congr. Biochem. Prague* 1988, p. 47.
7. Houghten, R.A., *Proc. Natl. Acad. Sci. USA* 1985, *82*, 5131–5135.
8. Dooley, C.T. Houghten, R.A., *Life Sci.* 1993, *52*, 1509–1517.
9. Houghten, R.A., *Curr. Biol.* 1994, *4*, 564–567.
10. Freier, S.M.; Konings, D.A.M.; Wyatt, J.R.; Ecker, D.J. *J. Med. Chem.* 1995, *38*, 344–352.
11. Merrifield, B.; I. *Life During the Golden Age of Peptide Chemistry*, Seeman, J., ed. Am. Chem. Soc., 1993.
12. A. Guyot in *Synthesis and Separations Using Functionalized Polymers*, Sherrington, D.C.; Hodge, P.; eds. John Wiley and Sons, Chichester, 1988, p 1ff.
13. Bayer, E.; Rapp, W. *Chem. Pept. Prot.* 1986, *3*, 3.
14. Brenner, S.; Lerner, R.A.; *Proc. Natl. Acad. Sci USA* 1992, *89*, 5381–5383.
15. Kerr, J.M.; Banville, S.C.; Zuckermann, R.N.J. *Am. Chem. Soc.* 1993, *115*, 2529.
16. Ohlmeyer, M.H.J.,; Swanson, R.N.; Dillard, L.W.; Reader, J.C.; Asouline, G.; Kobayashi, R.; Wigler, M.; Still, W.C.; *Proc. Natl. Acad. Sci USA* 1993, *90*, 10922–10926.
17. Grinsrud, E.P. in *Detectors for Capillary Chromatography*, Hill, H.H, and McMinn, D.G., eds. John Wiley and Sons, New York, 1992 pp 83–107.

18. Nicolaou, K.C.; Xiao, X-Y.; Parandoosh, Z.; Senyei, A.; Nova, M.P. *Angew. Chem. Int. Ed. Engl.* 1995, *43*, 2289–2291.
19. Fodor, S.P.A.; Read, J.L.; Pirrung, M.C.; Stryer, L.; Lu, A.T.; Solas, D.; *Science*, 1991, *251*, 767–773.
20. Frank, R.; *Tetrahedron*, 1992, *42*, 9217–9232.
21. Schwabacher, A.W.; Shen, Y.; Johnson, C.W. *J. Am. Chem. Soc.* 1999, *121*, 8669–8670.
22. Curran, D.P. *Chemtracts–Organic Chemistry*, 1996, *9*, 77–87.
23. Janda, K.D.; Gravert, D.J. *Chem. Rev.* 1997, *97*, 489–510.
24. Drewry, D.H.; Coe, D.M.; Poon, S. *Med. Res. Rev.* 1999, *19*, 97–148.
25. Keating, T.A.; Armstrong, R.W. *J. Am. Chem. Soc.* 1996, *118*, 2574–2583.
26. Brown, S.D.; Armstrong, R.W. *J. Am. Chem. Soc.* 1996, *118*, 6331–6332.

Case Study 4-1
A γ-Turn Mimetic Library: Development and Production

**Petr Kocis, James B. Campbell, Richard B. Sparks,
and Richard Wildonger**
Zeneca Pharmaceuticals, Wilmington, Delaware

ABSTRACT

Development of a γ-turn mimetic combinatorial library is described. The library chemistry was developed starting with "pilot" and "model" compounds synthesized in a parallel mode and produced using a split- and-mix protocol to assemble a 125 compound model library. The supporting experimental data provide a foundation for the potential production of combinatorial libraries containing several tens of thousands of compounds.

Chemicals

1. N-α-Fmoc-*N-t*-Boc-L-tryptophan, Bachem
2. Fmoc-L-proline, Bachem
3. Fmoc-4-(2-aminoethyl)-1-carboxymethylpiperazine (stored as dihydrichloride), Neosystem Laboratoire
4. Fmoc-L-aspartic acid β-*t*-butyl ester, Bachem
5. Fmoc-5-aminovaleric acid, Bachem
6. (1*R*)-(+)-Camphanic acid, 98%, Aldrich
7. 1-Methylcyclopropane-1-carboxylic acid, Aldrich
8. (3-Pyridyl)thiazole-4-carboxylic acid, Maybridge
9. Isovaleraldehyde, 97%, Aldrich
10. Benzaldehyde, redistilled, 99.5+%, Aldrich
11. 2-Methoxyphenylisocyanate, 99%, Aldrich

12. 4-Formylphenoxyacetic acid, 99%, Acros
13. Glycolaldehyde dimer, crystalline, Aldrich
14. (4-Chlorophenoxy)-1,3-dimethylpyrazole-4-carboxylic acid, Bionet
15. Trimethylorthoformate, anhydrous, 99.8%, Aldrich
16. Sodium cyanoborohydride, 95%, Aldrich
17. Sodium borohydride, 99%, Aldrich
18. Acetonitrile, anhydrous, 99.8%, Aldrich
19. [O-(7-Azabenzotriazol-1-yl)-1,1,3,3-tetramethyluronium hexafluorophosphate], HATU, Perseptive Biosystems
20. 1-Methylimidazole, NMI, redistilled, 99+%, Aldrich
21. N,N'-diisopropylethylamine, DIEA, 99%, Aldrich
22. Dimethylformamide, DMF, Fisher
23. Methylene chloride, DCM, Fisher
24. Trifluoroacetic acid, TFA, 99+%, Aldrich
25. Methanol, Fisher
26. Nitrophenylisothiocyanate-O-trityl, NPIT, synthesized according to reference 8
27. Tentagel Rink Amide Resin (4-(2′, 4′-dimethoxy-Fmoc-aminomethyl)phenoxy resin, Rapp Polymere)
28. Piperidine, 99%, Aldrich
29. 4-Methylmorpholine, redistilled, 99+%, Aldrich
30. Tetrakis(triphenylphosphine)palladium(0), 99%, Aldrich
31. Acetic acid, Glacial, Fisher
32. 6-Hydroxy-2,5,7,8-tetramethylchroman-2-carboxylic acid, 97%, Aldrich
33. 5-Methyl-2-pyrazinecarboxylic acid, 98%, Aldrich
34. 1-Adamantaneacetic acid, 98%, Aldrich
35. 2,3,4-Trimethoxybenzoic acid, 98+%, Aldrich
36. 3-Iodobenzoic acid, 98%, Aldrich
37. Alloc-4(S)-cis-amino-Fmoc-(S)-proline
38. 1-H-pyrazole-[N,N'-bis($tert$-butoxycarbonyl)]carboxamidine

Equipment and Supplies

1. Disposable syringes, all-polypropylene/polyethylene, 3 mL, 5 mL, 10 mL, Aldrich
2. Syringe pressure caps, Aldrich
3. Needles, 20 gauge, 1.5 in., Becton Dickinson
4. Polyethylene frits, cut from polyethylene stock (36″ × 36″ × 1/8″), Bel-Art Products
5. IKA KS501 digital shaker, Fisher

6. Barnstead/Thermolyne LabQuake shaker, Fisher
7. Eppendorf pipetors, 10 µL, 2–20 µL, 100 µL, 200 µL, 1 mL, 5 mL, Fisher
8. Custom manifold (see Fig. 3)
9. Brinkmann Dispensette, 1–5 mL, Fisher
10. Target* DP* 2 mL screw thread vials, Fisher
11. Microliter inserts with polypropylene feet for sample vials, Hewlett Packard
12. Zymate
13. BenchMate
14. System V
15. EasyLab and Master Lab Station are registered trademarks of Zymark Corporation, Hopkinton, MA
16. Teflon and Nomex are registered trademarks of E. I. DuPont de Nemours, Wilmington, DE
17. Bio-Chem is a registered trademark of Bio-Chem Valve Inc., Boonton, NJ
18. Chemraz is a registered trademark of Greene Tweed and Co., Kulpsville, PA
19. Visual Basic and Windows NT are registered trademarks of Microsoft Corporation, Redmond, WA.
20. HP8453 UV/VS spectrophotometer

I. INTRODUCTION

Turns have been recognized as having a very important role in molecular recognition. Many studies have demonstrated the medicinal relevance of β and γ turns and their mimetics [1,2]. Our studies, as well as a recently published study [3], showed that 4-*cis*-aminoproline exhibits features of a γ-turn mimetic. Another feature of this molecular scaffold is convergent functional groups, which allows one to display important functional groups and structural features in a diaxial mode [4] and hence mimic a discontinuous epitope. A combinatorial library comprising several tens of thousands of compounds was synthesized based on 4-*cis*-aminoproline as a molecular scaffold (Scheme 1). Both manual and automated synthesis techniques were used in the production of the library. The production of such a library was preceded by thorough development of chemistry for "pilot" and "model" compounds (vide infra) and a model library including analysis, a quality control evaluation of the model library, and appropriate feedback process revision.

Whereas pilot and model compounds were synthesized using a multiple

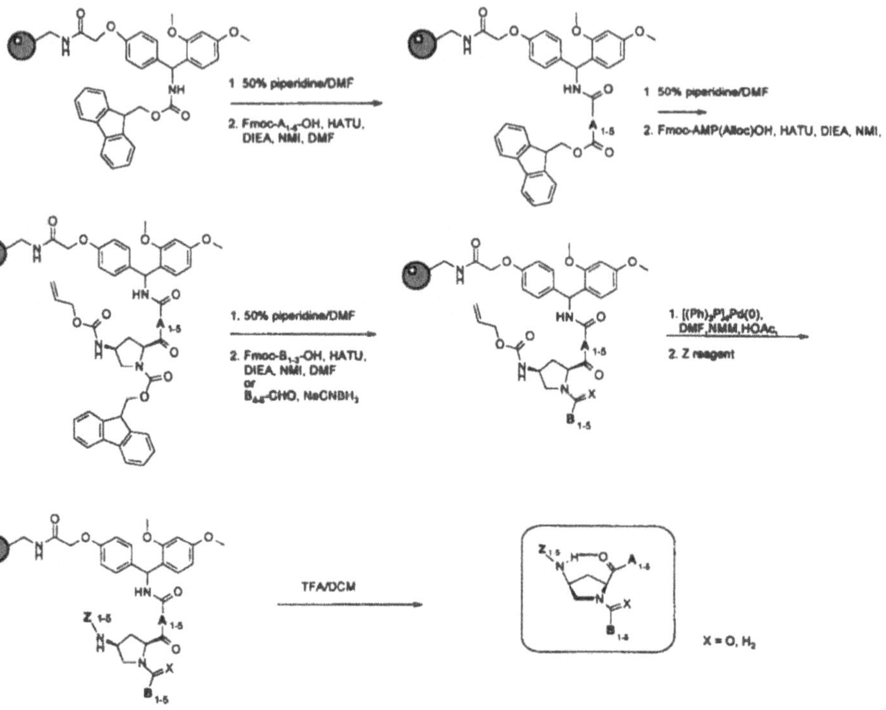

Scheme 1   Solid phase synthesis of a γ-turn mimetic library.

parallel approach, the model library was synthesized in combinatorial fashion using a split-and-mix method [5,6]. Thus, a simple yet expedient syringe technique was used for the synthesis of pilot and model compounds. The model library of 125 compounds, which will be described, was synthesized using both syringe technique and the Zeneca automated synthesizer developed in a collaborative effort between Zymark Corporation and Zeneca Pharmaceuticals, Inc. (U.S.) [7]. A significant supporting level of automation [8–10] is critical to the effective implementation of high-throughput methods of synthesis. However, while there are clear areas where automation can increase drug candidate synthesis throughput, there remain some critical issues that need to be considered for optimal application.

Exploratory synthesis of five initial pilot compounds containing substituents in all randomization positions of the 4-cis-aminoproline scaffold was done initially to indicate the overall feasibility of the synthetic approach and to gain insight into the capacity to functionalize each of the randomization positions

(Fig. 1). Model compounds were then prepared as an experimental "matrix mapping" of chemical feasibility, reactivity, purity, stereoelectronic properties, and many other factors. Molecular diversity and medicinally relevant properties were factored into the selection of diversity reagents for the synthesis of model compounds.

Based on the results from model compound synthesis and analysis, preview of theoretical mass distribution, molecular weight redundancy, consideration of positional isomers, representatives from the chemical feasibility range, product complexity, and other factors, a model library was designed to test library synthesis in a combinatorial (split-and-mix) mode, accounting for all aspects of manual and automated processes. This applied approach usually has revealed hidden logistic aspects of handling large numbers of reagents and handling parallel and serial steps in a timely manner. The model library synthesis also helps to organize the planning of all steps for larger library generation that include data management.

II. GENERAL PROCEDURE

A. General Multiple Parallel Procedure (Three Model Compounds)

For manual synthesis, the vessels were made of disposable syringes [11] fitted with polyethylene frits and Luer caps (Fig. 2). The assembled reactors could be agitated by either shaking (IKA or LabQuake) or by sparging with argon.

1. First Randomization Position

Tentagel rink amide resin (150 mg, 0.21 mmol/g; Rapp Polymere) was added to a 10-mL reactor (disposable syringe), swollen in DMF, and washed 3 × 1.8 mL DMF (custom manifold and Brinkmann Dispensette). Piperidine/DMF (1:1, 1.8 mL) was aspirated to the resin, syringe was Luer-capped and gently rotated for 10 min. Ninhydrin testing [12] with few beads indicated the presence of primary amine. The resin was then washed with DMF (10 × 1.8 mL). The piperidine/DMF filtrates were collected along with the washes and the loading of the resin was measured by UV fulvene reading at 302 nm. Loading was found to be 0.21 mmol/g.

The deprotected resin was then divided into three equal batches of 0.3 mL each in a 3-mL syringe reactor using a slurry transfer technique (Fig. 3). In three separate 2-mL vials, three Fmoc amino acids (A1, A2, A3, Fig. 1) (5 equivalents) were dissolved in 0.75 mL DMF and activated with HATU (5 equivalents), NMI (5 equivalents), and DIEA (10 equivalents). These solutions were added to the resin-loaded syringes by (1) placing the plunger back on the syringe and dispens-

Position A

A1   A2   A3   A4

A5

Position B

B1   B2   B3   B4   B5

Position Z

Z1   Z2   Z3   Z4   Z5

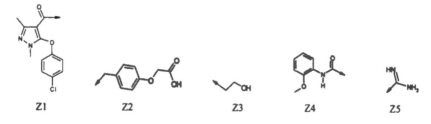

Figure 1 Building blocks for diversity positions A, B, and Z. Arrows indicate the connectivity to the corresponding diversity position on 4-*cis*-aminoproline scaffold.

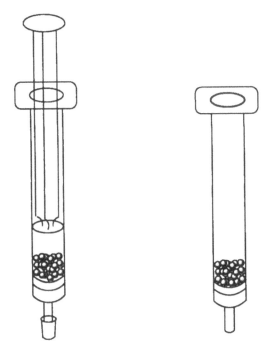

Figure 2 Syringe reactors.

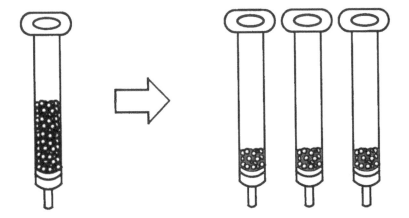

Figure 3 Slurry transfer technique.

ing residual DMF, (2) installing a syringe needle on the Luer tip of the syringe, (3) aspirating the activated acid solution, and (4) replacing the needle with a cap.

The reactors were typically shaken on a rotating LabQuake shaker or IKA benchtop shaker overnight. Ninhydrin testing on a few beads from each of the three reactors indicated absence of primary amine (complete reaction). The resins were washed with 10×0.6 mL DMF.

A small aliquot (50 beads) of each compound-resin was transferred to an empty 1-mL Isolute tube (Jones chromatography). These aliquots were washed with DCM (10×1 mL) and transferred to a suitable glass cleavage vessel. Trifluoroacetic (100 µl, 95%) acid in water was added to each aliquot of compound-resin. After 90 min, most of the cleavage solution was evaporated. To extract the cleaved compound, 50 µL methanol was added, and after 5 min the extracts were decanted with the aid of a gel loading tip, leaving the beads behind. Typically, 1 µL of the extracts was analyzed by liquid chromatography/mass spectroscopy (LC-MS). LC purities were generally required to be >98% at this stage.

2. Scaffold Attachment

After Fmoc deprotection with 50% piperidine in DMF for 10 min and DMF wash, the scaffold (Alloc-4(S)-cis-amino-Fmoc-(S)-proline, 3 equivalents) was coupled to each of the three compound-resins using the HATU activation procedure (3 equivalents). After obtaining a negative ninhydrin test and DMF wash, subsequent LC-MS analysis of a cleaved aliquot allowed for evaluation of chemical feasibility and purity at this stage. LC purities were generally required to be >95%.

3. Second Randomization Position

After Fmoc deprotection as described above, three diverse carboxylic acids (B1, B2, B3) were coupled to the three compound-resins using the HATU procedure. NPIT [13] testing was used to monitor completeness of the reaction (the presence of secondary amine). Subsequent LC-MS analysis of a cleaved aliquot at this stage permitted quality evaluation of the combination of the scaffold with the corresponding diversity reagents in the R1 and R2 randomization positions. LC purities were generally required to be >95% at this stage as well.

4. Third Randomization Position

The Alloc group was removed using the following procedure: the compound-resins were first washed with 3×0.5 mL mixture of DMF, AcOH, and NMM (10:2:1) and sparged with argon for 15 min in 0.5 mL of the same mixture prior to addition of 7.5 mg of [tetrakis(triphenylphosphine)]palladium(0). The compound-resins were protected from light with aluminum foil and sparged for

3 h. Ninhydrin testing of a few beads indicated presence of primary amine. The resin was washed 6 × 1 mL of DMF, 10 × 1 mL of DCM, and 3 × 1 mL of DMF. Three diverse carboxylic acids (Z1, Z2, Z3) were then coupled to each of the three deprotected compound-resins using the HATU procedure. After completion of the reaction and subsequent wash, LC-MS analysis of a cleaved aliquot allowed for quality evaluation of the completely assembled model compound. LC purities were generally required to be >85% at this stage.

B. Split-and-Mix Procedure (Model Library Synthesis)

1. Fmoc Deprotection, Resin Loading

Tentagel MB (2.3 g, Rapp Polymere) was added to a 40-mL reactor on the Zymark prototype and washed 3 × 1 min (sparge) with 20 mL DMF. 25 mL of 50% piperidine/DMF was added and the suspension sparged with nitrogen for 10 min. Ninhydrin testing of a few beads indicated the presence of primary amine. The resin was washed 20 × (1-min sparge and 20-mL DMF wash). The piperidine/DMF filtrates were collected along with the washes and the loading of the resin was determined by UV fulvene reading at 302 nm. Loading was found to be 0.21 mmol/g.

2. A Coupling

The resin was divided (using a slurry transfer technique) into five equal batches of 2.25 mL each in a 10-mL syringe reactors (Fig. 2) fitted with a polyethylene frit, a plunger, and an end cap. Five diverse carboxylic acids (A1–5) (5 equivalents) were activated separately in DMF (1.125 mL) with HATU (5 equivalents), NMI (5 equivalents), and DIEA (10 equivalents) and aspirated to the resin-loaded syringes. A syringe needle was used to draw up the activated acid solution and was replaced with the Luer cap to seal the reactor. The reactors were typically shaken on a rotating LabQuake shaker or a IKA benchtop shaker overnight. Ninhydrin testing of a few beads from each of the five reactors indicated absence of primary amine (complete reaction). The resins were washed with DMF (10 × 5 mL).

3. Fmoc Deprotection, Resin Loading

All five compound-resins were combined and added to a 40-mL reactor on the Zeneca/Zymark prototype automated synthesizer, 25 mL of 50% piperidine/DMF was added, and the suspension was sparged for 10 min. Ninhydrin testing of a few beads indicated the presence of primary amine. Since there were a few secondary amines present, NPIT [13] analysis was also done and indicated presence of secondary amines. The resin was washed with DMF (20 × 20 mL).

4. Scaffold Coupling

Alloc-4(*S*)-*cis*-amino-Fmoc-(*S*)-proline (3 equivalents) was activated in DMF (4 mL) with HATU (3 equivalents), NMI (3 equivalents), and DIEA (6 equivalents) and added to the combined and deprotected resin. The suspension was sparged intermittently overnight. Ninhydrin testing of a few beads indicated absence of primary amine (complete reaction). NPIT analysis indicated absence of secondary amine (complete reaction). The resin was washed with DMF (10 × 25 mL). 25 mL of 50% piperidine/DMF was added and the suspension sparged for 10 min. NPIT analysis of a few beads indicated the presence of secondary amine. The resin was washed with DMF (20 × 20 mL).

5. Coupling B Randomization Position

The resin was divided (using a slurry transfer technique) into five equal batches of 1.5 mL each in a 10-mL syringe reactor fitted with a polyethylene frit, a plunger, and an end cap. Three batches were used for carboxylic acid coupling (Scheme 1) and the remaining two batches were used for reductive alkylation (Scheme 2; see next paragraph). Three carboxylic acids (5 equivalents) were activated separately in DMF (1.125 mL) with HATU (5 equivalents), NMI (5 equivalents), and DIEA (10 equivalents). The appropriate aldehyde was aspirated to each of the resin-loaded syringes. The reactors were typically shaken on a rotating LabQuake shaker or an IKA benchtop shaker overnight. NPIT testing

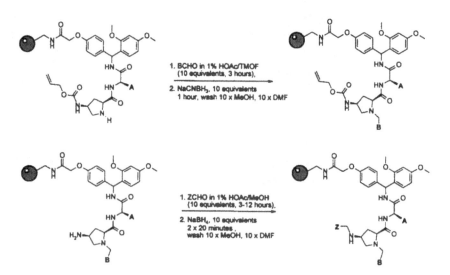

Scheme 2 Reductive alkylation in B and Z diversity positions.

of a few beads from each of the three reactors indicated absence of secondary amine (complete reaction). The resins were washed with DMF (10 × 5 mL).

6. Reductive Alkylation B Randomization Position

The remaining two syringes were washed with trimethylorthoformate (TMOF) (3 × 5 mL). Two aldehydes (10 equivalents) were dissolved separately in 1% AcOH/TMOF (8 mL) and the appropriate aldehyde was added to each of the resin-loaded syringes. The reactors were typically shaken on a rotating LabQuake shaker or a IKA benchtop shaker for 1 h. Sodium cyanoborohydride (10 equivalents) was added as follows: The aldehyde solution was dispensed from the syringe directly into a vial containing the solid sodium cyanoborohydride, during which the majority of the solid reagent dissolved (sometimes some cloudiness remained), and the reaction mixture was drawn back into the syringe and shaken for 1 h. The resins were then washed with methanol (10 × 5 mL) and DMF (20 × 5 mL). NPIT testing of a few beads from each of the reactors indicated absence of secondary amine (complete reaction).

7. Alloc Deprotection

All five compound-resins were combined and added to a 40-mL reactor on the Zymark prototype. The resin was washed 4 × 20 mL (77% DMF, 15% glacial acetic acid, and 7% N-methylmorpholine). Without draining the last wash and while sparging with nitrogen, tetrakis(triphenylphosphine)]palladium(0) (1.63 equivalents) was added and the suspension was sparged for 3 h. The reactor was wrapped in aluminum foil to protect the reaction mixture from light. Ninhydrin testing of a few beads indicated presence of primary amine. The resin was than washed for 1 min (sparge) with DMF (6 × 20 mL), DCM (10 × 20 mL), and DMF (3 × 20 mL).

8. Coupling Z Randomization Position

The resin was divided (using a slurry transfer technique) into five equal batches of 1.5 mL each in a 10-mL syringe reactor fitted with a polyethylene frit, a plunger, and an end cap. One of the batches was used for carboxylic acid coupling, another one was used for reaction with isocyanate, two of the batches were used for reductive alkylation, and the remaining fifth batch was reacted with a guanylating reagent. A carboxylic acid (5 equivalents) was activated separately in DMF (1.125 mL) with HATU (5 equivalents), NMI (5 equivalents), and DIEA (10 equivalents) and added to the resin-loaded syringe. The reactor was typically shaken on a rotating LabQuake shaker or a IKA benchtop shaker overnight. Ninhydrin testing of a few beads indicated absence of primary amines. The compound-resin was washed with DMF (20 × 5 mL).

9.   Reductive Alkylation Z Randomization Position

Two of the five batches were washed with methanol (3 × 5 mL). Two aldehydes (10 equivalents) were dissolved separately in 1% AcOH/methanol (8 mL) and added to the corresponding resin-loaded syringes. The reactors were typically shaken on a rotating LabQuake shaker or a IKA benchtop shaker for 3–12 h (depending on aldehyde). The compound-resins were then drained and the plunger removed. The open syringe was placed on a manifold where fresh methanol (5 mL) was added and nitrogen sparge was initiated. While sparging, sodium borohydride (10 equivalents) was added as a solid directly to the opened syringe against the stream of nitrogen. After 20 min of sparging, and additional 10 equivalents of sodium borohydride was added and the sparging continued for an additional 20 min. The compound-resins were washed with methanol (10 × 5 mL) and DMF (20 × 5 mL). Kaiser testing of a few beads from each of the reactors indicated the absence of primary amine (complete reaction).

10.   Reaction with Isocyanate

2-Methoxyphenylisocyanate (5 equivalents) was dissolved in DMF (8 mL) with diisopropylethylamine (5 equivalents) and added to the resin-loaded syringe. The reactor was shaken on a IKA benchtop shaker overnight. The compound-resin was then washed with DMF (20 × 5 mL). Kaiser testing of a few beads from the reactor indicated the absence of primary amine (complete reaction).

11.   Guanylation Z Randomization Position

The remaining batch of compound-resin was washed with acetonitrile (3 × 5 mL). 1-H-pyrazole-[N,N'-bis(*tert*-butoxycarbonyl)]carboxamidine (7.5 equivalents) was dissolved in acetonitrile (8 mL) and aspirated into the resin-loaded syringe. The reactor was shaken on a IKA benchtop shaker overnight. The compound-resin was then washed with DMF (20 × 5 mL). Kaiser testing of a few beads from the reactor indicated absence of primary amine (complete reaction).

12.   Cleavage

Resins from different batches were suspended in 95% TFA/water for 90 min and filtered. The resin was then reextracted with DCM. The DCM extracts were combined with the TFA filtrates and evaporated to afford a 125-member combinatorial library in the form of five mixtures of five compounds each.

III.   COMMENTARY: FEATURES AND FUNCTION OF THE PROTOTYPE SYNTHESIZER

A prototype synthesizer (Fig. 4) is mounted on a sheet of anodized aluminum (22 × 42 × ¼ in.). The entire apparatus can be conveniently accommodated

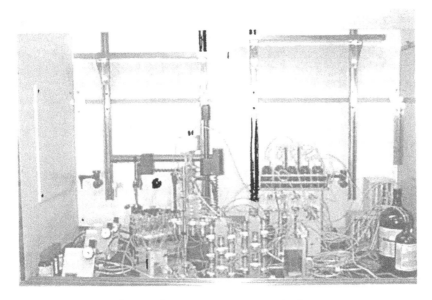

Figure 4 Zymark prototype synthesizer.

within our standard fume hoods. The modified BenchMate arm, located toward the back, can access the cannula wash, argon supply, diversity rack, and reaction vessel rack. Liquid reagents are metered and delivered by a modified Zymark Master Lab Station. The Master Lab Station (MLS) additionally meters solvent for the cannula wash function. Also located on the platform are the valves and plumbing for effecting parallel and serial delivery of reagents or solvents, as well as the plumbing and hardware necessary to divert reaction contents either to individual vessels for collection or to waste as appropriate. All wetted surfaces are Teflon, polyphenylene sulfide (PPS), glass, or stainless steel. All fluid routing is controlled by appropriately placed and operated Bio-Chem three-way solenoid valves that were built to custom specifications. In addition, all enclosures housing electrical circuits are maintained under an inert atmosphere of nitrogen, thus permitting the use of extremely flammable solvents, including diethyl ether.

The Zymark MLS provides liquid metering to the parallel and serial delivery functions, as well as to the cannula wash (Fig. 5). Ten 125-mL common reagent/solvent bottles are arrayed in a rack on top of the MLS, and these are kept under an inert blanket of argon or nitrogen. If larger solvent reservoirs are desired, they may be placed elsewhere in the hood. The MLS has been modified so that several of its resident firmware functions have been disabled and are now handled externally via a local controller. This allows more direct user control of certain timing functions. A standard 12-position rotary valve is used to select the

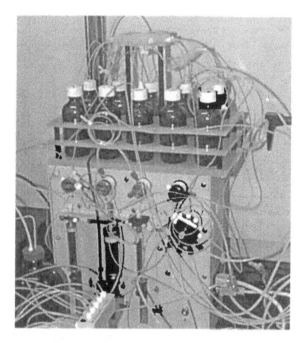

Figure 5   Zymark Master Lab Station.

appropriate common reagent or solvent and to divert the flow to either the parallel
or serial delivery system.

A key to the proper functioning of the synthesizer is inherent in the design
of the reaction vessels themselves (Fig. 6). The core features of the reaction
vessels are the three threaded access ports, the topmost (B) of which allows
reagent/solvent introduction from both parallel and serial sources; and the middle
and bottom ports, which, depending on gas flow direction, enable the sparge and
clear functions. Thus, positive gas pressure applied to the middle port (C) allows
gas to bubble through the filter frit to gently agitate the reactor contents. This
function is under software control and may be intermittent if so desired. Likewise,
positive gas pressure applied to the bottom port (D) causes the contents of the
reactor vessel to be cleared through port C either directly to waste or to collection
tubes. Each reactor is also equipped with a threaded septum cap (A) for robotic
access. Regardless of the liquid source, all liquids enter the reactor vessel itself
through an outer concentric tube with radially arrayed pinholes at its bottom,
thus giving a "showerhead wash" to the reactor vessel walls. This equipment
has been modified to accept 10 reaction vessels up to 100 mL in volume or to
accept up to two reactor vessels of 250-mL, 500-mL, or 1000-mL capacity. The
1000-mL vessel is depicted in Fig. 7.

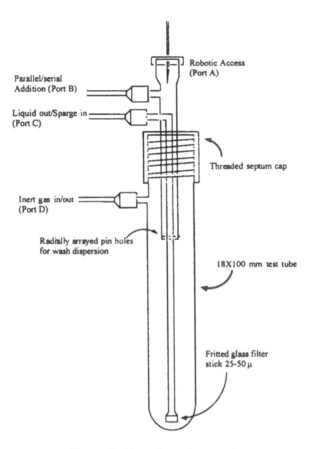

Figure 6   Zymark reaction vessel.

The prototype synthesizer is fitted with a heating/cooling block for conducting reactions at temperatures other than ambient. The design of this block (Fig. 8) permits both cooling to $-75°C$ and heating to $150°C$, as well as providing a reflux zone through which cooled liquid can be circulated at -10 to $-15°C$. The block is fabricated from two blocks of anodized aluminum separated by Nomex insulation. The outside of the block is insulated with Nomex as well and finally clad with stainless steel. The bottom of the block is fitted with strip resistance heaters held in place with a layer of ceramic electrical insulation, and the entire assembly is mounted on a 1-in. Teflon block with indexing holes.

The contents of the reactor vessels may be cleared either directly to waste or diverted to individual test tubes for further processing following cleavage from the resin. In the former case, one of three (aqueous inorganic, nonhalogenated

Figure 7 A 1000-mL large-scale reactor vessel.

organic, or halogenated organic) waste streams is software selectable. The 10-position collection rack is a modified test tube rack and mounted on its back is the valving for the waste stream selection. Since positive inert gas pressure is used to clear the reactor vessels, continuing that gas flow after delivery of the reactor contents to the collection rack provides an intermediate level inert atmosphere over the tubes if desired.

The diversity reagents are delivered by a BenchMate arm that has been fitted with a syringe hand (Fig. 9). The robot arm has relinquished its normal motion control firmware functions to a local controller which handles syringe functions as well. Software calls to deliver diversity reagents to specified reactor vessels initiate a complex series of moves that closely mimic good manual syringe technique.

The prototype synthesizer is controlled by a Zymark System V Controller operating in tandem with two local controllers. These are running the Zymate Easy Lab V2.24 operating system.

Flexibility in operational capability of the prototype synthesizer was con-

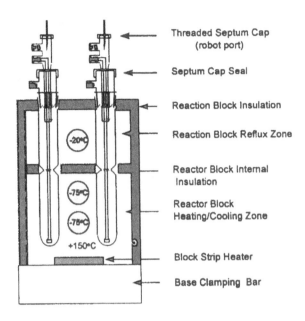

Figure 8   Heating/cooling block.

sidered to be a much higher priority than user friendliness. A prime consideration was the ability of the user to have control over functional operations even at the level of energizing single valves. Thus, at the lowest level there are routines that permit more precise control. Only six fundamental functions are required to have proper operation of the synthesizer. All other subprocess routines are derived from these six core functions.

1.   Serial delivery
2.   Parallel delivery
3.   Robotic delivery
4.   Sparge/agitate
5.   Clear to waste
6.   Clear to collect

IV.   COPYRIGHTS AND TRADEMARKS

Zymate, BenchMate, System V, EasyLab, and Master Lab Station are registered trademarks of Zymark Corporation, Hopkinton, MA. Teflon and Nomex are regis-

Figure 9 Diversity reagent robot.

tered trademarks of E. I. DuPont de Nemours, Wilmington, DE. Bio-Chem is a registered trademark of Bio-Chem Valve Inc., Boonton, NJ. Chemraz is a registered trademark of Greene Tweed and Co., Kulpsville, PA.

REFERENCES

1. Gardner, B., Nakanishi, H., and Kahn, M. *Tetrahedron 49*, 3433–3448 (1993).
2. Chen, S., Chrusciel, A.R., Nakanishi, H., et al. *Proc. Natl. Acad. Sci. USA 89*, 5872–5876 (1992).
3. Curran, T.P., Chandler, N.M., Kennedy, R.J., and Keaney, M.T. *Tetrahedron Lett. 37*, 1933–1936 (1996).
4. Kocis, P., Issakova, O., Sepetov, N.F., and Lebl, M. *Tetrahedron Lett. 36*, 6623–6626 (1995).
5. Furka, A., Sebestyen, F., Asgedom, M., and Dibo, G. *Int. J. Pept. Prot. Res. 37*, 487–493 (1991).

6. Lam, K.S., Salmon, S.E., Hersh, E.M., Hruby, V.J., Kazmierski, W.M., and Knapp, R.J. *Nature 354*, 82–84 (1991).
7. Campbell, J.B., and Wildonger, R.A. *ISLAR '96 Proc*. 127–141 (1996).
8. Zuckerman, N., Kerr, J.M., Siani, M.A., and Banville, S.C. *Int. J. Pept. Prot. Res. 40*, 497, (1992).
9. Sugawara T., Kato S., and Okamoto S.J. *Automation Chem. 16*, 33, (1994).
10. DeWitt, S.H., and Czarnik A.W. *Curr. Opin. Biotech. 6*, 640 (1995).
11. Krchnak, V., and Vagner J. *Pept. Res. 3*, 182–193, (1990).
12. Kaiser, E. *Anal. Biochem*, 34, 595, (1970).
13. Chu, S.S., and Reich, S.H. *Bioorgan. Medic. Chem. Lett. 5*(10), 1053–1058 (1995).

Case Study 4-2
Production of an α-Ketoamide-Containing Library via Convergent Parallel Synthesis

Alan P. Kaplan and Carmen M. Baldino
ArQule, Inc., Woburn, Massachusetts

ABSTRACT

By utilizing the ideas of convergent, multi-step synthesis in the combinatorial synthesis arena, ArQule has been able to build increasingly more complex small molecule libraries with enhanced biases towards certain biological targets. One particular library we have produced is a series of α-ketoamide-containing molecules that were designed to be inhibitors of serine and cysteine proteases. More specifically, a 1600 compound library containing a diamine scaffold that has been differentially functionalized with an α-ketoamide warhead and an aminomethyleneoxazolone (AMO) was produced in a parallel, solution-phase method using ArQule's Automated Molecular Assembly Plant (AMAP™). Herein we describe the methods that were used to produce this library.

Chemicals

1. Methyl 3-trifluoromethylbenzoylformate
2. Methyl benzoylformate
3. Methyl 4-bromobenzoylformate
4. Methyl 2,4-diflurobenzoylformate
5. Methyl 4-nitrobenzoylformate
6. Methyl 4-*tert*-butylbenzoylformate
7. Methyl 3-methylbenzoylformate

8. Methyl 3-methoxybenzoylformate
9. Methyl 3-fluorobenzoylformate
10. Methyl 4-methoxybenzoylformate
11. *N,N'*-Dimethyl 1,2-ethylenediamine
12. *N,N'*-Dimethyl 1,3-propanediamine
13. *N,N'*-Dimethyl 1,6-hexanediamine
14. 2-Thienyl ethoxymethyleneoxazolone (EMO)
15. 2-Napthyl EMO
16. 4-Biphenyl EMO
17. 3-Tolyl EMO
18. 4-Trifluoromethylphenyl EMO
19. 2-Furyl EMO
20. 2-Chlorophenyl EMO
21. 2-Tolyl EMO
22. 4-*tert*-Butylphenyl EMO
23. 3-Methoxyphenyl EMO
24. 2,4-Dichlorophenyl EMO
25. 3-Nitrophenyl EMO
26. 4-Bromophenyl EMO
27. 1-Naphthyl EMO
28. 3-Furyl EMO
29. 3,4-Methylenedioxylphenyl EMO
30. 3-Pyridyl EMO
31. 4-Tolyl EMO
32. 4-Chlorophenyl EMO
33. 4-Nitrophenyl EMO
34. Piperazine
35. Homopiperizine
36. 1,4-Diaminobutane
37. 1,3-Cyclohexanebis(methylamine)
38. 1,3-Diaminopropane
39. Methanol
40. Dimethylformamide (DMF)
41. Acetonitrile

Equipment and Supplies

1. Tecan fluid handler
2. Gilson fluid handler
3. Zymark robotic weighing and dissolution unit
4. Tomtec fluid handler
5. Custom 96-well aluminum reaction blocks fitted with 1-mL glass vials

6. Savant centrifugal solvent concentrator
7. Shimadzu HPLC with autosampler
8. Fisons triple-quad mass spectrometer

I. INTRODUCTION

As the field of combinatorial and parallel synthesis has progressed, one goal has been to diverge from peptide-and nucleotide-based libraries to more traditional organic molecules that are biased toward specific biological targets [1]. As part of ArQule's Mapping Array™ program, a library of diamines that have been selectively bis-functionalized with a series of α-ketoesters and a selection of other electrophiles has been produced. Previous work has shown that α-ketoamides are inhibitors of both serine and cysteine proteases [2,3]. The aim for production of these libraries was to use these compounds to generate new lead compounds for a variety of proteolytic targets. This report will comment on the production of a 1600-compound library that contains a diamine tethering an α-ketoamide and an aminomethyloxazolone (AMO) (Scheme 1) [4].

That we are able to selectively acylate a single amino group of a diamine can be explained by the proposed mechanism by which the reaction occurs (Scheme 2). After initial acylation of the first amine, the second, free amine is transiently protected by its interaction with the highly electrophilic α-keto group

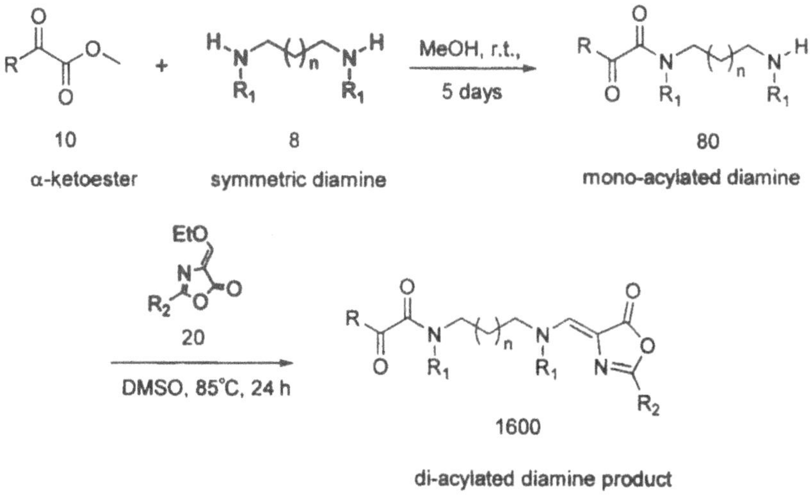

Scheme 1 General scheme of the α-ketoamide/AMO-containing library.

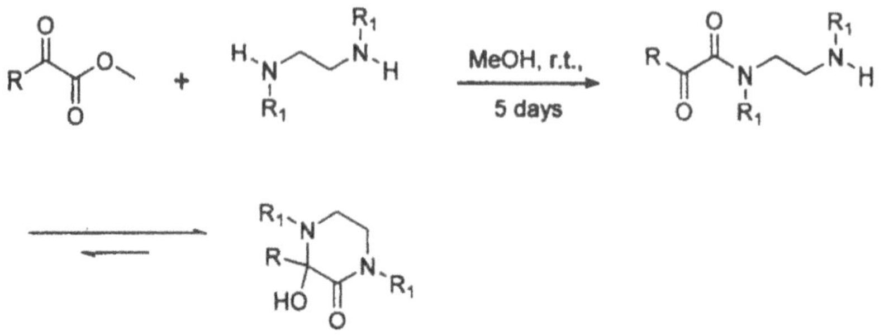

Scheme 2 Proposed mechanism of monoacylation of diamines.

to form a hemiaminal. While this process is clearly reversible, the equilibrium is sufficiently toward the hemiaminal to prevent further reaction with an additional α-ketoester. In the case of primary diamines, where the second amine of the hemiaminal could still be acylated, the amine is protected from further acylation by presumed steric constraints. Whether the hemiaminal is formed in an inter-or intramolecular fashion is unknown. One could postulate that the extent of intermolecular formation is dictated by the structure of the diamine that was used in the reaction. A flexible diamine, such as N,N'-dialkyl-1,2-ethylene-diamine, could easily form a six-member cyclic hemiaminal, whereas the free amine of a diamine such as piperizine more likely will be protected in an intermolecular fashion. After isolation of the initial monoacylated diamine, the second electrophile, which is more reactive than the initial α-ketoester, is reacted with the hemiaminal to yield the desired bis-functionalized product in high yield and purity.

Production of this library required the synthesis of many of the building blocks, as they were not available from commercial sources. The 10 α-ketoesters were synthesized at ArQule using the method developed by Wasserman [5] (Scheme 3), and the 16 EMOs were also produced at ArQule using procedures that were reported by Stammer [6] (Scheme 4). The 8 diamines were all commercially available and were used without further purification.

II. GENERAL PROCEDURE: SYNTHESIS

Many of the library syntheses at ArQule are run using custom-made reaction blocks that are designed to map directly to a 96-well microtitre plate. Each well in the reaction block was fitted with a 1-mL glass vial that acts as the reaction vessel for each individual reaction. The standard practice at ArQule is to leave

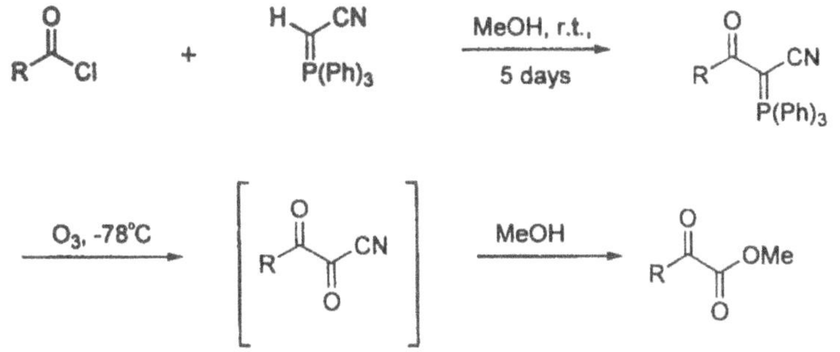

Scheme 3   General synthesis of the α-keto esters.

columns 1 and 12 blank in our production run; therefore, we run only 80 reactions per reaction plate. The standard reaction scale for the production of a library is 50 μmol. Stock solutions of each of the α-ketoesters were made in methanol to a final concentration of 0.25 M using a Zymark robotic weighing and dissolution station. The Zymark system was subsequently used to make all the other stock solutions for the production run. Using a Tecan fluid handler, each of the 10 individual α-ketoester solutions were dispensed in 200-μL aliquots into a reaction plate such that an entire column, 8 individual samples, contained the same α-ketoester. Across each row of the reaction plate was added 200 μL of a 0.25 M methanol solution of the 8 diamines. This 10 × 8 cross was repeated on 19 subsequent reaction plates to give 20 replicates of blocks of 80 individual products (Fig. 1). The reactions were capped and allowed to sit at 25°C for 5 days. After that, the reactions were uncapped, and the solvent was gradually evaporated upon standing at 25°C for 24 h. The resulting crude, monoacylated material was treated with 200 μL of acetonitrile (if the diamine used was a secondary amine)

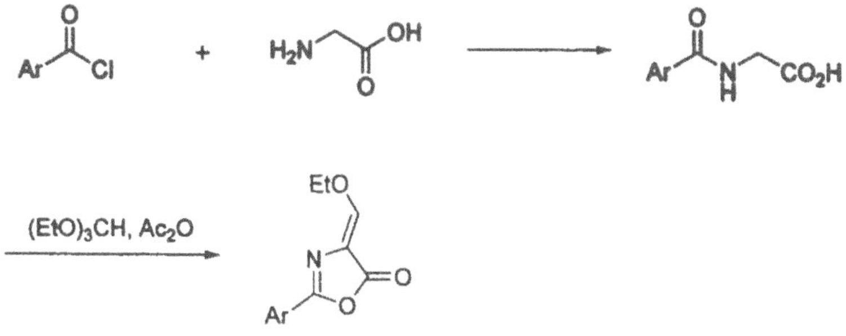

Scheme 4   General synthesis of the ethoxymethyleneoxazolones (EMOs).

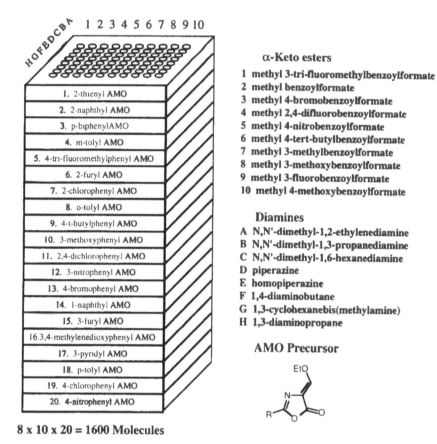

α-Keto esters

1 methyl 3-tri-fluoromethylbenzoylformate
2 methyl benzoylformate
3 methyl 4-bromobenzoylformate
4 methyl 2,4-difluorobenzoylformate
5 methyl 4-nitrobenzoylformate
6 methyl 4-tert-butylbenzoylformate
7 methyl 3-methylbenzoylformate
8 methyl 3-methoxybenzoylformate
9 methyl 3-fluorobenzoylformate
10 methyl 4-methoxybenzoylformate

Diamines

A N,N'-dimethyl-1,2-ethylenediamine
B N,N'-dimethyl-1,3-propanediamine
C N,N'-dimethyl-1,6-hexanediamine
D piperazine
E homopiperazine
F 1,4-diaminobutane
G 1,3-cyclohexanebis(methylamine)
H 1,3-diaminopropane

AMO Precursor

Array contents:

1. 2-thienyl AMO
2. 2-naphthyl AMO
3. p-biphenyl AMO
4. m-tolyl AMO
5. 4-tri-fluoromethylphenyl AMO
6. 2-furyl AMO
7. 2-chlorophenyl AMO
8. o-tolyl AMO
9. 4-t-butylphenyl AMO
10. 3-methoxyphenyl AMO
11. 2,4-dichlorophenyl AMO
12. 3-nitrophenyl AMO
13. 4-bromophenyl AMO
14. 1-naphthyl AMO
15. 3-furyl AMO
16. 3,4-methylenedioxyphenyl AMO
17. 3-pyridyl AMO
18. p-tolyl AMO
19. 4-chlorophenyl AMO
20. 4-nitrophenyl AMO

8 x 10 x 20 = 1600 Molecules

Figure 1   Array layout for the library.

or DMF (if the diamine used was primary). We have found that primary amines
and secondary amines behave differently with the α-ketoamide chemistry; hence,
we need to examine different reaction conditions for the different diamines. The
blocks were heated to 50°C to ensure complete dissolution of the intermediate
products. To all 80 reactions on an individual plate was added a 200-μL aliquot
of a 0.25 M acetonitrile solution of one of the 20 EMOs. Each of the subsequent
19 plates had a different EMO added to each reaction well, giving a total of
1600 unique reactions. The reaction solutions were heated with vials capped at
a temperature setting of 85°C for 24 h. The solvent was removed from the reaction

via a Savant centrifugal concentrator. The product residues were dissolved in 1.6 mL of DMSO.

III. POSTSYNTHESIS QC ANALYSIS

A deliberate pattern of compounds was sampled using a Gilson 215 fluid handler for submission to quality control (QC). The QC sample pattern that was used required that each individual reagent used in the array be analyzed a minimum of twice to ensure the quality and fidelity of the original starting material. In total, 35% (560 samples) of the products were examined. The sample purity was determined by UV-HPLC analyses using a reverse-phase chromatography with an acetonitrile/water (containing 0.1% trifluoroacetic acid) gradient on a Shimadzu autosampler HPLC system. The analysis indicated the presence of both the E and Z isomers of the desired products (Fig. 2). The combined average purity of the two isomers for the entire library was 88%. Further product confirmation was obtained via electrospray ionization (ESI)-MS, which clearly showed the expected $(M + H^+)$ for each compound. We further confirmed the identity of 5% of the library (80 samples) by HPLC-MS to explicitly determine that the major component of the product samples corresponded to products of the desired molecular weight. The HPLC-MS also confirmed the expected presence of both the E and Z isomers of the product. While we cannot state with complete confidence the geometry around the double bond, previous crystal structures that we have obtained with other AMO-derived compounds indicates that the major product of acylations with an EMO has the Z configuration. Subsequent aliquots of the library were made using a Tomtec 96-channel fluid handler to create the "daughter plates" that were then shipped to ArQule's collaborators for biological screening.

To test our initial premise of designing these compounds, the products of this library were submitted to ArQule's biology department for screening against a variety of proteases. The list of serine proteases included thrombin, factor Xa, trypsin, and plasmin, whereas the cysteine proteases include papain and cruzain.

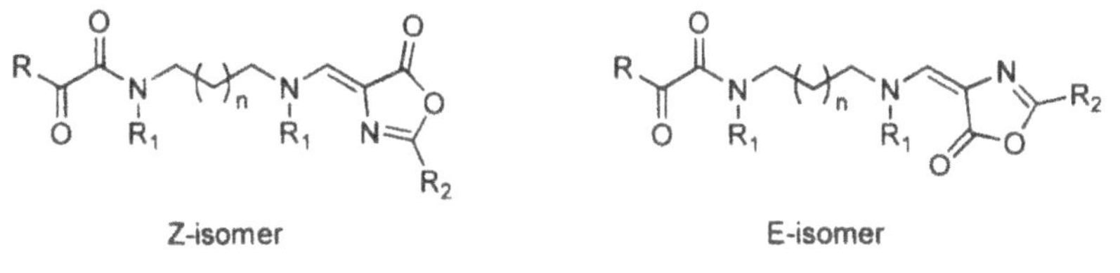

Z-isomer   E-isomer

Figure 2   Isomeric products of the EMO addition.

Several of the products from this array showed selective inhibition with some K_i values in the low to submicromolar range (1; A. Sheldon and M. Cleveland, personal communication, 1997; H. Cheng and M. Cullinane, personal communication, 1999.)

IV. COMMENTARY

The successful synthesis of this library, as well as all the libraries produced at ArQule, depends heavily on the process chemistry performed on the "front end." All of the reagents used in this library needed to be qualified in such a fashion that the final reaction conditions were sufficiently flexible that all 1600 compounds were obtained in good yields and purities. The success of doing efficient parallel synthesis requires that reagents that will not work effectively in the library process must be discovered during this qualification phase before entering the production phase so that they can be excluded from the production run.

REFERENCES

1. D. Coffen; C.M. Baldino; M. Lange; R.F. Tilton; C. Tu, *Med. Chem. Res.*, 1998, *8*: 206–218.
2. B. Munoz; C.-Z. Giam; C.-H. Wong, *Biorgan. Medic. Chem.*, 1994, *2*: 1085–1090 and B.E. Maryanoff; X. Qiu; K.P. Padmanabhan; A. Tulinsky; H.R. Almond Jr.; P. Andrade-Gordon; M.N. Greco; J.A. Kauffman; K.C. Nicolaou; A. Liu; et al., *Proc. Natl. Acad. Sci. USA*, 1993, *90*: 8048–8052.
3. Z. Li; A.C. Ortega-Vilain; G.S. Patil; D.L. Chu; J.E. Foreman; D.D. Eveleth, J.C. Powers; *J. Med. Chem.*, 1996, *39*: 4089–4098.
4. C.M. Baldino; S. Casebier; J. Caserta; G. Slobodkin; C. Tu; D.L. Coffen, *Synlett*, 1997, May: 488–490.
5. H.H. Wasserman; W.-B. Ho, *J. Org. Chem.*, 1994, *59*: 4364–4366.
6. J.M. Bland, C.H. Stammer, K.I. Varughese; *J. Org. Chem.*, 1984, *49*: 1634–1636.

Case Study 4-3
Manual Solid Phase Synthesis of a Fully Encoded, 46,656-Member Benzimidazole Library

David Tumelty, Lun-Cong Dong, Kathy Cao, Lanchi Le, and Michael C. Needels
Affymax Research Institute, Palo Alto, California

ABSTRACT

The manual synthesis of an encoded chemical library composed of substituted benzimidazoles is described. Three monomer sets composed of diverse amines and aldehydes were combined in a mix-and-split protocol to produce the final $36 \times 36 \times 36 = 46{,}656$-member library.

Chemicals

1. Tentagel HL NH$_2$ resin (0.36 mmol/g) (presieved to get a uniform bead size of 110 μm, Rapp Polymere GmbH, Tubingen, Germany)
2. Methanol (MeOH)
3. Dichloromethane (DCM)
4. Di-*tert*-butyl dicarbonate [(Boc)$_2$O]
5. 9-Fluorenylmethylchloroformate (Fmoc-Cl)
6. *N*, *N*-Diisopropylethylamine (DIEA)
7. Ninhydrin test reagents (phenol/ethanol, ninydrin/ethanol, KCN/pyridine, purchased as premade solutions from PE Applied Biosystems, Foster City, CA)
8. *N*, *N*-Dimethylformamide (DMF)
9. Piperidine

10. Diethyl ether (Et_2O), anhydrous
11. Twelve different *N*-Alloc-*N*-[(dialkylcarbamoyl)methyl]glycine tag monomers (synthesized in-house)
12. 1,3-Diisopropylcarbodiimide (DIC)
13. 1-Hydroxybenzotriazole hydrate (HOBt)
14. Trifluoroacetic acid (TFA)
15. 1-Methyl-2-pyrrolidinone (NMP)
16. 5-(4-Formyl-3,5-dimethoxyphenoxy)butyric acid (commercially available as the BAL linker from Perseptive Biosystems, Framingham, MA)
17. *p*-[(R,S)-α-[1-(9H-Fluoren-9-yl)-methoxyformamido]-2,4-dimethoxy-benzyl]-phenoxyacetic acid, known as the Fmoc-acid-labile linker (Calbiochem-Novabiochem, San Diego, CA)
18. Seventy-one primary amines (various suppliers or synthesized in-house)
19. Tetrahydrofuran (THF)
20. Sodium cyanoborohydride, 1.0 *M* in THF
21. Acetic acid, glacial
22. 4-Fluoro-3-nitrobenzoic acid
23. Acetic anhydride
24. Acetonitrile, HPLC grade
25. Water, HPLC grade
26. Tetrabutylammonium fluoride (TBAF), 1.0 *M* in THF
27. Azidotrimethylsilane (TMS-azide)
28. Tetrakis(triphenylphosphine)palladium(0) [$Pd(PPh_3)_4$]
29. *O*-(7-Azabenzotriazol-1-yl)-1,1,3,3-tetramethyluronium hexafluoro-phosphate (HATU) (Perseptive Biosystems, Milford, MA)
30. Tin(II) chloride dihydrate
31. Sodium diethyldithiocarbamate trihydrate
32. Thirty-five aldehydes (various suppliers or synthesized in-house)
33. Trimethylorthoformate (TMOF), anhydrous

Equipment and Supplies

Access to standard laboratory equipment and glassware is assumed.

1. Electronic label maker, with plastic-coated tape
2. 2-L glass synthesis vessel (Chemglass, Vineland, NJ)
3. Solvent resistant, 1-L polyethylene wash bottles (VWR, San Francisco, CA)
4. Overhead stirrer with Teflon stirrer blade (Chemglass)
5. Polypropylene reaction vessels (RVs), with appropriate filters and

caps (15-, 25-, and 75-mL capacity) (Alltech Associates, San Jose, CA)

6. Syringe pressure caps (Aldrich)
7. Disposable wooden stirrers (VWR)
8. 2× Vacuum manifold and glass chamber (24-port manifolds with Teflon needle inserts were used) (Alltech Associates)
9. 2× Medium-pressure vacuum pumps (VWR)
10. Polyethylene disposable transfer pipets (7-mL capacity) (VWR)
11. Disposable glass vials and screw-caps with Teflon liners (20- and 30-mL capacity) (VWR)
12. Temperature-controlled dry-block heater (holds six aluminum blocks, each block designed to hold six 20- or 30-mL glass vials) and thermometer (VWR)
13. Eppendorf Repeater pipettor (with different sizes of disposable Combi-syringes)
14. Variable volume pipetors (various brands, 0.1-, 0.25-, and 1.0-mL capacity) and corresponding tips
15. Large desiccator (VWR)
16. High-vacuum pump (VWR)

I. INTRODUCTION

This work was part of an ongoing program at Affymax aimed at synthesizing novel organic molecules by solid phase combinatorial chemistry for primary library screening. We have developed several chemistries that use nucleophilic aromatic substitution as a key transformation in the synthetic route to introduce diverse sets of monomers into recognized pharmacophores [1–3], and the benzimidazole library synthesis described here is a good example of this strategy. The route shown in Scheme 1 was reduced to practice by optimizing the reaction conditions for each step using a variety of different solid phase bound intermediates. Several other solid phase routes to related benzimidazole-based structures have also been published [4–7].

Critical early steps in the development stage of this chemical route were the demonstration of its compatibility with Affymax's encoding technology [8–10] and that a wide variety of monomers could be incorporated at each of the three diversity positions (amines for R1 and R2, aldehydes for R3). The encoding technology enables single beads to be used for assay, as the benzimidazoles released after cleavage can be identified from the mixture of chemical "tags" left on the bead. This technology has previously been described in detail [10]. The structure of a generic tag, as well as those of the linkers used in this chemistry, are shown in Fig. 1.

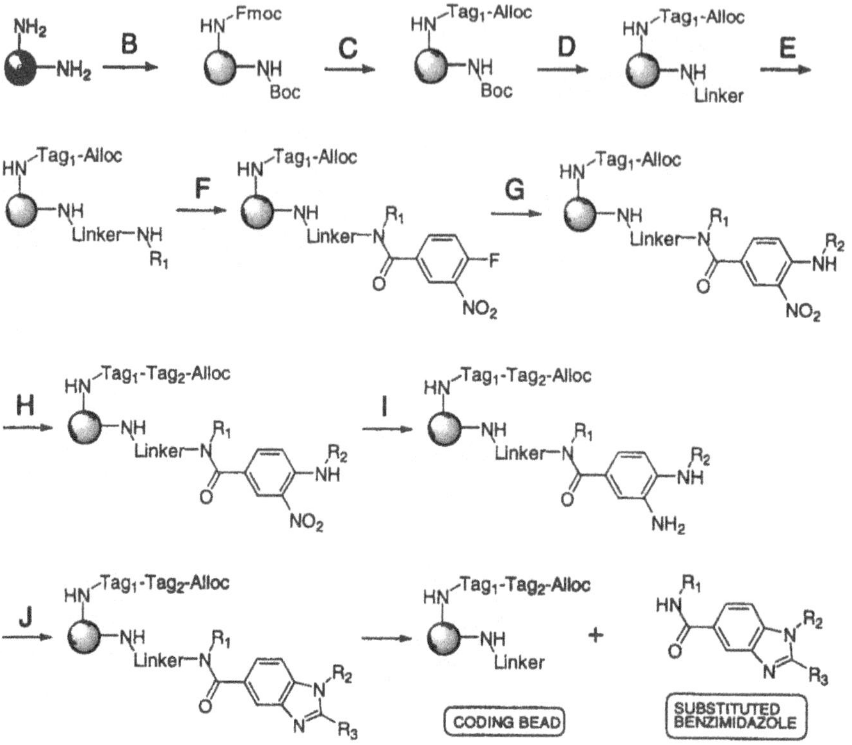

Scheme 1   Encoded benzimidazole library synthesis.

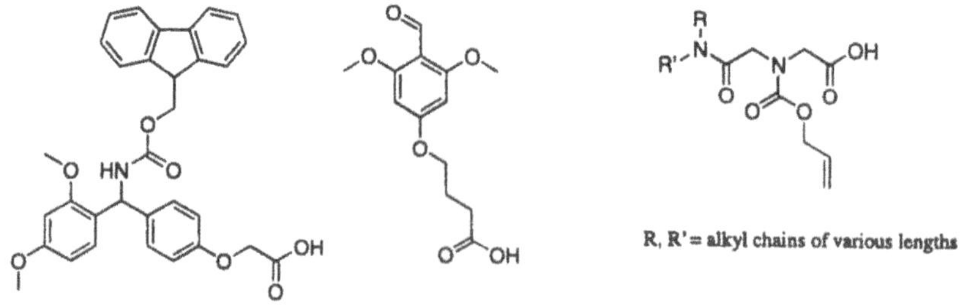

Fmoc-acid-labile linker   BAL linker   Generic *N*-Alloc-Tag monomer

Figure 1   Structures of linkers and tag monomers.

A key part of combinatorial chemistry development for this library was the "rehearsal" of potential monomers to ensure that they could be incorporated into the synthetic route and would lead to the correct final products. Considering that the final products were to be released for assay from a single bead with no subsequent purification steps, it was essential to identify monomers, which could give high product yields across a range of different starting intermediates. Numerous such model intermediates were thus used to test potential amine building blocks for successful incorporation at the R1 and R2 diversity positions (reductive amination and nucleophilic displacement, respectively), and aldehydes were screened for their ability to participate in the final cyclization reaction to incorporate R3. The rehearsal process thus involved numerous small-scale syntheses on resin, followed by acidolysis and analysis of the products by HPLC and LC/MS (and nuclear magnetic resonance in some cases). Overall, more than 300 monomers were judged suitable for library production. Description of the exact composition of the final 71 amines and 35 aldehydes chosen for the production of this library is subject to some proprietary restrictions, but several commercially available monomers that were used are shown in Fig. 2.

Prior to actual library production, numerous tagged, single compounds were synthesized by taking small batches of resin through the complete synthetic scheme, using single, known monomers for R1, R2, and R3 and incorporating the two tagging steps. At various stages in this process, the fidelity of both the benzimidazole intermediate and tag assembly on the beads was assessed. To achieve this, the product was first released by acidolysis from a batch of beads (5 mg) and examined by LC-MS. At the same time, *single*-bead samples from the same resin batch were subjected to acidolysis and examination by LC-MS. The larger scale product provided a useful standard to aid in the analysis of the single-bead LC-MS data. The single beads were also subject to decoding. The use of these analytical tools for the analysis of combinatorial libraries during synthesis and as a QC check for completed libraries has been described [11,12].

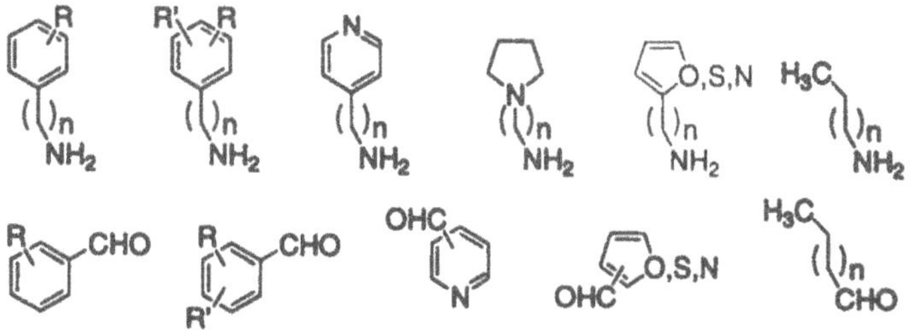

Figure 2 Generic structures of some commercial amines and aldehydes used in library production.

This process can be time consuming but was essential to give confidence that the encoded chemical route was working effectively during library production, especially after the mix-and-split steps.

We used slight variations of the protocol shown in Scheme 1 to enable R1 and R3 to be unsubstituted, i.e., a hydrogen atom rather than a side chain was present at these positions. One of the initial 36 resin batches (with only the first set of tags attached) was coupled with a Fmoc–acid-labile linker, which resulted in an unsubstituted primary carboxamide on final acid cleavage. In addition, one of the final R3 monomers was trimethylorthoformate, which adds a formyl equivalent to the substituted o-phenylenediamine starting intermediates under the described reaction conditions, resulting in 2(H)-benzimidazole products.

We chose to produce the actual library by manual synthesis as this enabled maximum flexibility in terms of the amounts of resin, reagent concentrations, and washing protocols that could be used. Initially, we synthesized 36 key tagged resin intermediates, each on a 2-g scale. Approximately one-third by weight of each of these resin intermediates was combined and taken onward to make the library as described below (two related libraries were also made from the remaining two-thirds of the resins).

The final library consisted of 36 pools with roughly 0.7 g in each; thus, theoretically around 450 single-bead copies of each compound should be present in the library. The final pools were kept separated so that the identity of the last aldehyde monomer was known (''spatial'' encoding), whereas that of the R1 and R2 amines could be identified by the subsequent decoding of an individual bead. After completion of the library, the chemical syntheses and tagging fidelity on beads in each pool were assessed by analyzing the products from single beads by LC-MS and decoding the corresponding bead, as outlined above [12]. It should be noted that unencoded libraries and single compounds (synthesized to confirm hits from screening of the library) can be synthesized using the same protocol as that shown in Scheme 1, except that the resin differentiation and tagging steps are omitted.

II. EXPERIMENTAL PROCEDURES

A. General Resin Handling Procedures

Reactions on large-scale batches of resin were carried out in a 2-L glass reaction vessel originally designed for solid phase peptide synthesis. The vessel consists of a cylindrical glass chamber, narrowed at the top, with a screw cap. At the bottom of the chamber is a medium-porosity glass filter and below this is a T-bore, Teflon stopcock attached to two side arms, the lower of which has a ground-glass joint for connection to a 2-L round-bottom flask that serves as a waste container. The other side arm is attached to a nitrogen source. Turning the stopcock enables either nitrogen to bubble through the reaction vessel (thus mixing any resin/solution that is present) or the solution in the vessel to drain into the waste flask. For the largest resin batches (72 g), the addition of an overhead paddle stirrer was required to ensure adequate resin/solvent equilibration during the washing and reaction steps. During draining, the waste flask was evacuated with a medium-vacuum pump to ensure fast and complete filtering of the resin in the reaction chamber. Prior to washes with ether, resin in this vessel was transferred to disposable 75-mL polypropylene tubes to facilitate its subsequent recovery (ether-dried resin was difficult to remove from the large glass vessel).

Smaller amounts of resin (i.e., 0.7 or 2.0 g) were washed in disposable polypropylene reaction vessels of various sizes (15, 25, or 75 mL as required), each equipped with two 20-μm filters. A fundamental, but critical, step was clearly distinguishing these RVs (or glass vials, see below) with solvent-resistant labels. This was especially vital as the labels determined which reagents were added to which resin and were essential for tracking resin batches through the entire synthetic scheme. In practice, this was most conveniently achieved with an electronic label maker using standard plastic-coated tape. For washing, solvents were added to the RVs via 1-L polypropylene wash bottles, the resin/solvent equilibrated by mixing with disposable wooden sticks for 30 s, then left to stand for 1 min before draining. The tubes were drained using a commercial vacuum manifold mounted on a rectangular glass chamber, which was evacuated by a medium-vacuum pump to ensure fast and complete filtering of the resin in the RVs. Twelve to eighteen RVs (depending on RV capacity) could be accommodated per manifold for simultaneous washing.

In addition to the washing protocols, the RVs were also used in several of the reaction steps. In this case, the RVs were removed from the manifold, the outlets plugged with syringe pressure caps, and the appropriate solutions added to the resins. The RVs were then placed in a test-tube rack for the duration of the reaction. Resin drying was generally carried out directly in the RVs, after resin washing, by placing the tubes in a large desiccator and evacuating with a high-vacuum pump. For some reaction steps, the resins were not dried down but were simply left solvated in the appropriate solvent prior to the addition of the next reagent.

When heating was required, the resins/solutions were transferred to glass vials with polypropylene disposable pipets, then the vials placed in aluminum blocks designed to hold six vials of 20 or 30 mL capacity (these were machined in-house from commercially available blocks). Six such blocks were accommodated in a thermostatically controlled oven for the duration of the reaction.

B. Resin Differentiation

Tentagel HL NH$_2$ resin (72 g, 27.4 mmol with respect to amine functionality) was placed in a 2-L glass reaction vessel and washed with MeOH (3 × 600 mL), with nitrogen bubbling and stirring from an overhead paddle. The resin was then washed with DCM (5 × 600 mL) and filtered to leave the resin solvated. Separately, Boc$_2$O (22.7 g, 104 mmol, 3.80 equivalents with respect to resin amine content) was dissolved in DCM (145 mL) and Fmoc-Cl (0.79 g, 3.1 mmol, 0.11 equivalent) was also dissolved in DCM (72 mL). Both solutions were mixed together and DIEA (37.5 mL, 216 mmol) was added to give a clear, colorless solution. DCM (600 mL) was added to the solvated resin and, with vigorous stirring and nitrogen bubbling, the Fmoc-Cl/Boc$_2$O/DIEA mixture was added to the resin. Stirring and nitrogen bubbling was continued for 4 h. The resin was then filtered and washed with DCM (3 × 600 mL), MeOH (3 × 600 mL), and DCM (3 × 600 mL).

The entire resin was washed in DMF (3 × 600 mL) then treated twice with piperidine (20% v/v) in DMF (800 mL) for 30 min, each time with vigorous stirring and nitrogen bubbling. The resin was then filtered and washed with DMF (3 × 600 mL), MeOH (3 × 600 mL), DCM (5 × 600 mL), and MeOH (3 × 600 mL). Finally, the resin was rinsed into four 75-mL polypropylene tubes with MeOH and washed with Et$_2$O (5 × 40 mL each). After filtering, the combined resin was transferred to a 250-mL crystallizing dish and dried overnight in vacuo.

C. Loading of the First Tags

The differentiated resin (72 g), with approximately eight-ninths of the available amines protected with the Boc group, was divided into 36 × 2-g portions and placed in separate, clearly labeled (RV1–RV36) 25-mL polypropylene RVs. Six stock solutions (A–F) were made up using six different alkylamine tags (20 mmol each) with DMF (70 mL). The encoding protocol for each RV was as follows using the resulting tag solutions A, B, C, D, E, and F;

The encoding mixtures were made up as follows from the individual stock solutions of tags A–F: for resins in RV1–6: 10 mL (2 mmol) of corresponding stock tagging solution A, B, C, D, E, or F only: for resins in RV7–21: 5 mL (1 mmol) from each of the two stock tagging mixtures, as indicated in Table 1; for resins in RV22–36: 3.33 mL (0.67 mmol) from each of three stock tagging mixtures, as indicated in Table 1.

To each of the 36 final tagging solutions (10 mL total) in a 20-mL glass vial was added DIC (0.33 ml, 2 mmol) and HOBt (0.32 g, 2 mmol) and the solution mixed for 20 min before being added to the corresponding RV. After 12 h, the resins were filtered, washed (DMF, DCM, MeOH, Et$_2$O, each 3 × 15 mL), and dried overnight in vacuo.

D. Coupling of the Acid-Labile Linkers

The resins (2 g each) in RV1–36 were carefully solvated with a solution of TFA (50% v/v) in DCM (20 mL) and left to stand for 1 h with occasional stirring.

Table 1   Encoding Mixtures

RV	1	2	3	4	5	6	7		10	11	12
Code	A	B	C	D	E	F	AB	E	AE	AF	BC
RV	13	14	15	16	17	18	19		22	23	24
Code	BD	BE	BF	CD	CE	CF	DE		ABC	ABD	ABE
RV	25	26	27	28	29	30	31		34	35	36
Code	ABF	ACD	ACE	ACF	ADE	ADF	AEF		BCF	BDE	BDF

The resins were drained, washed with DCM (5 × 20 mL), and washed with DIEA (10% v/v) in DCM (20 mL, 2 × 20 min). Each resin was then washed with DCM (3 ×20 mL), NMP (3 × 20 mL), filtered, and left solvated.

RV1 only: Fmoc–acid-labile linker (1 g, 3.6 mmol, approx. 6 equivalents), DIC (0.29 mL, 1.8 mmol), and HOBt (0.28 g, 1.8 mmol) were dissolved in NMP (8 mL), then stirred for 30 min before being added to the resin and left for 24 h. After this time, the resin was filtered, washed (NMP, DCM, MeOH, Et$_2$O, each 3 × 15 mL), and dried overnight in vacuo.

RV2–36: BAL linker (34 g, 127 mmol, approx. 6 equivalents over calcu-lated free amine on each resin), DIC (20.5 mL, 127 mmol), and HOBt (19.8 g, 127 mmol) were dissolved in NMP (made up to 350 mL) and stirred at room temperature for 30 min to give a dark brown solution. The solution was filtered through a medium sinter under reduced pressure to remove the fine white diiso-propylurea side product that had formed during the activation reaction. Equal portions of the resulting solution (10 mL) were added to RV2–36, mixed with the resin, then left to stand, with occasional stirring for 24 h. After this time, the resins were filtered, washed (NMP, DCM, MeOH, Et$_2$O, each 3 × 15 mL), and dried overnight in vacuo.

E.   Reductive Amination of BAL Linker with R1 Amines

RV2–36 only: The resins (2 g) were transferred by spatula from the 25-mL poly-propylene tubes to 30-mL glass vials. Separately, 35 amines (5 mmol, approx. 6.25 equiv. over resin-bound aldehyde) were dissolved in MeOH (0.3% v/v) in THF (10 mL), the solutions used to wash residual resin from the tubes and wash-ings added to the respective resins in the glass vials. The resin/solution was heated at 50°C for 5 h; then a solution of 1 M NaCNBH$_3$ in THF (5 mL, 5 mmol) was added, along with a little acetic acid (0.045 mL), and the mixture left at 50°C for an additional 12 h. After this time, the resins were transferred to 25-mL polypropylene tubes with MeOH washes and washed with MeOH, DMF, DCM, MeOH (each 2 × 20 mL), then Et$_2$O (3 × 20 mL), and dried overnight in vacuo.

F.   Coupling of 4-Fluoro-3-Nitrobenzoic Acid Scaffold to
Resins

RV1 pretreatment: The resin (2 g) was treated with piperidine (20% v/v) in DMF (15 mL, 2 × 15 min) to remove the Fmoc group, then washed (DMF, DCM, MeOH, Et$_2$O, each 3 × 15 mL) and dried overnight in vacuo.

All of the dry RV1–36 resins (2 g in 25 mL RV) were solvated with NMP (2 × 15 mL washes). Separately, 4-fluoro-3-nitrobenzoic acid (80 g, 432 mmol, total of about 20 equivalents with respect to amine available for coupling per

RV) and DIC (34 mL, 216 mmol, 10 equivalents per RV) were dissolved in NMP (400 mL) with vigorous stirring in a 2-L beaker for 30 min. The resulting solution was filtered in a fine-sinter funnel under vacuum to remove the fine white diisopropylurea side product, transferred to a 1-L measuring cylinder, and made up to 540 mL with NMP. Subsequently, a 15-mL aliquot of the resulting clear yellow solution was added to each resin. The resin was left standing in the solution at room temperature for 18 h with occasional stirring. After this time, the resins were washed with NMP, DCM, MeOH, then Et$_2$O (3 × 20 mL each), and dried overnight in vacuo.

1. Mix-and-Split Procedure

After successful decoding and analysis of the acid-cleaved products, 0.7 g of each of the 36 RV resins (25.2 g total) was combined by washing with NMP, then extra NMP was added (to 800 mL) in a 2-L glass reaction vessel and mixed thoroughly for 30 min with mechanical stirring and nitrogen bubbling. Following further washes with DCM and MeOH (3 × 250 mL each), the resin was washed into two 75-mL polypropylene tubes with MeOH and finally washed with Et$_2$O (7 × 40 mL each). The combined resin was transferred to a 250 mL crystallizing dish and dried overnight in vacuo.

G. Nucleophilic Displacement of Fluorine by R2 Amines

The dry, mixed resin (about 26 g in total) was divided equally among 36 × 15 mL polypropylene RVs (approx. 0.7 g in each). Separately, 36 amines (5 mmol, approx. 24 equivalents over the resin-bound fluorine) were dispensed or weighed into 20-mL glass vials and DIEA (0.44 mL, 2.5 mmol) added along with NMP (4 mL). This mixture was then transferred to the tubes containing the corresponding resin, mixed, and the slurry transferred back to the corresponding 20-mL vials. Upon resin settling, some of the amine/NMP solution was used to recover any remaining resin from the tubes and the washings returned to the corresponding vials, which were then heated at 50°C for 12 h. After this time, the slurry was washed into labeled 25-mL RVs with NMP and filtered. The resins were further washed with NMP, DCM, MeOH, then DCM (3 × 20 mL each time) before being filtered and left solvated.

H. Addition of the Second Set of Tags

Two separate solutions were first prepared for the Alloc group deprotection step;
Solution A: 1 M TBAF in THF (45 mL, 45 mmol) was mixed with TMS-N$_3$ (18 mL, 136 mmol) and diluted to 250 ml with DCM.

Solution B: Pd(PPh$_3$)$_4$ (3 g, 2.6 mmol) was dissolved in DCM (120 mL) to form a clear yellow solution.

To each resin in RV1–36 was added 6 mL of solution A, followed by 3 mL of solution B. The resulting light brown solution and resin were rapidly mixed and allowed to stand for 1 h at room temperature, with occasional stirring. Each resin was then washed with DCM, MeOH, DCM and NMP (3 × 15 mL each time), filtered, and the resin left solvated.

Six new tags were used to encode the amines that had just been used in the *ipso* displacement reaction (see Commentary, VII). Accordingly, to each of the 36 tagging mixtures (0.14 mmol, made up to 4 mL with NMP) in a 20-mL glass vial was added DIEA (0.044 mL, 0.25 mmol), followed by solid HATU (0.048 g, 0.125 mmol). The solutions (approx. 3.3 equivalents of activated tags with respect to resin-bound secondary amine exposed by Alloc removal) were stirred for 3 min to give an orange solution, then added to the appropriate resin, mixed thoroughly, and left to stand for 12 h at room temperature with occasional stirring. After this time, the resins were transferred to 15-mL RVs, filtered, and washed with NMP, DCM, MeOH, and finally Et$_2$O (4 × 10 mL each time) before being dried in vacuo overnight.

I. Second Mix-and-Split Step and Reduction of the Nitro Group

Each of the 36 RV resins (about 26 g total) was combined by washing with NMP into a 2-L glass reaction vessel. The total resin mass was washed with more NMP (2 × 250 mL), with mixing by nitrogen bubbling, filtered, and left solvated. Separately, tin(II) chloride dihydrate (68 g, 300 mmol, approx. 35 equivalents with respect to resin-bound nitro groups) was dissolved in NMP with vigorous stirring to give a final volume of 300 mL (1 M solution). The solution was added to the resin and mixed by nitrogen bubbling only for 12 h at room temperature. The resin was filtered and washed with NMP (6 × 250 mL) then treated with a 0.2 M solution of sodium diethyldithiocarbamate in NMP containing 5% DIEA (v/v, 250 mL total) for 1 h to remove excess Pd and Sn salts. The resin was filtered and washed with NMP (6 × 250 mL), DCM and MeOH (5 × 200 mL each). Finally, the resin was rinsed into two 75-mL polypropylene tubes with MeOH and washed with Et$_2$O (7 × 40 mL each). After filtering, the combined resin was transferred to a 250-mL crystallizing dish and dried overnight in vacuo.

J. Benzimidazole Formation with R3 Aldehydes

The dried resin from above (about 26 g in total) was divided equally among 36 × 15 mL polypropylene RVs (approx. 0.7 g in each), washed with NMP (2 × 10 mL), filtered, and left solvated. Separately, 35 aldehydes (3 mmol, approx. 15

equivalents with respect to resin-bound aniline) were dispensed or weighed into labeled 20-mL glass vials and dissolved with NMP (5 mL). The final aldehyde concentration was about 0.6 M in each case. In a separate vial, a solution of TMOF (7% v/v) in NMP (5 mL) was made up to provide the reactant for the 36th RV. Each mixture was then transferred to the tubes containing the corresponding resin, mixed, and the slurry transferred back to the corresponding 20-mL vials. Upon resin settling, some of the aldehyde (or TMOF)/NMP solution was used to recover any remaining resin from the tubes and the washings returned to the vials, which were then heated at 50°C for 12 h. After this time, the slurry was washed into labeled 25-mL RVs with NMP and filtered. The resins were further washed with NMP, DCM, MeOH, then Et_2O (5 × 20 mL each time) before being dried overnight in vacuo.

III. COMMENTARY (BY SECTION)

B. The resin differentiation procedure was modified from an in-house protocol (S. Raillard, personal communication, 1998). After reaction with the protecting group mixture, a ninhydrin color test on the resin revealed no free amines. A sample of this resin (0.1 g) was used to determine the differentiation ratio (i.e., ratio of Boc-protected amines to Fmoc-protected amines). A measured amount of the differentiated resin was treated with piperidine in DMF and the absorbance of the released dibenzylfulvene-piperidine adduct measured at 302 nm. This enabled the loading of Fmoc-protected amines to be calculated at 0.037 mmol/g. The resulting resin was treated with TFA in DCM to remove the Boc group, neutralized with DIEA in DCM, then treated with excess Fmoc-Cl/DIEA in DCM to reprotect all of the available amines. Upon Fmoc removal from this resin, the loading was estimated to be 0.300 mmol/g. The differentiation ration was thus calculated to be approximately 8:1 for ligand/tag amines. After Fmoc removal from the resin, a strong positive (blue coloration) showed the presence of free amines.

    C. After tag coupling and resin drying, five single-bead samples from each RV were subjected to the decoding procedures [11,12] to ensure that the tags had been properly coupled and could be read before continuing with library production.

    D. The dry resins swell considerably on addition of the TFA/DCM solutions. After linker coupling to RV1, a negative ninhydrin test was found. In contrast, when coupling the BAL linker to RV2–36, a slight blue tinge indicating almost complete acylation was found. The result of this test was unchanged even after double-coupling of some test resin batches with additional linker.

    E. The reductive amination procedure was adapted from an in-house protocol [13]. The resin should be washed first with MeOH after the reduction step to remove excess salt precipitates and facilitate adequate draining in the subsequent washes.

F. Acylation with the symmetrical anhydride of the acid is recommended. Coupling with DIC/HOBt or HATU activation leads to variable amounts (depending on the reaction time) of fluorine displacement by the alcohols present [1-hydroxybenzotriazole (HOBt) or 1-hydroxy-7-azabenzotrizole (HOAt), respectively] during these reactions, as previously reported [14]. Also, a small sample (5 mg) of each dried resin was acetylated with a mixture of acetic anhydride (0.1 mL, 1 mol) and DIEA (0.17 mL, 1 mmol) in DMF (1 mL) for 30 min, then filtered, washed, and dried (as above). Each resin was treated with a solution of TFA with water (5% v/v) added (0.5 mL) for 30 min, the solution filtered, evaporated to dryness, dissolved in a mixture of acetonitrile/water (1 : 1, 0.1 mL) then subjected to both LC and LC-MS to characterize the products. As before, single-bead samples from each RV were subjected to the decoding procedures to ensure the fidelity of decoding after the above chemical procedures.

G. The final amine concentration for displacement was about 1 M in each case. The displacement has also been successfully carried out using DMF or dimethylsulfoxide as solvent.

H. The coupling of the second set of tags required HATU activation to successfully acylate the exposed secondary amine from the first tag(s). We determined in model systems that no acylation of the resin-bound, substituted *ortho*-nitroanilines occurred during this tagging step, presumably due to the expected strong deactivation of the secondary aniline by the *ortho*-nitro group. The second set of tag mixtures was prepared in a similar way to the first set (see Table 1) except that six new, structurally different tags were used (e.g., tags U, V, W, X, Y, and Z). These correspond directly to tags A–F for the purposes of tag mixture composition as determined by Table 1. The final volume of tagging mixtures was also reduced to 4 mL for the 0.7 g of resin in each RV (compared to 10 mL for 2 g of resin used for the first set of tags). Single-bead samples from each RV were subjected again to the decoding procedures to ensure that both sets of codes could be read before proceeding.

I. The 36 resin batches were combined only after successful decoding of several beads (5 from each RV). The sodium diethyldithiocarbamate in NMP wash was used to remove excess Pd and Sn salts via chelation, as previously reported [9,15].

J. On initial addition of the aldehydes, the resins turn a bright yellow typical of the formation of the Schiff's base (i.e., most likely an imine between the primary aniline and the aldehyde is the first species formed). We discovered that benzimidazole formation could, in fact, be achieved by using an adequate excess of aldehyde alone. In our hands, using 15 equivalents in a concentration of around 0.6 M was optimal. Using less than 5 equivalents gave increased amounts of unreacted phenylenediamine starting material. To be more precise, a 2-CF$_3$ benzimidazole side product (formed from the reaction of the starting material with TFA during cleavage) was the species actually observed. If the excess was greater than approx. 35 equivalents, side products with mass spectra corresponding to undefined adducts began to appear, in addition to the expected products. Overall, with the aldehydes chosen and under the conditions described, we were confident that the correct final products were being formed with HPLC purities of at least 85% based on our extensive rehearsal reactions.

As an aside, during the production of this library, a similar route to benzimidazoles using aldehydes with the addition of an organic oxidant was described [7]. Considering the protocol described above, we have speculated as to the origin of the "oxidative" equivalents required for completion of the cyclization reaction. They may derive either from dissolved oxygen in the solvents or the excess aldehyde itself. Perhaps most likely, simply heating to 50°C was sufficient to drive the thermal cyclization of the imine intermediates formed by the initial addition of the aldehyde.

REFERENCES

1. D Tumelty, MK Schwarz, MC Needels. *Tetrahedron Lett 39*: 7467–7470, 1998.
2. MK Schwarz, D Tumelty, MA Gallop. *Tetrahedron Lett 39*: 8397–8400, 1998.
3. MK Schwarz, D Tumelty, MA Gallop. *J Org Chem 64*: 2219–2231, 1999
4. GB Phillips, GP Wei *Tetrahedron Lett 37*: 4887–4890, 1996.
5. J Lee, D Gauthier, RA Rivero. *Tetrahedron Lett 39*: 201–204, 1998.
6. Q Sun, B Yan. *Bioorg Med Chem Lett 8*: 361–364, 1998.
7. JP Mayer, GS Lewis, C McGee, D Bankaitis-Davis. *Tetrahedron Lett 39*: 6655–6658, 1998.
8. Z-J Ni, D Maclean, CP Holmes, MM Murphy, B Ruhland, JW Jacobs, EM Gordon, MA Gallop. *J Med Chem 39*: 1601–1608, 1996.
9. ZJ Ni, D Maclean, CP Holmes, MA Gallop. *Meth Enzymol 267 (Comb Chem)*: 261–272, 1996.
10. D Maclean, JR Schullek, MM Murphy, Z-J Ni, EM Gordon, MA Gallop. *Proc Natl Acad Sci USA 94*: 2805–2810, 1997.
11. WL Fitch. *Annu Rep Comb Chem Mol Divers 1*: 59–68, 1997
12. KC Lewis, WL Fitch, D Maclean. *LC-GC 16*: 644, 646, 648–649, 1998.
13. AK Szardenings, TS Burkoth, GC Look, DA Campbell. *J Org Chem 61*: 6720–6722, 1996.
14. GA Morales, JW Corbett, WF DeGrado. *J Org Chem 63*: 1172–1177, 1998.
15. SA Kates, SB Daniels, F Albericio. *Anal Biochem 212*: 303–310, 1993.

Case Study 4-4
Protease Substrate Specificity Mapping Using Membrane-Bound Peptides

Richard A. Laursen, Caizhi Zhu,* and Yongjun Duan†
Boston University, Boston, Massachusetts

ABSTRACT

Chromophore-bearing peptides are synthesized on small membrane disks. When disks are incubated in a solution containing a protease, peptide bonds are cleaved releasing the chromophore, whose rate of formation can be measured. Synthesis and analysis of a library consisting of 400 dipeptide substrates are described.

Chemicals

1. *N*-Fmoc–amino acids (*N*-9-fluorenylmethoxycarbonyl-amino acid-OH)
2. *N*-Fmoc-AC (*N*-Fmoc–aminocaproic acid)
3. HOBT (1-hydroxybenzotriazole)
4. TBTU (2-(1-benzotriazol-1-yl)-1,1,3,3,-tetramethyluronium tetrafluoroborate)
5. Dimethylformamide (DMF, distilled from P_2O_5)
6. DIC (diisopropylcarbodiimide)
7. Piperidine
8. Pyridine

Current affiliation:
* Joslin Diabetes Center, Harvard Medical School, Boston, Massachusetts.
† Symbollon Corporation, Framingham, Massachusetts.

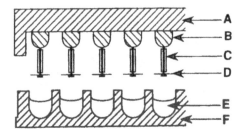

Figure 1 Peptidase activity assay device. (A) Pin holder block constructed from a flexible, 96-well polyvinyl microplate whose wells (B) are filled with wads of Parafilm or with beeswax, into the bottoms of which are stuck stainless-steel straight pins partially covered by a Teflon tubing sleeve (1 cm long) (C). The peptide-membrane disk (D) is impaled on the pin and placed between the pinhead and the sleeve, the space between which is about 0.5 mm to allow the disk to tilt. Aliquots of the enzyme solution (E) are placed in the wells of a clear polystyrene microplate (F). Reaction is initiated by lowering the disks into the solutions. At intervals, the pin block assembly is raised and the absorbance of the solutions in the microplate (F) read in a microplate reader.

9. *N*-Methylmorpholine
10. Reagent L (88% trifluoroacetic acid, 5% water, 5% dithiothreitol/2%, triisopropylsilane; freshly prepared)
11. FITC (fluorescein isothiocyanate)
12. Peptide synthesis membrane (polypropylene to which is bonded poly-aminoethylmethacrylamide; 0.22-μm pore size, 300–400 n mole/cm^2 amino groups; obtained as a development product from Millipore Corp.)

Equipment and Supplies

1. Clear, polystyrene microplates (e.g., Corning Costar no. 3797) for use with a microplate reader.
2. Disposable vinyl microplates (e.g., Corning Costar no. 2797) for construction of membrane disk holder described in Fig. 1.
3. Molecular Devices V_{max} kinetic microplate reader (or similar device of other manufacturer)
4. 1/4-in. (~6-mm) paper punch for making membrane disks
5. 20-mL screw-cap vials (liquid scintillation vials)
6. 2.0-mL polypropylene microcentrifuge tubes
7. Fine-tipped drafting pen filled with India ink

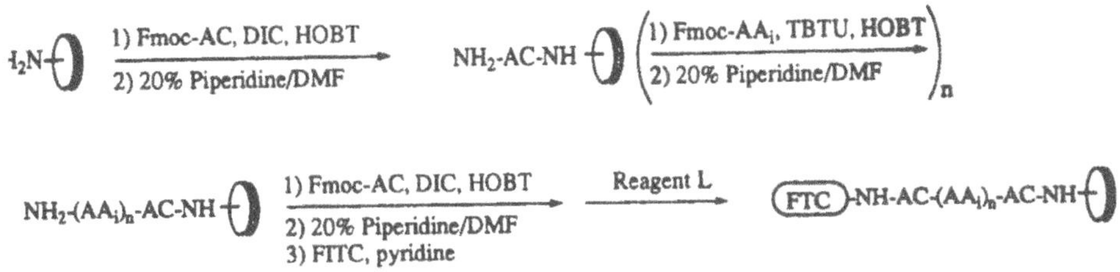

$$H_2N-\!\!\bigcirc \xrightarrow[\text{2) 20\% Piperidine/DMF}]{\text{1) Fmoc-AC, DIC, HOBT}} NH_2\text{-AC-NH}-\!\!\bigcirc \left(\xrightarrow[\text{2) 20\% Piperidine/DMF}]{\text{1) Fmoc-AA}_i, \text{TBTU, HOBT}} \right)_n$$

$$NH_2\text{-(AA}_i)_n\text{-AC-NH}-\!\!\bigcirc \xrightarrow[\substack{\text{2) 20\% Piperidine/DMF} \\ \text{3) FITC, pyridine}}]{\text{1) Fmoc-AC, DIC, HOBT}} \xrightarrow{\text{Reagent L}} \boxed{\text{FTC}}\text{-NH-AC-(AA}_i)_n\text{-AC-NH}-\!\!\bigcirc$$

Scheme 1 Synthesis of a dipeptide on a membrane disk: AC, ϵ-aminocaproic acid, AA, amino acid; FTC, the fluoresceinylthiocarbamyl chromophore.

I. INTRODUCTION

Proteases play important roles in many biological processes, both in the modification of precursor proteins after synthesis and the degradation of proteins. Definition of protease substrate specificity is essential for designing inhibitors and drugs to control these processes, as well as for understanding the details of proteolytic events in general. Proteolytic specifity is generally determined by the nature of the amino acids on either side of the scissile bond and/or, in many cases, residues more remote from the scissile bond. Numerous approaches to the assessment of protease specificity have been explored, but many of these either require time-consuming analyses of individual reactions or of complex mixtures of peptides, or provide information only about the specificity for one side of the scissile bond. The method described here allows simultaneous analysis of the hydrolysis of hundreds of peptides, which are easily synthesized on membrane disks, and systematic structure–function study of protease specificity [1].

Peptides are synthesized on individual disks of peptide synthesis membrane, consisting of a porous polypropylene core that supports a polyaminoethylacrylamide matrix. The substrate assembly consists of a spacer to separate the peptide from the polymer matrix; the peptide, which can be of any desired length but is illustrated here with a dipeptide; a second spacer; and a chromophore, in this case, fluorescein. The synthesis of the substrate is outlined in Scheme 1.

Synthesis of peptides (e.g., 400 dipeptides) utilizes a split-synthesis approach in which batches of disks are derivatized in a single vessel and then sorted and recombined into other batches for subsequent reactions. The result is a collection of disks, each of which bears a peptide of known sequence. The disks are attached to a pin assembly, which allows each disk to be immersed in a small volume of protease solution, as shown in Fig. 1. If the peptide is hydrolyzed by the protease, the N-terminal half of the peptide bearing the chromophore is re-

(FTC)-NH-AC-AA$_1$-AA$_2$-AC-NH⟨ ⟩ $\xrightarrow{\text{Protease}}$ (FTC)-NH-AC-AA$_1$-COO⁻ + NH$_2$-AA$_2$-AC-NH⟨ ⟩

Scheme 2 Cleavage of a membrane-bound peptide to release the FTC chromophore-bearing fragment into solution.

leased into the solution (Scheme 2) where it can be monitored spectrophotometrically after the disk is removed from the solution. The disk is then replaced in the solution for further reaction. In this way, the kinetics of hydrolysis of each peptide can be measured.

II. GENERAL PROCEDURES

A. Synthesis of Dipeptide Substrates

Synthesis involved six steps: sequential addition of a spacer, first amino acid, second amino acid, spacer, and chromophore, and removal of side chain protecting groups (Scheme 1). The addition of spacers and chromophores was done on batches of 400 (or multiples thereof), and attachment of individual amino acids on batches of 20. Membrane disks with a diameter of 6 mm were punched from sheets using a paper punch and lettered with India ink to indicate a particular dipeptide sequence. *Step 1.* Four hundred membrane disks were placed in a 20-mL screw-cap vial and washed with 3 × 12 mL of dimethylformamide (DMF). To the vial was added 2.5 mL of a solution of 0.2 M *N*-Fmoc-ε-aminocaproic acid, 0.2 M diisopropylcarbodiimide (DIC), and 0.2 M *N*-hydroxybenzotriazole (HOBT). The mixture was shaken at room temperature for 2 h. The spacer solution was removed by aspiration, and the disks were washed with 3 × 12 mL of DMF. After removal of the Fmoc protecting group by reaction with 8 mL of 20% piperidine in DMF for 15 min, the disks were washed 5 times with DMF and several times with methanol, and were dried under vacuum. (Complete removal of piperidine is necessary to prevent premature deprotection in the next step.) *Step 2.* The disks were sorted into batches of 20 corresponding to the first amino acid. Each batch of 20 was placed in a 2.0-mL microcentrifuge tube and to each tube was added 150 μL of 0.2 M Fmoc-amino acid, 0.19 M TBTU, 0.2 M HOBT, and 0.6 M *N*-methylmorpholine in DMF. After 1 h, the amino acid solution was aspirated off, the disks were washed with 3 × 1 mL of DMF and transferred to a 20-mL vial, and the Fmoc group was removed from the combined disks in 8 mL of 20% piperidine in DMF for 15 min. The disks were washed with DMF and methanol and then dried as in step 1. *Step 3.* The disks were resorted into batches of 20 corresponding to the second amino acid, and the second amino acid was coupled and deprotected as in step 2. *Step 4.* All 400 disks were combined and the second spacer was added as in step 1. *Step 5.* The chromophore was added by incubation with 3 mL of fluorescein isothiocyanate (50 mg/mL) in pyridine-DMF (1:9) at 50°C for 1.5 h, after which the disks were washed with DMF and finally acetone until the washings were no longer fluorescent. *Step 6.* Side-chain protective groups were removed by treatment with 8 mL of reagent L (88% trifluoroacetic acid, 5% dithiothreitol, 5% water, 2% triisopropylsilane) for 3 h, after which the disks were washed thoroughly with methanol and ether, dried under vacuum, and stored in a refrigerator.

Large-scale synthesis was performed as above on multiple batches of disks or on membrane strips (e.g. 2×3 cm) from which could be punched 10 or more disks. In this case, the reaction vessels and reagent volumes must be increased to permit adequate mixing and coverage of membranes by reagents.

B. Assay of Peptidase Activity

Membrane disks were attached to a 96-pin block described in Scheme 1. For the simultaneous assay of 400 peptides, the disks were arranged in rows of 10, on 5 blocks, with 80 disks per block. To each of 80 corresponding wells of 5 microplates was added 250 mL of enzyme solution containing 5–100 µg of protease (depending on its activity) in an appropriate buffer. In the case of porcine pancreatic elastase, each well contained 100 µg in 250 µL of 50 mM Tris buffer, pH 8.0. The reaction was initiated by placing the pin block over the microtiter plate and immersing the disks in the enzyme solution. The reactions were generally carried out at 37°C. For kinetics studies, the disks were removed at intervals from the enzyme solutions and the absorbance at 490 nm of the liberated chromophore was read on a Molecular Devices V_{max} microplate reader, after which the disks were replaced and incubation was continued. After several readings, the data could be plotted to reveal hydrolysis rates for each of the peptides.

III. RESULTS

The results of simultaneous hydrolysis by porcine pancreatic elastase of 400 dipeptides derived from the 20 common amino acids is shown in Table 1. In this case, it is seen that cleavage occurs primarily after P_1 alanine, as expected from the known specificity of elastase, but also after cysteine, which was unexpected. It is seen also that there is some preference for cleavage of peptides with small or neutral polar side chains at positon P_1'.

Table 1 Relative Rates of Hydrolysis of Membrane-Bound Dipeptides for Porcine Pancreatic Elastase[a]

$P_1 P_1'$	F	Y	W	L	M	I	V	H	C	A	P	G	T	S	Q	N	E	D	K	R
F	2	2	1	1	2	2	1	3	3	2	2	2	2	2	2	3	2	3	2	1
Y	1	2	1	2	2	3	1	3	3	3	2	4	3	2	3	3	2	3	3	3
W	1	1	1	1	1	1	1	1	2	2	1	1	1	1	1	1	1	1	1	1
L	1	2	1	0	2	0	1	5	5	4	3	3	3	3	5	4	2	3	4	2
M	2	2	1	2	3	2	2	6	7	4	3	3	3	3	4	3	3	3	3	3
I	1	2	1	1	3	1	1	4	7	8	2	3	2	3	6	6	2	3	2	2
V	1	1	2	1	4	2	2	4	9	9	3	3	3	4	6	6	3	2	4	2
H	4	3	1	2	4	2	4	5	8	5	4	4	4	3	4	5	3	4	3	3
C	6	7	2	7	**15**	4	7	**16**	**23**	**31**	7	**11**	**12**	**12**	**30**	**30**	**12**	**11**	6	6
A	4	4	2	6	**14**	6	7	**11**	**21**	**21**	8	5	9	**12**	**14**	**16**	6	9	**10**	7
P	1	2	1	2	2	2	2	3	8	3	2	3	3	3	2	2	2	3	4	2
G	2	2	1	2	3	2	3	5	5	4	3	2	3	3	4	5	2	3	3	3
T	2	2	1	3	5	2	3	2	8	7	2	3	3	3	2	2	2	3	4	2
S	2	1	1	2	2	2	2	3	6	3	2	2	2	2	3	3	2	3	3	2
Q	3	3	2	4	3	3	4	5	**11**	4	3	3	3	3	4	3	3	3	3	3
N	3	3	1	3	3	3	3	5	8	3	3	4	3	3	4	3	3	3	3	3
E	1	2	1	3	3	2	3	3	8	3	2	2	4	2	3	3	3	3	3	2
D	2	2	1	2	2	2	3	3	6	5	2	2	2	2	3	3	2	2	2	2
K	2	2	1	2	3	2	3	2	7	4	2	3	1	3	3	3	3	2	5	2
R	1	2	1	2	2	1	2	3	6	3	2	2	3	2	3	3	2	2	3	2

[a] $\Delta A_{490}/min$ ($\times 10^4$) values after incubation at 37°C; larger values are underlined and/or bolded for emphasis.

IV. COMMENTARY

The analysis described above is not limited to dipeptides. Peptides of any desired length may be synthesized using the procedures described and used to identify amino acid residues remote from the scissile bond that contribute to substrate specificity. Using this technique, peptide substrates with increasing rates of hydrolysis can be detected by systematic elongation of the sequence. For example, the best dipeptide substrate is elongated at the N terminus with each of the 20 amino acids to determine the best tripeptide substrate (out of 20 tripeptides), and so on. In the case of elastase, a pentapeptide sequence was found that is hydrolyzed about 50 times faster than one of the fastest dipeptides [2].

REFERENCES

1. Y Duan, RA Laursen. *Anal Biochem 216*: 431–438, 1994.
2. C Zhu. Solid phase protease substrate synthesis and studies of elastase and ginger protease-II specificity. PhD Dissertation, Boston University, 1999.

Case Study 4-5
Characterization of the Epitope Specificity of Antibodies Using Membrane-Bound Peptides

Richard A. Laursen and Zhengxin Wang*
Boston University, Boston, Massachusetts

ABSTRACT

Peptides of defined sequence are synthesized in a 96-spot array on polypropylene-based membrane sheets using a device that permits both synthesis and quantitative assessment of biological activity. The utility of this technique is illustrated with the identification of an epitope for a monoclonal antibody.

Chemicals

1. *N*-Fmoc–amino acids (*N*-9-fluorenylmethoxycarbonyl–amino acid–OH)
2. HOBT (1-Hydroxybenzotriazole)
3. TBTU (2-(1-benzotriazol-1-yl)-1,1,3,3-tetramethyluronium tetrafluoroborate)
4. Dimethylformamide (DMF, distilled from P_2O_5)
5. Piperidine
6. *N*-Methylmorpholine
7. Reagent L (88% trifluoroacetic acid, 5% water, 5% dithiothreitol, 2% triisopropylsilane; freshly prepared)

---

* *Current affiliation*: Rockefeller University, New York, New York.

8. Peptide synthesis membrane (polypropylene to which is bonded poly-aminoethylmethacrylamide; 0.22-mm pore size, 300–400 nmol/cm^2 amino groups; obtained as a development product from Millipore Corp.)

9. Reagents for immunoassay (see text)

Equipment and Supplies:

1. Clear, polystyrene microplates (e.g., Corning Costar no. 3797) for use with a microplate reader.
2. Multispot synthesis and assay apparatus (Fig. 2)
3. Single and multipipetors

I. INTRODUCTION

The past 15 years has seen the development of numerous techniques and strategies for the synthesis of large numbers of peptides for the assessment of various types of biological activity. In fact, it can be said that modern combinatorial chemistry originated with multipeptide synthesis. For many applications, particularly structure–function studies, it is convenient to work with a set of peptides that are displayed as a defined array on a surface. It is especially beneficial if the peptides can be synthesized without the need for expensive instrumentation. In the method described here, peptides of any desired sequence are synthesized as discrete spots on a membrane matrix consisting of an inert polypropylene core with a polyacrylamide matrix polymerized around it [1] (Scheme 1). Synthesis reactions are confined to discrete zones by means of a device that effectively separates each zone from the other [2]. The utility of this approach is demonstrated by identification of the antigenic determinant (epitope) in the protein antigen of a monoclonal antibody [3].

The membrane peptide synthesis method was developed concurrently with the multispot method of Frank [4], which is similar but uses a cellulose sheet. However, unlike the polypropylene membrane, the cellulose sheet is physically more fragile, will not stand up to multiple probing, and is not amenable to quantitative assays.

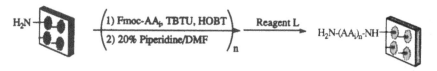

Scheme 1   Synthesis of peptide spots on a membrane.

II. GENERAL PROCEDURES

A. Peptide Synthesis

Peptide synthesis was carried out in the wells of the 96-well peptide synthesis device shown in Fig. 1. Membrane sheets (11.5 × 7.5 cm) were mounted in the

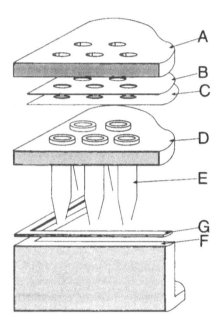

Figure 1 Cutaway schematic of the multispot peptide synthesis and assay apparatus. The device consists of two aluminum plates (A and D) between which is sandwiched the peptide synthesis membrane (C). Ninety-six holes are drilled in the top and bottom plates following the format of a standard microtiter plate, with 3/16-in-diameter holes in the top plate and 15/64-in. holes in the lower. A polyethylene microcentrifuge tube (E, 6 × 35 mm), with its bottom cut off at an angle, is inserted in each hole in the lower plate. A seal is formed around each membrane spot by compression of the membrane between the top of the microcentrifuge tube and a sheet of 1/64-in. silicone rubber (B) when plates A and D are bolted together (there are 12 1/4-in. bolts, not shown, spaced evenly around the perimeter of the plate). The plates are bolted onto a polypropylene vacuum chamber (F) sealed by a neoprene rubber gasket (G). Waste drains from the bottom of the vacuum chamber into a vacuum flask. In the binding assay mode, the unit is assembled in the same way, except that the vacuum chamber is removed and the upper plate assembly placed in a 96-well microplate in such a way that each of the 96 microcentrifuge tubes (E) is in a separate well. Assay solutions are washed from the membrane to the microplate wells by application of pressure to the holes in the plate above the membrane.

device and a small hole was punched in the membrane in each well using a syringe needle to permit rapid flow during washing steps. Amino acids were attached sequentially by manually pipetting activated amino acids into each well. Common washing and Fmoc deprotection steps were done using an 8-tip multipipetor to hasten the process. Activated amino acid solutions were prepared immediately before synthesis and contained 0.2 M Fmoc–amino acid, 0.19 TBTU, 0.2 M HOBT, and 0.6 M *N*-methylmorpholine. If kept cold (0–4°C), these solutions could be used within an 8-h work period. *Step 1*: To each well was added 10 µL of activated amino acid. Slower reacting amino acids (Ile, Val, Thr, trityl-Cys) were added first so that they would have a longer time in which to react. The time required for adding 96 amino acids is about 15 min. Reaction was allowed to proceed for 30 min after addition of the last amino acid, and the amino acid solution was removed by aspiration into the vacuum manifold of the device. *Step 2*: Each well was then washed with 3×100 µL of DMF, after which (*Step 3*) 25 µL of 20% piperidine in DMF was added. Deprotection was allowed to proceed for 10 min, the solution was aspirated, and the wells were washed (Step 4) with 7×100 µl of DMF. Steps 1–4 were then repeated until peptides of the desired length were constructed. After the final DMF step, the wells were washed with 100 µL of ether and the membrane was removed from the apparatus for deprotection of side chains.

B. Side Chain Deprotection

Side chain deprotection was accomplished by removing the membrane sheet from the synthesis apparatus and placing it in Reagent L (88% trifluoroacetic acid/5% water/5% dithiothreitol/2% triisopropylsilane), or similar cleavage cocktail, in a polyethylene or polypropylene tray for 2 h, after which the sheet was washed with 3×50 mL of methanol, air-dried, and then vacuum-dried. Sheets were placed in plastic bags or plastic wrap and stored in a refrigerator.

C. Enzyme-Linked Immunosorbent Assay (ELISA)

The membrane-bound peptides were subjected to an ELISA essentially as described by Wang and Laursen [1]. The membrane bearing synthetic peptides was washed with Tris/NaCl (0.02 M Tris-HCl, 0.5 M NaCl, pH 7.5), blocked with Tris/NaCl containing 10% ovalbumin and sequentially incubated with a monoclonal antibody (10 µg/mL) and protein A–alkaline phosphatase conjugate (from Sigma; diluted 1:2000 in Tris/NaCl containing 0.05% Tween-20 and 1% ovalbumin), with washing steps in between. Incubations and washings were done in a small plastic tray. The membrane was then mounted in the peptide synthesis device to detect bound antibodies by alkaline phosphatase–catalyzed hydrolysis of *p*-nitrophenyl phosphate, and the absorbance of *p*-nitrophenolate was read at

405 nm in a Bio-Tek BT2000 MicroKinetics microplate reader (or similar device). Prior to reprobing of the membrane with a different antibody, bound antibody and protein A–alkaline phosphatase were removed from the peptides by sonication for 30 min in an aqueous solution, prewarmed to 60°C, containing 0.1 M sodium phosphate/0.1% 2-mercaptoethanol/1% sodium dodecyl sulfate, and then several times with Tris/NaCl.

III. RESULTS

The example given here comes from a study designed to identify and characterize the epitopes in human Ha-*ras* protein recognized by specific antibodies [3]. The strategy in this study involves synthesizing all of the possible octapeptide sequences (1–8, 2–9, 3–10, . . ., 182–189). Each peptide is synthesized, without using a cleavable linker, as a discrete spot on the membrane, using the device shown in Fig. 1. The membrane sheets are removed from the synthesis device, and side-chain protective groups are removed with a trifluoroacetic acid cleavage cocktail. At this point, several sheets can be processed together in a single vessel. After the membrane sheets have been washed, they are incubated in solutions containing individual monoclonal antibodies (mAbs), followed by the protein A–alkaline phosphatase conjugate. The sheets are remounted in the synthesis/assay device and *p*-nitrophenyl phosphate solution is added to each well. Those peptides to which antibodies and protein A conjugate have bound cause hydrolysis of the substrate to form yellow *p*-nitrophenolate ion, which is washed into the wells of microplates and read on a microplate reader.

Figure 2 shows the results of such an assay with mAb-10, one of several mAbs that were studied [3]. In this case, only five octapeptides, encompassing residues 28–40 (FVDEYDPTIEDSY), bound to mAb-10, as determined by the production of *p*-nitrophenolate. From these data (and confirming data from studies with smaller peptides) it can be deduced that the smallest epitope recognized by mAb-10 is the tetrapeptide sequence YDPT.

IV. COMMENTARY

An outstanding characteristic of these membrane sheets is their toughness. Sheets could be reused many times for the analysis of several mAbs by stripping the mAb–protein A conjugate from the peptides, a process that requires prolonged sonication in warm detergent solution. The membranes remained intact through as many as 20 cycles, involving a total of 10 h of sonication.

Most of the DMF used in washing steps during synthesis can be replaced by ether, which is less expensive.

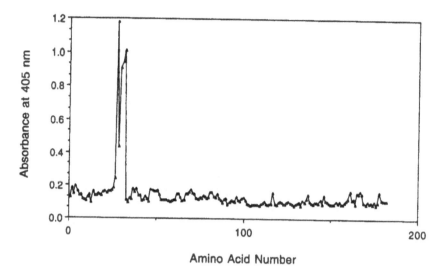

Figure 2 Epitope mapping of human Ha-*ras* protein against a monoclonal antibody (mAb-10). All 180 possible overlapping octapeptides covering the sequence of human Ha-*ras* protein were synthesized on two membrane sheets. The sheets were incubated with mAb-10 and, after washing, with protein A–alkaline phosphatase conjugate in the synthesis/assay device (Scheme 1). Bound antibodies were detected by alkaline phosphatase–catalyzed hydrolysis of *p*-nitrophenyl phosphate. The numbers along the *x* axis correspond to the positions of amino acid residues in the sequence of human Ha-*ras* protein. The *y* axis indicates the absorbance at 405 nm of *p*-nitrophenol formed during ELISA.

REFERENCES

1. Z Wang, RA Laursen. *Peptide Res.* 5: 275–280, 1992.
2. Z Wang, RA Laursen, Apparatus and process for effecting sequential peptide synthesis, U.S. Patent 5,188,733, issued Feb. 23, 1993.
3. Z Wang, WP Carney, RA Laursen. *J. Peptide Res.* 50: 483–492, 1997.
4. R Frank. *Tetrahedron 48*: 9217–9232 (1992).

Case Study 4-6
Rapid Parallel Synthesis Utilizing Microwave Irradiation

Brian M. Glass and Andrew P. Combs
DuPont Pharmaceuticals Company, Wilmington, Delaware

ABSTRACT

Many solid-phase reactions require higher temperatures and/or longer reaction times to drive these two-phase reactions to completion, in comparison with analogous solution phase reactions. Microwave irradiation provides a useful, expedient, and inexpensive tool to heat reactions and thus facilitates the rapid parallel synthesis of large combinatorial libraries. Here we demonstrate the use of microwave irradiation and the safety-catch linker to synthesize large libraries of diverse amide products.

Chemicals

1. Sulfamylbutyryl (Safety-Catch) resin
2. BOC-protected amino acids
3. Acid chlorides
4. Benzotriazol-1-yloxytrispyrrolidinophosphonium Hexafluorophosphate (PyBOP).
5. Diisopropylethylamine (DIEA)
6. Trifluoroacetic acid (TFA)
7. Bromoacetonitrile
8. 0.25 M stock plate containing 88 diverse primary and secondary amines in DMSO
9. dimethylformamide (DMF), dichloromethane (DCM), MeOH, *N*-methylpyrolidinone (NMP), dimethylsulfoxide (DMSO)

Equipment and Supplies

1. LabQuake rotary shaker
2. Solvent-resistant wash bottles
3. Polyfiltronics PKP Unifilter 2000 filter plates
4. Polyfiltronics Uniplate 2000 collection plates
5. Tomtec Quadra 320 96 channel pippetor
6. Sharp Carousel 1100-W commercial microwave
7. Tomtec vacuum manifold

I. INTRODUCTION

Synthesis of large libraries (10^2–10^5) of discrete compounds has become the most prevalent method used within the pharmaceutical industry for the generation of compound collections for high-throughput screening purposes. The main reason for this is that validation of screening "hits" from these libraries requires substantially less time than deconvolution of actives derived from libraries of complex mixtures. Thus, the availability of rapid, convenient, and cost-effective methods for the production of discrete compound libraries is highly desirable. The use of 96-well filter plates has become common practice and allows workstation approaches to library production. We report here microwave irradiation [1–12] as an efficient and rapid method for reaction heating utilizing a commercial microwave oven. This process not only allows for the rapid heating of samples, thus reducing reaction times, but also provides a low-cost alternative to conventional heating methods. The method is demonstrated by the synthesis of libraries of amide products using the Kenner safety-catch approach [13].

Library Synthesis Procedure

Method for the synthesis of an 880-member discrete compound library.

II. EXPERIMENTAL PROCEDURE

BOC-Ala was coupled to 1.0 g of 4-sulfamylbutyryl resin 1 using 5 equivalents of the amino acid, 5 equivalents of PyBOP, and 15 equivalents of DIEA dissolved in DMF (Scheme 1). The resin was mixed on a rotary shaker for 18 h and then washed with 10 mL of DMF, DCM, and MeOH 3 times each. The BOC group was removed from resin 2 by adding 15 mL of a 50% TFA/DCM solution and mixing for 30 min. The resin was washed with 10 mL of DCM and MeOH 3 times each and dried under vacuum. The amino acid–bound resin was acylated

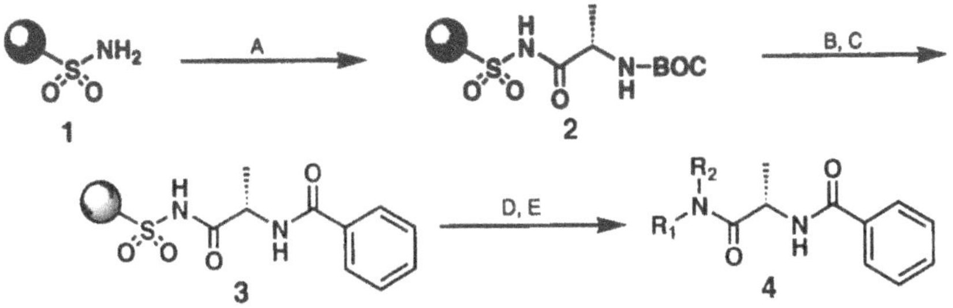

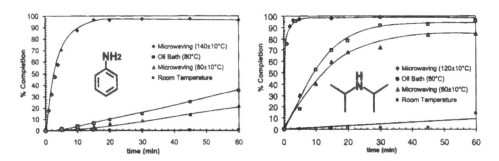

Scheme 1 (A) Boc-Ala, PyBOP, DIEA; (B) 50% TFA/DCM; (C) benzoyl chloride, DIEA; (D) BrCH$_2$CN, DIEA, NMP; (E) R$_1$R$_2$NH, DMSO.

with 5 equivalents of benzoyl chloride and 10 equivalents of DIEA in a solution of DCM for 1 h. Resin **3** was washed with 10 mL of DCM, DMF, and MeOH 3 times each and dried under vacuum.

Linker activation and cleavage: To resin **3** was added 10 mL of NMP, 10 mmol of bromoacetonitrile, and 2 mmol of DIEA (Scheme 1). The resin was mixed on a rotary shaker for 18 h and washed with 10 mL of NMP, DMF, and DCM 3 times each. The activated resin was dried under vacuum and cleaved with a 10 mM solution of an amine in DMSO.

Ellman et al. reports that cleavage with 10 mM benzylamine has a $t_{1/2}$ of less than 5 minutes; however, hindered and nonnucleophilic amines require heating [14]. Diisopropylamine and aniline were used to investigate cleavage rates of activated resin **3** under various temperature conditions over 1 h (Fig. 1) using both microwave and traditional oil bath heating. It was determined that microwave heating did not accelerate reaction rates over traditional heating; however, it provides a very convenient and rapid method of heating. In this case, even

Figure 1 Cleavage rate studies.

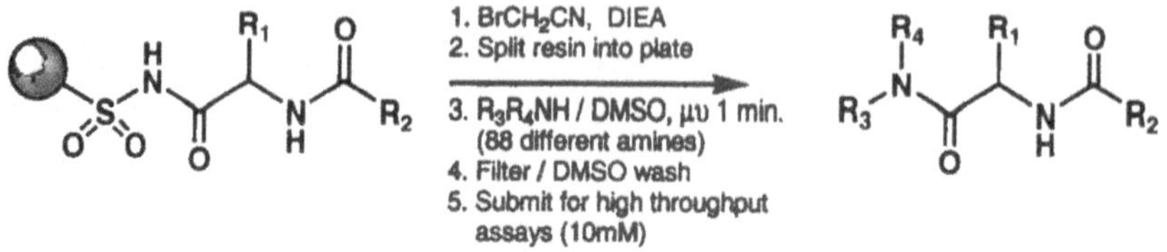

1. BrCH$_2$CN, DIEA
2. Split resin into plate

3. R$_3$R$_4$NH / DMSO, $\mu\upsilon$ 1 min.
   (88 different amines)
4. Filter / DMSO wash
5. Submit for high throughput
   assays (10mM)

Scheme 2   Rapid parallel synthesis utilizing microwave irradiation in 96 well plates.

cleavage with aniline could be accomplished within 15 min by microwaving to a temperature of 140°C.

Ten different BOC-protected amino acids were coupled to 1.0 g each of 4-sulfamylbutyryl resin. Each resin was then deprotected and acylated with a different acid chloride to yield approximately 1 g of 10 different resins. (Procedure same as above.)

Linker activation and cleavage (Scheme 2):   To each resin was added 10 mL of NMP, 10 mmol of bromoacetonitrile, and 2 mmol of DIEA. Resins were mixed on a rotary shaker for 18 h and washed with 10 mL of NMP, DMF, and DCM 3 times each. Each resin was then suspended in a 3:2 DCM/DMF solution and split between 88 wells of a Polyfiltronics filter plate using a Tomtec Quadra 96-channel pipetor. Resins were washed with DCM and dried under vacuum. Then 230 µL of DMSO was pipetted into each well of the filter plates using a 96-channel pipetor. Finally, 20 µL (0.8 equivalent of amine per well) of a diverse set of primary and secondary amines was transferred from a stock microtiter plate to the reaction plate using a 96-channel pipetor (Fig. 2). The plate of amines chosen in this experiment consisted of a diverse set of primary and secondary

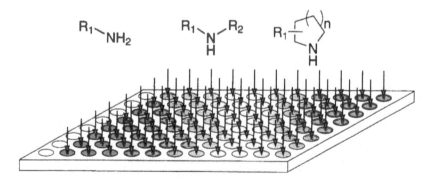

Figure 2   Addition of 88 diverse amines.

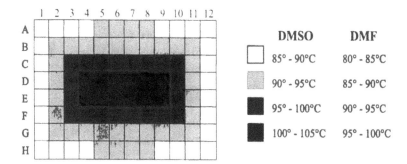

Figure 3 Temperature gradients within a microwave-heated microtiter plate—1 mL per well heated continuously for 1 min at 1100 W in a conventional microwave.

aliphatic amines. Anilines were not used due to the instabilities of the filter plates to the higher temperatures. Modified plates are in development to make workable these increased temperatures. Sets of four plates were then microwaved for 60 s to a temperature of 80°C in a commercial microwave. Temperature gradients exist between wells of the heated microtiter plate but do not present a problem with this chemistry (Fig. 3). The solutions were allowed to cool for 30 min and drained into a collection plate using a vacuum manifold. The resin was washed with an additional 250 μL of DMSO and filtered into the same collection plate to give 500 μL of a 10 mM DMSO solution for immediate biological screening.

III. POSTCLEAVAGE ANALYSIS

A row from each plate was spotted on a TLC plate with a 12-channel pipetor and eluted in a TLC chamber with 1:1 hexane/ethyl acetate to rapidly determine library purity. Ninhydrin staining was used to detect the presence of small amounts of unreacted amine. Daughter plates are made using a 96-channel pipetor for flow probe ^{1}H-NMR and electrospray mass spectrometric analysis. The remainder of the solution (500 μL of a 10 mM DMSO stock) was submitted for biological screening.

IV. COMMENTARY

We have used the procedure described above to synthesize large, diverse-amide libraries. Purities determined by thin-layer chromatographic analysis (starting amines were undetectable by ninhydrin stain) and ^{1}H-NMR were on the average greater than 90%. Liquid chromatography–mass spectrometry analysis identified more than 95% of the desired products.

REFERENCES

1. S Caddick, Microwave Assisted Organic Reactions, *Tetrahedron* 1995, *51*, 10403–10432.
2. CR Strauss, RW Trainor. *Aust J Chem* 1995, *48*, 1665–1692.
3. AG Whittaker, DMP Mingos, The application of microwave heating to chemical syntheses, *J. Microwave Power Electromagn Energy* 1994, *29*(4), 195–219.
4. G Majetich, R Hicks, The use of microwave heating to promote organic reactions, *J. Microwave Power Electromagn Energy* 1995 *30*(1), 27–45.
5. AK Bose, BK Banik, N Lavlinskaia, M Jayaraman, MS Manhas, MORE chemistry in a microwave, *Chemtech Sept 1997*, 18–24.
6. IC Cotterill, AY Usyatinsky, JM Arnold, DS Clark, JS Dordick, PC Michels, YL Khmelnitsky, Microwave assisted combinatorial chemistry, synthesis of substituted pyridines, *Tetrahedron Lett* 1998, *39*, 117–1120.
7. AP Combs, S Saubern, M Rafalski, PYS Lam, Solid supported aryl/heteroaryl C-N cross-coupling reactions, *Tetrahedron Lett* 1999, *40*,1623–1626.
8. M Larhed, G Lindeberg, A Hallberg, Microwave assisted Suzuki coupling on solid-phase, *Tetrahedron Lett* 1996, *37*, 8219–8222.
9. AML Hoel, J Nielsen, Microwave-assisted solid-phase Ugi four-component condensations, *Tetrahedron Lett* 1999, *40*, 3941–3944.
10. HM Yu, ST Chen, KT Wang, *J Org Chem* 1992, *57*(18), 4781–4784.
11. H Chang, C Gao, I Takeuchi, Y Yoo, J Wang, PG Schultz, XD Xiang, RP Sharma, M Downes, T Venkatesan, Combinatorial synthesis and high throughout evaluation of ferroelectric/dielectric thin-film libraries for microwave applications *Appl Phys Lett* 1998, *72*(17), 2185–2187.
12. CG Blettner, WA Konig, W Stenzel, T Schotten, Microwave-assisted aqueous Suzuki cross-coupling reactions, *J Org Chem* 1999, *64*, 3885–3890.
13. GW Kenner, JR McDermott, RC Sheppard. The safety catch principle in solid phase peptide synthesis, *Chem Commun 1971*, 636–637.
14. JB Backes, AA Virgilio, JA Ellman. *J Am Chem Soc* 1996, *118*, 3055–3056.

Chapter 5
High-Throughput Purification Strategies

Mark J. Suto
DuPont Pharmaceuticals Research Laboratories, San Diego, California

I. INTRODUCTION

The use of high-speed solution phase parallel synthesis has emerged as one of the best methods for the production of discrete compound libraries. This is due to both the advances made in automation and the development of high-speed purification and detection methods. In the past, solid-phase techniques were the methods of choice because they could provide cleaner products than multistep reactions. However, in some cases, the development time required outweighed the benefits achieved.

As the emphasis shifted to the use of solution phase parallel synthesis, it was quickly realized that the success of this approach was not dependent on the ability to synthesize the compounds but to purify them. This chapter outlines several methods that have been developed to avoid the "purification" bottleneck.

II. BACKGROUND

Medicinal chemistry has typically focused on the synthesis and purification of single discrete compounds. This was required to provide the quantity of material needed (>100 mg range) and to ensure that the chemical purity of the compound was sufficient for use in various biological assays. With the development of high-throughput assays, it became evident that larger numbers of compounds were required; however, also smaller quantities of each could be used.

Combinatorial chemistry or the simultaneous synthesis of large numbers of compounds or libraries was developed to meet the increased screening demand. The early libraries were peptide-based and did not provide suitable drug candidates (oral bioavailability and stability). However, these techniques were then applied to the synthesis of small "drug-like" molecules, resulting in a totally new approach to medicinal chemistry and drug discovery.

The early synthetic procedures for combinatorial libraries were derived from peptide chemistry and thus the first small-molecule libraries were prepared on an inert solid-phase material or resin. The compounds could either be evaluated in biological assays either attached to the resin or be cleaved from the resin and then tested (although this is highly dependent on the type of assay used). It was found that the use of solid-phase chemistry provided many advantages in that excess reagents could be used to drive the reactions to completion and then simply washed away when the reactions were complete. Cleavage from the resin resulted in relatively pure compounds.

The transition from synthesizing peptides on a solid phase to small drug-like molecules was found not to be straightforward. The synthesis of heterocycles in particular usually required a great deal of solid-phase chemistry development time (3–6 months). Reactions that worked in solution did not always work to the same degree on a solid phase and many reactions did not work at all. In some cases, the required development time for solid-phase chemistry outweighed the perceived advantages of this approach. One additional disadvantage to solid-phase chemistry was the difficulty in identifying or deconvoluting the active compounds from the assay. The production of large numbers of compounds using solid-phase chemistry usually was most effective when used to produce mixtures of compounds. Screening of these mixtures then required that additional techniques or technology be used to identify the active compounds. However, when the chemistry did work, large numbers of compounds (thousands) could be synthesized and made available for evaluation.

Many chemists involved in drug discovery were hesitant to embark on this new resin-based approach mainly due to the unfamiliarity and difficulty in developing the solid-phase chemistry. A potential way to circumvent the use of solid-phase chemistry would be to use more traditional solution chemistry in a high-throughput or automated fashion. However, if the chemistry were not high yielding, the products derived from this approach would be of insufficient purity for biological evaluation. This would be due to the presence of excess reagents or byproducts in the final compounds (as mentioned, this is not the situation with solid-phase chemistry). Some groups developed very high-yielding chemistries that could be used to prepare large numbers of individual compounds in solution [1]. The individual yields in a multistep synthesis would need to exceed 95% for this to work. This could be quite difficult, and it was found that developing the high-yielding solution chemistry could be as time consuming as that required for solid-phase approaches.

III. HIGH-THROUGHPUT PURIFICATION

The problem that now confronted many medicinal chemists was how to blend solution phase methods, with which they were familiar, with the developing field of high-throughput synthesis or combinatorial chemistry. The resolution would be to develop methods for high-throughput purification, as this was the limiting step associated with solution phase approaches. An initial tactic used to circumvent the purity issue was the development of parallel liquid–liquid extraction methodologies that could be used to remove unreacted starting materials [2]. In addition, novel liquid–liquid extraction methods, such as the use of fluorous reagents, were under development [3].

Many groups began developing "hybrid techniques" by combining solid-phase technology with solution phase synthesis. For example, multicomponent condensations are very useful in a combinatorial approach. They are able to generate large numbers of products very quickly but they are usually not high yielding. To obtain clean products from a multicomponent reaction, the final reaction mixture was treated with a solid-phase hydroxy Wang resin [4]. The final products are in essence captured on the resin. Successive washing of the resin to remove starting materials and byproducts and then treating the resin with acid provided the final "purified" products (Scheme 1).

A wide variety of related reports soon followed, including the use of other scavenger or capture approaches, high-throughput purification (high-speed silica gel or reverse phase approaches), reagent capture and recognition approaches as well as the use of solid-phase reagents. The common theme of all of these was to provide methodology in which solution-phase chemistry could be used in a parallel or automated fashion to afford purified compounds suitable for biological evaluation.

Removal of reagents or starting materials such as amines, isocyantes and acid chlorides that typically are used in excess to drive reactions to completion, can be done with resin bound electrophiles, nucleophiles and bases [5]. The solid-phase "scavengers" are added to the reaction once it is complete, the reaction is filtered and concentrated to provide only the desired purified products. Unreacted starting materials are bound to the resin (Scheme 2).

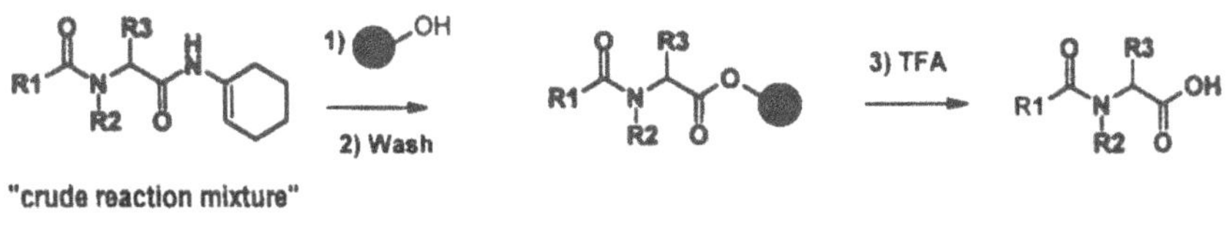

"crude reaction mixture"

Scheme 1

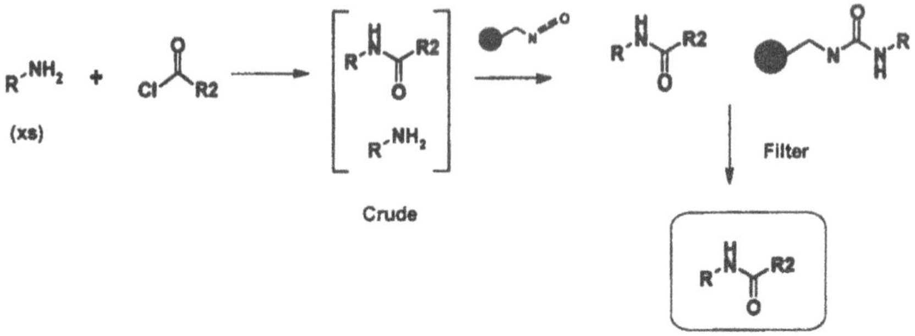

Scheme 2

Reagents such as hydrides or coupling reagents (Figure 1) can also be attached to a solid-phase and used in more traditional solution phase chemistry [6,7]. Once again the reactions are simply filtered to provide the desired products.

The reagents described work for a variety of reactions. However, as one tries to proceed through a series of synthetic steps it was found that chemistry specific scavengers would be more useful. The use of designed reagents to address certain types of chemistries was reported (molecular reactivity and recognition) [8]. The technique was used in a series of reaction to effect the synthesis and purification of a variety of compounds. Several excellent reviews have been recently published illustrating the versatility and applicability of this approach [9,10].

IV. SUMMARY

Following are specific examples on the use of a variety of scavengers or purification techniques as they have been applied to medicinal chemistry problems. These examples illustrate but a small example of a growing field.

Linclau and Curran developed fluorous scavengers and techniques that can be used to enhance the effectiveness of liquid-liquid extractions. The procedure described in this report involves the development of a fluorous-labeled amine.

Figure 1   Solid-phase reagents.

The amine can be used to scavenge unreacted reagents and the fluorous label allows for select extraction. In summary, a urea library was prepared by treating an amine with excess isocyanate. The excess isocyanate, which is needed to drive the reaction to completion, must now be removed to obtain the desired urea in sufficient purity. The compound was "purified" by treating the reaction mixture with a fluorous-bound amine described above, partitioning the organic layer with a fluorous solvent and separating the organic layer to provide the desired urea in high purity.

The processes can be automated, but some care must be taken in performing the liquid–liquid extractions as the volume of the phases (organic and fluorous) can change upon mixing. This makes it difficult to perform a consistent volume-directed liquid–liquid extraction.

A key step to this technique is to effect complete removal of the quenched fluorous product. In this case, the quenched urea was very polar and the standard fluorous reagent (39 fluorine atoms) did not have the proper partition coefficient. It was found that doubling the number of fluorine atoms in the scavenger (larger fluorine tagged amine) provided quenched ureas (scavenged isocyanate) with the proper partition coefficient.

Several other techniques have been developed to remove unreacted reagents or starting materials, including the use of ion exchange resins in the preparation of amide libraries (Suto) or the use of polymer-supported reagents (Parlow, Dice, and South). The latter approach was used in the preparation of a series of α-ketoamides and involves the development and combination of a series of resin-bound reagents to address a specific multistep synthesis. The synthesis contains three chemical reactions, two physical manipulations, and four different solid-phase reagents/scavengers. The technique uses designed sequestering agents (polymer-assisted solution phase synthesis, PSAP) to produce a series of keto-amides in excellent purity and yields with no additional intermediate purification. Almost all of the compounds prepared were more than 90% pure and many exceeded 95%.

The ion exchange resins (Suto) provide a one-pot procedure for promoting the reaction and removing starting materials. The chemistry described is the formation of an amide utilizing an acid chloride and an amine. Slightly less than one equivalent of the amine is used and the reaction is carried out in the presence of a basic ion exchange resin. The resin acts as a base in this situation. Next, the reaction is quenched with a small amount of water. The remaining acid chloride is converted to the carboxylic acid and the basic resin, then scavenges and thus purifies the mixture. Simple filtration and concentration provides the desired amides in high yields. The ability to obtain resins that vary in their pH and their cost makes this an economical and easy-to-use approach.

In summary, there now exist several different strategies for producing quality libraries in a solution phase approach. The purities obtained are comparable to solid-phase techniques. However, the time required to develop the procedures and the cost are much less prohibitive.

REFERENCES

1. Arqule Inc., Medford, Massachusetts. See also Case Study 4–2.
2. S. Cheng, D.D. Comer et al, *J. Am. Chem. Soc.*, 1996, *118*, 2567–2573.
3. D.P. Curran and S. Hadida, *J. Am. Chem. Soc.*, 1996, *118*, 2531–2532.
4. T.A. Keating and R.W. Armstrong, *J. Am. Chem. Soc.*, 1996, *118*, 2574–2583.
5. S.W. Kaldor, M.G. Siegel, et al, *Tet. Lett.*, 1996, *37*, 7193–7196.
6. S.W. Kaldor and M.G. Siegel, et al, *Curr. Opin. Chem. Biol.*, 1997, *3*, 101–106.
7. R.J. Booth and J.C. Hodges, *J. Am. Chem. Soc.*, 1997, *119*, 4882–4886.
8. D.L. Flynn, J.Z. Crich *et al*, *J. Am. Chem. Soc.*, 1997, *119*, 4874–4881.
9. D.L. Flynn, R.V. Devraj et al, *Curr. Opin. Drug Dis. Dev.*, 1998, *1*, 41–50.
10. D.L. Flynn, R.V. Devraj et al, *Med. Chem. Res.*, 1998, *8*, 219–243.

Case Study 5-1
Fluorous Scavenging

Bruno Linclau
University of Southampton, Southampton, England

Dennis P. Curran
University of Pittsburgh, Pittsburgh, Pennsylvania

ABSTRACT

This section describes the application of a fluorous quenching strategy for the fully automated parallel synthesis of a small urea library. Both the urea synthesis and the quenching procedure occur in solution phase and the final workup consists of fluorous-organic liquid–liquid extraction. A brief discussion concerning extraction properties of fluorous urea derivatives is included.

Chemicals

1. FC-72 (commercially available, 3M)
2. Tetrahydrofuran
3. *N,N*-Bis[3-[tris(2-perfluorohexylethyl)silyl]propyl]amine **3** (see ref. 1)
4. *o*-, *m*-, *p*-Trifluoromethylbenzylamine (commercially available)
5. *o*-, *m*-, *p*-Bromophenyl isocyanate (commercially available)

Equipment and Supplies

1. Hewlett Packard 7686 solution phase synthesizer
2. Reaction vials and "high-recovery" vials (commercially available, Hewlett Packard, HP)
3. Vacuum oven with temperature control

I. INTRODUCTION

Diarylureas are good inhibitors of cAMP-stimulated transepithelial Cl secretion across several epithelial cells. In collaboration with Dr. Ashvani K. Singh from the Department of Cell Biology and Physiology, we synthesized several urea libraries via automated parallel synthesis using the recently developed fluorous quenching method [1].

The use of fluorous methods in combinatorial chemistry was introduced a few years ago [2]. These methods rely on the fact that organic compounds become soluble in a fluorous phase when a perfluorinated tag is attached to the compound. By selective tagging of the desired (or undesired) products, purification by fluorous-organic liquid–liquid extraction can be accomplished. Fluorous solvents are immiscible in water and most organic solvents [3]. This has led to several solution phase reaction and purification strategies wherein selective tagging occurs before, during, or after the desired transformation is completed. The last case represents fluorous quenching (Scheme 1).

To obtain complete extractive removal of the quenched—now fluorous—product, a critical number of fluorines must be present. In this regard, the readily available tris(perfluorohexylethyl)silyl group [$(C_6F_{13}CH_2CH_2)_3Si$—] (F_{39}) and related groups are suitable tags that effectively convert small (MW < 200) nonpolar molecules into fluorous ones, resulting in efficient removal of the adducts from the organic phase by fluorous extraction [4]. However, larger and more polar substrates linked to an F_{39} tag either require repetitive extraction or cannot even be completely removed from the organic phase. For fluorous quenching purposes, it is an absolute requirement that the quenched fluorous product be efficiently removed from the organic phase. Urea groups are quite polar, and we found that an F_{39} tag is not large enough to effectively extract the fluorous urea into the

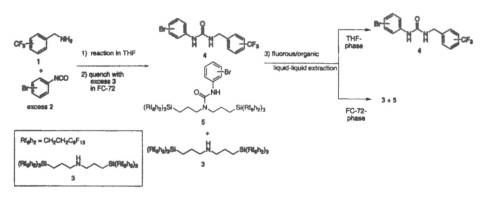

Scheme 1

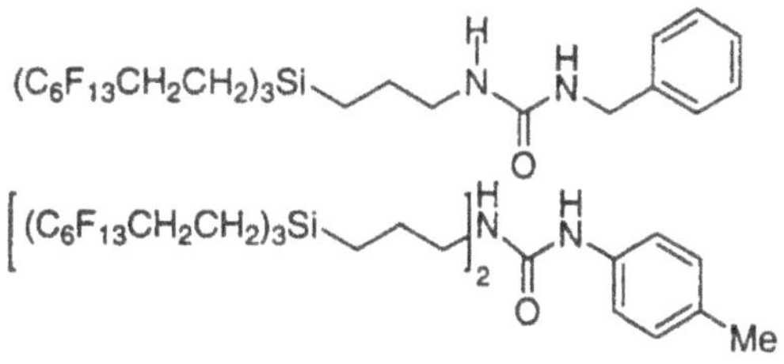

	P_{CHCl_3}	P_{CH_3CN}	P_{CH_3OH}	P_{THF}
6 $(C_6F_{13}CH_2CH_2)_3Si$	0.27	1.0	0.27	-
7 $[(C_6F_{13}CH_2CH_2)_3Si]_2$	27	>50	>100	11

Figure 1 Partition coefficients for two fluorous ureas.

very apolar fluorous phase by liquid–liquid extraction. For example, the fluorous urea derivative **6** has a low partition coefficient ($P = c_{fluorous\ phase}/c_{organic\ phase}$) against all organic solvents tested (Fig. 1). When two F_{39} tags are used as in **7**, the partition coefficient is significantly increased compared to **6** and is sufficiently high for effective extraction. Even in tetrahydrofuran (THF), the partition coefficient lies within a useful range for liquid–liquid extraction. As a result, the F_{78} amine **3** was chosen as a suitable scavenger for isocyanates (Scheme 1).

II. GENERAL PROTOCOL

Urea products **4** were formed by reaction of amine reagents **1** with an excess of isocyanate reagents **2**. After a short reaction period, fluorous quencher **3** was added, which transformed the excess **2** into fluorous adduct **5**. Fluorous/organic liquid–liquid extraction resulted in the removal of the fluorous products **3** and **5**, and pure urea products **4** were obtained after evaporation of the solvent. Figure 2 summarizes the results of a typical 3 × 3 library.

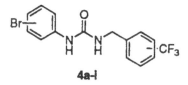

4a-l

Yield range	73–96 %	average 86 %
HPLC-purity	92.3–98.5 %	average 96.0 %
sym-urea contamination	0.2–4.4 %	average 2.0 %
Fluorous contamination	0.2–0.5 mol%	average 0.3 mol %

Figure 2 Results of the 3 × 3 library of ureas.

III. GENERAL PROCEDURE

Both the preparation of the reagent solutions and the reaction are performed on the robot, using two separate operations:

A. Preparation of the Reagent Solutions

For preparing the reagent solutions, an excess of each reagent was manually added to a dry empty HP vial, and the vial was tightly capped. The robot was programmed to transfer a certain amount of the pure reagent to another dry, empty, capped vial, immediately followed by dilution with THF, resulting in 1.5 mL of 0.5 M solution. The following amounts were used to program the robot: 107 μL of *p*-trifluoromethylbenzylamine; 107 μL of *m*-trifluoromethylbenzylamine; 105 μL of *o*-trifluoromethylbenzylamine; 94 μL of *m*-bromophenyl isocyanate; 93 μL of *m*-bromophenyl isocyanate. *p*-Bromophenyl isocyanate is a solid, and a 0.5 M solution was prepared by weighing 159 mg in a dry, empty vial that was immediately tightly capped. The solid was subsequently dissolved by dispensing 1.59 mL of THF. Fluorous amine 3 is not solid but is too viscous to be transferred with a needle. Solutions were prepared by transferring 0.5 mL FC-72 from a "stock vial" to each of the 9 vials containing 110 mg of 3. We choose to use "stock vials" of FC-72 rather than having a FC-72 solvent bottle to minimize contact of the FC-72 with the Teflon tubing in the robot.

B. Reaction-Quenching-Extraction Procedure

The following types of vials were listed in the vial table: amine reagent vials, isocyanate reagent vials (each charged with reagent solution), reaction vials (high recovery, empty)(9), fluorous amine solution vials, FC-72 stock vials, and fluorous waste vials.

1. Reaction

The robot was programmed first to dispense 0.35 mL of THF in each of the 9 reaction vials. Then 0.1 mL (0.05 mmol) of the (variable) amine solution was added to the reaction vial. The needle and sample loop were rinsed with THF (2.5 mL) immediately after each addition, followed by vortexing of the vial for 0.5 min at medium speed. Subsequently, 0.15 mL (0.075 mmol, 1.5 equivalent) of the (variable) isocyanate solution was added to the reaction vials in such a way that 9 different reaction combinations are obtained (rinse). The reaction vials were mixed again for 0.5 min at medium speed and allowed to stand for 30 min at room temperature.

2. Quenching

Fluorous amine solution (0.55 mL, 0.05 mmol) was transferred to each reaction vial. The fluorous amine solution vial was rinsed with 0.1 mL of FC-72, followed by transfer of the 0.1 mL to the reaction vial as well. The needle was rinsed again with THF (2.5 mL) and all reaction vials were vortexed for 30 min, upon which they were allowed to stand for 30 min at room temperature.

3. Extractions

The extractions were performed by transferring the (bottom) fluorous phase to a fluorous waste vial (see commentary below). At first, 0.68 mL was transferred. New FC-72 was added to each reaction vial, followed by mixing for 5 min. The phases were allowed to settle for 30 min, whereafter the bottom fluorous phase (0.6 mL) was again transferred to a fluorous waste vial. This step was repeated 3 more times.

Finally, evaporation on the robot of the organic solvent yielded nine different ureas **4a–i**, which were subsequently dried in a vacuum oven at 50–60°C prior to analysis. The yields and purities of the products from a typical nine-compound library are contained in Table 1.

Table 1 Results of the 3 × 3 Library of Ureas

Entry	No.	R¹	R²	Urea yield (%)	HPLC purity[a] (%)	%F[b]
1	4a	4-Br	4-CF₃	87	98.0 (0.5)	0.5
2	4b	4-Br	3-CF₃	80	98.5 (0.4)	0.3
3	4c	4-Br	2-CF₃	90	98.4 (0.2)	0.4
4	4d	3-Br	4-CF₃	79	95.7 (2.4)	0.3
5	4e	3-Br	3-CF₃	93	96.5 (1.7)	0.3
6	4f	3-Br	2-CF₃	85	96.8 (0.9)	0.3
7	4g	2-Br	4-CF₃	73	92.3 (4.4)	0.2
8	4h	2-Br	3-CF₃	96	94.0 (3.6)	0.2
9	4i	2-Br	2-CF₃	90	93.6 (3.6)	0.3

[a] Waters Nova-Pack C18 3.9 × 150 mm. Flow rate 1.5 mL/min. Detection at 210 nm, solvent, acetonitrile 40–60% in water. The numbers in parentheses refer to the amount of symmetrical urea.
[b] Mol%, determined in the ¹⁹F-NMR spectrum, relative to the product-CF₃ group.

IV. ANALYSIS: DETERMINATION OF TRACES OF FLUOROUS COMPOUNDS

Analysis of the obtained ureas was performed by reversed-phase high-performance liquid chromatography (HPLC), and all compounds show a purity of greater than 92%. The main impurities are the corresponding symmetrical ureas, formed by hydrolysis of the parent isocyanate. The levels of *sym*-urea impurities are clearly dependent on the starting isocyanate, suggesting that the starting material was already contaminated with the hydrolysis products. (Some commercially available isocyanates contain a solid impurity, presumably the *sym*-urea derivative, and have to be filtered before use.)

Fluorous impurities could not be detected from the ^{1}H-NMR- or HPLC spectrum. However, by using ^{19}F-NMR, it was possible to detect small quantities present in the organic phase. Quantification was achieved by incorporating a trifluoromethyl group in the reagent, which could be used as internal standard. The signal of the CF_3 group in the product was integrated against the CF_3 group of the fluorous product to determine the mol % of fluorous impurity. Because there are six fluorous chains present in the impurity, its CF_3 signal counts for 18 fluorines. As a result, very low levels could be detected. Using a carefully tuned spectrometer, we found that the level of impurity was lower than 0.5%, demonstrating that the F_{78}-tagged urea was effectively removed via fluorous-organic liquid–liquid extraction.

V. COMMENTARY

Concerning the extractions, the volume to be transferred has to be set in the program. There is no monitoring of the meniscus. This can be a problem when the volumes of the phases change after mixing, which is the case with THF and FC-72. In a separate experiment, we found that mixing of pure THF and FC-72 resulted in a volume increase of the THF phase, but mixing of a concentrated solution of fluorous urea in FC-72 (100 mg in 0.7 mL) with THF resulted in a volume increase of the fluorous phase. We noticed that the changes in phase volume differed for different compounds, which made optimization of the extraction volume setting somewhat difficult. This probably accounts for the different yields obtained for the ureas; however, the average yield obtained (86%) was satisfactory.

We also noted that some urea products formed in other libraries were not soluble in the reaction solvent (THF). However, this did not affect the fluorous quenching and extraction. The organic solid always was suspended in the (upper) THF phase, so that no clogging of the robot needle occurred. The yields for these insoluble ureas were even above average, presumably because no urea product was removed when some THF accidentally was taken away. This is an added advantage of the fluorous quenching method because insoluble product would be removed if a filtration-based purification technique were used.

REFERENCES

1. B Linclau, AK Singh, DP Curran. *J Org Chem 64*: 2835–2842, 1999.
2. A Studer, S Hadida, R Ferrito, S-Y Kim, P Jeger, P Wipf. *Science 275*: 823–826, 1997; DP Curran. *Angew Chem Int Ed Engl 37*: 1174–1196, 1998.
3. IT Horvath, J Rabai. *Science 266*: 72–75, 1994; B Cornils. *Angew Chem Int Ed Engl 36*: 2057–2059, 1997.
4. A Studer, P Jeger, P Wipf, DP Curran. *J Org Chem 62*: 2917–2924, 1997; A Studer, DP Curran. *Tetrahedron 53*: 6681–6696, 1997.

Case Study 5-2
Polymer-Assisted Solution Phase Synthesis of α-Ketoamides

John J. Parlow, Thomas A. Dice, and Michael S. South
Pharmacia Corporation, St. Louis, Missouri

ABSTRACT

A parallel solution phase synthesis of an α-ketoamide library is described. The two-step synthesis is accomplished using polymer-assisted solution phase (PASP) (Fig. 1) synthesis techniques as the sole method of purification to afford α-keto-amide products in high purity.

Chemicals

1. 4-Phenoxyphenethylamine
2. 3,5-Dimethoxybenzylamine
3. Benzylamine
4. 3-Ethoxypropylamine
5. Mandelic acid
6. 4-(Trifluoromethyl)mandelic acid
7. 4-Methoxymandelic acid
8. 4-Bromomandelic acid
9. 2-Hydroxy-3-methylbutyric acid
10. 2-Hydroxyisocaproic acid
11. (R)-(−)-Hexahydromandelic acid
12. 1-Hydroxybenzotriazole
13. Diethylenetriamine
14. Sodium thiosulfate pentahydrate
15. Amberlyst A-21 (obtained from Aldrich Chemical Company)

Figure 1   A polymer-assisted solution phase synthesis of α-ketoamides.

16. Amberlyst A-26 (Cl⁻) (obtained from Alfa Æsar Company)
17. 1,1,1-Triacetoxy-1,1-dihydro-1,2-benziodoxol-3(1H)-one   (obtained from Lancaster Synthesis Inc.)
18. Merrifield's resin (chloromethylated styrene-divinylbenzene copolymer, 2% cross-linked, obtained from Aldrich Chemical Company)
19. PS-carbodiimide resin (obtained from Argonaut Technologies)

Equipment and Supplies

1. 2-dram glass vials with Teflon-lined cap as reaction vessel
2. 25-mL volumetric flask
3. 48-position custom aluminum rack to hold 2 dram vials
4. Lab-Line orbital shaker (set to 200 rpm)
5. Custom 48-well Bohdan block to hold fritted vessels and 10-mL collection vessels
6. 15-mL fritted vessels with 20-μm inserted frit (P.J. Cobert, 944195) for Bohdan block
7. 10-mL Kontes conical vials (749001-0010) used in Bohdan block for collection
8. 48-position custom nitrogen blow-down manifold
9. Automated weighing workstation (by Bohdan Automation) to tare and gross 10-mL conical vials
10. Vacuum oven
11. Eppendorf brand 1-mL pipet

I. INTRODUCTION

As part of a medicinal program at Searle, an α-ketoamide library was prepared using PASP parallel synthesis techniques. The synthesis included three chemical transformations involving two physical manipulation steps. The first step involved the preparation of the α-hydroxyamides, as shown in Scheme 1. The α-hydroxy acids **1** were reacted with hydroxybenzotriazole **2** and the polymer-bound carbodiimide **3** to afford the α-hydroxy-activated esters **4**. A limiting

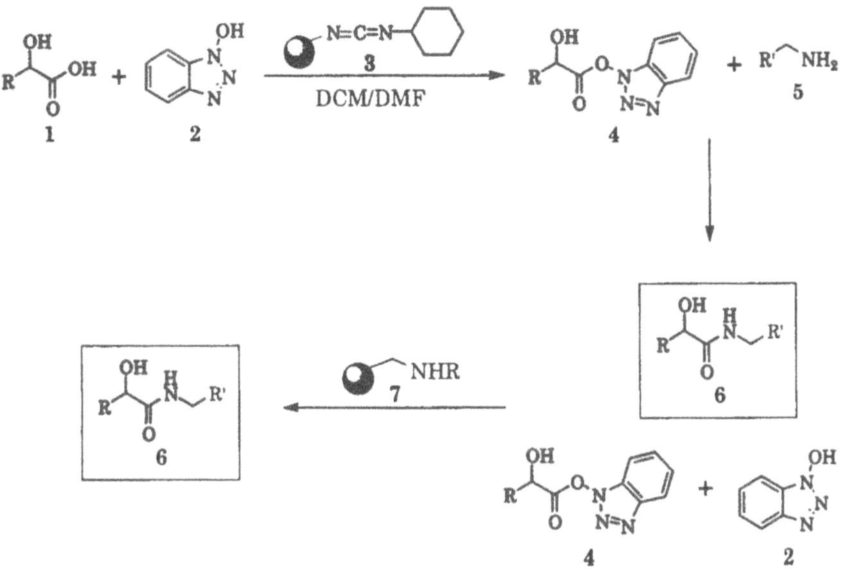

Scheme 1

amount of the amine **5** was added to each reaction vessel. After total consumption of amine **5**, the product mixtures contained the desired α-hydroxyamides **6**, excess α-hydroxy activated esters **4**, and hydroxybenzotriazole **2**. The reactant-sequestering nucleophilic polyamine resin **7** was added to react with the excess α-hydroxy-activated esters **4**, forming insoluble polymer-bound amides. The basic polyamine resin **7** also sequestered the hydroxybenzotriazole byproduct **2**, leaving the desired α-hydroxyamides **6** in solution. Simple filtration and rinsing with dichloromethane yielded a filtrate whereupon evaporation of the solvents left purified products **6** from each parallel reaction.

The second step of the synthesis involved preparation of the α-ketoamides, as shown in Scheme 2. α-Hydroxyamide **6** was oxidized with excess Dess-Martin periodinane reagent **8** to drive the reaction to completion. Upon completion of the reaction, a mixed-resin bed, containing the reducing thiosulfate resin **12** and the basic Amberlyst A-21 resin **13**, was added to the product mixture. The thiosulfate resin **12** reduced the excess Dess-Martin reagent **8** and the I^{III} species **11** to the sequesterable 2-iodobenzoic acid (I^{I} species). In addition, the thiosulfate resin **12** sequestered most of the resulting 2-iodobenzoic acid, with typically less than 5% of the acid remaining. The Amberlyst A-21 resin **13** was used as a base to sequester the remaining 2-iodobenzoic acid. Filtration, rinsing with dichloromethane, and concentration afforded purified α-ketoamide products **9**. (Fig. 2)

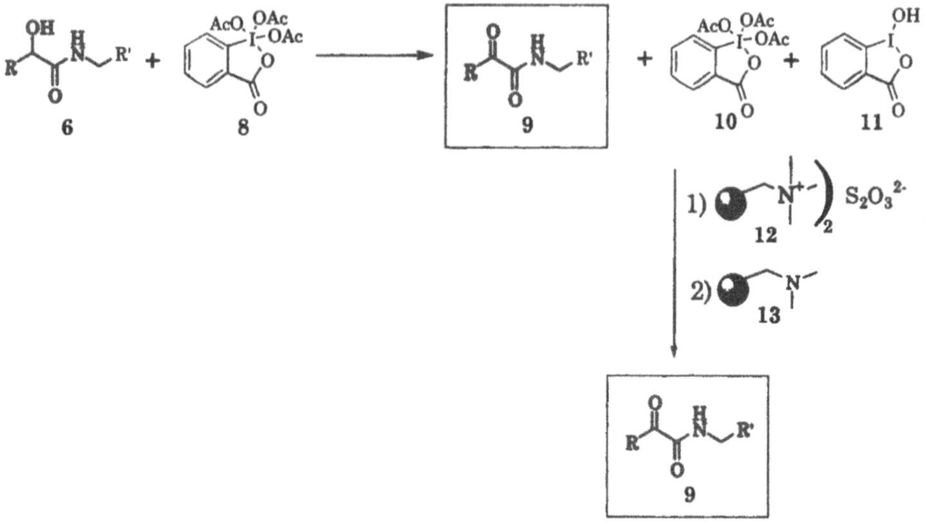

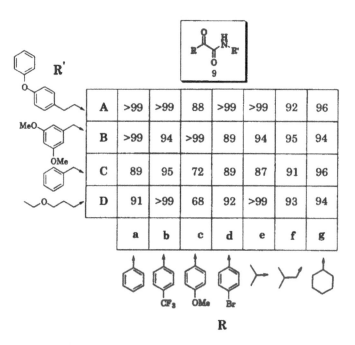

Scheme 2

R'	a	b	c	d	e	f	g
A	>99	>99	88	>99	>99	92	96
B	>99	94	>99	89	94	95	94
C	89	95	72	89	87	91	96
D	91	>99	68	92	>99	93	94

R

Figure 2 HPLC purities (%) for α-ketoamide products 9.

II. GENERAL

The reactions were run in 2-dram glass reaction vessels with Teflon-lined caps and placed in a 48-position custom aluminum rack. The filtrations were conducted through a Bhodan customized reaction block containing 15-mL polypropylene fritted vessels with a 20-μm inserted frit. The eluent was collected in 10-mL Kontes conical vials designed to fit in the collection tray of the Bhodan reaction block. The tare and gross weighing of the conical vials were conducted on a Bhodan automated weighing workstation to afford mass returns for each product. The conical vials containing the final products were dried in a high-vacuum oven with no heat prior to weighing. ^{1}H-NMR spectra were recorded using 300-MHz NMR spectrometers. HPLC purities were determined with a Hewlett Packard HP1100 equipped with a series 1100 MSD and an ODS Eclipse XDB-C18 3.5-μm 30 × 2.1 mm C18 column, eluting with a gradient system of 0:100 to 95:5 acetonitrile-H_2O with 0.1% TFA over 6 min at 1 mL/min, and detected by UV at 254 nm using a diode array detector. GC purities were determined with an HP6890 equipped with an MSD utilizing a capillary column (0.25 mm i.d., 5% phenylmethylsiloxane, 0.25-μm film thickness, 30 m, temperature program from 100°C to 300°C at 20°C/min). Reported yields are unoptimized, with emphasis on purity rather than quantity.

A. Preparation of Thiosulfate Resin (12)

Amberlyst A-26 (Cl$^-$) (1.0 meq/mL) (~1100 mL, ~1.1 mol) was packed in a column and flushed with deionized water. Twelve liters of a 1 M solution of sodium thiosulfate pentahydrate in deionized water was passed through the resin-packed column. The resin was rinsed with deionized water until the eluent was neutral. The resin 12 was removed from the column and rinsed 4 times with tetrahydrofuran, 3 times with diethylether, and dried *in vacuo*. Anal. Obsd.: N, 4.81%, 3.4 mmol/g; S, 11.18%, 3.5 mmol/g; Cl, 0.15%; Na, 0.19%.

B. Preparation of Polyamine Resin (7)

Merrifield's resin (2% cross-linked, 2.69 mmol/g) (1 kg) was added to diethyl-enetriamine (used as solvent) and the mixture heated at 100°C for 4 h. The polymer was filtered and successively rinsed 2 times with 10% triethylamine in dimethylformamide, once with dimethylformamide, 4 times with 10% triethylamine in tetrahydrofuran, 3 times with tetrahydrofuran, and 3 times with methanol. The resin 7 was then dried under vacuum to a constant weight. Anal. Obsd: Cl, 0%; N, 6.57%, 4.69 mmol/g.

C. Washing of Amberlyst A-21 Resin (13)

Amberlyst A-21 resin **13** was rinsed 3 times with dimethylformamide, 3 times with dichloromethane, soaked in dichloromethane for 1 h, rinsed with dichloromethane 2 times, rinsed 3 times with tetrahydrofuran, 3 times with diethyl ether, and dried in vacuo.

D. General Procedure A. Preparation of Amides 6 from α-Hydroxy Acids 1

1-Hydroxybenzotriazole **2** (120 mM, 80:20 dichloromethane-dimethylformamide) (1 mL, 0.12 mmol) was added manually to a slurry of the α-hydroxy acid **1** [mandelic acid, 4-(trifluoromethyl)mandelic acid, 4-methoxymandelic acid, 4-bromomandelic acid, 2-hydroxy-3-methylbutyric acid, 2-hydroxyisocaproic acid, (R)-(-)-hexahydromandelic acid) (0.11 mmol), and PS-carbodiimide **3** (2.00 mmol/g) (174 mg, 0.348 mmol)] in a 2-dram vial with 1 mL of dichloromethane. The vials, held in a 48-position custom rack, were capped, the rack turned sideways, and the suspension was agitated on an orbital shaker for 30 min. The amine **5** (4-phenoxyphenethylamine, 3,5-dimethoxybenzylamine, benzylamine, 3-ethoxypropylamine) in dichloromethane (100 mM) (1 mL, 0.10 mmol) was added manually and the suspension was agitated laterally on an orbital shaker for 1.5 h. Upon completion of the reaction, polyamine resin **7** (4.69 mmol/g) (0.10 g, 0.469 mmol) was added manually and the suspension laterally agitated on an orbital shaker for 16 h. The reaction solution from the 2-dram vials was filtered through the Bhodan custom block with the fritted vessels and the polymer was rinsed with five 1-mL portions of dichloromethane into a 10-mL conical vial. The combined filtrate and washings were evaporated under a stream of nitrogen to afford the pure products **6**. These products were characterized by GC-MS and carried on to the next step.

E. General Procedure B. Preparation of α-Ketoamides 9 from α-Hydroxyamides 6

The Dess-Martin periodinane reagent **8** [1,1,1-triacetoxy-1,1-dihydro-1,2-benziodoxol-3(1H)-one] in dichloromethane (125 mM) (1 mL, 0.125 mmol) was added manually to the α-hydroxy acid **6** in a 2-dram vial. The vials, contained in a 48-position custom rack, were capped and the resulting solution was agitated on an orbital shaker for 2 h. Upon completion of the reaction, the Amberlyst A-21 resin **13** (4.8 mmol/g) (0.20 g, 0.96 mmol) and the polymer-bound thiosulfate resin **12** (1.7 mmol/g) (0.50 g, 0.85 mmol) was added manually, followed by addition of 1 mL of dichloromethane, and the suspension was agitated laterally on an orbital shaker for 16 h. The reaction solution from the 2-dram vials were filtered through the Bhodan custom block with the fritted vessels and the polymers were rinsed with five 1-mL portions of dichloromethane into a 10-mL conical vial. The combined filtrate and washings were evaporated under a stream of nitrogen and placed in a high-vacuum oven (no heat) to afford the pure products **9**.

9Aa: General procedure B; mass recovery (yield): 26.0 mg, (75%); ^{1}H-NMR (CDCl$_3$) ppm 2.94 (t, 2H), 3.69 (m, 2H), 6.99–7.37 (m, 10H), 7.51 (t, 2H), 7.68 (t, 1H), 8.36 (dd, 2H); HPLC purity (retention time): >99% (4.3 min); HRMS calcd for C$_{22}$H$_{19}$NO$_3$ (M$^+$ + H) 346.1443, found 346.1449.

9Ab: General procedure B; mass recovery (yield): 27.7 mg, (67%); ^{1}H-NMR (CDCl$_3$) ppm 2.94 (t, 2H), 3.69 (m, 2H), 7.00–7.37 (m, 10H), 7.77 (d, 2H), 8.47 (d, 2H); HPLC purity (retention time): >99% (4.8 min); HRMS calcd for C$_{23}$H$_{18}$F$_3$NO$_3$ (M$^+$ + H) 414.1317, found 414.1283.

9Ac: General procedure B; mass recovery (yield): 28.6 mg, (76%); ^{1}H-NMR (CDCl$_3$) ppm 2.92 (m, 2H), 3.67 (m, 2H), 3.92 (s, 3H), 6.96–7.36 (m, 11H), 8.12 (bs, 1H), 8.42 (dd, 2H); HPLC purity (retention time): 88% (4.3 min); HRMS calcd for C$_{23}$H$_{21}$NO$_4$ (M$^+$ + H) 376.1549, found 376.1540.

9Ad: General procedure B; mass recovery (yield): 29.7 mg, (70%); ^{1}H-NMR (CDCl$_3$) ppm 2.92 (bt, 2H), 3.68 (m, 2H), 6.99–7.37 (m, 10H), 7.67 (d, 2H), 8.25 (d, 2H); HPLC purity (retention time): >99% (4.7 min); HRMS calcd for C$_{22}$H$_{18}$BrNO$_3$ (M$^+$ + H) 424.0548, found 424.0533.

9Ae: General procedure B; mass recovery (yield): 25.2 mg, (81%); ^{1}H-NMR (CDCl$_3$) ppm 1.16 (d, 6H), 2.87 (t, 2H), 3.60 (m, 3H), 6.97–7.36 (m, 10H); HPLC purity (retention time): >99% (4.2 min); HRMS calcd for C$_{19}$H$_{21}$NO$_3$ (M$^+$ + H) 312.1600, found 312.1596.

9Af: General procedure B; mass recovery (yield): 28.2 mg, (87%); ^{1}H-NMR (CDCl$_3$) ppm 0.98 (d, 6H), 2.22 (m, 1H), 2.82 (d, 2H), 2.88 (t, 2H), 3.59 (m, 2H), 6.97–7.39 (m, 10H); HPLC purity (retention time): 92% (4.4 min); HRMS calcd for C$_{20}$H$_{23}$NO$_3$ (M$^+$ + H) 326.1756, found 326.1757.

9Ag: General procedure B; mass recovery (yield): 30.3 mg, (86%); ^{1}H-NMR (CDCl$_3$) ppm 1.31 (m, 5H), 1.84 (m, 5H), 2.86 (t, 2H), 3.41 (m, 1H), 3.58 (m, 2H), 6.97–7.39 (m, 10H); HPLC purity (retention time): 96% (4.7 min); HRMS calcd for C$_{22}$H$_{25}$NO$_3$ (M$^+$ + H) 352.1913, found 352.1919.

9Ba: General procedure B; mass recovery (yield): 20.3 mg, (68%); ^{1}H-NMR (CDCl$_3$) ppm 3.82 (s, 6H), 4.54 (d, 2H), 6.43 (s, 1H), 6.51 (s, 2H), 7.50 (bs, 1H), 7.53 (m, 2H), 7.67 (m, 1H), 8.38 (d, 2H); HPLC purity (retention time): >99% (3.4 min); HRMS calcd for C$_{17}$H$_{17}$NO$_4$ (M$^+$ + H) 300.1236, found 300.1240.

9Bb: General procedure B; mass recovery (yield): 19.9 mg, (54%); ^{1}H-NMR (CDCl$_3$) ppm 3.82 (s, 6H), 4.54 (d, 2H), 6.44 (s, 1H), 6.50 (s, 2H), 7.45

(bs, 1H), 7.78 (d, 2H), 8.52 (d, 2H); HPLC purity (retention time): 94% (4.1 min); HRMS calcd for $C_{18}H_{16}F_3NO_4$ (M$^+$ + H) 368.1110, found 368.1107.

9Bc: General procedure B; mass recovery (yield): 24.8 mg, (75%); ^{1}H-NMR (CDCl$_3$) ppm 3.82 (s, 6H), 3.93 (s, 3H), 4.52 (d, 2H), 6.42 (s, 1H), 6.51 (s, 2H), 6.99 (d, 2H), 7.48 (bs, 1H), 8.48 (d, 2H); HPLC purity (retention time): >99% (3.4 min); HRMS calcd for $C_{18}H_{19}NO_5$ (M$^+$ + H) 330.1341, found 330.1355.

9Bd: General procedure B; mass recovery (yield): 21.8 mg, (58%); ^{1}H-NMR (CDCl$_3$) ppm 3.82 (s, 6H), 4.52 (d, 2H), 6.43 (s, 1H), 6.50 (s, 2H), 7.44 (bs, 1H), 7.68 (d, 2H), 8.32 (d, 2H); HPLC purity (retention time): 89% (3.9 min); HRMS calcd for $C_{17}H_{16}BrNO_4$ (M$^+$ + H) 378.0341, found 378.0332.

9Be: General procedure B; mass recovery (yield): 22.0 mg, (94%); ^{1}H-NMR (CDCl$_3$) ppm 1.18 (d, 6H), 3.66 (septet, 1H), 3.81 (s, 6H), 4.44 (d, 2H), 6.41 (s, 1H), 6.45 (s, 2H), 7.29 (bs, 1H); HPLC purity (retention time): 94% (3.2 min).

9Bf: General procedure B; mass recovery (yield): 22.5 mg, (81%); ^{1}H-NMR (CDCl$_3$) ppm 0.99 (d, 6H), 2.22 (m, 1H), 2.86 (d, 2H), 3.81 (s, 6H), 4.43 (d, 2H), 6.41 (s, 1H), 6.45 (s, 2H), 7.29 (bs, 1H); HPLC purity (retention time): 95% (3.6 min); HRMS calcd for $C_{15}H_{21}NO_4$ (M$^+$ + H) 280.1549, found 280.1563.

9Bg: General procedure B; mass recovery (yield): 25.3 mg, (83%); ^{1}H-NMR (CDCl$_3$) ppm 1.33 (m, mH), 1.84 (m, 5H), 3.43 (m, 1H), 3.81 (s, 6H), 4.43 (d, 2H), 6.41 (s, 1H), 6.44 (s, 2H), 7.25 (bs, 1H); HPLC purity (retention time): 94% (3.9 min); HRMS calcd for $C_{17}H_{23}NO_4$ (M$^+$ + H) 306.1705, found 306.1726.

9Ca: General procedure B; mass recovery (yield): 14.3 mg, (60%); ^{1}H-NMR (CDCl$_3$) ppm 4.61 (d, 2H), 7.52 (m, 8H), 7.55 (t, 1H), 8.41 (d, 2H); HPLC purity (retention time): 89% (3.3 min); HRMS calcd for $C_{15}H_{13}NO_2$ (M$^+$ + H) 240.1025, found 240.1029.

9Cb: General procedure B; mass recovery (yield): 13.7 mg, (45%); ^{1}H-NMR (CDCl$_3$) ppm 4.62 (d, 2H), 7.38 (m, 6H), 7.78 (d, 2H), 8.52 (d, 2H); HPLC purity (retention time): 95% (4.0 min); HRMS calcd for $C_{16}H_{12}F_3NO_2$ (M$^+$ + H) 308.0898, found 308.0935.

9Cc: General procedure B; mass recovery (yield): 19.5 mg, (72%); ^{1}H-NMR (CDCl$_3$) ppm 3.93 (s, 3H), 4.60 (d, 2H), 7.00 (d, 2H), 7.37 (m, 6H), 8.49 (d, 2H); HPLC purity (retention time): 72% (3.3 min); HRMS calcd for $C_{16}H_{15}NO_3$ (M$^+$ + H) 270.1130, found 270.1134.

9Cd: General procedure B; mass recovery (yield): 17.7 mg, (56%); ^{1}H-NMR (CDCl$_3$) ppm 4.61 (d, 2H), 7.37 (m, 6H), 7.66 (d, 2H), 8.30 (d, 2H); HPLC purity (retention time): 89% (3.9 min); HRMS calcd for $C_{15}H_{12}Br_1NO_2$ (M$^+$ + H) 320.0111, found 320.0117.

9Ce: General procedure B; mass recovery (yield): 11.6 mg, (57%); ^{1}H-NMR (CDCl$_3$) ppm 1.19 (d, 6H), 3.67 (septet, 1H), 4.52 (d, 2H), 7.35 (m, 6H); HPLC purity (retention time): 87% (3.0 min); HRMS calcd for C$_{12}$H$_{15}$NO$_2$ (M$^+$ + H) 206.1181, found 206.1192.

9Cf: General procedure B; mass recovery (yield): 17.3 mg, (79%); ^{1}H-NMR (CDCl$_3$) ppm 0.99 (d, 6H), 2.21 (septet, 1H), 2.87 (d, 2H), 4.50 (d, 2H), 7.35 (m, 6H); HPLC purity (retention time): 91% (3.5 min); HRMS calcd for C$_{13}$H$_{17}$NO$_2$ (M$^+$ + H) 220.1338, found 220.1346.

9Cg: General procedure B; mass recovery (yield): 21.7 mg, (88%); ^{1}H-NMR (CDCl$_3$) ppm 1.34 (m, 5H), 1.85 (m, 5H), 3.45 (m, 1H), 4.51 (d, 2H), 7.36 (m, 6H); HPLC purity (retention time): 96% (3.8 min); HRMS calcd for C$_{15}$H$_{19}$NO$_2$ (M$^+$ + H) 246.1494, found 246.1502.

9Da: 7 general procedure B; mass recovery (yield): 15.4 mg, (65%); ^{1}H-NMR (CDCl$_3$) ppm 1.28 (m, 5H), 1.90 (m, 2H), 3.55 (m, 4H), 7.50 (m, 6H); HPLC purity (retention time): 91% (2.7 min); HRMS calcd for C$_{13}$H$_{17}$NO$_3$ (M$^+$ + H) 236.1287, found 236.1297.

9Db: General procedure B; mass recovery (yield): 15.7 mg, (52%); ^{1}H-NMR (CDCl$_3$) ppm 1.30 (m, 3H), 1.92 (m, 2H), 3.57 (m, 6H), 7.76 (d, 2H), 7.84 (bs, 1H), 8.48 (d, 1H); HPLC purity (retention time): >99% (3.6 min); HRMS calcd for C$_{14}$H$_{16}$NO$_3$ (M$^+$ + H) 304.1161, found 304.1170.

9Dc: General procedure B; mass recovery (yield): 19.3 mg, (73%); ^{1}H-NMR (CDCl$_3$) ppm 1.28 (t, 3H), 1.90 (quintet, 2H), 3.55 (m, 6H), 3.92 (s, 3H), 6.97 (d, 2H), 7.76 (bs, 1H), 8.44 (d, 1H); HPLC purity (retention time): 68% (2.8 min); HRMS calcd for C$_{14}$H$_{19}$NO$_4$ (M$^+$ + H) 266.1392, found 266.1388.

9Dd: General procedure B; mass recovery (yield): 17.0 mg, (54%); ^{1}H-NMR (CDCl$_3$) ppm 1.29 (t, 3H), 1.90 (quintet, 2H), 3.56 (m, 6H), 7.65 (d, 2H), 7.80 (bs, 1H), 8.27 (d, 1H); HPLC purity (retention time): 92% (3.4 min); HRMS calcd for C$_{13}$H$_{16}$BrNO$_2$ (M$^+$ + H) 314.0392, found 314.0396.

9De: General procedure B; mass recovery (yield): 1.8 mg, (9%); ^{1}H-NMR (CDCl$_3$) ppm 1.02 (m, 3H), 1.24 (m, 6H), 1.83 (m, 3H), 3.55 (m, 6H); HPLC purity (retention time): >99% (2.4 min).

9Df: General procedure B; mass recovery, (yield): 11.4 mg, (53%); ^{1}H-NMR (CDCl$_3$) ppm 0.98 (d, 6H), 1.27 (t, 3H), 1.82 (quintet, 2H), 2.20 (m, 1H), 2.82 (d, 2H), 3.48 (m, 6H), 7.58 (bs, 1H); HPLC purity (retention time): 93% (2.9 min); HRMS calcd for C$_{11}$H$_{21}$NO$_3$ (M$^+$ + H) 216.1600, found 216.1619.

9Dg: General procedure B; mass recovery (yield): 11.4 mg, (53%); ^{1}H-NMR (CDCl$_3$) ppm 1.31(m, 8H), 1.84 (m, 8H), 3.47 (m, 6H), 7.56 (bs, 1H); HPLC purity (retention time): 94% (3.3 min); HRMS calcd for C$_{13}$H$_{23}$NO$_3$ (M$^+$ + H) 242.1756, found 242.1760.

Case Study 5-3
Ion Exchange Resins for the Preparation of Amide Libraries

Mark J. Suto
DuPont Pharmaceuticals Research Laboratories, San Diego, California

ABSTRACT

The use of ion exchange resins to facilitate the synthesis and purification of small-amide-based libraries is described. The resins have utility as both a solid-phase base and as a means to perform "in situ" purification of the reaction mixtures. The technique has been used to prepare a variety of heterocyclic-based amide libraries for lead optimization. Several examples are presented.

Chemicals

1. 3,4-Dimethoxybenzoyl chloride
2. 4-Cyanobenzoyl chloride
3. 3,5-Dichlorobenzoyl chloride
4. Furfurylamine
5. 4-(Methylthio)aniline
6. Ethyl acetate
7. Amberlyst 21 ion exchange resin
8. Amberlyte IRA-68 ion exchange resin
9. 2-Chloro-4-trifluoromethyl pyrimidine acid chloride

Equipment and Supplies

1. 96-well 1-mL microtiter plate
2. 8-channel pipet

3. Branson 1210 benchtop sonicator
4. Custom-fritted funnel (1–2 mL)
5. Test tubes (18 × 150 mm)
6. BenchMate (Zymark Corp.)

I. INTRODUCTION

The formation of amides has been a fundamental reaction in the preparation of combinatorial libraries. This is due to their ease of synthesis and the availability of a variety of amines and carboxylic acids. However, the formation of amides in combinatorial chemistry was usually performed on a solid-phase resin or required the use of an aqueous workup, the latter being difficult to perform in a high-throughput synthetic manner. These synthetic or purification strategies were required to obtain compounds of sufficient purity for biological evaluation.

As part of a lead optimization program, a series of heterocyclic amides were considered necessary to fully investigate the structure–activity relationship of a lead series. The compounds needed to be of high purity (>90%) because dose–response studies would be completed. In addition, large numbers of compounds were anticipated. However, the targeted compounds were not amenable to solid-phase approaches, so that an alternative method of synthesis and purification was needed.

Ion exchange resins have been in use in organic chemistry for quite some time as a methods for delivering "bound" acids or bases. Therefore, an investigation into the use of ion exchange resins for the preparation of libraries in a parallel synthetic manner was undertaken [1]. The approach was to use a basic resin to assist in the formation of an amide from an acid chloride and an amine where it would serve as a base to neutralize the HCl formed and minimize the amount of amine needed. It could also scavenge or remove unreacted materials (hydrolzed acid) to "purify in situ" the reaction mixtures (Scheme 1). Simple filtration and

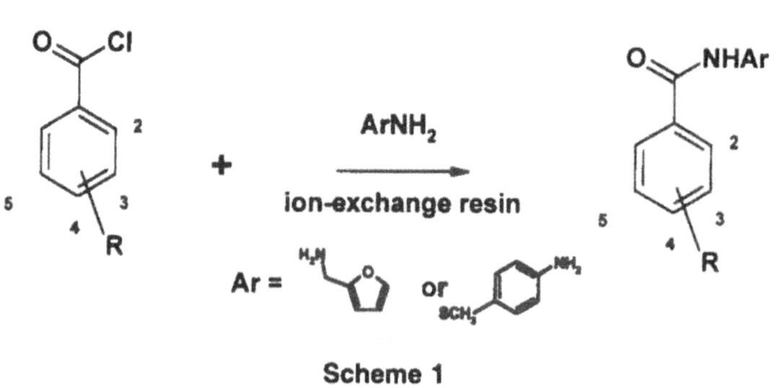

Scheme 1

concentration provided the final products. Different types of resins (Amberlyst, Amberlyte, and Dowex resins) that varied in their pH and composition, and a variety of solvents (CH_2Cl_2, EtOAc, THF, and DMF) were investigated. Listed below are several examples of the use of basic ion exchange resins for the preparation of focused libraries of compounds for lead optimization [2].

II. GENERAL PROCEDURE FOR SCHEME 1

To a glass centrifuge tube or related vessel was added Amberlyte IRA68 ion exchange resin (0.05 g), followed by the corresponding amine (4 or 5, 0.0475 mmol) and ethyl acetate (6 mL). Then a solution of the acid chloride (1, 2, or 3, 0.05 mmol) in 6 mL of ethyl acetate was added. The reaction was shaken for 18 h; then 100 μL of water was added and shaking continued for an additional 0.5 h. The mixture was filtered through a fritted funnel (1- or 2-mL fritted disk) and the solvent concentrated in vacuo. The material was analyzed by reversed-phase HPLC (isocratic, acetonitrile-water 65:35, 1% trifluoroacetic acid). The yield of the six compounds prepared ranged from 93% to 100% and were more than 99% pure.

In addition, the solvent could be decanted from the resin, the resin washed with additional ethyl acetate, and the solvent combined and concentrated.

The approach described can also be applied to the preparation of amides in a microtiter format, as shown in Scheme 2.

III. GENERAL PROCEDURE FOR SCHEME 2

To the appropriate number of wells of a 96-well 1-mL microtiter plate was added Amberlyst 21 ion exchange resins (5–10 beads) followed by 200 μL of ethyl acetate. Then, to each well containing resin was added 22.4 μmol of an amine (in 100 μL of ethyl acetate; in this particular example, anywhere from 40 to 80

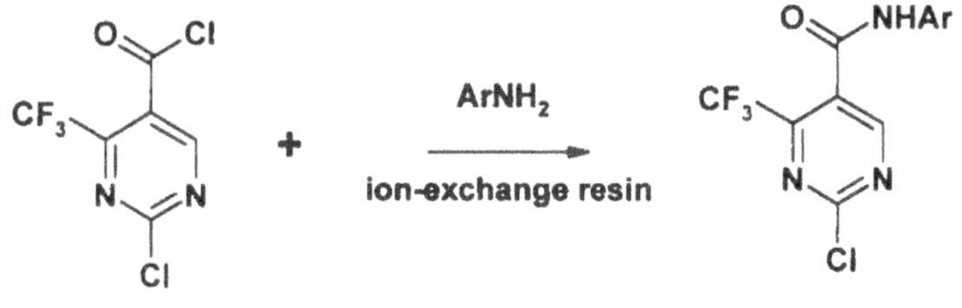

Scheme 2

amines were used) followed by a solution of 2-chloro-4-trifluoromethylpyrimidine acid chloride (24.6 μmol in 100 μl ethyl acetate). The plate was capped and placed in a benchtop sonicator for 0.25 h. The plate was removed and to each reaction well was added 30 mL of water using an 8-channel pipet and then the plate was sonicated for an additional 0.1 h. The resin was allowed to settle to the bottom of each well, and 280 μL of the solvent was removed and transferred to a tarred test tube (the tubes were prepared with BenchMate). The solvents were removed in vacuo, the tubes reweighed, and the products solvated to a concentration of 5 mg/mL. A percentage of each plate was analyzed by HPLC as described above.

IV. COMMENTARY

Historically, ion exchange resins have been used in a variety of chemistries and represent one of the first types of solid-phase reagents reported. Their use in the preparation of amides and related compounds as illustrated here can be quickly adapted to most organic laboratories with little or no additional capital investment. However, they can be used with automated synthesizers as well because they can be handled and filtered easily. The cost of the resins is also an advantage over more typical solid-phase resins currently in use. Recently, a large number of papers have appeared illustrating the use of both acidic and basic ion exchange resins in parallel library synthesis.

   One last point of interest is the apparent selectivity achievable with ion exchange resins. As shown in Scheme 2, a 2-chloropyrmidine acid chloride was selectively reacted with a variety of amines to form the amide. Using Amberlyst 21 resin, no substantial reaction at the 2-chloro position was observed. This was the case for a wide variety of pyrimidines. However, if Amberlyst 27 was used, the reactions proceeded but purity of the final products was much lower. Amberlyst 27 is slightly more basic than Amberlyst 21, and it is believed that this difference resulted in additional byproducts or reactions.

REFERENCES

1. Work performed at Signal Pharmaceuticals, San Diego, CA.
2. M.J. Suto, et al *Tetrahedron*, 1998, *54*, 4141–4150.

Chapter 6
Automating High-Throughput Organic Synthesis

Paul D. Hoeprich, Jr.*
Hewlett Packard/Agilent, Wilmington, Delaware

I. INTRODUCTION

The pharmaceutical research community has been facing acute pressures over the past few years to become more productive. Driven by the need for more efficient approaches to reduce ''time to market'' for new molecular entities (NMEs), research efforts have become focused on overall throughput and productivity in synthesis of new compounds, screening for desired biological activity, and information management of the research enterprise.

During this same time period, the success of molecular biology and genetics has generated large numbers of potential therapeutic targets. In many cases, these new targets have been or will be incorporated into automated high-throughput screening activities. The preparation of new and diverse drug molecules, however, has not kept pace with developments in molecular biology. To achieve parity, a new experimental chemical synthetic logic has emerged, known as *combinatorial chemistry*. The latter can be described as a set of synthetic tools for generating, simultaneously, a multitude of chemical compounds rapidly and efficiently. The net effect has been to increase greatly the range of molecular diversity available to the medicinal chemist.

Perhaps the greatest power of the combinatorial method is that the contributory elements have been to a large degree automated. Given that statement, it is

---
Current affiliation: Genicon Sciences, San Diego, California.

important to look at automation and what this term really means. Historically, automation refers to acting or operating with little or no external influence, manual intervention, or control. In the medical/scientific arena, the concept has been applied to rendering more palatable those tasks that are repetitive, well defined, and relatively simple, e.g., weighing, pipetting, simple measurements (pH, optical density (OD), etc.), physical manipulation (plate transfer, test tube movement, etc.), and the like. Extending these basic principles to chemical synthesis, the areas of bioploymer synthesis are hallmarks of success. Peptide and oligonucleotide synthesis automation has been successful because the steps are highly repetitive, i.e., one deals with a single class of "building block" and one is usually making either a peptide bond or a phosphodiester bond. The problem is considerably more complex when extending principles of automation to general organic synthesis. Now one must factor in unlimited building blocks, a vast array of bond formation chemistries, a vast assortment of chemical reagents, extremes in reaction conditions, intermediate isolation, characterization and purification, final product isolation, and so forth.

The goal becomes a challenge to design equipment that incorporates features most important to synthetic organic chemists. A set of basic criteria is listed that has been culled from my own experience and from others in the instrumentation field and those actually practicing science. What follows is that general list:

> Modular design for solution phase and solid-phase approaches
> Inert materials compatible with the wide range of reagents used in organic synthesis
> Inert atmosphere for storage and delivery of air-sensitive reagents
> Multiple reaction vessels with individual temperature control
> Gentle, effective agitation
> Simple-to-use yet powerful graphical user interface (GUI) operating software
> "Manual" intervention without compromising automated run

I think it is safe to say that there is not a single solution, i.e., one "box," that will affordably and adequately fill these needs. Nonetheless, in the early 1990s, several commercial efforts were launched to develop, manufacture, and sell "turn-key" or one-box automated synthesis products. At one point, 3–4 years ago, there were some 19 companies offering "automated" synthesis platforms for general organic synthesis; presently, there are half that number. The impetus for this flurry of activity was derived from the initial excitement outlined above describing the combinatorial approach to synthesis.

Conceptually, the idea of automating organic synthesis is not new. In 1976, Professor Clifford Leznoff wrote of his pioneering work describing solid-phase synthesis of insect pheromones that it "has the potential for being adapted to an automated procedure" [1]. At the root of automation of chemical synthesis is an enhanced appreciation and acceptance of the solid-phase synthesis principle as an enabling technology for drug discovery. The concept, originally conceived and demonstrated by R. B. Merrifield nearly 40 years ago [2,3], has endured and flourished within the realm of biopolymer synthesis, especially in the area of peptide synthesis. The principle itself is straightforward in scope and concept: an insoluble, inert support "bead" is rendered chemically active by specific functionalization (e.g., halogenation); a "linker" entity is appended to the support via the functional group, offering both a means for chemical elaboration and facilitating selective removal of final product; chemical synthesis ensues (sequential addition of synthetic elements and requisite facilitating agents), resulting in creation of a compound; and finally, the product compound is released into solution. All soluble reagents can be used in excess to drive reactions to completion and, ideally, furnish high yields. These excess reagents and byproducts are separated from the support-bound target molecule by simple repetitions of washing and filtration. Stepwise syntheses are further simplified and accelerated because all reactions are performed in the same reaction vessel, thereby avoiding losses associated with intermediate transfer, purification, etc. Finally and most importantly, all of these operations are amenable to automation.

Early on, it was realized that the repetitive steps of solid-phase peptide synthesis could be automated. Two years after his historic report [2] "Solid Phase Synthesis of a Tetrapeptide," Merrifield [4] and later Merrifield and John Stewart [5] reported the design and construction of a device that was capable of automating the stepwise assembly of protected amino acids on polystyrene beads to give peptides. Final deprotection and cleavage of the peptide from the solid support was accomplished independent of the instrument. Of note in these early accomplishments were the use of chemically inert materials, e.g., glass for the single reactor and "Teflon" (polytetrafluoroethylene, PTFE) tubing for delivery reagents and solvents through a rotary valve; all solutions and solvents were pumped through the system using a diaphragm pump. The ordered series of synthetic events was programmed using a "drum" programmer that was literally a drum containing a series of plastic pins that tripped switches activating appropriate solenoid valves. This fundamental instrument concept was commercialized shortly thereafter in 1972 by Beckman Instruments [6]. It is remarkable that several system elements described above coincided with the idealized list at the outset of this chapter, e.g., the use of inert materials, simple programming of events, and so forth.

Among the many challenges associated with automating more general organic synthetic chemistry that for so many years has been done with round-bottomed flasks, reflux condensers, dropping funnels, etc., is the requirement that chemists can comfortably transfer their experience to a new platform or series of platforms (modules). This argues that chemistry done in solution be accommodated as a high priority. But pressures for enhanced "productivity" utilizing solid-phase principles as well make practice of both approaches an added requirement.

A necessary initial step toward addressing these challenges is to devise an automated platform that facilitates adaptation of classical solution methods and allows practice of the same or similar chemistry to a nonhomogenous phase. This sort of flexibility provides a means by which "rehearsal" or optimization of chemistry can occur before ramping up to levels of "high throughput," e.g., producing libraries of compounds (>1000 members) in a combinatorial/parallel manner using either solution or solid phase approaches.

Given a proper, functional, and robust hardware platform(s) or modules, real flexibility in chemistry is made possible by system operating software. All instrument operations should be controlled by an operating software system run through a graphical user interface, most likely on a standalone PC. The requirements of the software are many:

Be user-friendly
Incorporate intuitive setup and use
Execute synthetic operations, e.g., reagent addition, resin washing, etc.
Allow flexible and varied synthesis setup
Archive executed procedures and runs
Import data from external reagent and compound databases

As such, as we shall see in the case studies below, a myriad of technologies have been developed to achieve high-throughput organic synthesis through automated and semiautomated approaches.

II. CASE STUDIES

Case Study 1

A 24-member pyrazole library was prepared on solid phase in a continuous three-step sequence. The initial reaction involved condensing a series of esters with an immobilized "keto" functionality (4-acteyl moiety from a resin-bound benzamide) generating a resin-bound 1,3-diketo species. Cyclization with a variety of substituted hydrazines following a protocol that ramped the reaction vessel temperature from −10°C to 90°C yielded a family of pyrazoles. These reactions required dry conditions (solvents), as well as transport of a strong organic base and reactive synthons (e.g., substituted hydrazines). The Solaris 530 organic synthesis system (PE Biosystem) combines high-speed liquid handling through eight liquid delivery tips with a reaction vessel module that holds up to 48 reactors under an inert environment that provides correct conditions for these activities.

Cleavage of the pyrazoles from the styrenic support was accomplished in a solution of trifluoroacetic acid (TFA) requiring a "sealed" system. The ability of the Solaris system to carry out a multistep reaction sequence in parallel with air-sensitive reagents and a temperature gradient is noteworthy. Unattended operation is enabled via a GUI-based operating system.

Case Study 2

This first of two contributions from Trega Biosciences presents a synthetic regimen that is largely manual in execution. In fact, where they chose to use "automation" was in fact merely the use of robotic liquid handling devices from Packard, Tomtec, and Spyder Instruments. Unlike the first case study, where reagents were loaded onto the instrument in a series of steps programmed into the operating software and the run initiated, the approach described here embodies significant "manual" manipulation. The result is indeed a high-throughput experience in terms of numbers of compounds produced but is marginally automated. Often, when one adds a robotic capability, e.g., simultaneous liquid pipetting, with a given process the tendency has been to assert that, the process is now "automated." As suggested at the outset of this chapter, automation has had greatest success when dealing with simple repetitive tasks. Given that, the use of the products from Packard, Tomtec, and Spyder in the preparation of 30,000 + isoquinolinones that facilitated reagent preparation, solvent removal, and solution transfers does suggest a practice that is semiautomated, i.e., a combination of manual procedure with those performed in an unattended manner. In terms of the chemistry, straightforward acylations, Schiff base formation, and anhydride condensation composed the synthetic operations. The noteworthy aspects of parallelism enabling the connotation of "high throughput" were the grouped "teabag" shaking (although not mentioned, the device used was probably a Parr hydrogenation shaker) and the gaseous HF treatment to cleave the quinolinones from their solid support.

Case Study 3

This paper describes a unique approach described as a "tilted centrifugation" (TC) for the purposes of this paragraph. Basically, this approach involves a centrifuge with 8 positions capable of holding a 96-well microtiter plate (~200 μl). Each well of the plate contains 5 mg resin added as a slurry in DMF. Rapid centrifugation simultaneously depletes each well of solvent and the resin remains behind. Interfacing the centrifuge with a liquid delivery system (Packard Multiprobe 104) and placing all operations under computer control, it is possible to link together several steps to enable preparation of tetrahydroisoquinolinones; the chemistry is the same as described in Case Study 2. The entire process can be carried out in an unattended manner and in a single run. Moreover, the tiresome and repetitive actions of reagent addition and washing have been "automated."

Case Study 4

This case study, like Case Studies 2 and 3, leverages a combination of manual and automated manipulation to achieve significant levels of throughput; these experiments were done entirely on solid phase utilizing Unisphere aminomethyl-ated polystyrene resin. Noteworthy in this approach are several unique technologies/techniques. High-throughput results were achieved by utilizing the "split-and-pool" method described 8 years ago by Lam et al. [9]. The variation in the IRORI method is that when the reaction sequence is complete, each reaction cartridge (holding 20 mg resin) has a unique compound; Lam's original split-and-pool approach provided one peptide per resin bead. The synthetic manipula-tions are carried out manually, i.e., reductive amination, amide bond formation, and Suzuki coupling, in a common container, e.g., round-bottomed flask or Teflon-capped bottle. The sorting of the "pooled" reaction cartridges or "Micro-Kans" from the common container is done by a gantry robotic device that reads a unique radio frequency–emitting tag (Rf tag) as determined by the Accutag system. Cleavage of biphenyl compounds from the solid support was done in parallel and deposited in deep-well microtiter plate. The overall process is con-trolled by the IRORI–Afferent Software package.

Case Study 5

This very creative study involves again a combination of several approaches, the net result of which is multifaceted and highly productive. The blend of solution chemistry at reduced temperature and inert conditions to generate a collection of methyl ketones followed by transfer to a new set of reaction vessels containing a "capture" resin effected the solution to solid-phase transition effortlessly. The closing chemistry step, Hurd-Mori reaction in the presence of $SOCl_2$, accom-plished cyclization with concomitant removal from the solid support. The instru-mentation, Argonaut Quest 210 SLN, enabled simultaneity, temperature control, and inert reaction environment and final product isolation. The reaction sequence was accomplished in parallel because of the instrumentation and the modularity designed and built in to the devices associated with the core synthesizer unit. Reaction of Grignard reagents with the Weinreb amide made from p-bromoben-zoic acid, in situ formation of methyl ketones, was done in solution at reduced temperature under an inert atmosphere. The reaction mixture was quenched with a sulfonic acid bound to polystyrene support, acidified with acetic acid, and then cannulated into an adjacent bank of reaction vessels each charged with PS-sulfo-nylhydrazide resin. Corresponding hydrazones were formed by the heating of each reaction vessel to 50°C for 4 h—effectively "capturing" only the desired or expected ketone from the previous step. After thoroughly washing the hydrazone resins, cyclization and release were achieved in the presence of $SOCl_2$ at 60°C for 5 h. The reaction mixtures were neutralized by filtration through the liquid–liquid extraction ChemElut Plus cartridges (diatomaceous earth pre-equilibrated with saturated Na_2CO_3) into scintillation vials. Concentration yielded pure 1,2,3-thiadiazoles.

Case Study 6

This work describes a multistep solution phase reaction sequence resulting in two classes of compounds—diamines and allylic amines. As in the above study, the starting material was a series of Weinreb amides. The latter were reacted with lithium aluminum hydride (LAH) or methyl Grignard to generate aldehydes or ketones, respectively. Reductive amination of the carbonyl compounds following by Boc removal afforded the diamine families. Using the Hewlett Packard 7686 solution phase synthesizer (no longer available), preparation of the carbonyl compounds was enabled by the syringe pump fluidics that permitted dropwise addition of a LAH/THF to a solution of aldehyde or MeMgBr/ether to Weinreb amide and allowed to react. The reaction mixture was diluted with ethyl acetate and 2 N HCl, mixed, and the phases allowed to separate. The upper layer (EtOAc) was removed using the syringe mechanism and transferred to a vial containing $NaSO_4$. The aqueous layer was extracted a second time with an additional aliquot of EtOAc; the combined organic phases were dried over $NaSO_4$, then aliquoted equally into two separate vials and concentrated under N_2 at 35°C. Reductive amination was carried out in methanolic solution with the appropriate amine for 7 h, followed by $NaCNBH_3$ for 2 h. Methanol was evaporated and the residue partitioned between EtOAc and H_2O, followed by drying and concentrating as described above. Wittig chemistry was performed on aldehydes by adding methyl(triphenylphosphranylidene) acetate in CH_3CN for 2 h at 80°C. The reaction mixture was collected and passed over a silica gel cartridge on board the instrument platform. Triphenylphosphine oxide and other reaction byproducts were not retained; the desired allylic amines were eluted with EtOAc/hexane into individual vials. Of note on this platform was the ability to do many manipulations common on the "bench," such as liquid–liquid extraction and workup protocols including "drying" over $NaSO_4$. These were easily accomplished through the operating system software. (Author's note: HP discontinued production of the 7686 synthesizer in late 1998 but still supports those units placed in the field.)

III. COMMERCIALLY AVAILABLE PRODUCTS

The most recent survey of users, symposia, and Web searches has led to Table 1, which identifies current vendors of automated synthesis equipment.

Table 1 Automated Synthesizers, Vendor List

Advanced Chem Tech	Louisville, KY	*www.peptide.com*
Argonaut Technologies	San Carlos, CA	*www.argotech.com*
Bohdan Automation	Vernon Hills, IL	*www.bohdaninc.com*
Charybdis Technologies	Carlsbad, CA	*www.charybtech.com*
ChemSpeed	Basel, CH	*www.chemspeed.com*
PE Biosystems	Foster City, CA	*www.pebio.com*
IRORI	San Diego, CA	*www.irori.com*
Rapp Polymere	Tubingen, Germany	www.rapp-polymere.com
Spyder Instruments	San Diego, CA	www.5z.com/spyder
Zinsser Analytic	Frankfurt, Germany	*www.zinsser-analytic*

A. Advanced ChemTech

This company is the oldest of the currently active vendors, beginning in earnest in the early 1980s with a singular focus on automation of solid-phase peptide chemistry. With the renewed interest in solid-phase approaches being applied to more general organic synthesis, their platforms were used initially to try reactions beyond amide bond formation. They presently offer a portfolio of products including reagents, resins, and instrumentation. In addition, they offer a handbook for automated synthesis that is available through their website (see Table 1).

B. Argonaut Technologies

Founded in 1994 with a focus on automating synthetic organic chemistry, both solid-phase and solution phase synthesis, this company offers the most complete and diverse line of synthesizer platforms. They have some seven different products for automated and semiautomated synthetic chemistry. A complete spectrum of applications is embodied in their product offerings, ranging from manual reaction setup to fully automated preparation of thousands of compounds. They also offer a selection of solid-phase resins for synthesis and solid-phase scavengers and reagents.

C. Bohdan Automation

Bohdan focuses on two areas: automated sample preparation and automated synthesis. The two areas have been integrated to provide a solution to high-throughput synthetic chemistry that readily enables solid- and solution phase chemistries, weighing and dispensing, reagent preparation, and compound dissolution. Their products are tied together and integrated by software of their own creation.

D. Charybdis

This southern California company builds synthesis technology around their reaction block system know as Calypso. Shaker technology integrated with a robotic platform (Cavro) makes up their automated synthesis capability. They have software that integrates synthetic activities with compound analysis, primarily LC and LC-MS, in a convenient and seamless manner.

E. Chemspeed

Chemspeed offers solutions to increase productivity in organic synthesis through commercialization of core technology developed at Roche in Basel, Switzerland. Their parallel synthesizer family is well designed and offers robustness that enables parallelization of nearly any synthetic reaction that can be carried out in solution. In addition, they offer an electronic lab journal capability to monitor/record laboratory events.

F. PE Biosystems (currently, Applied Biosystems)

The Solaris 530 organic synthesis system is a high-speed robotic system that leverages a proven liquid-handling platform (Tecan). The multiprobe (eight liquid handling tips) fluidic arm enables high-speed transfers and deliveries to reaction vessels housed in a proprietary manifold that maintains inertness and temperature control. One can address up to 48 reaction vessels. An off-line incubation/mixing module makes the robotic platform available for continued syntheses, i.e., reagent additions, etc., thereby enhancing overall throughput.

G. IRORI

This combinatorial chemistry approach is a collection of components that if utilized as an ensemble can provide multimilligram quantities of discrete compounds based on a variation of the split-and-pool method. At the heart of their method is a radiofrequency encoding technique to track/follow the progress of solid-phase synthetic chemistry. Their system, AccuTag-100, facilitates the generation of compounds. Split and pooling activities are monitored by Synthesis Manager software that controls AutoSort-10K System and cleavage of finished molecules from the solid phase is carried by AutoCleave-96 workstation. The reactors are known as "Kans" and come in two sizes depending on the scale of synthesis.

H. Rapp Polymere

Known primarily for their polyoxyethylene-polystyrene-grafted copolymers, this small German company has produced two products for high-throughput chemistry. The APOS 1200 system is a highly modularized system that allows parallel synthesis of individual compounds. Each reaction module can be controlled with respect to reaction time and temperature and can be interfaced with a Tecan 75 liquid-handling system. An independent module or reaction block, called Syrem, is also offered by the company.

I. Spyder Instruments

This company has developed a new method to automated solid-phase synthesis that is based on centrifugation of "tilted" microtiter plates the details of which are best described at their web site, given in Table 1. In brief, centrifugation is used to separate liquid and solid phases before, during, and after a reaction sequence, thus allowing the steps of a reaction sequence to be automated.

J. Zinsser Analytic

This Frankfurt, based company has built a modular-based system, Sophas M, that has great flexibility. Reaction volumes can range between 96 well plates to 25-mL reaction vessels. The operating software is based on Windows NT and can import data from any database.

IV. CONCLUSIONS

The flexible automated and semiautomated organic synthesis technology platforms described above are based on past experiences and approaches spanning nearly 40 years, beginning with the efforts of Merrifield. A fuller dimensionality of more general chemical synthesis is being realized by drawing on a legacy of man's desire to make chemicals, beginning with dyes in the Egyptian, Roman, and Phoenician era, to the alchemy of the Middle Ages and burst of activity in the nineteenth century, leading to modern organic synthesis, which began to take form in the early twentieth century.

In order to move to fast, high-throughput, and more inclusive synthetic organic chemistry, development of integrated technology platforms that feature instruments and techniques designed to follow the course of an organic reaction along with hardware that can prepare, store, add, mix, separate, and analyze chemical compounds is necessary. The result has been an impressive portfolio of unique instruments, many of which are described in the case studies.

I think that we are now entering a period of a *mature beginning* in what perhaps is best described as a genuine "paradigm shift" [7] toward automating synthetic organic chemistry. Of the 3 million or so known organic compounds, only 2000–3000 are available commercially in pure form. "If the chemist wants any one of the other three million, or an of the infinite number of possible compounds so far unknown, he is going to have to make it from something that is available—in other words, he must *synthesize* it" [8]. There will always be a need for synthetic organic chemistry.

The role of automation and high-throughput approaches will continue to advance one's ability to "synthesize it." By mediating control of environmental conditions (through fluidics and temperature regulation), understanding compatibility issues, and integrating information-handling capabilities and molecular design approaches, a chemist can now create ensembles of compounds reflecting maximum chemical and structural diversity for screening as potential drug candidates.

REFERENCES

1. Leznoff, C.C., and Fyles, T.M. 1976. *J. Chem. Soc. Chem. Commun.*, 251–252.
2. Merrifield, R.B. 1962. *Fed. Proc. 21*: 412.
3. Merrifield, R.B. 1963. *J. Am. Chem. Soc. 85*: 2149–2154.
4. Merrifield, R.B. 1965. Nature 207: 522–23 and 1965. *Science 150*: 178–185.
5. Merrifield, R.B., Stewart, J.M., and Jernberg, N. 1966. *Anal. Chem. 38*: 121–140.
6. J.W. Stewart, personal communication.
7. Kuhne, T. *The Structure of Scientific Revolutions*. University of Chicago Press, 1970.
8. Allinger, N. et al. *Organic chemistry*. Worth Publishers, New York, 1971.
9. Lam, K.S. et al. 1991. *Nature 354*: 82–84.

Case Study 6-1
Automated Robotic Parallel Synthesis of a Small-Molecule Library Using Multistep Reactions on Solid Support

Surendrakumar Chaturvedi,* Peter V. Fisher, Kenneth M. Otteson, and B. John Bergot†
Applied Biosystems, Foster City, California

ABSTRACT

Robotically based parallel solid-phase organic synthesis (SPOS) is attaining widespread acceptance for production of discrete molecules on the multimilligram scale. Herewith we describe the synthesis of a 24-member pyrazole library prepared via sequential Claisen condensation and hydrazine cyclization on support-loaded 4-acetylbenzoic acid using an automated robotic workstation.

MATERIALS

Chemicals

Esters

1. E1—methyl benzoate
2. E2—methyl 2-thiophenecarboxylate
3. E3—methyl 4-bromobenzoate

---

* *Current affiliation*: (Corresponding author) Orchid Biosciences, Princeton, New Jersey.
† Retired.

4. E4—methyl 4-(trifluoromethyl)benzoate
5. E5—methyl 2-furanocarboxylate
6. E6—methyl 4-methoxybenzoate
7. E7—methyl 4-(diethylamino)benzoate
8. E8—methyl nicotinate

All esters were supplied by Aldrich Chemicals (Milwaukee, WI), with the exception of 2 and 7, which were obtained as the corresponding carboxylic acids thence methylated with diazomethane under usual conditions.

Hydrazines

1. N1—phenylhydrazine
2. N2—methylhydrazine
3. N3—2-hydrazinopyridine (hydrochloride salt)

Hydrazines (Aldrich Chemicals) were used as received with the exception of 3, which was liberated from its bishydrochloride salt by neutralization with sodium hydroxide solution (see below).

Resin

N-Fmoc-protected Rink amide [4-(2′,4′-dimethoxyaminomethyl)phenoxy] DVB-polystyrene support was obtained from Applied Biosystems and used as supplied.

Miscellany

Potassium t-butoxide, 1,3-dimethyl-2-imidazolidinone (DMEU), hydroxybenzotriazole (HOBT), diisopropylcarbodiimide (DIC), 4-acetylbenzoic acid, acetic acid, and dimethylacetamide (DMA) were obtained from Aldrich Chemicals. N, N-Dimethylformamide (DMF) and dichloromethane (DCM) were from Burdick and Jackson (Muskegee, MI). The Fmoc deprotection reagent, 20% piperidine in DMF (v/v), and the cleavage reagent, trifluoroacetic acid (TFA), were obtained from Applied Biosystems.

Stock Solutions

1. 1 M potassium t-butoxide in DMEU: in a dry 250-mL flask equipped with a magnetic stirrer, rubber septum, and an argon gas inlet was placed potassium t-butoxide (9 g, 80 mmol) in dry DMEU (80 mL); the suspension was stirred until complete dissolution (~3 h) and used immediately thereafter.
2. 1 M hydrazine solutions: phenylhydrazine (4 mL, 40 mmol) and methylhydrazine (2.1 mL, 40 mmol) were diluted to 40 mL with dry DMA; 2-pyridylhydrazine dihydrochloride (5.8 g, 32 mmol) was neutralized with NaOH (2.5 M, 12 mL) and then diluted to 40 mL with DMA.

3. 30% aqueous acetic acid: 30 mL acetic acid was diluted to 100 mL with deionized water.

Preparation of 4-Acetylbenzoic Acid Amide Resin

Fmoc–amide resin was deprotected by stirring about 10 g support for 2 h in 100 mL 20% piperidine/DMF, washed copiously with DMF then DCM, and dried in vacuo overnight. The resultant amino resin (9 g, 0.65 mmol/g, 5.85 mmol) was suspended in 50 mL dry DMF for 10 min to swell support, then supplied with 50 mL "preactivated" 4-acetylbenzoic acid solution [4-acetylbenzoic acid (3.0 g, 17.6 mmol) and HOBT (2.6 g, 19.3 mmol). The product was dissolved in 50 mL DMF, then supplied with DIC (3.0 mL, 19.3 mmol) over a 0.5-h period with stirring. The reaction mixture was shaken for 24 h, filtered by suction, washed in succession with 3 × 100-mL portions of water, methanol, DMF, and DCM, then dried in vaccuo for 5 h. A Kaiser test showed virtually quantitative coupling to the resin.

EQUIPMENT

All synthetic manipulations, except as noted above, were carried out on the So-laris model 530 organic synthesizer manufactured by Applied Biosystems (Foster City, CA). HPLC analysis was performed using a system comprising the Series 200 quaternary pump, Series 200 autosampler, model 235C diode array UV detector, under TurboChrom control, all supplied by Perkin-Elmer (Norwalk, CT). Chromatography was conducted on 4.6 × 50mm Targa C18 columns from Higgins Analytical (Mt. View, CA) using a 0.1 M sodium acetate buffer (pH 4.5) and acetonitrile. Mass spectrometry was conducted on a PE-Sciex (Foster City, CA) API-100 mass spectrometer equipped with atmospheric pressure CI source using direct injection via a syringe pump supplied by Harvard Apparatus (St.-Laurent, Quebec, Canada), and sampling in the positive-ion mode. Evaporation of cleavage solutions was performed on a Speed-Vac Plus concentrator from Savant (Farmingham, NY).

I. INTRODUCTION

As part of the growing armamentarium of new technologies being applied to modern drug discovery processes [1,2], the use of automated devices for carrying out heretofore manual ("benchtop") synthetic organic chemical reactions has become an invaluable tool in the past 4–5 years [3]. In particular, the adaptation of classical organic reactions to a *solid-phase* (polymer-supported) methodology

[4] has greatly enhanced the ability of the medicinal chemist to produce broad arrays of compounds (typically characterized as "small molecules," molecular weights ≤500 amu) quickly with minimum postsynthetic workup. To illustrate the power of this technique, we have selected a target set of molecules—suggested by a recently published paper [5,6] on the solid-phase synthesis of substituted pyrazoles and isoxazoles—for this case study, using the Solaris model 530 organic synthesizer to conduct all synthetic manipulations. The process (Scheme 1) requires initial coupling of the 4-acetylbenzoic acid precursor to deblocked Rink Fmoc–amide resin off-line, followed by a Claisen-type condensation under basic conditions of a series of 8 esters; finally, the β-ketoester intermediates are treated under reflux conditions with 3 (unsymmetrical) substituted hydrazines to afford the 24 final products after acid cleavage from the support. Products are then analyzed by high-performance liquid chromatography (HPLC) and structures confirmed by mass spectrometry (MS). It is beyond the scope of this case study to exhaustively discuss detailed operation of the Solaris workstation; interested readers are referred to the manufacturer for more information.

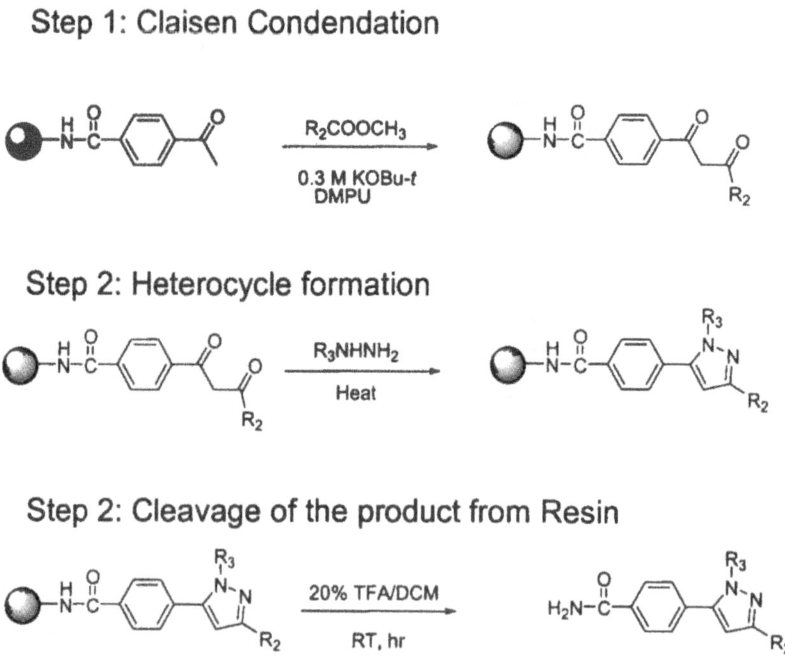

Step 1: Claisen Condendation

Step 2: Heterocycle formation

Step 2: Cleavage of the product from Resin

Scheme 1 Outline of organic reactions carried out on the automated synthesizer for the pyrazole synthetic process.

II. THE SYNTHESIZER INSTRUMENT

A. Brief Description of Solaris Model 530 Organic Synthesizer

Figure 1 and Table 1 depict and summarize, respectively, the overall instrument functionalities. Actual chemistries take place within reaction vessels (10 mL), attached to the reaction block using threaded neck fittings. Reagents and solutions are delivered into or removed from the block through machined pipetting or vacuum ports, respectively (a glass fritted filtration tube placed in each vessel allows vacuum tip entry and liquid removal while retaining the solid support). For retention of the *liquid phase*, the pipetting tips can access the filtration tube port and take up liquid by displacement of the syringe pump plungers, resulting in transfer to other containers on the deck. Figure 2 illustrates details of the reaction block and vessel assembly. The reaction block has provisions for supra-ambient reactions, where a coolant is passed through a "condensing zone" laid out within

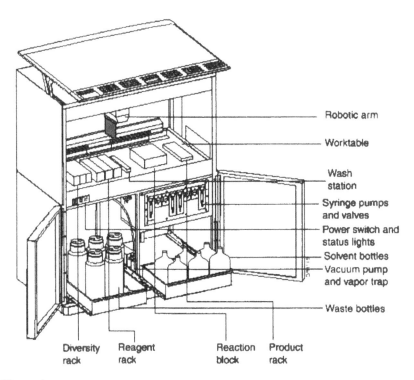

Figure 1   Schematic rendition of the Model 530 organic synthesizer showing key components and their location.

Table 1 Summary of Subsystems and Components Composing the Solaris Model 530 Organic Synthesizer

Status lights	Aspirating and coaxial tips
Racks	Syringe pumps
Reaction block	Distribution valves
Reaction block holders	Waste system
Work tables	Inert gas system
Robotic arm	Sensors
Vacuum pump and vapor trap assembly	Interlock system
Plumbing system	Optional offline incubation workstation

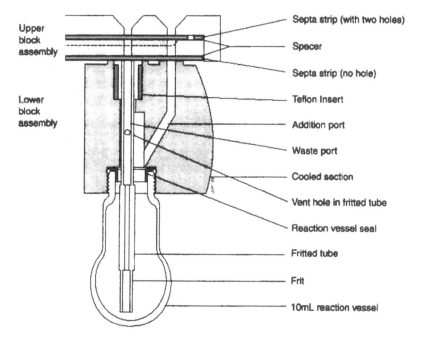

Figure 2 Cross-sectional drawing of a portion of the reaction block detailing reaction vessel attachment and components therein, as well as structural features of the block. Reagents and solvents are introduced by the pipetting tips through the *addition port*; liquid waste is removed by the vacuum tips engaging the *fritted tube* by way of the *waste port*; inert atmosphere is preserved by a double *septum strip* arrangement located between the upper and lower block assemblies; and vapor condensation during heating of the reaction vessels is provided by internal tubing (not shown) conveying chilled liquid within the *cooling zone*.

the block—simulating conventional refluxing—as flask contents are subjected to high-temperature conditions, as described below.

There is also an off-line incubation workstation available, consisting of computer-controlled heating/cooling/stirring capability for the reaction block when certain chemistry manipulations not requiring the robot are being performed. A solid-state module mounted atop a shaker platform can heat reaction vessel solutions with agitation, and a recirculating chiller can provide the necessary condensing temperatures; the same solid-state module can also supply cooling to flask contents through embedded cooling coils served by a separate refrigeration unit.

B. Solaris Software

The software for the 530 organic synthesizer is accessed by the chemist through a graphical user interface (GUI), which allows the user to specify instrument settings and preferences; write "scripts," which instruct the robotic arm assembly to perform requisite operations to carry out synthetic reactions; acquire a solvent/reagent volume usage profile; and operate the device in manual or automated mode. Other functions of the GUI allow the user to lay out work tables; specify system solvent type and location; monitor hardware and sensor operations; write subroutines, which are special command functions; and compose sample lists, which allow batchwise addition of reagents.

III. GENERAL PROCEDURE

A. The Synthesis Process

The flow diagram below (Fig. 3) illustrates the process for performing a synthesis on the 530 organic synthesizer. The process includes planning the synthesis, preparing the reagents and hardware components, and entering the corresponding settings into the software. Key blocks in this flow diagram are described below.

B. Reaction Scheme

Synthetic reactions conducted on the Solaris are recapitulated in Scheme 1 and reaction precursors are summarized in Fig. 4: preloaded resin bearing the 4-acetylbenzoyl starting substrate and distributed in 24 vessels is first reacted with each from a series of eight methyl esters (E1–8) under basic conditions to effect Claisen formation of 1,3-diketo intermediates; these, in turn, are treated with each from a series of three substituted hydrazines (N1–3) under elevated temperature

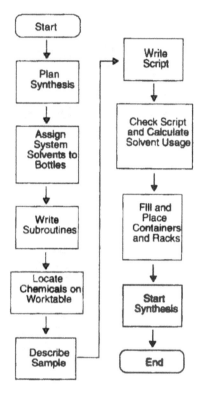

Figure 3   Flow diagram of a typical synthesis process when using the Solaris synthesizer as outlined in the 530 user's manual.

to produce the desired pyrazole heterocycles.* Finally, products are removed from the resin under acid cleavage conditions and transferred to their respective product vials.

It is important to note that, in general, all reactions to be conducted on the Solaris were first evaluated on the bench scale to obtain optimal conditions for the expected synthetic outcome. For example, a representative ester (methylbenzoate) was chosen for determining satisfactory Claisen condensation conditions; then a substituted hydrazine (phenylhydrazine) was used to obtain an effective process for condensative ring closure to the pyrazoles. When conditions are estab-

---

* The use of asymmetrically substituted hydrazines lead to the production of mixtures of N^1- and N^2-substituted N-aryl- and N-alkylpyrazoles; see Refs. 5 and 6 for additional comments.

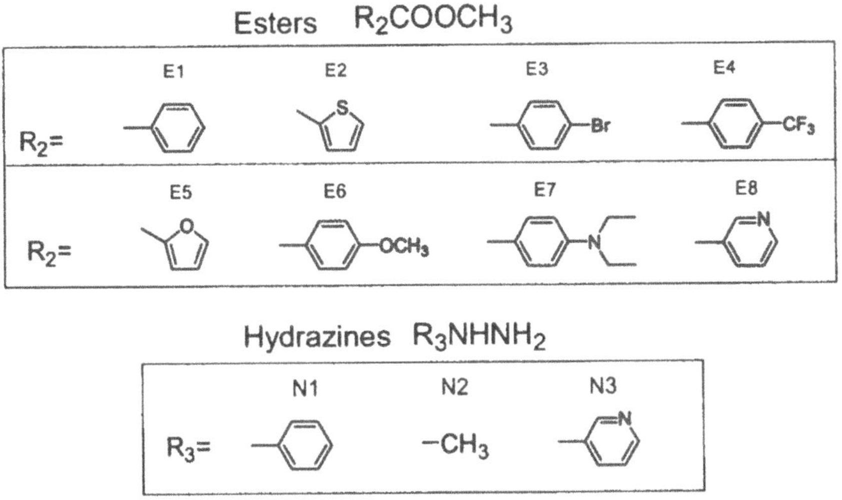

Esters R₂COOCH₃

Hydrazines R₃NHNH₂

Figure 4 Ester and hydrazine precursors ("diversity elements") used in the demonstration synthesis.

lished to the satisfaction of the chemist, these are then "translated" to an appropriate "script" for execution by the automated device (see below).

C. Preparation of Reagents

Table 2 summarizes the formulation of the synthesis precursors (diversity elements), solutions, and disposition of starting solid support to prepare the Solaris for a 24-vessel library matrix (8 × 3), laid out in the following manner:

Vessel no.	Ester	Hydrazine
1	E1	N1
2	E2	N1
3	E3	N1
4	E4	N1
5	E5	N1
6	E6	N1
7	E7	N1
8	E8	N1
9	E1	N2
10	E2	N2

continued on p. 179

Table 2 Formulations of All Starting Materials and Reagents Used in the Claisen Demonstration Pyrazole Synthesis, Including Container Types

Reagent	Wt/vol	Diluent	Container
Esters			
E1: methylbenzoate	2 mL	Neat	10-mL tube
E2: methyl-2-carboxythiophene	2 mL	Neat	10-mL tube
E3: methyl 4-bromobenzoate	2 mL	Neat	10-mL tube
E4: methyl 4-(trifluoromethyl)benzoate	2 mL	Neat	10-mL tube
E5: methyl 2-furanocarboxylate	2-mL	Neat	10-mL tube
E6: methyl 4-methoxybenzoate	2-mL	Neat	10-mL tube
E7: methyl 4-(diethylamino)-benzoate	2 mL	Neat	10-mL tube
E8: methylnicotinoate	2 mL	Neat	10-mL tube
Others			
Potassium *t*-butoxide	9 g	80 mL DMEU	4 × 60 mL bottle, 15 mL ea
1,3-Dimethyl-2-imidazolidinone (DMEU)	60 mL	Neat	4 × 60 mL bottle, 15 mL ea
30% Acetic acid	90 mL	DI water, to 300 mL	4 × 60 mL bottle, 30 mL ea
20% TFA	40 mL	Dichloromethane, to 200 mL	4 × 60 mL bottle, 40 mL ea
Hydrazines			
N1: Phenylhydrazine	4 mL	DMA, to 40 mL	4 × 10 mL tube, 10 mL ea
N2: Methylhydrazine	2.1 mL	DMA, to 40 mL	4 × 10 mL tube, 10 mL ea
N3: 2-Pyridinylhydrazine bis-HCl		12.5 mL 2.5 M NaOH, then to 40 mL w/DMA	4 × 10 mL tube, 10 mL ea
Resin			100 mg/reaction flask, preloaded w/4-acetylbenzoic acid, ~ 0.6 mM/g

Abbreviations are defined in the Materials section of this case study.

Vessel no.	Ester	Hydrazine
11	E3	N2
12	E4	N2
13	E5	N2
14	E6	N2
15	E7	N2
16	E8	N2
17	E1	N3
18	E2	N3
19	E3	N3
20	E4	N3
21	E5	N3
22	E6	N3
23	E7	N3
24	E8	N3

Each vessel ("well") will have a unique compound assigned to it as a consequence of the manner in which individual components are distributed. Tubes and bottles containing synthesis reagents and precursors are then placed on the deck in supplied racks. (*Note*: the potassium *t*-butoxide solution must be used within 2–3 hr. of preparation, as a fine powder is thrown down on longer standing, causing fouling of the pipetting tips.) Reaction vessels are charged individually with the starting resin by placing tared vessels on a microbalance and transferring resin with a (plastic) spatula to the specified weight.

D. The Script

After the synthetic process is defined (see above), reagent formulations prepared, the work table laid out, necessary subroutines written, the chemist then composes the run script, which entrains the entire sequence of events controlling the synthesis that is to be executed on board. It is noteworthy to point out that a large proportion of operations consist of resin washings, which again illustrate both the utility of conducting organic synthesis on solid supports and the power of a robot to execute repetitive tasks over several hours or, indeed, overnight. "Pauses" are placed in the script to prompt the chemist that an off-line event is necessary or for inspection of the reaction vessels before continuing with the next phase of the synthesis. Note that the "Cleavage and Collection" script module uses a *Weave* function, which directs the instrument to execute two or more commands sequentially (in this case *Transfer* and *Clean*) in groups of four vessels for most effective use of the four pairs of tips. For optimal productivity, reagents common to all flask positions in the reaction block should be apportioned among four containers per reagent, as up to four pipetor tips can enter bottles and be filled simultaneously, greatly decreasing robot arm/tip travel when compared to filling four tips from one bottle one tip at a time. It also is necessary to include a tip cleaning step between different reagent additions (or product transfers) to minimize cross-contamination. For best results, scripting of reactions should be guided by previous manual "rehearsals," where by reagent type and quantity used, solubilities, length of reaction time, temperatures required, and the like are established. Of course, one can as well use the Solaris for "process development," where by variables are built into the script according to the particular facet of a reaction under investigation. However, the example used in this case study for automated synthesis was first worked out at the bench, then translated to the Solaris. After scripting is completed, it is strongly suggested that the file be run in *Simulation*, where the chemist can match the Logfile output (not shown) with the intended operation, to check for any entry errors, command omissions, verification of prospective volume usages, etc. This is a very useful feature, particularly when one's synthetic script requires extensive washes, for which it is essential that adequate solvent volumes be supplied before launching the script in "real time."

E. Executing the Synthesis

After all reagents and solutions have been placed in their respective rack positions on the deck (work table), the script is initiated by double-clicking on the run icon on the toolbar. Following addition of the hydrazines, the reaction block is removed from the deck and placed into the off-line incubation workstation for completion of the cyclization to the desired heterocyclic pyrazoles. First, the recirculating chiller temperature is set at $-10°C$, and chiller hoses are connected to the reaction block; the chiller is then run for 1 h until the desired temperature is attained. Next, the heater is programmed to reach 90°C over 0.5 h, then held at the set point for 14 h. Finally, the shaker is programmed to stir at 150 rpm for the length of the incubation, and the device is started. After incubation is completed, the reaction block is removed from the incubation workstation, flask contents allowed to cool to ambient temperature, and the block is restored to its position on the deck for completion of the synthesis script.

F. Sample Workup

After the synthesis products have been transferred from the reaction vessels to the product rack, tubes containing product in the cleavage solution are transferred to the Speed-Vac, evaporated to dryness, then reconstituted to 5 mL in acetonitrile for subsequent HPLC and MS analysis.

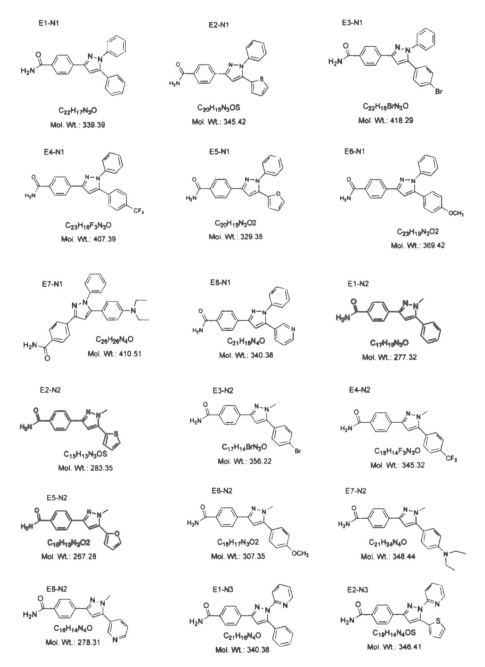

Figure 5 Structures, molecular formulas, and molecular weights of the 24 expected pyrazoles from the solid-phase demonstration synthesis. The alphanumeric designation for compounds refers to both the ester (E) and hydrazine (N) precursors used for each respective synthetic elaboration (see key in Figure 5).

E3-N3

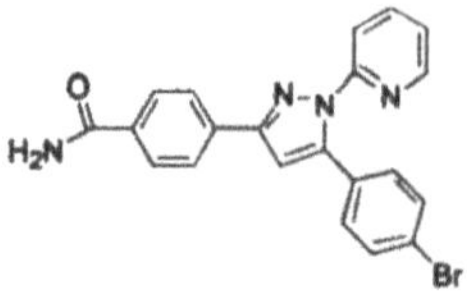

C$_{21}$H$_{15}$BrN$_4$O
Mol. Wt.: 419.27

E4-N3

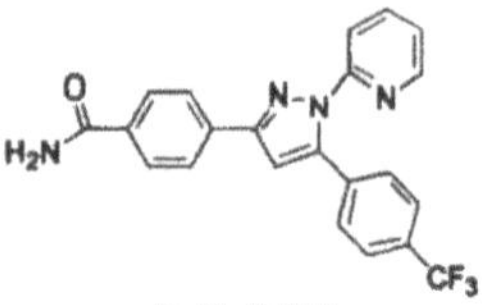

C$_{22}$H$_{15}$F$_3$N$_4$O
Mol. Wt.: 408.38

E5-N3

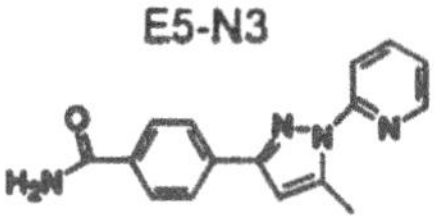

C$_{16}$H$_{14}$N$_4$O
Mol. Wt.: 278.31

E6-N3

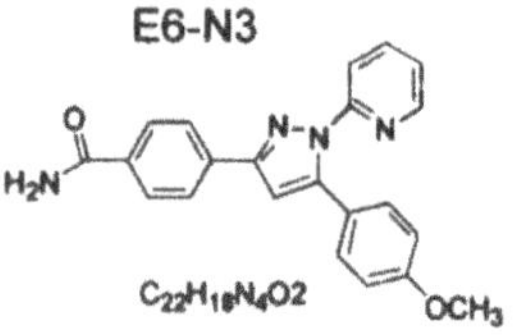

C$_{22}$H$_{18}$N$_4$O2
Mol. Wt.: 370.40

E7-N3

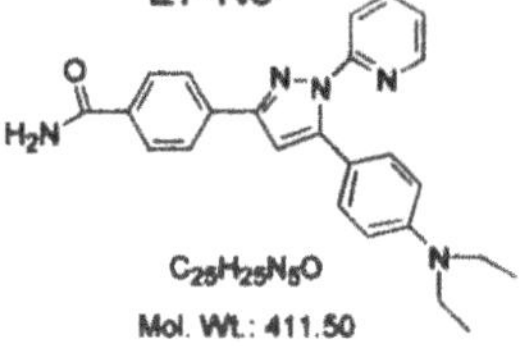

C$_{25}$H$_{25}$N$_5$O
Mol. Wt.: 411.50

E8-N3

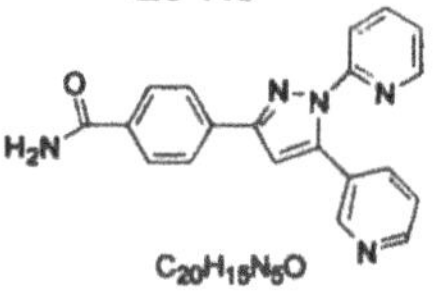

C$_{20}$H$_{15}$N$_5$O
Mol. Wt.: 341.37

Figure 5 Continued

IV. RESULTS

Expected products, their structures, molecular formulas, and theoretical molecular weights are shown in Fig. 5. Table 3 summarizes the relevant analytical data for vessels 1–24. Both regioisomers of those pyrazoles where chromatographic resolution was attained were combined to arrive at an aggregate purity of products. In those cases of demonstrably lower purity (entries 2–7), MS showed the presence of the uncyclized ketohydrazone as the principal byproduct. It is not obvious why this particular series of coupled hydrazines cyclized incompletely, as there seemingly is very little difference in steric bulk of this substituted hydrazine when compared to entries 17–24, using a hydrazine of comparable spatial aspect.

Figure 6 shows a typical chromatogram (A) of a regioisomeric pair (entry

Table 3 Summary of Analytical Data for Reaction Product Entries 1–24[a]

Entry	Mol. formula	HPLC purity (%)	Mol wt, theoretical	Mol wt, experimental
1 (E1-N1)	$C_{22}H_{17}N_3O$	94.9	339.14	339.39
2 (E2-N1)	$C_{20}H_{15}N_3OS$	41.0[b]	345.09	345.42
3 (E3-N1)	$C_{22}H_{16}BrN_3O$	41.9[c]	417.07	417.29
4 (E4-N1)	$C_{23}H_{16}F_3N_3O$	55.2[d]	407.12	407.39
5 (E5-N1)	$C_{20}H_{15}N_3O_2$	47.3[e]	329.12	329.35
6 (E6-N1)	$C_{23}H_{19}N_3O_2$	58.0[f]	369.15	369.42
7 (E7-N1)	$C_{26}H_{26}N_4O$	31.2[g]	410.21	410.51
8 (E8-N1)	$C_{21}H_{16}N_4O$	96.7	340.13	340.38
9 (E1-N2)	$C_{17}H_{15}N_3O$	95.9	277.12	277.32
10 (E2-N2)	$C_{15}H_{13}N_3OS$	91.8	283.08	283.35
11 (E3-N2)	$C_{17}H_{14}BrN_3O$	89.0	355.05	355.22
12 (E4-N2)	$C_{18}H_{14}F_3N_3O$	93.4	345.11	345.32
13 (E5-N2)	$C_{15}H_{13}N_3O_2$	80.1	267.10	267.38
14 (E6-N2)	$C_{18}H_{17}N_3O_2$	92.2	307.13	307.35
15 (E7-N2)	$C_{21}H_{24}N_4O$	52.8	348.20	348.44
16 (E8-N2)	$C_{16}H_{14}N_4O$	95.7	278.12	278.31
17 (E1-N3)	$C_{21}H_{16}N_4O$	95.3	340.13	340.38
18 (E2-N3)	$C_{19}H_{14}N_4OS$	96.4	346.09	346.41
19 (E3-N3)	$C_{21}H_{15}BrN_4O$	95.6	418.06	418.33
20 (E4-N3)	$C_{22}H_{15}F_3N_4O$	95.3	408.11	408.32
21 (E5-N3)	$C_{19}H_{14}N_4O_2$	92.9	330.11	330.35
22 (E6-N3)	$C_{22}H_{18}N_4O_2$	89.1	370.14	370.30
23 (E7-N3)	$C_{25}H_{25}N_5O$	64.8	411.20	411.36
24 (E8-N3)	$C_{20}H_{15}N_5O$	97.6	341.12	341.35

[a] Purities are expressed as area percent, and molecular weights determined by MS are corrected for M + 1 molecular ions.
[b] Ketohydrazone = 28.7%.
[c] Ketohydrazone = 49.4%.
[d] Ketohydrazone = 37.6%.
[e] Ketohydrazone = 21.3%.
[f] Ketohydrazone = 20.3%.
[g] Ketohydrazone = 36.9%.

12) and (B) a mixture of regioisomeric pyrazoles and their respective 3-ketohydrazone intermediates (entry 4). Lower pyrazole purities of entries 7, 15, and 23 were ascribable to a less successful Claisen reaction of ester E7; retention resin samples from each Claisen reaction were cleaved and examined by HPLC, and the product resulting from ester E7 was significantly lower in purity (~65%) than the other analogs (80–90%, data not shown).

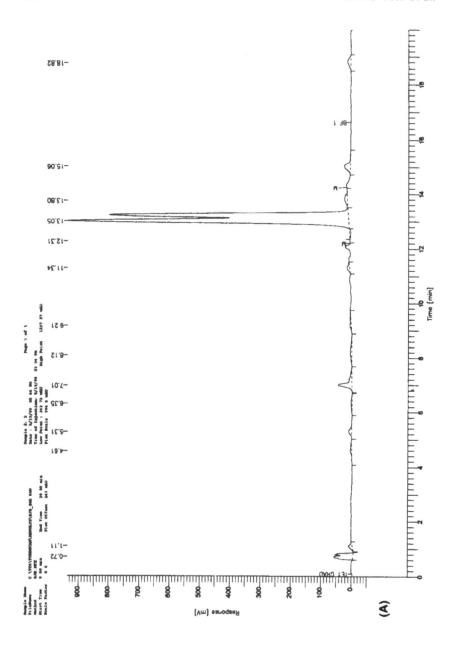

(A)

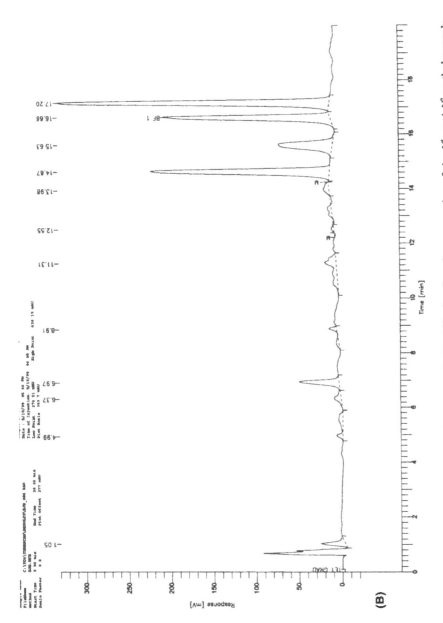

Figure 6 HPLC chromatograms of (A) pyrazole E4–N2, showing clean separation of the N^1- and N^2-methylpyrazole regioisomers; and (B) pyrazole E3–N1 regioisomers along with the faster eluting, incompletely cyclized 3-ketohydrazone intermediate pair.

Results obtained here and those published by Marzinik and Felder [5,6] were not strictly comparable, as the latter also included a C-alkylation step of the 1,3-diketo intermediates. In our hands, this reaction went poorly at the bench and was not included in the overall automated chemistry scheme. Also, for our synthesis, overall yields were not established, though only trace amounts of both the starting 4-acetylbenzoic amide and the intermediate diketo compounds were noted in the chromatograms, suggesting reasonably good conversion to final product (save for entries 7, 15, and 23, for reasons noted above).

V. CONCLUSION

Satisfactory execution of a multistep synthesis on the solid phase has been demonstrated using an automated robotic synthesizer under full computer control. Reaction conditions included the use of a labile strong base, protracted heating, and volatile organic acid cleavage. The data suggest some areas for reaction optimization, which can be accommodated both by manual (bench scale) techniques and by the Solaris automated synthesizer. For best results in translating solid-phase reactions described in the literature to an automated platform, performance of rigorous bench level rehearsals is recommended.

REFERENCES

1. P. Campbell, Ed. *Nature 384* (Suppl 6604): 1–26, 1996.
2. E.M. Gordon, M.A. Gallop, D.V. Patel. *Acc. Chem. Res. 29*: 144–154, 1996.
3. R. Baum, Ed. (Special Report) *Chem. Eng. News Feb. 12*: 28–73, 1996.
4. N.K. Terrett, M. Gardner, D.W. Gordon, R.J. Koblyecki, J. Steele. *Tetrahedron 51*: 8135–8173, 1995.
5. A.L. Marzinzik, E.R. Felder. *Tet. Lett. 37*: 1003–1006, 1996.
6. A.L. Marzinzik, E.R. Felder. *Molecules 2*: 17–30, 1997.
7. *User's Manual*, Solaris 530 Organic Synthesis System, PE Corp, 1998.

Case Study 6-2

Solid-Phase Parallel Synthesis of Large Tetrahydroisoquinolinone Library

Michal Lebl and Jaylynn C. Pires
Spyder Instruments, Inc., San Diego, California

Christine Burger, Yidong Ni, and Yuewu Chen
Trega Biosciences, Inc., San Diego, California

ABSTRACT

A library of 30,816 tetrahydroisoquinolinones was prepared using a combination of "tea-bag" synthesis and synthesis in microtiterplates with "surface suction" separation of solid and liquid phases.

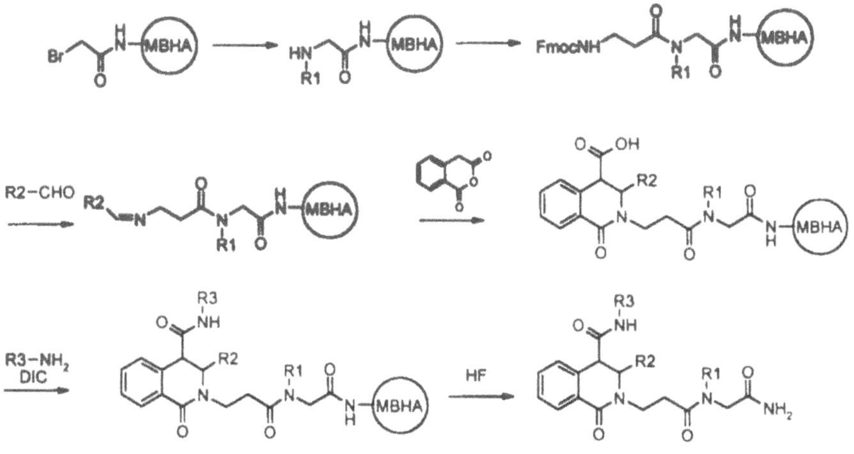

Scheme 1

Chemicals

1. Bromoacetic acid
2. β-Alanine
3. Benzaldehyde
4. Salicylaldehyde
5. 1-Naphthaldehyde
6. 2-Chloro-5-nitrobenzaldehyde
7. 2-Chloro-6-fluorobenzaldehyde
8. 2-Cyanobenzaldehyde
9. 2-Imidazolecarboxaldehyde
10. *o*-Anisaldehyde
11. 2-Naphthaldehyde
12. 2-Pyridinecarboxaldehyde
13. 2-Quinolinecarboxaldehyde
14. Piperonal
15. 3,5-bis(trifluoromethyl)benzaldehyde
16. 5-Nitrovanillin
17. 3-Methylbenzaldehyde
18. 3-Nitrobenzaldehyde
19. 3-Phenoxybenzaldehyde
20. 3-Thiophenecarboxaldehyde
21. 4-(3-Dimethylaminopropoxy)benzaldehyde
22. 4-(Dimethylamino)benzaldehyde
23. 4-(Methylthio)benzaldehyde
24. Trifluoro-*p*-tolualdehyde
25. 4-Biphenylcarboxaldehyde
26. 4-Bromobenzaldehyde
27. 4-Hydroxybenzaldehyde
28. *p*-Tolualdehyde
29. 4-Propoxybenzaldehyde
30. 6-Methyl-2-pyridinecarboxaldehyde
31. 3,4,5-Trimethoxybenzaldehyde
32. 4-Ethylbenzaldehyde
33. *N*-(2-Aminoethyl)pyrrolidine
34. Pyrrolidine
35. *N*,*N*-Dimethylethylenediamine
36. 2-Methoxyethylamine
37. Cyclohexylamine
38. 1-Methylpiperazine
39. Tetrahydrofurfurylamine

40. Benzylamine
41. 3-Methylbenzylamine
42. 1-(3-Aminopropyl)imidazole
43. Tyramine
44. 2-Chlorobenzylamine
45. 4-Chlorobenzylamine
46. 3-(Trifluoromethyl)benzylamine
47. Cyclopropylamine
48. Propylamine
49. 4-(Aminomethyl)pyridine
50. Allylamine
51. Morpholine
52. 3-(Aminomethyl)pyridine
53. 2-Thiophenemethylamine
54. 4-(2-Aminoethyl)morpholine
55. 4-(Ethyleminomethyl)pyridine
56. 3-Methoxybenzylamine
57. 2-(4-Methoxyphenyl)ethylamine
58. 2,3-Dimethoxybenzylamine
59. 2,4-Dichlorophenethylamine
60. N,N-Diethyl-1,3-propanediamine
61. 3-Dimethylaminopropylamine
62. 1-(2-Aminoethyl)piperidine
63. N,N,N',N'-Tetraethyldiethylenetriamine
64. Isoamylamine
65. Methylisobutylamine
66. 3-Ethoxypropylamine
67. Thiomorpholine
68. (R)-$(-)$-1-Cyclohexyethylamine
69. Isonipecotamide
70. N,N-Diethyl-N'-methylethylenediamine
71. 1,2,3,4-Tetrahydroisoquinoline
72. N,N,N'-Triethylethylenediamine
73. β-Alanine ethyl ester hydrochloride
74. 1-(3-Aminopropyl)-2-pipecoline
75. Ethyl-1-piperazinecarboxylate
76. 4-(Trifluoromethyl)benzylamine
77. 1-Benzylpiperazine
78. 1-(2-Furoyl)piperazine
79. L-Leucine methyl ester hydrochloride
80. 4-Bromopiperidine hydrobromide

81. 1-Bis(4-fluorophenyl)methylpiperazine
82. 4-(1-Pyrrolidinyl)piperidine
83. 3,3′-Dipicolylamine
84. Neopentylamine
85. *N,N*-Di-*N*-butylethylenediamine
86. Ethyl-3-aminobutyrate
87. 3-Butoxypropylamine
88. *N*-(3-Aminopropyl)morpholine
89. 4-*tert*-Butylcyclohexylamine
90. 4-Amino-2,2,6,6-tetramethylpiperidine
91. 2-Amino-5-diethylaminopentane
92. *N*-(2-Aminoethyl)-*N*-ethyl-*m*-toluidine
93. Dodecylamine
94. Acethydrazide
95. Methylhydrazinocarboxylate
96. 2-(2-Aminoethyl)pyridine
97. 2-(Aminomethyl)-1-ethylpyrrolidine
98. Piperazine
99. Diisopropylcarbodiimide
100. Bromophenol blue
101. HATU
102. *p*-Methylbenzhydrylamine resin
103. Trimethylorthoformate (TMOF)
104. Homophthalic anhydride
105. *N*-Hydroxybenzotriazole
106. *N,N*-Dimethylaminopyridine

Equipment and Supplies

1. Eppendorf Pipettman (1 mL)
2. Multichannel pipetor
3. GeneVac centrifugal evaporator
4. Yaskawa robotic platform
5. 96-channel distribution manifold
6. 96-channel suction manifold
7. Sealing device for sealing tea-bags
8. Platform shaker
9. Polypropylene bottles of various sizes
10. Automatic pipetor Multiprobe 104
11. 96-channel pipetor Quadra 96
12. Polypropylene HF chamber
13. HF apparatus

I. INTRODUCTION

As part of an ongoing drug discovery effort a procedure for the synthesis of large numbers of small heterocyclic molecules was needed. We have found the combination of manual synthesis of intermediates in "tea-bags" [1] followed by the synthesis in the wells of deep-well microtiterplates to be a very convenient method for preparation of large libraries. Hundreds of intermediates from tea-bags can be distributed into tens of thousands of wells which are processed on the robotic platform. The washing is achieved by "surface suction" [2]. However, it is not necessary to use the robot for the last step of the synthesis; manual washing is merely more tedious and time consuming. The cleavage from the solid support was accomplished by exposing the resin in the plates to gaseous hydrogen fluoride (HF) [3]. An example is the modification of the process presented earlier for the synthesis of mixture library [4].

II. GENERAL PROCEDURE

Solid support (p-methylbenzhydrylamine resin, 1.1 mmol/g, 130 μm, Chem-Impex, Wood Dale, IL) was distributed into polypropylene (mesh polypropylene, 100 mesh) bags (1.1 g) and sealed. Resin was swollen in dimethylformamide (DMF), and 780 bags were shaken for 20 min in 50% piperidine/DMF to neutralize the resin. The bags were then washed 4 times with DMF, once with 0.3 M HOBT/DMF, and then 4 more times with DMF (the last wash contained 0.01% bromophenol blue for monitoring the subsequent acylation step [5]).

A. Attachment of Bromoacetic Acid and Formation of N-Substituted Glycine

A solution of bromoacetic acid in DMF (1 M and 8.7 equivalents) was added to DIC (1.2 M) and the mixture was shaken in a polypropylene flask for 1–2 h. After disappearance of the blue coloration, the bags were washed 3 times with DMF and separated into groups of 30 bags. Each group was washed twice with dimethylsulfoxide (DMSO) and the solution of an amine (1 M) in DMSO was added. The bags were shaken for 18–24 h, washed once with DMSO, and then 5 times with DMF (last wash containing bromophenol blue).

B. Coupling of β-Alanine

A solution of Fmoc–β-alanine was added (0.3 M and 3 equivalents) to DIC (0.3 M, 3 equivalents), HOBT (0.3 M, 3 equivalents), and a 1/10 equivalent of

DMAP. After disappearance of the blue coloration in bags not containing tertiary amine groups (these bags remain blue), the bags were washed 3 times by DMF and 4 times with dichloromethane. The bags were then laid out in the hood to air-dry. When the resin was dry, a small sample of resin from each amine group was taken, cleaved with HF, and analyzed. Only the tea-bags that had the expected material in a purity of 90% or greater were carried through to the next step (Fmoc deprotection). To remove the Fmoc group from the β-alanine, a solution of 50% piperidine in DMF was added. After 20 min of incubation, the bags were washed with DMF 5 times and separated into groups of 24 (3 amines in the R1 position did not pass quality control, total number of processed bags dropped to 690).

C.   Schiff Base Formation and Cyclization

The appropriate aldehyde solutions (0.8 M in DMF) were combined with an equal volume of 1.6 M trimethylorthoformate (TMOF) in DMF and then were added to each group of 24 bags. After 3 h incubation, the liquid was removed and two washes with 0.2 M TMOF/DMF were performed. A solution of homophthalic anhydride (0.4 M in DMF and 5 ×) with DIEA (0.03 M) was added to each bottle and bags were shaken overnight. The liquid was removed and six washes with DMF were performed, followed by three washes with t-butyl methyl ether. At this stage the bags were dried again, and a sample from each bag was taken, placed into a well of a microtiterplate, and exposed to gaseous HF for 2 h. Samples were extracted and analyzed by LC-MS. Only the bags containing expected product in purity better than 85% (evaporative light-scattering detection) were taken to the next step of synthesis (48 bags were removed and 642 bags were processed further).

D.   Coupling of Amine

In this step the content of each bag was divided into 48 wells of a deep-well microtiter plate. A set of "master plates" of 48 amine solutions (1 M in DMF) was created in deep-well microtiter plates by automatic pipetting from stock solutions (Multiprobe 104, Packard Canberra, Meriden, CT). A solution of HATU (0.3 M in DMF) was added to all wells of each plate by multichannel pipetting (Quadra 96, Tomtec), and after 20 min incubation, the appropriate amine solutions were added to each well by pipetting from the master plates (the HATU solution remains in the wells during this reaction). After overnight incubation of closed microtiter plates on the shaker, the solution was removed by surface suction and incubation with HATU and amine solution was repeated overnight, once again. The solution was removed by surface suction (Fig. 1) and microtiter plates

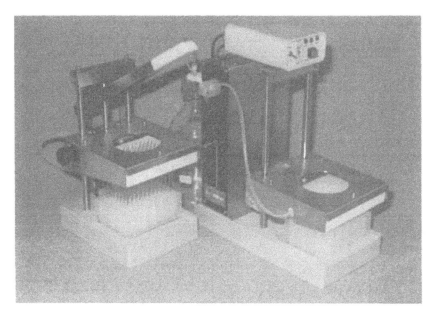

Figure 1 Manual 96-channel delivery (right) and suction (left) manifolds used for surface suction washing of solid phase distributed in microtiter plates.

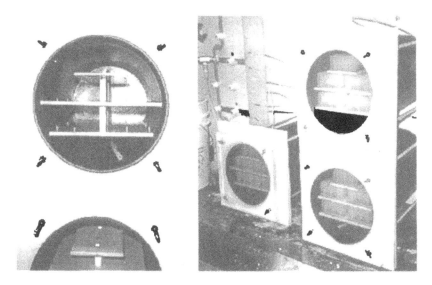

Figure 2 Polypropylene chambers used for cleavage of compounds from benzhydrylamine resin by exposure to gaseous hydrogen fluoride. The window is made of polymethacrylate covered with polypropylene foil.

Table 1 Purities of Tetrahydroisoquinolinones Synthesized by Combination of Tea-bag and Surface Suction Techniques[a]

Plate classification	Number of plates	MS (%)	HPLC (%)
QC passed	263	90.52	80.64
QC failed	58	48.63	78.49
Total	321	82.17	80.21

[a] MS, average percent of wells in the plate in which the correct molecular weight was identified; HPLC, average purity of 12 wells analyzed in each plate. QC = Quality Control.

were washed with DMF 10 times and with 50% *tert*-butyl methyl ether/DCM 4 times.

E. Cleavage from the Solid Support

The plates were air-dried for 2 days and placed in a polypropylene chambers (Fig. 2). The chamber was flushed with nitrogen for 30 min and then filled with gaseous hydrogen fluoride. After 2 h at room temperature, the chamber was flushed with nitrogen overnight, and plates were removed and placed in the desiccator. After overnight evacuation the plates were placed onto the table of the Multiprobe 208 (Packard Canberra) and solid support was extracted by repeated (4 times) addition and removal of 165 µL of acetic acid into the individual wells of microtiter plate. The extracts were transferred to deep-well polypropylene microtiter plates and evaporated in the Gene Vac, or lyophilized.

F. Postcleavage Analysis

All wells were analyzed by flow injection MS and one row of each plate was analyzed by HPLC with evaporative light-scattering detector (ELSD). The results are given in Table 1.

III. COMMENTARY

Separation of solid and liquid phases is the major problem in automation of solid-phase synthesis. We have found as a most convenient the use of "surface suction" method. This method does not require the use of any porous material for separation of liquid and solid phases. The manifold of 96 needles is slowly lowered against the surface of the liquid with continuous suction so that the liquid is shaved from the surface without disturbing the settled resin beads. Therefore, the needles can go very close to the layer of sedimented particles without removing them from the mixture. Obviously, this method requires that the solid phase sediment in the washing step. Even though it is convenient to perform this step robotically, it is feasible even for the manual synthesis as the suction and delivery manifolds are available commercially.*

REFERENCES

1. Houghten R.A. *Proc. Natl. Acad. Sci. USA* 1985, *82*, 5131.
2. Krchnák V, Weichsel A. S., Lebl M., Felder S. *Bioorg. Med. Chem. Lett.* 1997, 7, 1013.
3. Lebl M., Krchnák V. Techniques for massively parallel synthesis of small organic molecules. In *Innovation and Perspectives in Solid Phase Synthesis and Combinatorial Libraries*, Epton R., Ed.,; Mayflower Scientific Limited: Birmingham, 1999, p.43.
4. Griffith M.C., Dooley C.T., Houghten R.A., Kiely J.S. Solid-phase synthesis, characterization, and screening of a 43,000-compound tetrahydroisoquinoline combinatorial library. In *Molecular Diversity and Combinatorial Chemistry. Libraries and Drug Discovery*, Chaiken I.M., Janda K.D., Eds.; American Chemical Society: Washington, DC, 1996; p. 50.
5. Krchnák V., Vágner J., Safar P., Lebl M. *Collect. Czech. Chem. Commun.* 1988, *53*, 2542.

---

* Spyder Instruments, Inc. (*http://www.5z.com/spyder*), Torviq (*http://www.torviq.com*).

Case Study 6-3
Solid-Phase Parallel Synthesis of a Large Array of Tetrahydroisoquinolinones

Michal Lebl and Jaylynn C. Pires
Spyder Instruments, Inc., San Diego, California

ABSTRACT

An array of 768 tetrahydroisoquinolinones was prepared using a tilted-plate centrifugation technique. This technique allows for parallel processing of shallow-well microtiter plates used as a reaction vessel. Automated instrument simplifies the tedious task of manual distribution of building blocks and plate washing.

Chemicals

1. *N,N*-Dimethylethylenediamine
2. Benzylamine
3. 3-(Trifluoromethyl)benzylamine
4. Cyclopropylamine
5. Propylamine
6. Acethydrazide
7. Methyl hydrazinocarboxylate
8. Piperazine
9. Benzaldehyde
10. 1-Naphthaldehyde
11. o-Anisaldehyde
12. 2-Pyridinecarboxaldehyde

13. 3-Phenoxybenzaldehyde
14. 4-(3-Dimethylaminopropoxy) benzaldehyde
15. 4-(Methylthio)benzaldehyde
16. 4-Biphenylcarboxaldehyde
17. 4-Bromobenzaldehyde
18. 6-Methyl-2-pyridinecarboxaldehyde
19. 3,4,5-Trimethoxybenzaldehyde
20. 4-Ethylbenzaldehyde
21. 1-(2-Aminoethyl)pyrrolidine
22. Pyrrolidine
23. *N,N*-Dimethylethylenediamine
24. *N,N,N',N'*-Tetraethyldiethylenetriamine
25. *N,N*-Diethyl-*N'*-methylethylenediamine
26. *N,N,N'*-Triethylethylenediamine
27. 1-(3-Aminopropyl)-2-pipecoline
28. 4-(Trifluoromethyl)benzylamine
29. Diisopropylcarbodiimide
30. Bromophenol blue
31. *N*-[(Dimethylamino)-1H-1,2,3-triasol[4,5-b]pyridylmethylene]-*N*-methylmethanammonium hexafluorophosphate *N*-oxide (HATU)
32. *p*-Methylbenzhydrylamine resin
33. Trimethylorthoformate (TMOF)
34. Homophthalic anhydride
35. *N*-Hydroxybenzotriazole (HOBt)
36. *N,N*-Dimethylaminopyridine (DMAP)
37. *N,N*-Dimethylformamide (DMF)
38. Di-isopropylethylamine (DEA)
39. Di-isopropylcarbodiimide (DIC)
40. Dimethylsulfoxide (DMSO)

Equipment and Supplies

1. Intelligent centrifuge Compas 768.2
2. Eppendorf Pipettman (1 mL)
3. Multichannel pipetor
4. Gene Vac centrifugal evaporator or lyophilizer
5. Platform shaker
6. Automatic pipetor Multiprobe 104 and 208
7. Polypropylene HF chamber
8. HF apparatus

I. INTRODUCTION

After the identification of a hit from large "screening" library, the logical next step is the "hit explosion"—synthesis of a large number of compounds with structure similar to the original which were not included in the original screening library. We have developed the technique of tilted centrifugation [1], which allows synthesis of an array of up to 768 members in one batch.

The principle of tilted centrifugation is shown in Fig. 1. A resin-containing vessel is attached in the tilted position at the perimeter of the centrifugal plate and spun. Resin, which has sedimented at the bottom of the vessel, does not remain at the bottom of the flask. As the surface of liquid supernatant moves, the solid support layer moves as well. If the speed of rotation is increased, the centrifugal force created by rotation (which depends on the radius of rotation and the speed) combines with gravitation and the resulting force causes the liquid surface to stabilize at an angle perpendicular to the resulting force vector. At the ratio of relative centrifugal force (RCF) to G of 3, the angle of the liquid surface will be about 61 degrees. If the speed is increased so that the ratio of these forces is more than 50, the situation is close to the RCF of infinity; therefore, the liquid (and resin layer) angle will be close to 90 degrees. The pocket created by the tilt now allows only solid phase to remain in the pocket and all of the liquid is expelled.

We have built the computer-driven centrifuge with eight positions for tilted microtiter plates. A 96-channel distributor connected to six-port selector valve performs the delivery of washing solvents and common reagents. The centrifuge

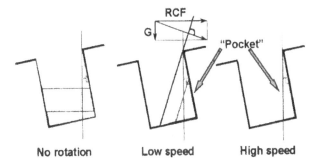

No rotation   Low speed   High speed

Figure 1 Principle of tilted-plate centrifugation. Formation of the pocket in the well of a tilted plate during centrifugation (direction: left to right). The solid support (lower layer) is collected in the pocket, while the liquid (upper layer) is expelled from the well. The liquid surface angle is perpendicular to the resulting force vector of the relative centrifugal force (RCF) and gravity (G).

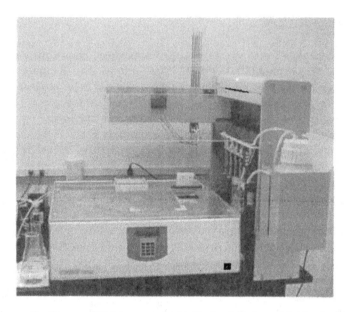

Figure 2   Compas 768.2 integrated with Packard Canberra Multiprobe 104.

was integrated with the Packard Multiprobe 104 liquid distribution system for the delivery of individual building blocks and reagents. Inclusion of the pipetting system allows us to perform the whole synthesis in a completely automatic regimen. Figure 2 shows the view of this instrument.

The synthesis is performed in the following way. A microtiter plate with a slurry of solid support distributed into it is placed on the perimeter of a rotor with a permanent tilt of 9 degrees. The rotor is rotated at the speed required for complete removal of the liquid portion of the well content. After stopping the rotation, the microtiter plate is placed (rotor is turned) under the multichannel (96-channel) liquid delivery head. The solvent selector valve is turned into the appropriate position and the washing solvent is delivered by actuating the syringe pump. This operation is repeated until all plates are serviced. The rotor is spun at the speed at which the liquid phase is just reaching the edge of the well, thus wetting all solid support in the "pocket," and after reaching this speed rotation is stopped. The cycle of slow rotation and stopping is repeated, thus gently mixing the slurry of solid support in the liquid phase. After shaking for the appropriate time, the plates are spun at high speed. The process of addition and removal of washing solvent is repeated for as many washes as are required. The plates are then consecutively placed under the array of 96 openings in the centrifuge cover, and appropriate building block solutions and coupling reagents are delivered by

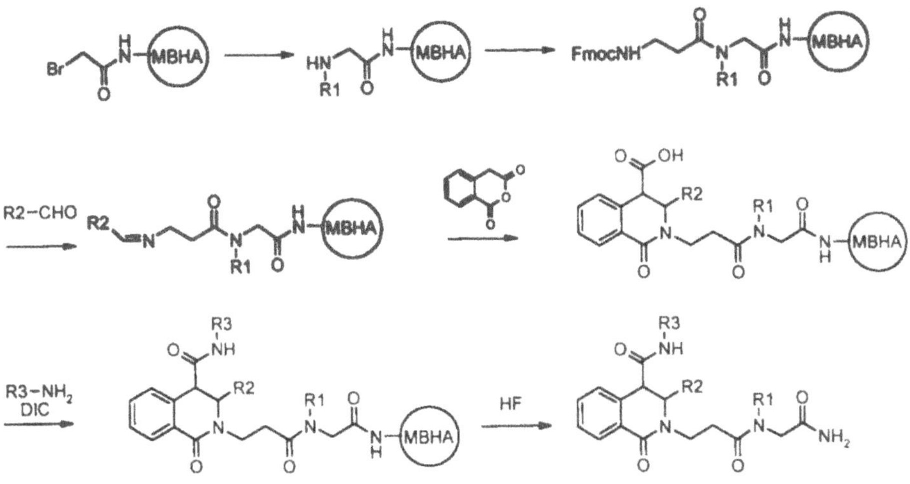

Scheme 1

pipetting (Multiprobe 104) through the openings from the stock solutions placed on the centrifuge cover. Alternatively, building blocks are delivered by manual pipetting with a multichannel pipetor from a trough or a prepared "master plate." This alternative is a faster option in the case where the number of building blocks used in the particular step is compatible with logical division of the microtiter plate into rows and columns (4,6,8,12), or when only one building block is distributed over the large part of the plate. When incubation at the elevated temperature is required, plates are removed from the centrifuge, stoppered with the cap mats, and incubated in the shaker oven. After the final wash and drying of the resin in the plate, cleavage can be performed by gaseous hydrogen fluoride (HF) as described in the communication about library synthesis [2].

As an example of the tilted-place centrifugation technique, we present here the explosion of the hit from large library synthesized by combination of tea-bag and surface suction techniques, presented earlier in this book [2]. The chemistry is illustrated in Scheme 1.

II. GENERAL PROCEDURE

The solid support (*p*-methylbenzhydrylamine resin, 1.1 mmol/g, 130 μm, Chem-Impex, Wood Dale, IL) was allowed to swell in DMF. The resin slurry was then distributed into the wells of eight polypropylene shallow-well microtiter plates (5 mg of resin per well). The microtiter plates were placed on the centrifugal rotor in a tilted position (9-degree tilt) and solvent was removed by centrifugation at 350 rpm in the Compas 768.2 (Spyder Instruments, San Diego, CA) intelligent centrifuge. The resin was neutralized with a 5% DIEA/DMF solution and washed 6 times with DMF (the last wash contained 0.01% bromophenol blue for monitoring the subsequent acylation step [3]).

A. Attachment of Bromoacetic Acid and Formation of N-Substituted Glycine

After six additional DMF washes, a solution of bromoacetic acid (1 M) and DIC (1.2 M) in DMF (100 µL/well) was added, and the mixture was oscillated for 2 h. After the disappearance of blue coloration, the plates were washed 6 times with DMF, 4 times with DMSO, and a solution of an amine (amines 1–8, 1 M) in DMSO was added. The plates were shaken overnight and then washed once with DMSO and 5 times with DMF (last wash containing bromophenol blue).

B. Coupling of β-Alanine

A solution of Fmoc–β-alanine in 0.3 M HOBT and DIC was added (100 µL, 0.3 M). After disappearance of the blue coloration in plate wells not containing tertiary amine groups (these wells remain blue), the plates were washed 4 times with DMF and a solution of 50% piperidine in DMF was added. After 20 min of incubation, the plates were washed 5 times with DMF.

C. Schiff Base Formation and Cyclization

The aldehyde solutions (0.8 M) were combined with an equal volume of 1.6 M TMOF/DMF and added to the appropriate wells of each plate (100 µL). After 3 h incubation, the liquid was removed and two washes with 0.2 M TMOF/DMF were performed. A solution of homophthalic anhydride (75 µL, 0.4 M in DMF) with DIEA (0.03 M) was added and plates were shaken overnight. The liquid was removed and six washes with DMF were performed.

D. Coupling of Amine

A solution of HATU (0.3 M in DMF, 75 µL) was added to each well. After 20 min incubation, a solution of an appropriate amine (amines 21–28, 1 M in DMF, 75 µL) was added. After overnight incubation of closed microtiter plates on the shaker, the solution was removed by centrifugation and the process of preincubation with HATU and incubation with amine solution was repeated and allowed to react overnight once again. The solution was removed by centrifugation and

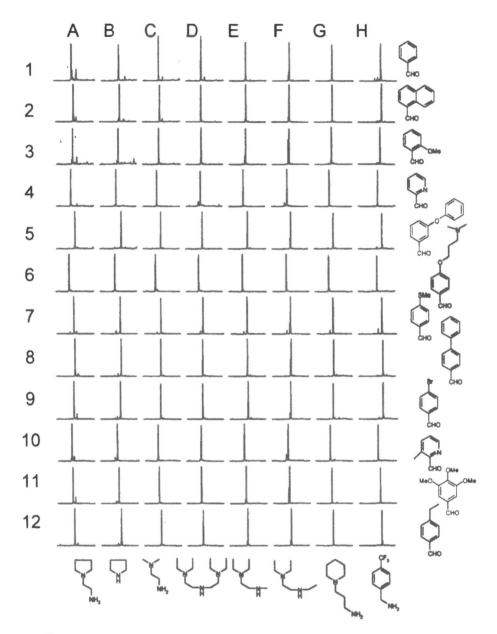

Figure 3 HPLCs of products from one plate from the synthesis of 768 tetrahydroisoquinolinones (R1 cyclopropylamine, R2 in rows, R3 in columns). Please note that products
are diastereoisomers—in some cases separated by HPLC (column F).

all microtiter plates were washed with DMF 12 times and with *tert*-butyl methyl ether 6 times.

E. Cleavage from the Solid Support

The plates were air-dried for 3 days and placed in a polypropylene chamber [4]. The chamber was flushed with nitrogen for 30 min and then filled with gaseous hydrogen fluoride. After 2 h at room temperature, chamber was flushed with nitrogen overnight and the plates were removed and placed in the desiccator. After overnight evacuation the plates were placed onto the table of the Multiprobe 208 (Packard Canberra, Meriden, CT) and the solid support was extracted by repeated (4 times) addition and removal of 150 µL of acetic acid into the individual wells of the microtiter plate. The extracts were transferred to deep-well polypropylene microtiter plates, the content was frozen, and acetic acid was removed in the lyophilizer.

F. Postcleavage Analysis

All wells were analyzed by LC-MS. The average purity of the prepared compounds in plates 1–6 was 87%. Plates 7 and 8, in which hydrazine derivative was used in the first step, did not produce acceptable products. The HPLC traces of the compounds synthesized in one plate (plate 4, R1 cyclopropylamine) are shown in Fig. 3.

III. COMMENTARY

We have described the new technique for very simple separation of solid and liquid phases in solid-phase synthesis that uses standard polypropylene microtiter plates as reaction vessels. This technique allows the building of very simple instruments in which all filtration problems are avoided and which can be completely automated. The obvious limitation of this technique is the requirement that the solid support must sediment in the given solvent system; however, solvents of higher density than the used solid support can always be diluted or evaporated prior to the application of tilted centrifugation. Our enthusiasm for the new technique is best illustrated by the fact that we quit our jobs at Trega Biosciences, where this technology was developed, and are now commercializing the new synthesizer in our company, Spyder Instruments, dedicated to supplying every chemist with his own Compas 768.2. Application of this technology for smaller scale, higher density (up to the level of synthesis on individual beads), or larger scale, lower density synthesis is obvious.

ACKNOWLEDGMENT

Part of this work (tilted-centrifugation technique development) was supported by SBIR NIH Grant IR43GM58981-01 to Spyder Instruments, Inc.

REFERENCES

1. Lebl M. *Bioorg. Med. Chem. Lett.* 1999, *9*, 1305.
2. Lebl M., Pires J., Burger C., Ni Y., Chem Y. in this book p. 187.
3. Krchnák V., Vágner J., Safar P., Lebl M. *Collect. Czech. Chem. Commun.* 1988, *53*, 2542.
4. Lebl M., Krchnák V. Techniques for massively parallel synthesis of small organic molecules. In *Innovation and Perspectives in Solid Phase Synthesis and Combinatorial Libraries*, Epton R., Ed.; Mayflower Scientific Limited: Birmingham, 1999, p. 43.

Case Study 6-4

High-Output Solid-Phase Synthesis of Discrete Compounds Using Directed Split-and-Pool Strategy

Rongshi Li and Xiao-Yi Xiao
ChemRx/IRORI, Discovery Partners International, San Diego, California

K. C. Nicolaou
Skaggs Institute for Chemical Biology, Scripps Research Institute, and University of California, San Diego, La Jolla, California

ABSTRACT

To demonstrate high-output solid-phase synthesis technology, the synthesis of a library of 96 discrete compounds using the IRORI AccuTag-100 combinatorial chemistry system is described. The library was synthesized by a sequence involving reductive amination, amide bond formation, and Suzuki coupling on Uni-Sphere aminomethylated polystyrene resin.

Chemicals

1. 4-(3,5-Dimethoxy-4-formylphenoxy)butyric acid (Bal-linker)
2. Diisopropylethylamine (DIEA)
3. Benzotriazol-1-yloxytrispyrrolidinophosphoniumhexafluorophosphate (PyBOP)
4. 1,2-Dichloroethane (DCE)

Professor Nicolaou is an advisor to Discovery Partners International, San Diego, California.

5. Sodium triacetoxyborohydride [NaBH(OAc)$_3$]
6. Glacial acetic acid (HOAc)
7. Potassium carbonate (K$_2$CO$_3$)
8. Tetrakis(triphenylphosphine)palladium (0) [Pd(PPh$_3$)$_4$]
9. 4-Methoxybenzylamine
10. Benzylamine
11. Cyclopentylamine
12. Cyclohexylamine
13. Allylamine
14. Tetrahydrofurfurylamine
15. 2-Bromophenylacetic acid
16. 3-Bromophenylacetic acid
17. 4-Bromophenylacetic acid
18. 4-Bromobenzoic acid
19. Phenylboronic acid
20. 4-Chlorophenylboronic acid
21. 4-Methoxyphenylboronic acid
22. 3-Nitrophenylboronic acid
23. Dichloromethane (DCM)
24. Tetrahydrofuran (THF)
25. Trifluoroacetic acid (TFA)

Equipment and Supplies

1. IRORI AccuTag-100 combinatorial chemistry system
2. IRORI MultiTag-96 tag dropper
3. IRORI dry-resin filler
4. IRORI MutiCap-96 capping station
5. IRORI AccuCleave-96 cleavage station
6. Platform shaker

I. INTRODUCTION

To demonstrate how to achieve high-output solid-phase synthesis using the split-and-pool strategy [1–4], a library of 96 discrete compounds (biphenyl amide, Scheme 1 and Fig. 1) was designed and synthesized by a total of 14 reactions using the IRORI AccuTag-100 combinatorial chemistry system. This system produced discrete compounds using the ''directed sorting'' split-and-pool technique [5] and radiofrequency (RF)–labeled MicroKans containing UniSphere amino-

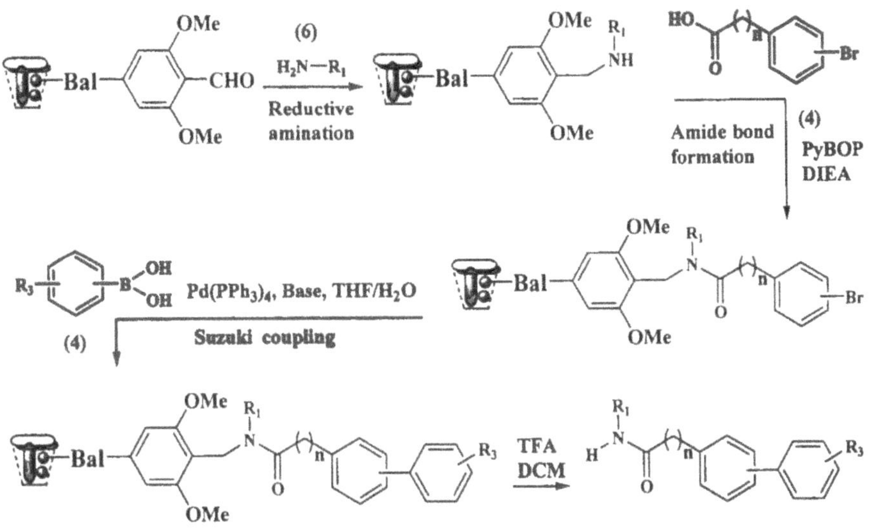

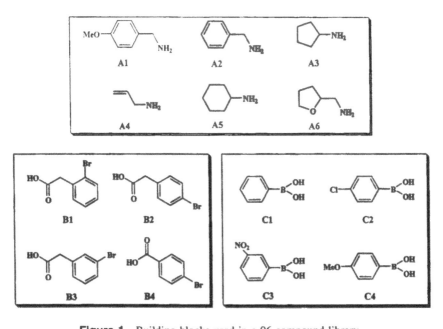

Scheme 1   Solid-phase synthesis of biphenyl amide library.

MeO——⟨ ⟩——CH₂—NH₂   ⟨ ⟩——CH₂—NH₂   ⟨cyclopentyl⟩—NH₂
       A1   A2   A3

CH₂=CH—CH₂—NH₂   ⟨cyclohexyl⟩—NH₂   ⟨tetrahydrofuran⟩—CH₂—NH₂
       A4   A5   A6

HO—⟨2-Br-phenyl⟩   HO—⟨4-Br-phenyl⟩
       B1   B2

HO—⟨3-Br-phenyl⟩   HO—⟨4-Br-benzoyl⟩
       B3   B4

⟨phenyl⟩—B(OH)₂   Cl—⟨phenyl⟩—B(OH)₂
       C1   C2

NO₂—⟨phenyl⟩—B(OH)₂   MeO—⟨phenyl⟩—B(OH)₂
       C3   C4

Figure 1   Building blocks used in a 96-compound library.

methylated polystyrene resin. A large library of discrete compounds (up to 10,000) can be produced using the IRORI AutoSort-10K microreactor sorting system to generate individual compounds in multimilligram quantities. At the end of the synthesis, 96 discrete compounds (up to 15 μmol per compound) were cleaved into 96-well plates directly using an AccuCleave-96 cleavage station. Combinatorial library product data, including structural information, plate/well locations, and synthetic history were archived using the IRORI Afferent Software Systems.

II. DEFINING BUILDING BLOCKS, SORTING AND ARCHIVING

Six amine building blocks (A1–A6) for reductive amination, four bromo-substitued carboxylic acids (B1–B4) for coupling, and four phenylboronic acids (C1–C4) for the Suzuki reaction step (Fig. 1) were defined using IRORI AccuTag-100 combinatorial chemistry system. This Synthesis Manager software records and assigns a code to each MicroKan in the first step and then directs the sorting of the MicroKans in subsequent steps. After each step, the Synthesis Manager reads each of the MicroKan RF tags and sorts them for the next reaction step. At the end of the synthesis, the Synthesis Manager archives the products in a format of either individual vials or 96-well.

III. BAL-LINKER ATTACHMENT

After 96 RF tags were loaded into MicroKans using a MultiTag-96 tag dropper, UniSphere aminomethylated polystyrene resin (20 mg/Kan, 1.10 μmol/g) was loaded using a dry-resin filler, then the MicroKans were capped simultaneously using a MutiCap-96 capping station. Under an argon atmosphere, a bottle containing 96 MicroKans was charged with DCM (1.5 mL/Kan), Bal-linker (3.0 eq), PyBOP (3.0 eq), and DIEA (6.0 eq). After shaking on a platform shaker for 24 h at room temperature, the MicroKans were washed with DCM (5 times) and dried under vacuum.

IV. REDUCTIVE AMINATION

Reductive amination of Bal-linker resin in 16 MicroKans was performed after sorting in a vial containing 4-methoxybenzylamine (A1, 10.0 eq), NaBH(OAc)$_3$ (5.0 eq), and HOAc (0.20 eq) in DCE (1.5 mL/Kan) at room temperature overnight. The MicroKans were washed with THF-H$_2$O (1:1) and THF (4 times) and dried under vacuum. The same procedure was followed for benzylamine (A2), cyclopentylamine (A3), allylamine (A4), cyclohexylamine (A5) and tetrahydrofurfurylamine (A6).

V. AMIDE BOND FORMATION

Coupling of 4-bromobenzoic acid (B4, 3.0 eq) to the corresponding resin-attached amine in 24 MicroKans was carried out after sorting in a vial containing PyBOP (3.0 eq) and DIEA (6.0 eq) in DCM (1.5 mL/Kan) at room temperature overnight. The MicroKans were washed with THF (3 times) and dried under vacuum. The same procedure was applied to 2-, 3- and 4-bromophenylacetic acids.

VI. SUZUKI COUPLING

The corresponding bromo-substituted resins in 24 sorted MicroKans were heated at 60°C for 2 days in a Teflon-capped bottle containing the corresponding boronic acid (25 eq), K_2CO_3 (50 eq), and $Pd(PPh_3)_4$ (0.20 eq) in THF-H_2O (4:1, 1.5 mL/Kan). The MicroKans were washed with THF-H_2O (4 times) and THF (4 times) and dried under vacuum.

VII. CLEAVAGE

The MicroKans were sorted into a MicroKan carrier, then placed into an Accu-Cleave-96 cleavage station. Ninety-six discrete compounds were obtained upon cleavage using 2 mL of 50% TFA in DCM for 2 h at room temperature.

VIII. PRODUCT CHARACTERIZATION

^{1}H-NMR (500 MHz, CD_2Cl_2) and electrospray mass spectrometry were used to characterize 6% of samples chosen randomly. Yields were determined using ^{1}H-NMR with an internal standard (TMS). Purity of the compound, consistent with ^{1}H-NMR results, ranged from 86% to 95% by HPLC (Table 1) [Column: C_{18}; mobile phase: A (H_2O), B (MeCN); gradient: 100–0% A and 0–100% B over 25 min; flow rate: 1.0 mL/min; detector: 254 nm] (Table 1).

Table 1 Sample Characterization Results

Compound	Structure	Yield (%)[a]	Purity (%)[b]
1	(structure, OMe)	36	95
2	(structure, NO$_2$)	45	90
3	(structure, OMe)	45	88
4	(structure, MeO)	55	92
5	(structure, OMe)	64	86
6	(structure, Cl)	45	94

[a] Overall yields were based on Unisphere aminomethylated polystyrene resin.
[b] For HPLC conditions, see text.

IX. CONCLUSION

The split-and-pool strategy on solid support is a very powerful synthetic tool for the generation of large numbers of discrete compounds. For instance, the synthesis of 10,000 discrete compounds requires only 40 reactions in a four-step sequence with 10 building blocks from each step, whereas it might require 40,000 reactions in a parallel synthesis (minimally 11,110 reactions).

ACKNOWLEDGMENT

The authors thank Dr. David Coffen for valuable discussions.

REFERENCES

1. X.-Y. Xiao, R. Li, H. Zhuang, B. Ewing, K. Karunaratne, J. Lillig, R. Brown, K.C. Nicolaou. *Biotechnol. Bioeng. (Comb. Chem.) 71*: 44, 2000.
2. K.C. Nicolaou, X.-Y. Xiao, Z. Parandosh, A. Senyei, M.P. Nova. *Angew. Chem. Int. Ed. Engl. 34*: 2289, 1995.
3. X.-Y. Xiao, M.P. Nova. In: S.R. Wilson, A.W. Czarnik, ed. *Combinatorial Chemistry: Synthesis and Applications*. New York, John Wiley & Sons, 1996, p. 135.
4. E.J. Moran, S. Sarshar, J.F. Cargill, M.M. Shahbaz, A. Lio, A.M.M. Mjalli, R.W. Armstrong. *J. Am. Chem. Soc. 117*: 10787, 1995.
5. X.-Y. Xiao, C. Zhao, H. Potash, M.P. Nova. *Angew. Chem. Int. Ed. Engl. 36*: 780, 1997.

Case Study 6-5
Multistep Synthesis of 1,2,3-Thiadiazoles Using "Catch and Release"

Yonghan Hu,* Sylvie Baudart,** and John A. Porco, Jr.†
Argonaut Technologies, San Carlos, California

ABSTRACT

1,2,3-Thiadiazoles were prepared using "resin capture," wherein ketones synthesized in solution were captured as resin-bound sulfonylhydrazones and released as 1,2,3-thiadiazoles with thionyl chloride (Hurd-Mori reaction). After cleavage, reaction mixtures were neutralized using liquid–liquid extraction cartridges to afford high-purity products.

Materials

Reagents required for the parallel "catch-and-release" synthesis of 1,2,3-thiadiazoles are outlined in Table 1.

Current affiliation:

*Genetics Institute, Cambridge, Massachusetts.

**Consultant, New York, New York.

† Department of Chemistry and Center for Streamlined Synthesis, Boston University, Boston, Massachusetts.

This chapter was adapted in part from Hu Y, Baudart S, Porco JA, Jr. *J Org Chem 64*: 1049–1051, 1999. © 1999 American Chemical Society.

Table 1 Materials Required

Material	Source	Property	Amount used
PS-TsNHNH$_2$ resin	Argonaut	2.4 mmol/g	1.0 g
MP-TsOH resin	Argonaut	1.45 mmol/g	10 g
N-Methoxy-N-methyl-p-bromobenzamide	Prepared (Lit.)^a	FW 244.09 d 1.434	1.08 mL
CH$_3$MgCl	Aldrich	3.0 M	465 μL
n-BuMgCl	Aldrich	2.0 M	695 μL
EtMgBr	Alfa Aesar	3.14 M	442 μL
iso-BuMgCl	Aldrich	2.0 M	695 μL
PhCH$_2$MgCl	Aldrich	2.0 M	695 μL
CH$_3$COOH	Fisher Scientific	FW 60.05 d 1.049	1.5 mL
SOCl$_2$	Aldrich	FW 118.97 d 1.631	3.5 mL

^{a} S Nahm, SM Weinreb. *Tetrahedron Lett* 22: 3815–3818, 1981.

Equipment and Supplies

1. Quest 210 SLN organic synthesizer (Argonaut Technologies) equipped with a lower Luer manifold
2. Bank-to-bank transfer cannulas (Argonaut Technologies)
3. Solid-phase extraction (SPE) rack (Argonaut Technologies)
4. Julabo F83-MW refrigerated recirculating chiller
5. ChemElut Plus 3-mL aqueous capacity (12-mL cartridge) Argonaut part. no. 900252 (pk/40).

I. INTRODUCTION

Many novel methodologies have been developed as a result of the increasing use of combinatorial solid-phase [1] and solution phase [2] synthesis to make compound libraries with potential biological and therapeutic significance. These include "catch-and-release" [3] and "resin capture" [4] strategies for the expedited workup and purification of compounds synthesized in solution. In this case study, we provide procedures for a catch-and-release synthesis of 1,2,3-thiadiazoles. Ketones are prepared in solution on bank A of the Quest 210 organic synthesizer, transferred to a sulfonylhydrazine resin in bank B to form resin-bound sulfonylhydrazones, and cleaved from the resin to afford 1,2,3-thiadiazoles (Scheme 1).

1,2,3-Thiadiazoles are an important class of biologically active compounds

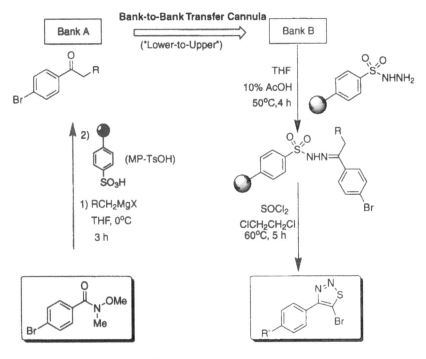

Scheme 1 1,2,3-Thiadiazoles prepared via "catch and release."

[5] as well as useful intermediates in organic synthesis [6]. For example, 4,5-bis-(4'-methoxyphenyl)-1,2,3-thiadiazole was found to be an active inhibitor of collagen-induced platelet aggregation in vitro [4a]. Many methods have been developed for the synthesis of 1,2,3-thiadiazoles [4d,4e], including the Hurd-Mori cyclization of α-methylene ketones employing *p*-toluenesulfonylhydrazone intermediates [7,8]. A gel-type polystyrene-sulfonylhydrazide resin (PS-Ts-NHNH₂), initially developed for carbonyl scavenging applications [9,10] was used to capture synthesized carbonyl compounds for 1,2,3-thiadiazole synthesis, in which case the Hurd-Mori cyclization constituted the cleavage step.

II. EXPERIMENTAL PROCEDURE

All parallel-synthesis transformations were performed on the Quest 210 organic synthesizer. A series of five Grignard reagents were reacted with a representative Weinreb amide in reaction vessels located in bank A. The tetrahedral intermediates generated were quenched with MP-TsOH resin to afford arylketones. The ketones were then transferred using a bank-to-bank transfer cannula to reaction vessels containing PS-TsNHNH$_2$ resin in bank B and captured to form polymer sulfonylhydrazones. After sulfonylhydrazone formation and Hurd-Mori cyclizative cleavage, excess thionyl chloride was neutralized in parallel utilizing Chem-Elut Plus liquid–liquid extraction cartridges [11–13] preloaded with saturated Na$_2$CO$_3$ and mounted on the Quest SPE rack. Final workup involved filtration and concentration of the products.

PS-TsNHNH$_2$ resin (200 mg, 2.4 mmol/g, 0.48 mmol) was added to 5-mL Teflon reaction vessels on bank A of the Quest 210. The reaction vessels containing the resin were then purged with nitrogen. On bank B of the Quest 210, N-methoxy-N-methyl-p-bromobenzamide (215 μL, 1.25 mmol) was added to five 5-mL Teflon reaction vessels with 3 mL dry THF. The reaction vessels on bank B were cooled to 0°C using a Julabo recirculating chiller. The appropriate Grignard reagents (1.38 mmol, 1.1 equivalents): CH$_3$MgCl (3.0 M, 465 μL), n-BuMgCl (2.0 M, 695 μL), EtMgBr (3.14 M, 442 μL), iso-BuMgCl (2.0 M, 695 μL), PhCH$_2$MgCl (2.0 M, 695 μL) were then added to the reaction vessels through the septum Luer plugs via syringe under nitrogen. Reaction mixtures were agitated at 0°C for 3 h.

While maintaining nitrogen gas flow, the Luer plugs were removed and to each reaction vessel was added 1 g (1.45 mmol/g) of MP-TsOH resin. The reaction mixtures were agitated for 10 min at 0°C, followed by addition of 0.3 mL of AcOH. Bank-to-bank transfer cannulas were attached to the Luer ports of reaction bank A and flushed with nitrogen. The cannulas were then attached to the male Luer fittings of the lower valve manifold of the adjacent reaction vessel positions on bank B. By toggling the reaction vessel drain valves of bank B to the open position, gas pressure was used to transfer the solution to the reaction vessels of bank A. The reaction vessels in bank A were then agitated at 50°C for 4 h under nitrogen. The vessels were cooled to room temperature, drained, and washed with THF (3x), hexane (2x), and dichloroethane (3x). To perform product cleavage, 2.3 mL of dichloroethane and 700 μL of SOCl$_2$ (9.6 mmol, 20 equivalents) were added to each reaction vessel and the reaction mixtures agitated for 5 h at 60°C.

Five liquid–liquid extraction cartridges (ChemElut Plus) [13] were mounted on the SPE rack. To each cartridge 2.5 mL of saturated Na$_2$CO$_3$ was added and the cartridges were allowed to soak for 10 min. The reaction mixtures (and three dichloroethane washes) were neutralized by filtration through the liquid–liquid extraction cartridges into scintillation vials. The solutions were concentrated to afford pure 1,2,3-thiadiazole products.

III. RESULTS AND DISCUSSION

The formation of support-bound sulfonylhydrazones from noncommercially available ketones was facilitated using resin capture wherein ketones synthesized in solution were captured as resin-bound sulfonylhydrazones (Scheme 1, Table 2). Five *p*-bromophenyl ketones were prepared in parallel on the Quest 210 organic synthesizer by reacting *N*-methoxy-*N*-methyl-*p*-bromobenzamide with a variety of Grignard reagents (THF, 0°C). The reaction mixtures were then quenched with a macroporous polystyrene–sulfonic acid resin (MP-TsOH) to decompose the tetrahedral intermediate [14]. Acetic acid (10% v/v) was added and the ketone solutions were directly transferred via cannula to reaction vessels containing PS-TsNHNH$_2$ resin. The sulfonylhydrazone formation was complete in 4 h at 50°C in the presence of acetic acid. After thionyl chloride cleavage (dichloroethane, 60°C, 5 h) and product purification (liquid–liquid extraction cartridges), 1,2,3-thiadiazoles were obtained in high chemical yield and purity. All products were characterized by GC [175°C for 3 min, ramped up to 300°C (20°C/min) for 5 min] and were found to have high purity (>94% GC area). The 1,2,3-thiadiazoles were isolated with chemical yields ranging from 59% to 98%. All compounds were characterized by ^{1}H- and ^{13}C-NMR (see spectroscopic data section). Bisaryl compounds similar to those shown in entry 5 are of great interest

Table 2 Thiadiazoles Prepared via Resin Capture of Ketones

Entry	Ketone	Thiadiazole	Yield (%)	GC purity (%)
1			98	100
2			82	94
3			77	97
4			59	97
5			67	98

because antithrombotic compounds have been found to bear aromatic substituents at both 4 and 5 positions of the 1,2,3-thiadiazole ring.

IV. SPECTROSCOPIC DATA

Gas chromatography, ^{1}H-NMR, ^{13}C-NMR, and MS (APCI) for 1,2,3-thiadiazole compounds synthesized are provided below:

Entry 1, 4-(4'-Bromophenyl)-1,2,3-thiadiazole: ^{1}H-NMR (300 MHz, CDCl$_3$): δ 8.65 (s, 1 H, =CH), 7.93 (d, 2 H, J = 8.7 Hz, Ar-H), 7.65 (d, 2 H, J = 8.7 Hz, Ar-H) ppm; ^{13}C-NMR (75 MHz, CDCl$_3$): δ 161.68, 132.24, 129.97, 129.63, 128.72, 123.50 ppm.

Entry 2, 4-(4'-Bromophenyl)-5-*n*-propyl-1,2,3-thiadiazole: ^{1}H-NMR (300 MHz, CDCl$_3$): δ 7.62 (m, 4 H, Ar-H), 3.02 (t, 2 H, J = 7.7 Hz, —CH$_2$—), 1.78 (m, 2 H, —CH$_2$—), 1.01 (t, 3 H, J = 7.4 Hz, —CH$_3$) ppm; ^{13}C-NMR (75 MHz, CDCl$_3$): δ 158.03, 153.12, 131.89, 130.34, 120.27, 123.00, 27.50, 24.95, 13.48 ppm; MS (APCI) showed [M +1]$^+$: 283.0 (calcd for C$_{11}$H$_{11}$N$_2$SBr: 282.1).

Entry 3, 4-(4'-Bromophenyl)-5-methyl-1,2,3-thiadiazole: ^{1}H-NMR (300 MHz, CDCl$_3$): δ 7.65 (m, 4 H, Ar-H), 2.71 (s, 3 H, —CH$_3$) ppm; ^{13}C-NMR (75 MHz, CDCl$_3$): δ 158.46, 146.55, 132.76, 131.91, 130.07, 123.02, 10.10 ppm.

Entry 4, 4-(4'-Bromophenyl)-5-isopropyl-1,2,3-thiadiazole: ^{1}H-NMR (300 MHz, CDCl$_3$): δ 7.65 (d, 2 H, J = 8.4 Hz, Ar-H), 7.56 (d, 2 H, J = 8.4 Hz, Ar-H), 3.51 (septet, 1 H, J = 6.6 Hz, —CH—), 1.39 (d, 6 H, J = 6.6 Hz, —(CH$_3$)$_2$) ppm; ^{13}C-NMR (75 MHz, CDCl$_3$): δ 161.39, 157.71, 131.92, 130.50, 130.34, 123.05, 26.85, 25.56 ppm.

Entry 5, 4-(4'-Bromophenyl)-5-phenyl-1,2,3-thiadiazole: ^{1}H-NMR (300 MHz, CDCl$_3$): δ 7.51 (m, 5 H, Ph-H), 7.44–7.33 (m, 4 H, Ar-H) ppm; ^{13}C-NMR (75 MHz, CDCl$_3$): δ 156.31, 151.07, 131.82, 131.60, 130.48, 129.83, 129.18, 129.13, 127.51, 123.18 ppm; MS (APCI) showed [M + 1]$^+$: 317.2 (calcd for C$_{14}$H$_9$N$_2$SBr: 316.2).

V. CONCLUSION

A multistep solution/solid-phase sequence for the synthesis of 1,2,3-thiadiazoles employing resin capture of ketones has been performed on the Quest 210 SLN organic synthesizer. Transfer of in situ–prepared ketones was accomplished using bank-to-bank transfer cannulas. Ketones were captured to the solid support as sulfonylhydrazones using PS-TsNHNH$_2$ resin. Cleavage of resin-bound sulfonylhydrazones with thionyl chloride afforded 1,2,3-thiadiazoles that were neutralized in parallel using ChemElut Plus cartridges.

REFERENCES

1. (a) M.A. Gallop, R.W. Barrett, W.J. Dower, S.P.A. Fodor, E.M. Gordon. *J. Med. Chem. 37*: 1233–1251, 1994. For a recent review, see (b) M.A. Thompson, J.A. Ellman. *Chem. Rev. 96*: 555–600, 1996.
2. (a) D.L. Coffen. Ed. Solution phase combinatorial chemistry, *Tetrahedron 54*: 1998. (b) S.W. Kaldor, M.W. Siegel. *Curr. Opin. Chem. Biol. 1*: 101–106, 1997.
3. For "catch and release" of amines, see: (a) M.G. Siegel, P.J. Hahn, B.A. Dressman, J.E. Fritz, J.R. Grunwell, S.W. Kaldor. *Tetrahedron Lett. 38*: 3357–3360, 1997. (b) A.J. Shuker, M.G. Siegel, D.P. Matthews, L.O. Weigel. *Tetrahedron Lett. 38*: 6149–6152, 1997. (c) Y. Liu, C. Zhao, D.E. Bergbreiter, D. Romo. *J. Org. Chem. 63*: 3471–3473, 1998.
4. For examples of "resin capture", see: (a) T.A. Keating, R.W. Armstrong. *J. Am. Chem. Soc. 118*: 2574–2583, 1996. (b) S.D. Brown, R.W. Armstrong. *J. Am. Chem. Soc. 118*: 6331–6332, 1996. (c) S.D. Brown, R.W. Armstrong. *J. Org. Chem. 62*: 7076–7077, 1997. (c) C. Chen, I.A. McDonald, B. Munoz, *Tetrahedron Lett. 39*: 217–220, 1998.
5. (a) E.W. Thomas, E.E. Nishizawa, D.C. Zimmermann, D.J. Williams. *J. Med. Chem. 28*: 442–446, 1985. (b) G.S. Lewis, P.H. Nelson. *J. Med. Chem. 22*: 1214–1218, 1979. (c) T.C. Britton, T.J. Lobl, C.G. Chidester. *J. Org. Chem. 49*: 4773–4780, 1984. For reviews on the chemistry of 1,2,3-thiadiazoles, see (d) E.W. Thomas. In: K.T. Potts, Vol. Ed. A.R. Katritzky, C.W. Rees, Series Eds. *Comprehensive Heterocyclic Chemistry*. London: Pergamon Press, 1984, Vol. 6, Part 4B, Chapter 4.24, p. 447. (e) E.W. Thomas. In: R.C. Storr, Vol. Ed. A.R. Katritzky, C.W. Rees, E.F.V. Scriven, Series Eds. *Comprehensive Heterocyclic Chemistry*. London: Pergamon Press, 1996, Vol. 4, Chapter 4.07, p. 289.
6. C. Rovira, J. Veciana, N. Santalo, J. Tarres, J. Cirujeda, E. Molins, J. Llorca, E. Espinosa. *J. Org. Chem. 59*: 3307–3313, 1994.
7. C.D. Hurd, R. Mori. *J. Am. Chem. Soc. 77*: 5359–5364, 1955.
8. (a) M. Fujita, T. Kobori, T. Hiyama, K. Kondo. *Heterocycles 36*: 33–36, 1993. (b) P. Stanetty, M. Kremslehner, M.J. Mullner. *Heterocyclic Chem. 33*: 1759–1763, 1996.
9. PS-TsNHNH$_2$ resin (1.8–2.5 mmol/g, 1% cross-linked polystyrene-*co*-divinylbenzene) is commercially available from Argonaut Technologies.
10. For reports on the preparation and use of sulfonylhydrazide resins, see (a) O. Galioglu, A. Akar. *Eur. Polym. J. 25*: 313–316, 1989. (b) D.W. Emerson, R.R. Emerson, S.C. Joshi, E.M. Sorensen, J.M. Turek. *J. Org. Chem. 44*: 4634–4639, 1979 (c) H. Kamogawa, A. Kanzawa, M. Kadoya, T. Naito, M. Nanasawa. *Bull. Chem. Soc. Jpn. 56*: 762–765, 1983.
11. Y. Hu, S. Baudart, J.A. Porco Jr. *J. Org. Chem. 64*: 1049–1051, 1999.
12. For examples of parallel workups employing liquid-liquid extraction cartridges, see

(a) C.R. Johnson, B. Zhang, P. Fantauzzi, M. Hocker, K.M. Yager. *Tetrahedron 54*: 4097–4106, 1998. (b) J.G. Breitenbucher, C.R. Johnson, M. Haight, J.C. Phelan. *Tetrahedron Lett. 39*: 1295–1298, 1998.

13. ChemElut Plus liquid–liquid extraction cartridges (3-mL aqueous capacity, 12-mL cartridge) were obtained from Argonaut Technologies (Part. No. 900252). The cartridges were preloaded with 2.5 mL saturated Na_2CO_3 for 10 min before use.

14. MP-TsOH resin (1.1–1.6 mmol/g, macroporous polystyrene-*co*-divinylbenzene) is commercially available from Argonaut Technologies.

Case Study 6-6
Automated Multistep Solution Phase Synthesis of Diamines and Allylic Amines

Fariba Aria
Kimia Corporation, Santa Clara, California

ABSTRACT

A series of Weinreb amides were transformed into aldehydes and ketones, which were then treated either with a series of amines to yield diamines (Scheme 1) or with triphenylphosphoranylidene to give allylic amines (Scheme 2).

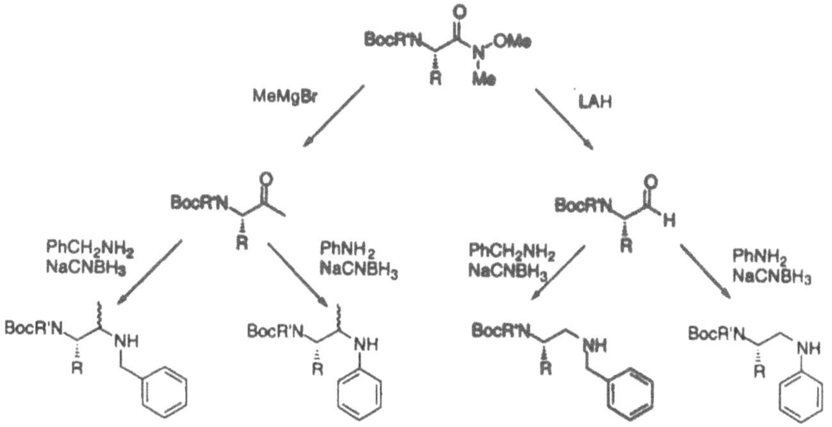

Scheme 1   Reductive amination.

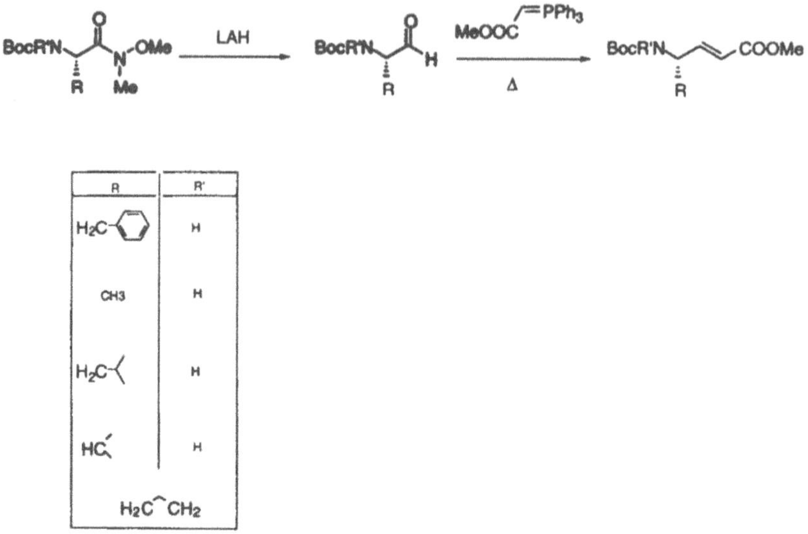

Scheme 2 Wittig reaction.

Chemicals

1. *N*-(*tert*-Butoxycarbonyl)L-alanine *N'*-methoxy-*N'*-methylamide
2. *N*-(*tert*-Butoxycarbonyl)L-valine *N'*-methoxy-*N'*-methylamide
3. *N*-(*tert*-Butoxycarbonyl)L-phenylalanine *N'*-methoxy-*N'*-methylamide*
4. *N*-(*tert*-Butoxycarbonyl)L-proline *N'*-methoxy-*N'*-methylamide
5. *N*-(*tert*-Butoxycarbonyl)L-leucine *N'*-methoxy-*N'*-methylamide
6. Lithium aluminum hydride 1 M in tetrahydrofuran
7. Methyl magnesium bromide 3 M solution in diethyl ether
8. Sodium cyanoborohydride
9. Methyl(triphenylphosphoranylidene)acetate
10. Aniline
11. Benzylamine
12. Tetrahydrofuran
13. Methanol
14. Water
15. Ethyl acetate
16. Acetonitrile
17. 1 N HCl
18. Sodium sulfate (anhydrous)

---

* This compound was prepared by coupling N (tBoc) Phe and *N*-methoxy-*N*-methylamine hydrochloride using isobutyl chloroformate and *N*-methylmorpholine.

Equipment

1. Hewlett Packard 7686 solution phase synthesizer
2. Computer equipped with ChemStation software
3. 2-mL vials
4. Syringes
5. Hewlett Packard silicagel cartridges

I. PROCEDURE

A. Synthesis of Diamines

The HP solution phase synthesizer was equipped with four bottles for solvents that were filled with water, 1 N HCl, ethyl acetate, and tetrahyrdofuran. The following were prepared manually:

Five vials of different Weinreb amides (0.4 mmol)
Two vials of 1 M solution of LiAlH₄ in tetrahydrofuran
Five vials of 3 M solution of MeMgBr in diethyl ether (200 μL)
Two vials of 2 M solution of NaBH₃CN in Methanol (1.2 mL)
Twenty vials of anhydrous sodium sulfate (1 mmol).
Two vials of a 1 M aniline solution in methanol (1.2 mL)
Two vials of a 1 M benzylamine solution in methanol (1.2 mL)

All of the vials were capped and placed at the designated place on the synthesizer's tray.

The rest of the procedure was done following an automated protocol. Tetrahydrofuran (500 μL) was dispensed in vials containing Weinreb amides and homogenized using the synthesizer's front mixer.

1. Preparation of Carbonyl Compounds

(a) Synthesis of Ketones. To each vial containing MeMgBr was added the previously made Weinreb amide solution (250 μL). Each reaction mixture was mixed, diluted with ethyl acetate (450 μL), and treated with 2 N HCl (450 μL) to obtain a clear liquid having two distinct layers. The upper layer was transferred into one of the vials containing the anhydrous sodium sulfate. The aqueous layer was then extracted with ethyl acetate (300 μL) and the extract added to the previous sodium sulfate vial. The combined dry solution was then transferred as two equal portions into two empty vials and concentrated at 35°C. After completion of this procedure a total of 10 vials containing 5 different ketones (5 ketones in duplicate vials) were produced.

(b) Synthesis of Aldehydes. To each of the five weinreb amide (250 μL) solutions was added the LiAlH₄ solution (120 μL) dropwise and then mixed. Each reaction mixture was diluted with ethyl acetate (450 μL) and treated with 2 N

HCl (450 µL) to obtain two distinct layers. The upper layer was transferred to a vial containing anhydrous sodium sulfate. The aqueous layer was extracted with ethyl acetate (300 µL) and the extract was added to the previous vial containing the sodium sulfate. The dry solution was transferred as two equal portions into two empty vials and concentrated at 35°C. After the procedure a total of 10 vials containing 5 different aldehydes were produced (5 aldehydes in duplicate vials).

2.   Reductive Amination

Methanol (200 µL) was dispensed into all 20 vials containing the ketones and aldehydes and mixed thoroughly. To one set of the duplicate vials containing the ketones and aldehydes was added a 1 M solution of aniline in methanol (300 µL) and to the other duplicate set was added a 1 M solution of benzylamine (300 µL). Each of the reactions were thoroughly mixed and left to react for 7 h. At the end of 7 h, each reaction mixture was treated with a 1 M solution of sodium cyanoborohydride in methanol (120 µL) for 2 h. The methanol was then evaporated and the resulting residue partitioned between ethyl acetate (800 µL) and water (400 µL). Each of the organic layers was transferred to vials containing sodium sulfate. Each of the water layers was extracted with ethyl acetate (400 µL) and combined with the previous vial containing the sodium sulfate. The combined dried organic layers were then transferred into an empty preweighted vial and concentrated. In this way 20 different diamines were generated in concentrations ranging from 31 to 79 µmol in a single run. The formation of these compounds was confirmed by LC-MS.

B.   Preparation of Allylic Amines

The followings was prepared manually:

> Five vials of Weinreb amides composed of the following amino acids: val, pro, ala, phe, leu (0.18 mmol)
> Five vials of solid methyl(triphenylphosphoranylidene)acetate (0.1 mmol)
> Two vials of acetonitrile (1.8 mL)
> Five vials of sodium sulfate (1 mm)

The automated part of this synthesis is described below.

1.   Synthesis of Aldehydes

The synthesis of the aldehydes was performed as described above with 0.18 mmol of starting Weinreb amide.

2.   Wittig Reaction

Acetonitrile (500 µL) was added to each vial containing the resulting aldehydes. Each solution was then transferred to the vials containing methyl(triphenylphosphoranylidene)acetate. The reaction mixtures were heated for 2 h at 80°C, the solvent evaporated, and the residues taken up with ethyl acetate/hexane (150 µL/ 150 µL). The supernatants were each purified over a silica gel cartridge using ethyl acetate/hexane (1:9) as eluant to separate triphenylphosphine oxide from the desired allylic amine. In this way five pure allylic amines were synthesized in yields ranging from 12% to 79% (22–142 µmol). The structures of these allylic amines were confirmed through NMR and LC-MS.

Part III
High-Throughput Synthesis of New Materials and Catalysts

Chapter 7
Synthesis and High-Throughput Evaluation of Complex Functional Materials Chips

Xiao-Dong Xiang
Lawrence Berkeley National Laboratory, Berkeley, California

I. INTRODUCTION

From the Stone Age to the Information Age, functional materials have always played an important role in the advancement of human society. Understanding and application of functional materials have dramatically changed our way of life during the past millennium. The discovery of the periodic table formed the basis of modern chemistry and materials science. Development of the quantum theory of solids further facilitated our understanding of the structure–property relationship of functional materials. In the last century, materials scientists have mainly focused on the understanding and application of simple materials formed by one or two elements of the periodic table. However, serendipitous discoveries of complex materials with fascinating properties, including high-temperature superconductivity, have begun to draw attention to more complex material systems. Our current knowledge of complex functional materials is miniscule considering the large universe of complex compounds that can be formed by different elements from the periodic table. J. C. Phillips estimated at the end of the 1980s that approximately 24,000 inorganic phases are known, of which 16,000 are binary and pseudobinary, and only 8000 are ternary and pseudoternary [1]. Due to the fact that given a desired functionality we still cannot theoretically predict a corresponding complex structure (that can be manufactured), experimental exploration will play a dominant role in materials research. To explore this large universe of complex systems in the new millennium, the conventional methods applied in the past to synthesize and test materials one at a time are antiquated. More efficient, thorough, systematic, and cost-effective methods must be adopted to meet the challenge.

There has long been much desire and effort to accelerate the rate of materials exploration. As early as the 1960s, materials physicists started using "composition spread" methods to explore metal alloy systems, especially superconductors. Miller and Shirn published a paper in 1967 [2] in which they discussed an Au-SiO$_2$ system using a cosputtering technique of Au and SiO$_2$. Geometrical arrangement of targets and substrate allows a continuously varying composition ratio of Au to SiO$_2$ deposited on the substrate. Electrical resistivity was measured as a function of weight percentage of Au. Similar methods were used by Hanak et al. to study the grain size effect in transition metal alloy superconductors [3] and by Sawatzky and Kay to study the Gd$_3$Fe$_5$O$_{12}$ system [4] in 1969. Hanak in 1970 described an approach using a one-cathode, multicomponent (two or three) sliced target cosputtering technique to generate binary or ternary composition spread [5]. In the paper, Hanak proposed to replace the traditional method of making one sample at a time with this approach in materials research. In fact, he conducted a large effort search of superconductors using this method in the RCA Laboratories during that period. Independently, Berlincourt at the U.S. Office of Naval Research (as the Director of the Physical Science Division) in 1973 proposed a national effort to conduct a search for high-temperature superconductors using this method and other automated screening techniques [6]. This proposal was circulated among the top physicists and materials scientists and sharply criticized. This approach was not further developed or applied to various materials systems at the time and was eventually forgotten by most scientists in the field. Many reasons may be attributed to this. Lack of support by funding agents to further develop the techniques around the approach may be the critical factor. The techniques used at the time were not effective enough to make any major impact in materials research. The narrowly focused application area (metal alloy superconductors) may also account for this. Even during the 10 years of worldwide high-T_c superconductor searches involving thousands of scientists after the discovery of cuprate superconductors, this method was barely used or mentioned at all.

During the late 1980s and early 1990s, stimulated by the discovery of high-T_c superconductors, scientists again started developing techniques to accelerate the materials discovery process. For example, robotic approaches were used to emulate and replace the human activity of bulk materials synthesis in Japan [7] and solution synthesis in the United Kingdom [8] of inorganic compounds in an effort to find new high-T_c superconductors.

During the 1980s, scientists in the field of biochemistry and the drug industry began developing various approaches to speed up the drug discovery process. These approaches are known as combinatorial chemistry and high-throughput screening [9–17]. Among these, spatially addressable libraries and gene chips inspired our recent work on combinatorial materials chips. We began an effort at Lawrence Berkeley National Laboratory (LBNL) in 1994 to develop systematic approaches and technologies on what I call "synthesis and high-throughput screening of combinatorial materials chips" [18]. Over the last 2 years, this approach has been widely adopted and further developed by scientists around the world in both academic and industrial research laboratories to explore different materials systems. The approach has been applied to a wide range of applications, including superconducting, magnetic, electronic, photonic materials, battery and fuel cell electrodes, and catalysis. In the following, we discuss the scientific and technical issues of the approach and review the achievements and status of the field.

II. GENERAL APPROACH

The application of a combinatorial materials chips approach to the discovery and optimization of materials is a multistep process. The first step involves the design of the materials chips. Some opponents and proponents of the approach often characterize the approach as "blind shotgun" and "Edisonian science." These characterizations all miss the point that the approach is simply an innovation in experimental methods and techniques, which allows thousands to millions of experiments to be carried out within the time frame of one experiment of conventional fashion. The aspect of human intervention in experiments has not been and should not be changed. As long as the experiments of exploratory nature are still needed in our scientific endeavors, i.e., the experiments are not only designed to confirm the prediction of our ingenious theorists, this type of approach will be of great value in science. In fact, timely availability of more complete, thorough, and systematic data on complex systems should improve and accelerate our understanding of the nature of complex systems. Efforts to catalog materials systems, e.g., catalog all ternary phase diagrams (and quaternary oxides, nitrides, and hydrides), including mapping of all useful properties analogous to the genome mapping project, may be technically possible but may not be realistic in the near future considering the economic cost and potential benefit. Therefore, good theoretical guidance and experimental design are always critical to ensure quick success of the experimental research. In fact, much more time of scientific talents freed from traditional time-consuming, repetitive, and tedious work of materials preparation can be put into the more intelligent portion of experiments. Specifically, materials chips may be designed to contain diverse discrete compositions aimed at exploring large segments of the composition landscape. It may also be designed to contain continuously varying compositions (such as ternary phase diagrams) aimed at identification of phase boundaries and isolation of important narrow-phase regions of interest based on our knowledge of a particular class of complex material systems. Mathematical strategies can be used to maximize the efficiencies of the experimental design.

The second step is the fabrication and synthesis of materials chips. Thin-film deposition techniques and new thin-film synthetic methods (as opposed to bulk synthesis) play an important role in this step. Processing parameters, such as temperature profiles and atmosphere in heat treatment, are also varied as are composition variables. Significant effort involving various characterizations, such as X-ray diffraction (XRD), Rutherford backscattering, Auger spectroscopy, atomic force microscopy, etc., performed on differently synthesized chips have been made to improve the quality and reliability of these materials chips. Continuous research in this area is needed to perfect the techniques.

The third step is high-throughput screening of the materials chips. Through rapid measurements of one (or several) relevant key physical property (or properties), a family of "lead" compositions and phase region with a desired property are identified. In this aspect, the screening techniques involved in the combinatorial materials chips approach are more challenging and sophisticated than those of combinatorial chemistry in the life sciences, where the luminescence tagging technique predominates. Characteristics ranging from superconductivity, electrical and dielectric properties, magnetic properties, thermal properties, and optical properties, each comprising many parameters, have to be classified. Fortunately, most of these techniques have been made available by physicists in the past, although many have to be further developed or improved to allow for imaging with higher spatial resolution. More importantly, structural studies often are critical and present the most challenging task.

In the next step of the process, a lead composition or narrow-phase region is further modified or optimized. Finally, a material of interest is synthesized in large thin films or bulk forms, and detailed structural and physical properties are characterized in the conventional manner. In some cases, actual devices are fabricated and tested for device performance using the refined materials. Conventional synthesis, especially single-crystal growth, hot-pressed ceramic processing and high-quality thin-film growth, and characterizations, will not be replaced but rather will be made more meaningful and effective on highly selected specimens. In addition to time and cost efficiencies, the environmental benefits of this approach are enormous—more than a six orders of magnitude reduction in chemical waste generation has been achieved compared with conventional methods.

III. SYNTHESIS METHODS

One of the distinctions between recent combinatorial materials chips technique and previous "composition-spread" through cosputtering technique is the use of combinatorial masks combined with multilayer thin-film deposition. The main consideration is to increase efficiency by using mathematically designed mask strategies. Other considerations are to simplify and reduce the cost of quality control and increase the reproducibility. In the cosputtering technique or any other codeposition techniques, the plume profile of each precursor target is usually different, which depends not only on the precursor composition but on details of the deposition condition, such as power, voltage, and atmosphere. Differences in sticking coefficients, resputtering effects for different precursors, and substrate temperatures further complicate the stoichiometric distribution. Therefore, it is often necessary to perform postdeposition profiling of stoichiometric distribution with various analytical techniques. These techniques do not always give accurate results for all precursors. Furthermore, they are very time consuming and have a very high cost of ownership. Our current approach uses precision masks to define the profile of each precursor combined with in situ monitoring of a uniform deposition profile and constant deposition rate. This method eliminates the need for routine mapping of the spatial distribution of compositions on the materials chips. The substrate is usually at room temperature or very low temperature during the layer-by-layer precursor deposition. Postannealing and heat treatment are used for diffusion of precursors and phase formation. This method dramatically reduces the cost and level of complexity of the experiment, which makes practice of the technique possible in typical small materials research laboratories.

Discrete materials chips are fabricated using different combinatorial mask configurations to effectively search through large compositional phase spaces. Figure 1 illustrates an example of a quaternary combinatorial masking scheme [19]. This masking scheme involves a series of n masks, which successively subdivides the substrate into a series of self-similar patterns of quadrants. The rth ($1 \leq r \leq n$) mask contains 4^{r-1} windows where each window exposes one-quarter of the area deposited with the preceding mask. Within each window there are an array of 4^{n-r} gridded sample sites. Each mask is used in up to four sequential depositions, each time rotated by 90 degrees. This process produces 4^n different compositional combinations by $4n$ precursor depositions and can be effectively used to survey a large number of materials with different compositions, each consisting of up to n elemental components where in each component is selected from a group of (up to four) precursors.

The implementation of masking schemes is accomplished using either photolithographic lift-off or physical shadow masks. The lithographic method is well suited for generating chips containing a high density of sites due to the high spatial resolution and alignment accuracy. Figure 2a is a photograph (taken under daylight) of a 1024-member materials chip designed to search for phosphors. This chip was deposited on a 1×1 in. Si substrate. The diversity of colors of different sites is attributable to the varied thickness and optical indexes of the films and reflects the complexity of the compositions.

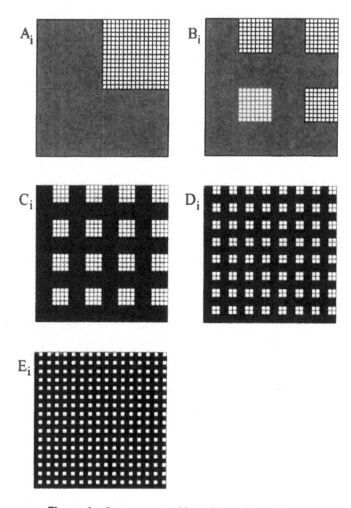

Figure 1 Quaternary combinatorial masking scheme.

The continuous phase diagram chips are fabricated by using linear masks that move at a controlled rate during each precursor deposition. For example, a ternary phase diagram is synthesized from gradient depositions of three precursors using high-precision in situ linear shutter systems (Fig. 3a) and subsequent ex situ postannealing. Compared to the previously used composition spread method by cosputtering of multiple targets [2–5], this technique can easily generate precisely controlled stoichiometric profiles within a small area. This advantage is crucial when various expensive single-crystal substrates need to be used

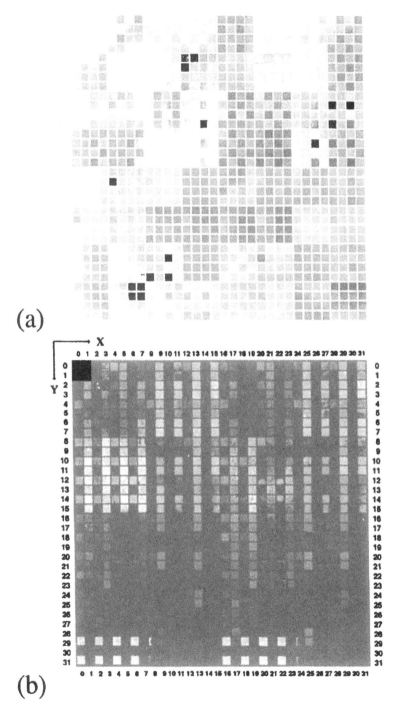

Figure 2 A 1024-member materials chip fabricated using quaternary combinatorial masking scheme.

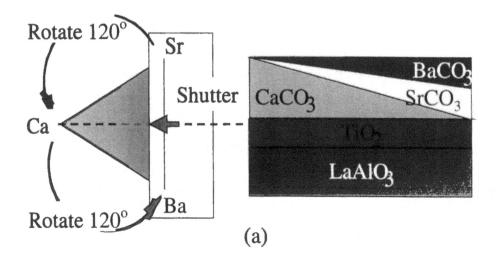

(a)

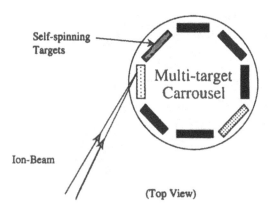

(b)

Figure 3 (a) Generation of ternary phase diagram using linear shutter system. (b) Multitarget carousel system.

(as in this study) for high-quality epitaxial film growth. Another advantage is that no complicated poststoichiometric determination is needed.

On the surface, this multilayer thin-film synthesis approach seems straightforward. However, it was not clear that the approach would generate meaningful crystalline compounds. Since the precursors are deposited sequentially as thin-film multilayers, nucleation could occur at each interface between precursors at elevated temperatures. Therefore, it is very likely that multiple binary phases will be formed instead of the desired multielement single phase. In fact, our group made many failed attempts at converting the precursors into crystalline compounds before an effective two-step annealing process was found [20]. This approach was inspired by the work of Johnson and colleagues at the University of Oregon on metal precursor interdiffusion [21]. It takes advantage of competition between interdiffusion and nucleation at the interfaces of precursor layers. There exists a critical thickness (a few tens to hundreds of angstroms for most oxides) for sequential precursor layers. Below this thickness, the interdiffusion process is dominant over the nucleation process at relatively low temperatures. Therefore, it is possible to interdiffuse the precursor layers at a relatively low temperature to form an intermediate amorphous state close in stoichiometry to the desired crystalline phase before the nucleation of thermodynamically stable phases starts. We found that most metal alloys or oxide compounds can be formed from multilayers of metal (or, in most cases, oxide) precursors through controlled thermal treatments. Careful postannealing processes are required to ensure uniform diffusion of precursors and subsequent phase formation. An extended period of low-temperature annealing under different atmospheres is often necessary for proper interdiffusion of metal or oxide precursors, stress release, and to prevent evaporation of nonreacted layers. High-pressure/high-vacuum ovens with controlled atmospheres are used for diffusion and crystal phase growth of the materials chips. Various analysis tools, Rutherford backscattering, X-ray, Auger spectroscopy, ion scattering spectroscopy, microprobe, etc., can be used to characterize the diffusion and crystalline growth of the films to optimize the layering sequences as well as the annealing procedures. Note that only a few sampling experiments are required for a large class of similar materials chips.

Volatile elements, e.g., Pb, will evaporate before the interdiffusion of other nonvolatile elements is even close to completion. Thus, for example, to make a $Pb(Zr, Ti)O_3$ chip an extra step has to be added. The nonvolatile precursors of Ti and Zr are annealed at 800°C for interdiffusion in vacuum (see Case Study 7-1, Fig. 5). Pb is deposited after they are uniformly alloyed and the PbO-alloy multilayer is heated at 500°C under 20 atm of Ar to diffuse PbO into the underlying metallic alloy (see Case Study 7-1, Fig. 6). High-pressure Ar is used primarily to prevent evaporation of PbO. Ar instead of O_2 is used because high-pressure O_2 will oxidize Pb^{2+} to Pb^{4+} at that temperature. When the layers are uniformly diffused, the film is heated in flowing oxygen at 800°C to form the crystalline perovskite phase.

For materials chips fabrication, proper selection of substrate materials is critical. Refractory single-crystal substrates are usually used to minimize substrate–sample reactions. Lattice matching and thermal expansion coefficients are often taken into consideration so as to fabricate epitaxial thin-film libraries in some cases. Common substrates that we have used include MgO, $LaAlO_3$, sapphire, and Si.

We have found that the two-step annealing method combined with substrate lattice matching, despite the fact that the crystalline compounds are formed ex situ rather than in situ, can be used to synthesize materials chips with high-quality epitaxial films (Fig. 4a). X-ray studies shown in Figure 4 demonstrate a good example. Figure 4b is a $\theta/2\theta$ XRD pattern for $Nd_{0.7}Sr_{0.3}MnO_3$ on $LaAlO_3$ (LAO) substrate, indicating that the film is (100) oriented. Even in logarithmic scale, we can hardly see any other impurity phase. Figure 4c shows ϕ scans of the (101) planes indicating that the film is in-plane aligned with the substrate. This capability of epitaxially growing entire integrated materials chips is important in many fields where material properties are closely tied to the crystalline quality of the films. This method can be used to synthesize metal alloys, oxides, even nitrides and other materials that are difficult to make by conventional bulk synthesis methods.

Previously we used pulsed laser deposition (PLD) and radiofrequency (RF) sputtering techniques for deposition of precursors. We recently developed an ultrahigh-vacuum (UHV) combinatorial ion-beam sputtering system that is more suitable to fabricated chips of different materials system including metal alloys, nitrides, hydrides, and oxides (even with air-sensitive or volatile precursors). Originally, this system is designed only for metal alloy systems. However, we realized later that this system is crucial for investigating many oxide systems, such as ferroelectric materials systems. For example, we found that layers of TiO_2 and ZrO_2 remain undiffused even at temperatures of 800°C for 2 days. In order to interdiffuse the nonvolatile components for $Pb(Zr, Ti)O_3$, elemental precursors of metallic Ti and Zr were used. The metal alloys interdiffuse more readily than their oxide counterparts. Therefore, materials chips of all metal alloys, nitrides, hydrides, and oxides systems can be fabricated using this system, where oxidation, nitration, and hydration are performed during the post-annealing steps.

Different from conventional thin-film deposition systems, the chamber is equipped with an automated eight-target carousel (Fig. 3b) and x-y precision in situ shutter system. Due to the high reflectance and thermal conductivity of metal targets, laser ablation is not suitable for metal deposition. Metal precursor targets, including air-sensitive precursors such as Ca, Sr, Ba, etc., are easily available in comparison with the oxide targets used in PLD, and metal precursors can be

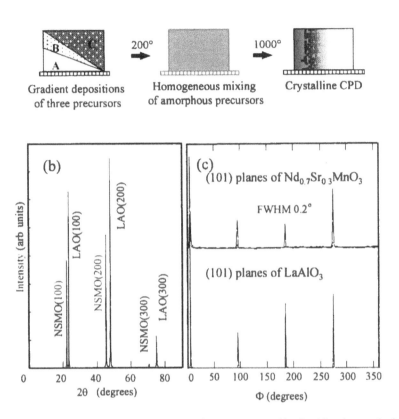

Figure 4 X-ray studies of two-step annealing system combined with substrate lattice matching.

easily diffused (critical for combinatorial synthesis). The chamber is maintained at UHV of better than 10^{-10} torr by a cryopump and Ti sublimation (or ion) pump suitable for metal alloy deposition. A load-lock chamber is installed for sample exchange without breaking the vacuum of the main chamber. During ion beam sputtering, deposition rates may vary as the beam current fluctuates. Thus, by feedback control of the deposition rate (through the beam current of the ion beam power supply) with the thickness monitor signal, we were able to keep a constant deposition rate that varies less than 0.001 Å/s. In order to calibrate and monitor the deposition rate before and during deposition, a thickness monitor with flexible feed-through was installed (it can be positioned either in front of substrate or at the side of substrate). Currently, the system is equipped with x-y shutters highly effective in the fabrication of the continuous phase diagram chips with quick turnaround. The system can also be easily equipped with a series of discrete combinatorial masks described above to perform in situ fabrication of discrete materials chips.

IV. HIGH-THROUGHPUT SCREENING

Once materials libraries containing thousands to millions of different composi- tions are fabricated, it is necessary to rapidly measure the particular material properties of interest. Although detailed characterization may require bulk sam- ples, quick screening of a few key properties of samples on the chips is crucial. In general, optical properties and the properties that can be derived from optical measurements are easier to screen. Our group at LBNL and that of Danielson at Symyx Technologies have used home-built optical systems to characterize the properties of luminescent materials libraries [22,23]. We have also developed an electro-optical coefficient measurement system to screen for better electro-optical materials, a magneto-optical (Kerr/Faraday rotation-based techniques) measure- ment system to map the surface magnetization and an optical measurement sys- tem to characterize the piezoelectric or strain effects.

A. X-ray Characterization

Often the characterization of specific properties of tiny samples on materials chips presents real scientific challenges. Structural characterization is probably the most difficult and currently still represents a time-consuming bottleneck in the combi- natorial materials library approach. Recently, Issacs at Lucent Technologies, Aeppli at NEC Research Institute, and their colleagues have made significant progress in applying the X-ray microbeam technique available at a number of synchrotron radiation facilities for direct structural characterization of materials chips [24]. X-ray microbeam techniques (spot size = $3 \times 20 \ \mu m^2$) have been applied to characterize the composition and structure of rare-earth activated $Gd(La,Sr)AlO_3$ phosphor thin films grown at LBNL by combinatorial synthesis (Fig. 5). Using X-ray fluorescence, X-ray diffraction, and near-edge X-ray ab- sorption spectroscopy, they measured the chemical composition, crystallographic structure, and valence state of the rare-earth activator atom Eu. These measure- ments demonstrate the power of X-ray microbeam analysis to nondestructively characterize as-grown combinatorial libraries. It even permits individual films to be mapped in exquisite detail, revealing the identity of secondary phases and compositional inhomogeneities within them. Although analogous instruments

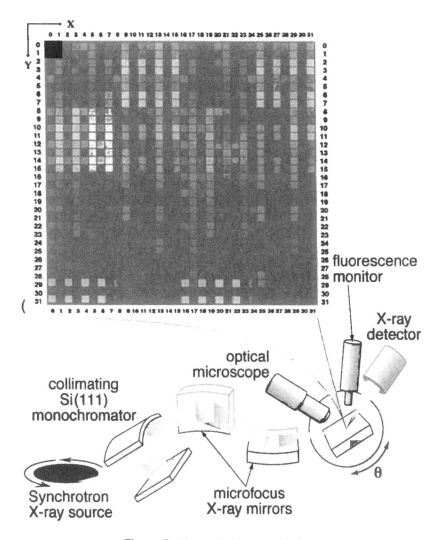

Figure 5 X-ray microbeam technology.

employing other probes, such as electrons, are conceivable, X-rays are unique both because their penetrating power allows the libraries to be examined under arbitrary environmental conditions and because their collimation and weak coupling nature allow extraordinarily precise quantitative analysis. Furthermore, the X-ray microprobe is nondestructive and can be applied directly to materials chips, as grown on their original substrates. Eventually, X-ray microbeams might even find use as in situ probes to optimize combinatorial synthesis. It is envisioned that combining an X-ray microfocus beam line at a third-generation synchrotron, like the Advanced Photon Source at Argonne National Laboratory, will allow complete structural and compositional characterization of 1000-element libraries on time scales of less than an hour. Efforts are underway to develop laboratory-based X-ray imaging techniques to allow routine characterization of materials chips.

B. Electrical Impedance Characterization

For screening libraries of electronic materials, quick, nondestructive, and quanti-tative characterization of the complex electrical impedance of material also pre-sents a challenging task. Characterization of electrical impedance is extremely important for many applications, including superconductivity, ferroelectrics/dielectrics, magnetoresistivity, and various semiconductor properties. For exam-ple, in the case of frequency-agile materials applications, three key properties—electric field tunability (nonlinear dielectric constant), dielectric constant, and loss tangent—need to be characterized at microwave frequencies. Conventional contact measurements using electrodes often yield misleading information due to interfacial effects, in addition to being destructive and difficult to use with materials libraries.

We have developed a novel scanning evanescent microwave probe (SEMP) capable of nondestructive and quantitative mapping of the complex electrical impedance (of any material) with submicrometer resolution [25]. The microwave frequency range was chosen because this is the relevant frequency range for most electronic applications. The SEMP is based on a probe made of a high-quality factor (Q) microwave coaxial resonator with a sharpened metal tip mounted on the center conductor (Fig. 6). The tip extends beyond an aperture formed on a thin, metal-shielding end wall of the resonator. The tip and the shielding structure are designed so that the propagating far-field components are shielded within the cavity, whereas the nonpropagating evanescent waves are generated at the tip. This feature is crucial for both high-resolution and quantitative analyses. In con-trast to the conventional antenna probes (a far-field concept), this probe does not emit significant energy (and therefore has very high Q to boost the sensitivity). Only when the tip is close to the sample will the evanescent waves on the tip interact with the material. The interaction gives rise to resonant frequency and Q changes of the cavity and consequently the microscopy of the electrical imped-ance. Similar shielding aperture tip structures can be implemented in all (either resonating or nonresonating) transmission line–type probes.

A 100-nm resolution on dielectric materials ($\lambda/10^6$) has been demonstrated [25]. Because the scanned tip is a part of a high-Q resonator, the SEMP has very high sensitivity ($\Delta f/f \sim 10^{-7}$ and $\Delta\varepsilon/\varepsilon \sim 10^{-4}$) [26]. More importantly, we have

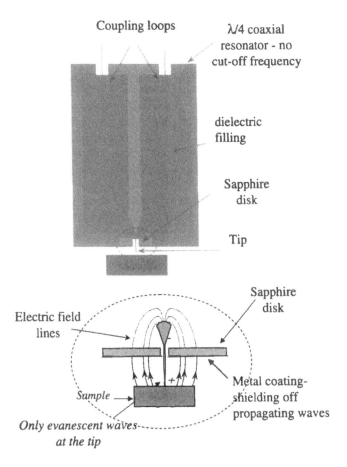

Figure 6   Scanning evanescent microwave microscope.

performed a theoretical near-field analysis that yielded analytical solutions for both insulating and conducting materials [26]. The theory enables quantitative local measurements of the complex dielectric constant of insulators or the conductivity of conductors. Novel feedback control methods have been developed to obtain reliable, nondestructive, and quantitative images of topography and physical properties simultaneously. The microscope is routinely used for quantitative characterization of linear and nonlinear dielectric constants of ferroelectrics and dielectrics [27,28]. A low-temperature version has also been developed to characterize superconducting materials [29].

V. APPLICATIONS

A. Superconductors

Superconductors are among the most fascinating materials both in terms of underlying physics and in their potential applications. It is not surprising that early applications of composition spread and automation techniques were all focused on superconductors. The discovery of copper oxide high-T_c superconductors is perhaps the most important materials discovery in the century. More than 10 years of extensive materials exploration efforts by thousands of scientist worldwide following the initial discovery by Bednorz and Mueller represent one of the largest scale materials explorations in modern history. It is therefore desirable to apply the combinatorial materials chips approach to search for new superconductors. We first demonstrated the feasibility of our technique to search for copper oxide superconductor [20]. In this study, we tried to answer the following question: if this technique were available 10 years ago, could we have used it to discover YBCO and BSCCO superconductors after the initial discovery of $La_{1-x}Sr_xCuO_4$?

A 128-member materials chip was generated by depositing precursors of $BaCO_3$, Y_2O_3, Bi_2O_3, CaO, $SrCO_3$, and CuO through a series binary masks [20]. The ratio of each precursor was kept at 1:1 to avoid much after-the-fact knowledge being used in design. After annealing at low temperature (200–400°C) and sintering at high temperature (840°C), sites containing $BiSrCaCuO_x$ and $YBaCuO_x$ were all found to be superconducting. Even though these sites do not have the correct stoichiometry of superconducting compounds and the sintering temperature was not optimal for individual superconducting compound synthesis, the study did successfully identify the superconducting sample sites.

Although the proof of concept of our technique was first demonstrated in copper oxide superconductors, using the approach to realistically search for new superconductors still represents a most challenging task. Not only are there few guidelines in terms of where to search, but we also have to implement sensitive screening techniques at cryogenic temperatures. We are currently implementing the SEMP and a magneto-optical imaging system at low temperatures to perform noncontact electrical and Meissner effect measurements.

B. Magnetoresistive Materials

Soon after the first demonstration, we used materials chips to identify a class of cobalt oxide magnetoresistive materials of the form $(La_{1-x}Sr_x)CoO_3$ [30]. Prior to this study, large magnetoresistances, potentially useful in magnetic reading head devices, were found only in Mn-based perovskites, $(La, R)_{1-x}A_xMnO_{3-d}$, where R is rare earth, and A is Ca, Sr, or Ba. The question arises whether these effects are unique to Mn-based perovskite oxides or can be found as an intrinsic property of other materials.

Using materials chips, we searched through simple perovskite ABO_3 and related $A_2 BO_4$ or $A_{n+1} B_n O_{3n+1}$ higher order structures, where A is (La, Y, rare earth)$^{3+}$ partially substituted with (Ca, Sr, Ba, Pb, Cd)$^{2+}$ and B is (Mn, V, Co, Ni, Cr, Fe). This has resulted in the discovery of a new family of Co-containing magnetoresistive materials. Magnetoresistance was found to increase as the size of the alkaline ion increased, in contrast to Mn-containing compounds, in which the magnetoresistive effect increases as the size of the alkaline earth ion decreases. An analysis of the effects of spin configuration and electronic structure on the magnetoresistive properties of the Co- and Mn-based compounds should help to elucidate the underlying mechanism of the magnetoresistance effect. Moreover, the discovery of diverse classes of the colossal magnetoresistance materials may help significantly in efforts to optimize these materials for eventual device applications.

C. Luminescent Materials

The technique quickly found applications in the optimization and discovery of luminescent materials since the screening is straightforward. We reported a systematic search of luminescent materials that led to the discovery of a number of rare-earth-doped refractory metal oxide phosphors, including $(Gd_{0.6}Sr_{0.4})Al_{1.6}O_{3+\delta}F_{1.8}:Eu^{2+}{}_{0.08}$(green), $LaAl_{3.1}O_{3+\delta}:Eu^{2+}{}_{0.08}$(blue), and $GdAl_{1.6}O_{3+\delta}:Eu^{3+}{}_{0.08}$(red) with photoluminescent quantum efficiencies of approximately 100%, 60%, and 100%, respectively [31]. Danielson et al. at Symyx Technologies reported a discovery of a novel one-dimensional luminescent oxide Sr_2CeO_4 [32]. This blue-white color phosphor has an emission maximum at 485 nm with a quantum yield of 50%. This work demonstrated that the combinatorial materials synthesis approach could actually be used to identify fundamentally new, unexpected structures with properties that arise from unusual mechanisms. We also identified a red phosphor $(Gd_{1.54}Zn_{0.46})O_{3-\delta}:Eu^{3+}{}_{0.06}$, with photoluminescent quantum efficiency of about 86% and a superior color chromaticity ($x = 0.656$, $y = 0.344$) compared to the state-of-the-art red phosphor $Y_2O_3:Eu^{3+}$ [33]. These phosphors also show excellent cathodoluminescent properties and hold great promise for applications in the display industry.

More recently, we came across an entirely unexpected novel blue-emitting luminescent composite material $Gd_3Ga_5O_{12}/SiO_x$ [34] while searching for phosphorescent materials using the integrated materials chip shown in Fig. 2a. This materials chip contains various silicate and gallate host materials doped with a number of different activators. The library design is based on the observation that some silicate- and gallate-based phosphors with relatively small band gaps have properties desirable for electroluminescent display applications. The photo-luminescent (PL) photograph of the thermally processed chip was taken under ultraviolet (UV) irradiation (centered on 254 nm). For more quantitative analysis of the libraries, a scanning spectrophotometer was used to measure the excitation and emission spectra of individual luminescent samples in the library. The highest PL intensity corresponded to the composition $Gd_3Ga_5O_{12}/(SiO_2)_{0.08}$. Our XRD studies showed no trace of crystalline SiO_2 or silicate in $Gd_3Ga_5O_z{:}(SiO_2)_y$, ($y \leq$ 0.5) bulk or thin-film samples. However, X-ray photoemission spectroscopy (XPS) and energy dispersive X-ray analysis (EDAX) confirmed the presence of SiO_2 in the sample. We conclude that the interfacial effect between $Gd_3Ga_5O_{12}$ and SiO_x is responsible for efficient blue photoluminescence in this composite material. This example clearly demonstrated that the integrated materials' chip approach can dramatically increase the probability of serendipitous discoveries.

D. Ferroelectric/Dielectric Materials

van Dover et al. at Lucent Technologies recently used the composition spread technique to discover new amorphous dielectric thin-film materials ($Zr_{0.15}Sn_{0.3}Ti_{0.55}O_{2.8}$) that may be used in the node capacitors of prototype dynamic random-access memories [35]. They used a three-target, off-axis, reactive cos-puttering system to generate amorphous ternary composition spreads on heated silicon substrates. The electrical properties were evaluated using a scanning Hg probe, widely employed in the semiconductor industry. More than 30 ternary composition spreads were evaluated; among them, the Ti-Sn-O system exhibited the highest figure of merit. This study is discussed in detail in Case Study 7-2.

Our group has been working on ferroelectric/paraelectric materials applica-tions [27,28]. Our emphases are on crystalline phase diagram mapping and modi-fications by doping. The ability to reliably map crystalline phase diagrams is of great significance in materials research. Structure–property relationships are the most fundamental issue addressed by materials research. Practically, very often the most interesting properties are found near narrow regions of phase boundaries. For example, many ferroelectric/paraelectric materials with superior performance properties are found near the morphotropical phase boundaries, such as PZT and PMN-PT, or near paraelectric to ferroelectric phase boundaries [36]. To locate these very narrow regions in a large-phase space, the conventional method of materials synthesis is particularly inadequate. As discussed in Case Study 7-1, we recently demonstrated that reliable crystalline phase diagram mapping could be achieved using our techniques. Many skeptical materials experts have long considered this capability impossible. We believe this will dramatically impact various areas of materials research.

VI. OTHER STUDIES

Many scientists and research groups have recently begun to work on other innovative ideas and technologies. Many interesting works in the areas of catalysis and polymer science have been reported. In this chapter we focus only on inorganic materials useful in the electronic and information industry applications. For works in the catalysis area, readers should refer to Chapter 8 in this volume as well a recent review by Janeleit and Weinberg [37]. At the March 1998 American Physical Society meeting, Koinuma of the Tokyo Institute of Technology reported on his group's unique combinatorial materials process approach. The concept comes from the realization that many important materials properties arise from different processing conditions, such as plasma energy, partial pressure of reactive gases, and temperature. Koinuma and his collegues demonstrated this concept in a study of hydrogenerated amorphous silicon for solar cell applications. They developed a combinatorial plasma-assisted chemical vapor deposition system. The plasma surface treatment condition of amorphous silicon is varied combinatorially to study the growth and properties of amorphous silicon. This group's combinatorial in situ superlattice engineering approach represents another exciting new possibility of combinatorial materials synthesis where by atomic layer-by-layer deposition is used to control the in situ crystal growth of novel artificially engineered compounds [38].

VII. CONCLUSION

The successful integration of the combinatorial materials chips approach with established conventional materials synthesis and characterization techniques will make possible new directions in materials research.

REFERENCES

1. J.C. Phillips, *Physics of High-Tc Superconductors*, Academic Press, New York, 1989.
2. N. Carl Miller and George A. Shirn, *Appl. Phys. Lett. 10*, 86 (1967).
3. Hanak, J.I. Gittleman, J.P. Pellicano, and S. Bozowski, *Phys. Lett. 30A*, 201 (1969).
4. E. Sawatzky and E. Kay, *IBM J. Res. Dev.* 696, Nov. (1969).
5. J.J. Hanak, *J. Mater. Sci. 5*, 964 (1970).
6. Ted Berlincourt, private communication (1973).
7. H. Yamauchi et al., *ISTEC J.* (in Japanese) *5*, 25 (1992).
8. S.R. Hall and M.R. Harrison, *Chem. in Br. 739*, Sept. (1994).

9. S. Tonegowa, *Nature 302*, 575 (1983).
10. J.K. Scott and G.P. Smith, *Science 249*, 386 (1990).
11. S.P.A. Fodor, J. L. Read, C. Pirrung, et al., *Science 251*, 767 (1991).
12. C. Tuerk and L. Gold, *Science 249*, 505 (1990).
13. K.S. Lam, S.E. Salmon, E.M. Hersh, et al., *Nature 354*, 82 (1992).
14. M.C. Needels et al., *Proc. Natl. Acad. Sci. U.S.A. 90*,10700 (1993).
15. B.A. Bunin, M.J. Plunkett, and J.A. Ellman, *Proc. Natl. Acad. Sci. U.S.A. 91*, 4708 (1994).
16. P.G. Schultz and R.A. Lerner, *Science 269*, 1835 (1995).
17. L.A. Thompson and J. A. Ellman, *Chem. Rev. 96*, 550 (1996).
18. X.-D. Xiang, *Annu. Rev. Mater. Sci. 29*, 149 (1999).
19. X.-D. Xiang and P.G. Shultz, *Physica C 282–287*, 428 (1997).
20. X.-D. Xiang, et al., *Science 268*, 1738–1740, (1995).
21. L. Fister, T. Novet, C.A. Grant, and D.C. Johnson, *Adv. Synth. React. Solids 2*, 155 (1994).
22. J. Wang, Y. Yoo, C. Gao, I. Takeuchi, X. Sun, X.-D. Xiang, and P.G. Schultz, *Science 279*, 1712 (1998).
23. E. Danielson, J.H. Golden, E.W. Mcfarland, C.M. Reaves, W.H. Weinberg, and X.D. Wu, *Nature 389*, 944 (1997).
24. E.D. Issacs, M. Kao, G. Aeppli, X.-D. Xiang, X. Sun, P. Schultz, M.A. Marcus, G.S. Cargill, and R. Haushalter, *Appl. Phys. Lett. 73*, 1820 (1998).
25. C. Gao, T. Wei, F. Duewer, and X.-D. Xiang, *Applied Physics Lett. 71*, 1817 (1997);
26. C. Gao and X.-D. Xiang, *Rev. Sci. Instr. 69*, 3846 (1998).
27. H. Chang, C. Gao, I. Takeuchi, Y. Yoo, et al., *Appl. Phys. Lett. 72*, 2185 (1998).
28. H. Chang, I. Tacheuchi, and X.-D. Xiang, *Appl. Phys. Lett. 74*, 1165 (1999).
29. I. Takeuchi, F. Duewer, and X.-D. Xiang, *Appl. Phys. Lett. 71*, 2026 (1997).
30. G. Briceno, H. Chang, X. Sun, P.G. Schultz, and X.-D. Xiang, *Science 270*, 273 (1995).
31. X. Sun, C. Gao, J. Wang, and X.-D. Xiang, *Appl. Phys. Lett. 70*, 3353 (1997).
32. E. Danielson, M. Devenney, D.M. Giaquinta, et al., *Science 279*, 837 (1998).
33. X. Sun, and X.-D. Xiang, *Appl. Phys. Lett. 72*, 525 (1998).
34. Jingsong Wang, Young Yoo, Chen Gao, Ichiro Takeuchi, Xiaodong Sun, X.-D. Xiang, Peter G. Schultz, *Science 279*, 1712 (1998).
35. R.B. van Dover, L.F. Schneemeyer, R.M. Fleming, *Nature 392*, 162 (1998).
36. Yuhuan Xu, *Ferroelectric Material and Their Applications*, North-Holland, Amsterdam, 1991.
37. B. Jandeleit, W.H. Weinberg, *Chem. Ind. 19*, 795 (1998).
38. Y. Mastsumoto, M. Murakami, Zhengwu Jin, A. Ohtomo, M. Lippmaa, M. Kawasaki, H. Koinuma, *Jap. J. Appl. Phy. 38*, 603 (1999).

Case Study 7-1
Studies of Ferroelectric/Dielectric Materials Using Combinatorial Materials Chips

Hauyee Chang and Xiao-Dong Xiang
Lawrence Berkeley National Laboratory, Berkeley, California

ABSTRACT

Mapping of phase boundaries and modification of dielectric/ferroelectric properties of ferroelectric/paraelectric materials are performed using continuous and discrete materials chips, respectively. Through the cases discussed here, we demonstrate the effectiveness of these techniques. Guided by good empirical observation and theoretical understanding, the approach could be 10,000 times faster than conventional approaches and highly effective and successful in real-world materials research.

I. EQUIPMENT AND SUPPLIES

1. Combinatorial deposition systems, including a high-vacuum pulsed-laser ablation and a ultrahigh-vacuum ion-beam sputtering system equipped with eight-target carousel and linear shutters.
2. Targets of $BaCO_3$, $SrCO_3$, $CaCO_3$, TiO_2, PbO, Ti, and Zr.
3. A 0.5-mm-thick (100) cut $LaAlO_3$ single crystal.
4. Scanning evanescent microwave probe (SEMP).
5. Optical systems for measuring the linear and quadratic electro-optic coefficients of thin films.

6. Software for laser beam scanning, shutter movement, and deposition rate control in deposition systems; SEMP software and electro-optic effects scans, written in Labview and C.

II. INTRODUCTION

Ferroelectric/paraelectric materials have been the focus of intense scientific research efforts due to their great technological importance. Examples of their important properties and relevant applications include the piezoelectric effect for transducer/actuator and microelectromechanical (MEM) device applications, electro-optic (e-o) effect for telecommunication applications, high dielectric constants and remanent polarization for microelectronic applications, electric field tunability at microwave frequency for radar and microwave electronic applications, and pyroelectric effect for infrared imaging. The combinatorial materials chips approach is especially effective in systematic studies where multiple material properties are screened over the same chip. In this case, a large number of ferroelectric/dielectric properties are often closely related. For instance, a material with high dielectric constant will usually have high piezoelectric constant and tunability. High tunability may also be related to large electro-optic coefficients. With abilities to screen all these different properties over the same material chip, we should be able to correlate systematic changes in properties as a function of composition.

There exists a large database of this class of materials that includes hundreds of compounds thanks to more than 50 years of continuous worldwide intensive research. The fundamental physics of these effects are also relatively well understood. However, subtle material modifications that dramatically improve performance characteristics are usually conducted in old-fashioned trial-and-error experiments.

For example, it was found that solid solutions of perovskite compounds with slightly different crystal structures often give rise to morphotropic phase boundary (MPB) compositions. These MPB compositions usually have dramatically enhanced piezoelectric coefficients, dielectric constants and other related properties. The best known example of such a compound is the $Pb(Zr, Ti)O_3$ (PZT) system in which the crystal structure changes from the Zr-rich ferroelectric rhombohedral structure to the Ti-rich ferroelectric tetragonal one at a Zr/Ti ratio of 0.53 [1]. Furthermore, it was found that for the MPB compounds with transition between relaxor (diffused and metastable phase that can be electrically driven from paraelectric to ferroelectric) and ferroelectric phase usually have even better properties. In addition to large dielectric and piezoelectric constants, they often have very large quadratic e-o coefficients, large microwave tunabilities, and low losses. The PLZT system is a perfect example. Addition of La to $Pb(Zr, Ti)O_3$, not only changes the ceramic from opaque to transparent at visible wavelength (critical to e-o applications), it also induces a region of relaxor-ferroelectrics near the MPB (instead of the original ferroelectrics-ferroelectrics MPB). PLZT 8.5: 65:35 composition has a quadratic e-o coefficient an order of magnitude higher than the slightly different composition 9:65:35 and at least two orders of magnitude higher than other known ferroelectric materials [1]. Identifying these narrow-phase regions in a binary and ternary system requires performing about 100 and 10,000 experiments, respectively. At least a few hundred candidate systems are worth exploring, which would take about 10,000 scientist work years. The continuous phase diagram chip technique discussed in the above review article is perfectly suited for this type of study. We will discuss several such experiments in this case study.

Another type of modification of the materials involves addition of small amounts of elemental or compound dopants. Properties such as dielectric and mechanical loss can improve dramatically with his type of modification. To perform this type of study, much more tedious and time-consuming experiments have to be performed than phase diagram mapping. The discrete materials chips technique is well suited for this type of exploration. We will also give an example in this case study to demonstrate the effectiveness of the technology.

III. PHASE DIAGRAM MAPPING

A. $(Ba_{1-x-y}Sr_xCa_y)TiO_3$ System

We made a continuous phase diagram chip to explore $Ba_xSr_yCa_{1-x-y}TiO_3$ (BSCT), where $0 \leq x$ and $y \leq 1$. The substrate is a piece of equilateral triangle–shaped $LaAlO_3$ with a height of 1 in. The bottom-most precursor layer consists of a uniformly 750-Å-thick TiO_2. A shuttle in the deposition system moving with constant velocity during deposition gave rise to the linear gradient thickness of $SrCO_3$ (0–1475 Å), $BaCO_3$ (0–1647 Å), and $CaCO_3$ (0–1225 Å) on top of the TiO_2 layer [Fig. 3(b)]. The substrate is rotated 120 degrees anticlockwise after the deposition of each layer of carbonates. This gives rise to a composition spread for a ternary phase diagram of BSCT. This chip was heated under flowing oxygen at 400°C for 24 h for mixing of the precursors, followed by 900°C for 1.5 h for crystallization. The fluorine from the precursors is removed during the annealing process.

Figures 1a and 1b are the ε_r and tan δ maps of the BSCT triangle. Along the *upper edge* of the ε_r map the composition is $(Ba_{1-x}Sr_x)TiO_3$ ($0 < x < 1$). Figure 2 is a line scan of ε_r versus composition along the upper edge, with a

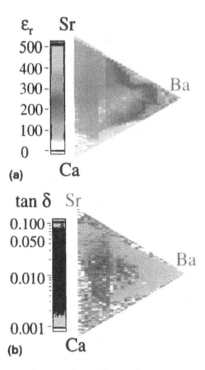

Figure 1   Dielectric properties mapping of Ba$_{1-x-y}$Sr$_x$Ca$_y$TiO$_3$ continuous-phase diagram chip.

maximum ε_r at Ba$_{0.7}$Sr$_{0.3}$TiO$_3$. This is consistent with previous studies on bulk samples where a pseudocubic paraelectric to tetragonal ferroelectric phase transition occurs at $x = 0.3$ at room temperature. This observation is significant in that we demonstrated that crystalline phase boundary could be reliably identified with a continuous material chip. We also measured the electro-optic coefficients of BST on the composition spread. Figure 3 shows the linear (γ) and quadratic (s) coefficients as a function of x. The nonzero value of the linear (γ e-o coefficient above $x = 0.3$ indicates that the transition from ferroelectric-paraelectric phase boundary (FPB) is not complete at $x = 0.3$. The diffusive nature of the boundary indicates the coexistence of ferroelectric and cubic phases in these films.

On the *left* and *lower edge* of the ε_r map where the composition is (Sr$_{1-x}$Ca$_x$)TiO$_3$ and (Ca$_{1-x}$Ba$_x$)TiO$_3$ ($0 < x < 1$), respectively, we found that ε_r decreases with increasing amounts of Ca. Similar trends have been observed previously in ceramic studies of such compositions [2,3]. Compositions of Ba$_{0.5-0.7}$Sr$_{0.3-0.5}$TiO$_3$

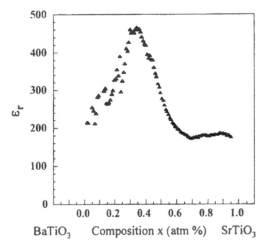

Figure 2   Dielectric constant vs. composition x of $Ba_{1-x}Sr_xTiO_3$. The composition with maximum dielectric constant coincides with phase boundary at $x = 0.3$ (pseudo-cubic paraelectric phase—tetragonal ferroelectric phase).

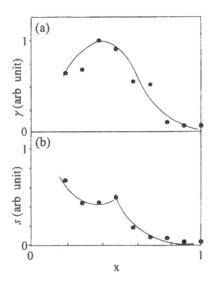

Figure 3   Graphs of measured (a) linear and (b) quadratic electrooptic coefficients vs. composition (x) for the $Ba_{1-x}Sr_xO_3$ composition gradient.

display the highest ε_r (≈ 500), SrTiO$_3$ the lowest ε_r (≈ 150) and BaTiO$_3$ has $\varepsilon_r \approx$ 400 at 1 GHz. Shown in the loss tan δ map, the value of tan δ for BaTiO$_3$ is 0.4, the same as that of the single-crystal value, whereas the value for SrTiO$_3$ is lowest as expected. Atomic force microscopy measurements of the triangle phase spread show that the average grain size is 150×150 nm^2 with no signs of off-stoichiometry outgrowths at the surface. Thus, we believe that the variations in both ε_r and tan δ of the phase spread are predominantly due to changes in composition of the thin films rather than extrinsic effects, such as differences in microstructures and strain in the film.

X-ray diffraction (XRD) patterns of large individual samples of various selected compositions from the BSCT triangle as well as doped BST showed that these films were (100) oriented and *epitaxial* with respect to the (100) LaAlO$_3$ single-crystal substrates. Rutherford backscattering (RBS) results on similar samples deposited on MgO indicated that the precursor layers have interdiffused. There are no significant differences in surface morphology or grain sizes among samples with different compositions in atomic force microscopy studies. These results suggest that the observed changes in physical properties (ε_r and tan δ) are not significantly affected by changes other than composition.

The grayish white regions in Figure 1b are the areas with tan δ values less than 0.02. Selecting regions of high ε_r with low tan δ from Figure 1a, we find that compositions within the regions of Ba$_{0.12-0.25}$Sr$_{0.35-0.47}$Ca$_{0.32-0.53}$TiO$_3$ have the lowest tan δ and are the best candidates for dielectric applications. These films have tan δ values less than 0.02 and ε_r between 130 and 160. From Figure 1, we find that the values of tan δ of this region are even lower than that of SrTiO$_3$, whereas the values of ε_r are the same or higher.

B. Pb(Zr$_{1-x}$Ti$_x$)O$_3$

Lead-containing perovskite compounds exhibit a myriad of interesting properties. A family of lead-based ferroelectrics, such as Pb(Mg$_{1/3}$Nb$_{2/3}$)O$_3$ (PMN), Pb(Zn$_{1/3}$ Nb$_{2/3}$)O$_3$ (PZN), Pb(Fe$_{1/2}$Ta$_{1/2}$)O$_3$ (PFT), Pb(Co$_{1/2}$W$_{1/2}$)O$_3$ (PCW), etc., are classified as relaxor ferroelectrics. Table 1 listed many complex relaxor compounds similar to PMN. These compounds have a perovskite structure (ABO$_3$) in which the B sites are occupied by two types of ions having chemical valences different from 4^+. They display diffuse paraelectric-ferroelectric phase transitions because of the fluctuations in the ratio of B$_1$ to B$_2$ ion concentration [1]. The diffused and metastable phases can be electrically driven from the paraelectric to ferroelectric phase. This property is extremely useful in many applications, including e-o devices in telecommunication and tunable microwave devices in microwave electronics and radar applications.

By adding one of these complex compounds to a PZT binary system, ternary solid solution with perovskite structure can be formed. Figure 4 shows two-

Table 1 Complex Metal–Ion Combination in Some Ternary Piezoelectric Ceramics

$Mg_{1/3}Nb_{2/3}$	$Co_{1/3}Nb_{2/3}$	$Li_{1/4}Ta_{3/4}$	$Mn_{1/2}Nb_{1/2}$	$Ni_{1/3}Sb_{2/3}$
$Zn_{1/3}Nb_{2/3}$	$Fe_{1/3}Nb_{2/3}$	$Cu_{1/4}Nb_{3/4}$	$Mn_{2/3}W_{1/3}$	$Ni_{1/3}Bi_{2/3}$
$Mg_{1/3}Ta_{2/3}$	$Fe_{1/3}Sb_{2/3}$	$Sb_{1/2}Nb_{1/2}$	$Mn_{1/3}W_{3/2}$	$Ni_{1/2}W_{1/2}$
$Zn_{1/3}Ta_{2/3}$	$Ni_{1/3}Nb_{2/3}$	$In_{1/2}Nb_{1/2}$	$Mn_{1/2}W_{1/2}$	$Fe_{1/2}Sb_{1/2}$
$Sb_{1/3}Nb_{2/3}$	$Sn_{1/3}Nb_{2/3}$	$Co_{1/2}W_{1/2}$	$Mn_{1/2}Ta_{1/2}$	$Sb_{1/2}Ta_{1/2}$
$Mn_{1/3}Nb_{2/3}$	$Li_{1/4}Sb_{3/4}$	$Cd_{1/2}W_{1/2}$	$Mn_{1/2}Sb_{1/2}$	$Al_{1/2}Te_{1/2}$
$Mn_{1/3}Ta_{2/3}$	$Li_{1/4}Nb_{3/4}$	$Co_{1/3}Nb_{2/3}$	$Mn_{1/3}Bi_{2/3}$	$In_{1/2}Te_{1/2}$
$Cd_{1/3}Nb_{2/3}$	$Fe_{1/2}Ta_{1/2}$	$Mg_{1/2}W_{1/2}$	$Mn_{1/3}Sb_{2/3}$	$Te_{1/3}Fe_{2/3}$
$Y_{1/2}Nb_{1/2}$	$Zn_{1/3}Ta_{2/3}$	$Ni_{1/5}Fe_{1/5}Nb_{3/5}$		

phase diagrams at room temperature for the ternary systems [1]. As in PZT, compositions near MPBs in the ternary system correspond to large values of dielectric and piezoelectric parameters, especially for compositions near the pseudo-cubic (PC)–tetragonal (T)–rhombohedral (R) transition. If the composition is on the PC side, the compounds should be a relaxor with large quadratic e-o effect and microwave tunability. However, these have not been characterized due to the tremendous amount of work involved if conventional materials synthesis methods are used. We believe that great opportunities exist in these materials systems to apply continuous-phase diagrams chips technique.

Synthesis of these materials is in itself a challenging task. The termodynamically more stable perovskite phases are usually contaminated with the more kinetically favorable but undesirable pyrochlore phases. Several methods of fabrication for both bulk and thin films have been developed to overcome this hurdle [4]. To overcome this difficulty, we developed a technique that allows the formation of pyrochlore-free lead pervoskite compounds. First, 200 Å of Zr and 150 Å of Ti (give a Zr/Ti ratio of $1:1$) was deposited and annealed at 800°C under 1 atm of Ar for 4 days to interdiffuse the metals (Fig. 5). Then 670 Å of PbO was deposited a top the alloyed thin film and heated under 10 atm Ar at 400°C for 4 days, to enable diffusion of Pb into the metal alloy (Fig. 6). The RBS spectra in Figs. 4 and 5 confirmed significant mixing of the different components. The film is then further annealed under flowing oxygen at 700°C for a half-hour to form the crystalline perovskite compound. XRD data showed only PZT peaks with no observable pyrochlore peaks.

Figure 7 is the SEMP scan of a $Pb(Zr, Ti)O_3$ composition spread from $PbTiO_3$ to $PbZrO_3$. The ε_r peaks at 48:52 (Ti/Zr) ratio, the well-known MPB in PZT [1]. This again demonstrated our ability to reliably map the phase boundaries in ferroelectric materials systems. Based on this success, we are now intensively mapping relaxor-PT-PZ or relaxor-PT-PN ternary systems with exciting results, which are beyond the scope of this case study.

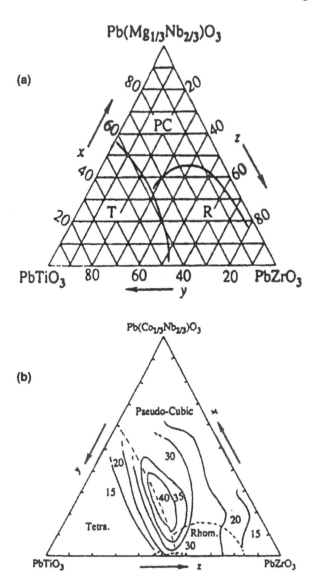

Figure 4 (a) Phase diagram at room temperature for the pseudoternary system $Pb(Mg_{1/3}Nb_{2/3})O_3$-$PbTiO_3$-$PbZrO_3$. Slid lines are morphotropic phase boundaries. PC is pseudo-cubic; T stands for tetragonal and R for rhombohedral. [After H. Ouchi, K. Nagano, and S. Hayakawa, *J. Am. Ceram. Soc. 48*, 630 (1965); H. Ouchi, K. Nagano, and S. Hayakawa, *J. Am. Ceram. Soc. 51*, 169 (1968).] (b) Phase diagram at room temperature for the pseudo-ternary system $Pb(Co_{1/3}Nb_{2/3})O_3$-$PbTiO_3$-$PbZrO_3$. Dotted lines are morphotropic phase boundaries. Numbers denote the percentage of K_p values at isopleths. [After T. Kudo, T. Yazaki, F. Naito, and S. Sugaya, *J. Am. Ceram. Soc. 53*, 326 (1970).]

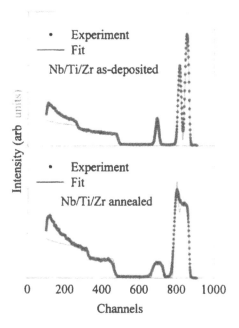

Figure 5 RBS studies of as-deposited metal precursor layers and annealed composite. The broadening of peaks indicate that each precursor is diffused throughout the composite.

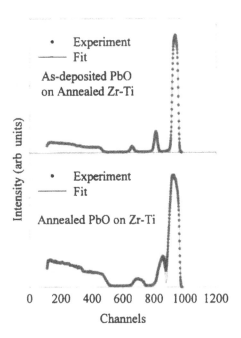

Figure 6 RBS studies of as-deposited Pb precursor layer and annealed composite. The broadening of Pb peaks indicate that Pb is diffused throughout the composite.

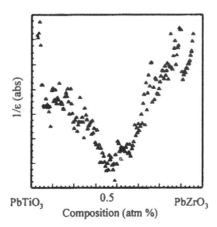

Figure 7 Inverse dielectric constant (arbitrary unit) vs. composition x of $PTi_{1-x}Zr_xO_3$. The composition with maximum dielectric constant coincides with morphotropic phase boundary at $x = 0.47$ (ferroelectric tetragonal-rhombohedral phase).

IV. DOPANTS STUDY

To facilitate study of the effects of different dopants on the dielectric properties of $(Ba_{1-x}Sr_x)TiO_3$ (BST), a chip consisting of four different stoichiometries of BST thin films ($x = 1.0, 0.8, 0.7$, and 0.5) was generated. These four hosts were doped with different combinations of up to three of nine different metallic elements, with each dopant added in excess of 1 mol % with respect to the BST host. The quaternary combinatorial masking scheme (see Chapter 7, Fig. 1) was used to generate $4^4 = 256$ different compositions in 16 steps. To fabricate the library, TiO_2 (870 Å) was deposited first to generate an array of 256 samples, each $650 \times 650 \ \mu m^2$, on a single-crystal (100) $LaAlO_3$ substrate. This was followed by deposition of BaF_2, SrF_2, and different dopants in the following sequence: C_2: Fe_2O_3(7 Å); C_1: W(5 Å); C_4: CaF_2(12 Å); D_2: Cr(4 Å); D_1: Mn_3O_4 (7 Å); D_4: CeO_2(12 Å); E_2: MgO(7 Å); E_1: Y_2O_3(10 Å); E_4: La_2O_3(12 Å); B_2: BaF_2(1640 Å); B_1: SrF_2 (270 Å) + BaF_2(1320 Å); B_4: SrF_2(410 Å) + BaF_2(940 Å); B_3: SrF_2(680 Å) + BaF_2(830 Å), where A, B, C, and D are the quaternary masks used in the deposition steps (see Chapter 7, Fig. 1), the subscript i indicates an $(i - 1) \times 90°$ clockwise rotation of the mask relative to the orientation of the mask, and the numbers in parentheses indicate the film thickness. The resulting dopant map of the library (for each host compound) is shown in Figure 8a. The dopant layers were sandwiched between TiO_2 and the fluoride materials to prevent excessive evaporation of dopants during subsequent annealing steps. The library was then heated at 400°C in flowing oxygen for 24 h to facilitate mixing of the precursors (TiO_2, BaF_2, and SrF_2) as well as diffusion of the dop-

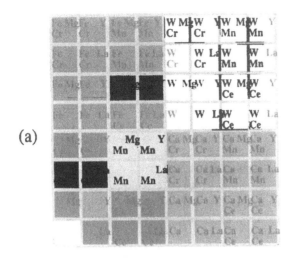

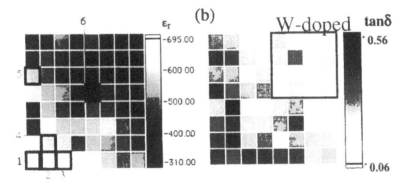

Figure 8 (a) Dopant (each element is 1 mol %) map; (b) dielectric constant and loss tangent maps of BTO.

ants. This was followed by a further annealing in flowing oxygen at 900°C for 1.5 h. The resulting high-density integrated materials chip of doped (Ba_xSrF_{1-x})-TiO_3 is of epitaxial quality, as confirmed by X-ray studies.

We used the SEMP to make nondestructive measurements of the dielectric properties of the thin-film library samples. The microwave impedance images (at 1 GHz) of the BTO-containing region of the library, consisting of 64 samples, is shown in Figure 8b; each square corresponds to a different thin-film sample site. The data have been converted to averaged ε_r and tan δ values for each site from images of resonant frequency and Q shifts, respectively.

The ε_r image of the library (Fig. 8b, upper), in which a lower dielectric constant is represented by a darker shade, shows that samples on the *upper right-hand quadrant* have a lower value of ε_r relative to pure BTO (marked sample 1 in Fig. 8b, lower). All these sites contain 1 mol % W, which indicates that doping with W lowers the dielectric constant of BTO. This trend is also observed in other films with different ratios of Ba and Sr. In general, most dopants decrease ε_r, except for a few instances like La (sample 2) and Ce (sample 3) doped samples. A comparison of samples that differ only in being doped with either La or Y but are otherwise identical reveals that the samples doped with the larger ion La have higher ε_r, than those doped with Y. This is evident in the alternating light and dark sites on the even columns of the ε_r image for samples with the BTO host (Fig. 8b). We have found that the differences between La-doped and Y-doped samples become less pronounced in hosts containing more Sr.

In the tan δ image (Fig. 8b, upper), a lighter shade implies a lower loss tangent (tan δ). Thus, all samples in the W-doped quadrant, which have reduced ε_r slightly, also have lower loss tangent in comparison with pure BTO. For example, the tan δ of W-doped BTO (sample 6) is reduced to 0.1 from 0.42 for undoped BTO (sample 1), whereas its ε_r is reduced to 406 from a value of 593 (BTO) at 1 GHz. In capacitance applications, tan δ (not ε_r) is the figure of merit. For tunable microwave devices, the figure of merit is given by $2(\Delta\varepsilon_r/\varepsilon_r)/\tan \delta$. This reduction in tan δ by doping with W is also observed in the other host materials studied here. Such a finding is especially important for the microwave application of these tunable dielectric materials.

The effects to dielectric constants and tan δ of both host materials by the addition of tungsten are summarized in Table 2. The tunabilities ($\Delta\varepsilon_r/\varepsilon_r$) of the tungsten-doped samples are not significantly lower than those of the pure materials. The results are summarized in Table 3, which demonstrates that the figures

Table 2   Dielectric Constant (ε_r) and Loss (tan δ) at 1 GHz, of the Highlighted Films (Fig. 8)[a]

Films	Sample	ε_r	tan δ
$BaTiO_3$	1	593	0.42
$BaTiO_3/La_2O_3$	2	695	0.35
$BaTiO_3/CeO_2$	3	634	0.45
$BaTiO_3/Y_2O_3$	4	576	0.54
$BaTiO_3/Fe_3O_4$, MgO	5	559	0.28
$BaTiO_3/W$	6	406	0.10
$SrTiO_3$		215	0.02
$SrTiO_3/W$		177	0.01

[a] Pure and doped samples of $SrTiO_3$ are included for comparison.

Table 3 Figure of Merits for Tungsten-Doped Materials

Films	Tunability (%)	ε_r at 0 kV/cm	tan δ at 0 kV/cm	FOM
$Ba_{0.5}Sr_{0.5}TiO_3$	56	1187	0.030	37
W-doped $Ba_{0.5}Sr_{0.5}TiO_3$	40	706	0.012	67
$Ba_{0.7}Sr_{0.3}TiO_3$	45	1106	0.034	25
W-doped $Ba_{0.7}Sr_{0.3}TiO_3$	39	1001	0.017	46

ᵃ The figure of merit improvement for tungsten-doped $(Ba,Sr)TiO_3$ is close to a factor of 2. Tunability is taken as the percentage change in dielectric constant from 0 kV/cm to 40 kV/cm. The measurements were done at 1 MHz.

of merit for tungsten-doped samples are significantly improved (by a factor of about 2). Based on this study, careful growth of high-quality film of W-BSTO by J. Horwitz at the Naval Research Laboratory has yielded a record low microwave loss (tan δ = 0.005) at room temperature. This improvement is of great significance in device applications.

V. CONCLUSION

As we demonstrated in this case study, combinatorial materials chips techniques allowed us to perform experiments up to 10,000 time more effectively and faster than conventional methods. Guided by good empirical observations and theoretical understanding, the approach has been demonstrated to be highly effective and successful in real-world materials research. However, this is only the beginning of a long journey with many challenges ahead. With continuous innovative efforts by scientists worldwide, this approach should be widely adopted in various fields of materials research and bring a revolution in materials research in the next millennium.

REFERENCES

1. Yuhuan Xu, *Ferroelectric Materials and Their Applications*, North-Holland, 1991.
2. G. Durst and M. Grotenuis, A.G. Barkow, *J. Am. Ceram. Soc. 33*, 133 (1950).
3. T. Mitsui and W.B. Westphal, *Phys. Rev. 124*, 1354 (1961).
4. Y.-F. Chen, T. Yu, J.-X. Chen, L. Shun, P. Li, and N.-B. Ming, *Appl. Phy. Lett. 66*, 148 (1996); Y.J. Song, Y. Zhu, and S.B. Desu, *Appl. Phys. Lett. 72*, 2686 (1998); Y. Lin, B.R. Zhao, H.B. Peng, B. Xu, H. Chen, F. Wu, H.J. Tao, Z.X. Zhao, and J.S. Chen, *Appl. Phys. Lett. 73*, 2781 (1998).

Case Study 7-2
Synthesis and Evaluation of Thin-Film Dielectrics Using a Composition Spread Approach

R. Bruce van Dover and Lynn F. Schneemeyer
Bell Laboratories, Lucent Technologies, Murray Hill, New Jersey

ABSTRACT

Synthesis of thin-film dielectrics with composition spreads in two dimensions is described. Automated rapid measurement allows key properties to be evaluated efficiently. The technique allows most of a pseudoternary oxide phase diagram to be meaningfully evaluated in about 24 h.

I. EQUIPMENT AND SUPPLIES

1. Turbo-pumped vacuum chamber (base pressure 10^{-7} torr) equipped with three US Gun II brand magnetron sputter guns oriented for 90-degree off-axis deposition; four 600-W radiofrequency (RF) power supplies for sputtering and substrate bias; O_2 and Ar gas supply with mass flow controllers (0–50 sccm).
2. 2-in-diameter metal sputter targets (e.g., Zr, Ti, and Sn).
3. 7×7 cm Si substrates, coated with 60 nm of TiN, to serve as a base electrode.
4. sample holder with provision for heating and RF bias.
5. probe station with x-y translation Hg probe.
6. Instrumentation for measuring capacitance (C), and current (I)–voltage (V) characteristics, specifically a Keithley 617 source/electrometer and an HP4174 impedance analyzer.

265

7. Software for controlling Hg probe position, V source, and I and C measurement, written specifically for this purpose in LabVIEW.

II. INTRODUCTION

New materials with a high dielectric constant and high breakdown field strength could replace thin-film SiO_2 as the dielectric in integrated-circuit capacitors and would be especially useful for increasing the bit density in dynamic random access memory (DRAM) [1]. As part of a Bell Laboratories effort to identify candidate materials, we fabricated pseudoternary oxides (i.e., oxides with three cations) and evaluated their electrical properties [2,3]. The films were deposited in a composition spread configuration using 90-degree off-axis cosputtering with three independent sources and RF power supplies. Evaluation of the films was by point-by-point measurement of the capacitance and breakdown voltage, the product of which is a useful figure of merit (FOM) [4].

III. GENERAL PROCEDURE

Films were deposited by reactive cosputtering using the experimental arrangement shown in Figure 1a. Three planar magnetron sputter guns equipped with metal targets are arranged at 90-degree intervals each approximately 2 cm from

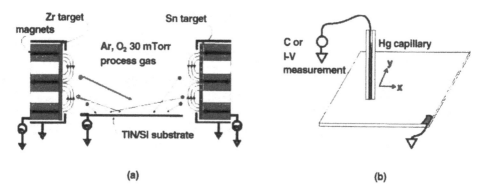

(a)   (b)

Figure 1 (a) Schematic illustration of the 90-degree off-axis sputtering configuration showing two (of three) magnetron sputter guns. A natural composition gradient is created by the diffusion limited arrival on the substrate of atoms or molecules sputtered from the target. (b) Schematic illustration of the Hg-capillary measurement technique. A capillary with an inner diameter of about 1 mm is filled with Hg and translated across the sample. Contact is made to the Hg and to a wire soldered to the base electrode (TiN) to make electrical measurements.

the side of a rectangular substrate. The Zr, Ti, and Sn guns were controlled independently and were typically run at 75, 125, and 15 W of RF power, respectively, in order to prepare a suitable fraction of the Zr/Ti/Sn/O pseudoternary phase spread. Depositions were performed at a total pressure of 30 mtorr, with flows of 15 sccm Ar and 10 sccm O_2. Substrates were Si wafers coated with 60 nm of TiN and were biased with 10 W of RF power during the deposition to promote surface mobility on the growing film, which we found to be important for growing high-quality films in the off-axis geometry. TiN, a metallic conductor used in conventional silicon integrated circuit fabrication, then served as the bottom electrode of the capacitors. The temperature of the substrate holder was held at about 200°C during the typical 20-min deposition time. The film thickness (typically 400–700 Å) and cation compositions were measured at selected spots by Rutherford backscattering (RBS) and interpolated using a model developed previously.

A. Electrical Characterization

The electrical properties of the resulting films were typically evaluated using a scanning Hg probe instrument schematically illustrated in Figure 1b. The area of the resulting capacitor was 4.05×10^{-3} cm^2. With proper attention this approach has proven reliable. The capacitance was measured using a commercial impedance analyzer with a frequency of 10 kHz and an alternating-current amplitude of 0.1 V. The results of this measurement are shown in Figure 2a as a function of position on the sample that was deposited at $T_s \sim 200$°C with 40% oxygen in the ambient sputtering atmosphere. I-V curves were measured by stepping the voltage, waiting 1 s, and measuring the resulting current with an electrometer. When the current jumped by more than 10-fold, or when it exceeded

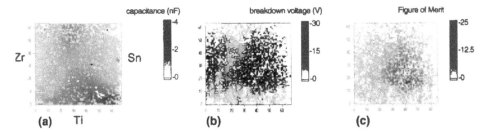

Figure 2 (a) Capacitance as a function of position. The x and y scales are in millimeters. The approximate positions of the sputter guns are as indicated. (b) Breakdown voltage as a function of position, for the same sample as in (a). (c) The figure of merit, defined as CV_{bd}/A, as a function of position. This data are obtained by multiplying the data of (a) and (b) point by point. The intensity scale is 0–25 µC/cm^2.

approximately 2×10^{-4} A/cm^2, the capacitor was deemed to have reached breakdown and the corresponding voltage is denoted V_{br}. The results of this measurement are shown in Figure 2b.

B. Figure of Merit

The capacitors were evaluated using the product $\varepsilon E_{br} \equiv CV_{br}/A$ as a FOM. This value physically corresponds to the maximum charge that can be stored on the capacitor and is typically in the range of 1–5 μC/cm^2. For example, in high-quality deposited SiO$_2$ films, $\varepsilon_r = 3.9$ and $E_{br} \sim 8$ MV/cm, so $\varepsilon_r\varepsilon_0 E_{br} \sim 2.8$ μC/cm^2. Using V_{br} as a measure of the breakdown voltage, we can evaluate the FOM point by point from the measurements taken with the Hg probe, as shown in Fig. 2c. This FOM, CV_{br}/A, is independent of dielectric thickness as long as ε_r and E_{br} are independent of thickness. The extent to which we can neglect film thickness effects is worthy of mention since the off-axis deposition technique being used to produce these compositional spreads necessarily produces films that vary somewhat in thickness over the substrate. Furthermore, the film thickness needed would depend on the design rules and other design considerations. However, in these and related amorphous films, thickness-dependent effects have been found to be relatively unimportant [5].

C. Composition Mapping

RBS measurements at a few points together with a model for the cation deposition rate as a function of distance from each gun allows mapping of the data onto a conventional ternary composition diagram, as shown in Fig. 3. Because more

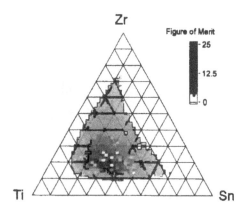

Figure 3 Data of Fig. 2c mapped onto a conventional ternary composition diagram, indicating the optimum compositions for the amorphous Zr-Sn-Ti-O dielectric thin-film system.

than one original data point can be mapped onto a single pixel, we have chosen the highest value in the set. The properties of thin films in the a-Zr-Sn-Ti-O system depend strongly on deposition conditions, particularly substrate temperature.

IV. COMMENT

High-throughout synthesis/evaluation of inorganic materials systems—specifically the continuous composition spread technique we have described—can successfully address well-defined problems where the range of compositions and processing parameters are constrained and where a suitable evaluation protocol can be defined. Success also requires that the behavior of the material be dominated by chemical composition rather than extrinsic factors (such as microstructure) that are inevitably highly process-dependent. In the present case, the material is intended to be used in thin-film form, so that the relevance of the synthetic technique was easy to establish with control runs of standard materials, such as Ta_2O_5. The deposition temperature could be constrained because it is evident that to make the desired amorphous films the substrate temperature should be fairly low. A few experiments were all that was needed to determine that 200°C would give results that reliably reflect the general trends, though certainly the properties depend on this parameter [3]. Since we wanted dielectric films with low leakage current, we reasoned that a relatively high oxygen partial pressure would help ensure fully oxidized cations and minimize traps and hopping sites caused by oxygen vacancies, and therefore settled on a standard 40% O_2 ratio (balance Ar, by volume).

REFERENCES

1. B El-Kareh, GB Bronner, and SE Schuster, *Solid State Technol.* May 1997, pp. 89–101; *The National Technology Roadmap for Semiconductors* (Sematech, 1994) p. 123. (also http://www.sematech.org/public/roadmap/ntrs94.pdf).
2. RB van Dover, LF Schneemeyer, and RM Fleming, *Nature 392*, 162 (1998); RB van Dover and LF Schneemeyer, *IEEE Electron Dev. Lett. 19*, 329 (1998).
3. LF Schneemeyer, RB van Dover, and RM Fleming, in *Ferroelectric Thin Films VII*, Materials Research Society, Pittsburgh, PA. Vol. 541, pp. 567–572.
4. D Gerstenberg. In: L Maissel and R Glang, eds. *Handbook of Thin Film Technology.* McGraw-Hill, New York, 1970, p.19–8.
5. Note that this is not the case for polycrystalline films, for which a strong dependence of ε on thickness is typically observed; see, e.g., C Basceri, SK Streiffer, AI Kingon, and RJ Waser, *J. Appl. Phys. 82*, 2497–2504 (1997).

Chapter 8
High-Throughput Screening of Catalyst Libraries

Richard C. Willson, David R. Hill, and Phillip R. Gibbs
University of Houston, Houston, Texas

I. INTRODUCTION

The recent flowering of high-throughput screening (HTS) for pharmaceutical discovery has been followed by application of these methods to materials and catalysts. While the development of HTS catalyst testing is in its early stages, there
is a broadening sense that high-throughput approaches may have an enormous
impact in this area.

Some of the roots of combinatorial materials chemistry can be traced back
to the 1970s and early 1980s. While at RCA, J. J. Hanak conducted early explorations into improving experimental efficiency by simultaneous testing of multiple
samples [1]. A radiofrequency co-sputtering method was used to deposit binary
and ternary mixtures of candidate superconducting materials. After each deposition, film thickness was measured as a rapid assay for composition as a function
of position. Superconductivity was screened in parallel using arrays of 50 gold
contacts predeposited along the substrate at composition intervals of about 2%.
The whole strip was tested in a cryogenic device, and raw test data were processed
and plotted by computer. As pointed out by Hanak, the availability of synthesis
methods covering a large range of multicomponent compositions, a simple, nondestructive measure of materials composition, and a rapid screen for material
properties of interest are the necessary elements for this ''multiple-sample concept'' to be realized. Despite its promise, the project was discontinued due to its
then-burdensome computational requirements [2].

The intervening period has seen enormous development of HTS methods
in pharmaceutical and life sciences applications, which are thoroughly reviewed

elsewhere in this volume. The recent resurgence of combinatorial methods for materials and catalyst applications can be traced to the publication by Xiang et al. [3] describing a combinatorial library of candidate superconducting ceramics. These workers produced a positional array of ceramic superconductor candidates by sequential sputtering of precursors through masks onto a flat-wafer substrate. The library was thermally processed to produce crystalline cuprate superconductor candidates. Each element in the library was then tested sequentially for resistance as a function of temperature. While no new superconductors were discovered, the paper demonstrated the screening of relatively complex materials in a parallel fashion.

During the last few years, many publications have appeared in the literature applying the combinatorial approach to catalyst research, as summarized in Table 1. This work has been summarized and critiqued in several recent reviews [4–8].

A. Early Combinatorial Catalysis

Several pioneering papers used classical methods of activity determination but took a combinatorial approach to catalyst synthesis (Table 1). Representative of these was the work of Menger et al. [9] on development of functionalized polymer catalysts having phosphatase activity. Polyallylamine was functionalized with various combinations of eight carboxylic acids to create randomly functionalized polymers bearing mixed amides. Additional diversity was provided by adding metal ions, and p-nitrophenylphosphate hydrolysis activity was measured by spectrophotometry. The most active formulation accelerated hydrolysis by more than 10,000-fold.

Case Study 8-1 by Kuntz et al. illustrates the latest development of the work of the Hoveyda group in iterative positional optimization for enantioselective catalysts. In addition to the new catalysts generated, this report is of interest for its practical guidance on highly productive methods of catalyst development.

B. Labeling vs. Positional Identification

An important aspect of any combinatorial assay is the "addressing" of each element in a catalyst library. There are two general methods used to accomplish this task: positional addressing by spatial location in the catalyst array, and labeling. Positional addressing involves identifying each element of a library, e.g., by cartesian coordinates on a grid for a rectangular two-dimensional array. Other two-dimensional array configurations may also be addressed in similar manner. One-dimensional arrays on thread-like supports have recently been proposed by Schwabacher et al. [10]. These offer an intriguing compatibility with Fourier analysis of patterns of variation across the array, and the reduced cost of charac-

Table 1 History of Combinatorial Catalyst Screening

Date	Area
	Combinatorial Concept Development
1970	Hanak introduces "multiple-sample concept" in screening superconducting alloys.
1984	Geysen introduces combinatorial libraries of peptides on polymeric supports [26].
1995	Xiang et al. synthesize library of superconducting ceramics on solid support.
	Combinatorial Catalysis (serial activity measurement)
1995	Menger et al. screen phosphatase activity among randomly functionalized polymers.
1996	Hill and Gall combinatorially synthesize polyoxometalates, test for oxidation activity [27].
1995	Liu et al. examine solid-phase synthesis of 2-pyrrolidinemethanol ligands as catalysts for enantioselective addition reactions.
1996	Burgess et al. examine metal catalysts for diazo compound asymmetrical C-H insertion.
1996	Cole et al. improve catalyst activity and selectivity by optimizing one position at a time from a core structure through iterative positional scanning [28].
1998	Sigman et al. show synergistic effects between components of Schiff base catalysts used for enantioselective asymmetrical Strecker reactions.
	IR Thermography
1996	Moates et al. screen gas phase heterogeneous catalytic oxidation reactions.
1998	Taylor and Merken screen liquid phase acylation catalysts on encoded beads.
1998	Reetz et al. screen enzymatic acylation and chiral epoxide catalyst.
1998	Holzwarth et al. examine heterogeneous catalysis of 1-hexyne hydrogenation and also the oxidation of iso-octane and toluene.
	Dyes
1998	Copeland and Miller use on-bead fluorescent pH indicators to monitor acylation catalysts.
1999	Crabtree et al. develop novel dyes for detection of alkene and imine hydrosilation activity.
1999	Mallouk et al. screen fuel cell electrocatalysts using fluorescent indicators.
	Selectivity Screens
1998	Senkan uses REMPI to selectively detect reaction products in reaction plumes.
1999	Cong et al. develop scanning mass spectrometer probe for catalyst library screening.
	Polymerization Catalysts
1999	Boussie et al. examine ethylene polymerization in multireactor device.
1999	Boussie et al. assess labeled ethylene polymerization catalysts by size of product mass.

terization by point detectors as the array is pulled past the sensor. Although spatial addressing is the simplest method for tracking individual catalysts, arrays are not compatible with all assay methods and may not be efficient for very large libraries. In such cases, monitoring is accomplished with labeling.

Labeling techniques involve attaching an identifier to each member of a library. Chemical encoding might be preferred for very large libraries to allow a single-pot assay. However, if the library is smaller it may be easier to generate an array than to encode and decode individual candidates. Taylor and Morken [11] chemically encoded catalyst beads prior to single-pot assay by infrared (IR) thermography by inclusion of a label in each bead. The most active catalyst beads were removed and decoded to determine the catalytic structure. Symyx Technologies used similar chemical labeling techniques for identifying active catalysts responsible for increased polymer growth in a single-pot ethylene polymerization. Also, physical identifiers, such as fluors and radiofrequency tags, are available commercially to track chemical compounds placed in microreactors.

C. Library Preparation

Combinatorial catalyst libraries have been prepared by a wide variety of procedures. Most are based on thin-film deposition or on solution-based methods, such as conventional impregnation, sol-gel procedures, and split-and-pool synthesis. Solution-based methods generally require at least partial automation, since library preparation can be laborious and time consuming when performed manually. Attempts to streamline library preparation have resulted in novel approaches, such as modified inkjet printers which print catalyst precursor solutions onto a support [12]. The most common means of automation for solution-based library preparation, however, is pipetting robots. Originally marketed for biochemical applications, robots significantly decrease operator effort in preparing catalyst libraries by conventional impregnation and sol-gel procedures. Compounds that are not easily prepared by solution-based methods are often prepared by thin-film deposition methods, such as thermal evaporation, radiofrequency sputtering, and pulsed-laser ablation. Akporiaye et al. [13] demonstrated the combinatorial synthesis of zeolites for use as catalysts and adsorbents in a hydrothermal synthesis autoclave capable of carrying out 100 syntheses in parallel. The entire phase diagram of the $Na_2O-Al_2O_3-SiO_2-H_2O$ system was reconstructed from a single experiment.

II. LIBRARY SCREENING

A. Infrared Thermography

We and our co-workers at the University of Houston demonstrated that IR thermography was effective for detecting catalytic activity using the oxidation of hydrogen to water as the model reaction [14]. Thermography takes advantage of the heat liberated by many reactions; an active catalyst causes the reaction to proceed at a higher rate on the catalyst surface, generating localized excess heat. Emitted IR radiation is detected by an IR camera and translated into a surface temperature reading. Active catalysts appear on the IR image as bright spots against a monochrome background. These proof-of-concept experiments revealed enhanced activity for formulations known from the literature to be good oxidation catalysts. A control was performed to account for any variation of emissivity with composition but was found not to be necessary in this case. Variations in emissivity, the ratio of radiated power of a real surface to that of an ideal black surface at the same temperature, could be misinterpreted as temperature variations if not corrected. However, modern digital image-processing techniques facilitate such corrections.

Maier and co-workers at the Max-Planck-Institüt fur Kohlenforschung extended this work to quantitation of small surface temperature rises by carefully accounting for varying emissivities within the catalyst array, and by eliminating a variety of optical barriers to measurement of small temperature differences [15]. They used this improved methodology to screen an array of 37 different amorphous microporous mixed-metal oxides for catalytic activity in the hydrogenation of 1-hexyne.

Reetz et al. [16] demonstrated IR thermographic screening of catalysts for solution phase reactions, including the enantioselective lipase-catalyzed acylation of 1-phenylethanol with vinyl acetate and enantioselective transition metal catalysis of ring-opening hydrolysis of epoxides with formation of diols. IR emission from the center of the wells of the microplate was monitored in each case.

Taylor and Morken [11] extended IR thermography to screening of catalysts on encoded polymer beads for activity in solution. Encoded polymer beads, each loaded with a single catalyst prepared by split-and-pool methods, were placed in a single pot of reactants in solution and monitored for enhanced activity. By adjusting the density of the solution, the researchers were able to float the beads above the IR-opaque solution and image the entire library with an IR camera. Catalysts most active toward an acylation reaction appeared as bright spots in the IR image. Further developments and greater methodological detail are given in Case Study 8-2.

IR thermography has emerged as a popular means of detecting catalytic activity, stability, poisoning resistance, and reactant selectivity. However, it does not provide information on catalyst product selectivity.

B. Local pH Indicators, Reactant Dyes, and Product-Specific Dyes

The activity of catalyst candidates can be directly indicated through the use of pH-sensitive indicators (for reactions that liberate or sequester protons), reactant-analog dyes that undergo color changes in the presence of an active catalyst (in analogy with chromogenic enzyme substrates), and/or dyes that react with characteristic functional groups on products to produce colored products.

Mallouk and co-workers developed a method for screening fuel-cell electrocatalysts for the electrooxidation of methanol using fluorescent pH indicators [12]. The liberation of protons by electro-oxidation of methanol induces a local pH drop in the solution surrounding an active catalyst, which is signaled by the glow of a fluorescent indicator when viewed under an ultraviolet light source. Catalysts with greater activity exhibit higher fluorescent intensities and can be readily identified by visual inspection.

The researchers generated a triangular library of catalyst spots on conducting carbon paper and submerged the array into an aqueous methanolic sodium sulfate electrolyte solution containing a fluorescent indicator. A three-electrode cell was used to screen for catalytic activity as an overpotential was induced in the solution. The experiment revealed catalysts that promoted electro-oxidation at lower overpotentials, as well as the potentials at which other catalysts displayed activity. The technique was subsequently improved by modifying an ink-jet printer to print metal salts on the conducting carbon paper, allowing the rapid generation of much larger catalyst libraries.

Bead-coupled fluorescent pH indicators have been used to determine acylation activity of bead-bound catalysts by parallel screening [17]. Fluorescence was observed in response to the generation of either alcohol or acetic anhydride, products of acyl transfer reactions. Several catalysts of known activities for acylation were first tested in solution and screened using fluorometry, revealing high fluorescent intensities for the more active catalysts. No fluorescence was detected in the absence of a catalyst. When catalysts were attached to beads along with the indicator, beads containing the most active catalysts fluoresced as expected, indicating that the method may be used for screening single-bead/single-catalyst libraries using fluorescence-activated bead sorting or fluorescence microscopy.

Reactive dyes allow simple and effective detection of catalytic activity. Crabtree and co-workers developed novel dyes for detection of alkene and imine hydrosilation activity [18]. These ferrocenyl-substituted dyes contain electron donor and acceptor groups, which interact through a reactive functionality that can be saturated by an active catalyst, diminishing the electronic connection between the groups. The result is a change in the color of the dye, which is visible to the human eye and is easily recorded by a digital camera for further analysis. Twelve hydrosilation catalysts were screened and the most active was identified as Wilkinson's catalyst, which is known to be an active hydrosilation catalyst. A cyclic palladium catalyst was also newly identified as being active. It is also possible to assess product formation and selectivity by product-selective dyeing, as has been described by workers at DuPont. A large body of functional group–specific dyes has been developed, many of which have been used in developing thin-layer chromatography plates.

C. Mass Spectrometry

Information on both catalyst activity and selectivity can readily be obtained by use of mass spectrometry. However, a single mass spectrometer can only analyze a product stream from one sample at a time, making the goal a "rapid sequential" assay as opposed to the truly parallel screening possible with some of the optical methods.

Symyx Technologies developed a device to automate and accelerate the sequential scanning of a 136-element library by mass spectrometry [19]. The researchers place the catalyst library on a flat support capable of translating in two dimensions. The array moves beneath a probe that feeds reactant gases to one catalyst at a time. The catalyst being tested is heated with a CO_2 laser, and a vacuum line samples the product gases directly above the catalyst and sends the product gas to the mass spectrometer. The remaining product gases are removed by vacuum to prevent contamination of the other catalysts on the array. The model reaction for the experiment was the oxidation of CO to CO_2. Experimental results identified active catalysts for this reaction that are known to be active from prior work.

D. Resonance-Enhanced Multiphoton Ionization (REMPI)

A parallel screen for selectivity in gas-phase reactions has been developed by Selim Senkan of UCLA [20]. Resonance-enhanced multiphoton ionization (REMPI) uses tunable lasers to selectively photoionize a preselected reaction product in the product gas streams leaving catalyst candidates, with the resulting photoions being detected by microelectrodes positioned just downstream of each candidate. Multiple products can be detected by repeating the measurement at a different wavelength, and selectivity calculated from the results.

This technique was demonstrated with catalysts for the dehydrogenation of cyclohexane to benzene. Each product benzene molecule was excited by a first photon and then ionized by a second photon, releasing an electron. The resulting photoions were detected nearly simultaneously by an array of microelectrodes positioned directly above each catalyst. As only benzene was ionized in this experiment, no information on product selectivity was generated.

Senkan and Ozturk [21] constructed a novel ceramic array microreactor system allowing the application of REMPI product detection to 16 samples (plus one control) simultaneously. Impregnated alumina pellets were held in small parallel channels. This reactor was used for rapid testing of 66 Pt/Pd/In ternary catalyst combinations for cyclohexane dehydrogenation activity. The REMPI technique shows great promise, as it is capable of providing parallel real-time analysis of both activity and selectivity for an entire catalyst library. However, the researchers point out that the method requires careful analysis of the spectral and photoionization properties of each product molecule to be detected.

E. Optical Absorbance

Optical absorbance has been used by Menger and co-workers at Emory University to detect catalytic activity for the dehydration of a β-hydroxyketone [9]. Increases in ultraviolet absorbance were monitored simultaneously for multiple-catalyst samples.

We are exploring imaging IR absorbance product detection as a means of parallel catalyst *selectivity* determination. Infrared absorption by a reaction product can be detected using an IR-sensitive camera with an appropriate narrow-bandpass filter [22]. To test the feasibility of the method, a polypropylene tube was sealed at each end with an IR-transparent polyethylene film and filled with air containing CO_2 at known concentrations. Carbon dioxide absorbs IR radiation near 4.2 μm, and imaging of the gas-filled tube through a 4.2 μm bandpass interference filter allowed direct observation of the concentration of this compound, a common undesired byproduct in partial oxidations. Since the same IR camera was used for both IR thermography and IR absorption imaging, these methods potentially could be combined to yield a compound HTS method. Related work has been reported by McFarland and Archibald [23], and work using fourier transform infrared microscopy is underway in the laboratory of Jochen Lauterbach at Purdue University.

F. Imaging Polarimetry

Polarimetry, the measurement of the optical activity of solutions by their rotation of linearly polarized light, is widely employed in the pharmaceutical and food industries. However, it suffers from relatively low throughput because measurements to date have been made on only one sample at a time. We have recently constructed an imaging polarimeter capable of following the optical activity of hundreds of samples simultaneously. The device utilizes a collimated, uniform light source, inlet and outlet polarizers, multisample array, and a CMOS camera fitted with a telecentric lens. Optical rotation is detected as an increase or decrease in transmitted light intensity detected after passage through the analyzing polarizer or as a shift in the analyzing polarizer position giving maximum intensity. We are currently working to increase the sensitivity of the method. We anticipate that, among other applications, imaging polarimetry may be useful in optimizing the enantioselectivity of synthetic reactions, whether classical or enzyme-catalyzed.

G. Polymerization Catalyst Screening

Development of methods for HTS of polymerization catalysts stands as a major analytical challenge. The complex assemblage of characteristics, such as molecular weight distribution, monomer incorporation, and branching, that determine a polymer's performance is difficult to assess in a highly parallel way. Most reported progress in this area comes from Symyx Technologies.

Boussie et al. [24] reported that polymerization by active catalysts exposed to ethylene gas could be detected visually by the growth of the polymer masses around labeled catalyst beads, and that active catalysts could be identified by recovering and decoding the chemical labels. Jandeleit et al. [7] reported the development of a high-pressure polymerization reactor with a series of 48 reaction chambers in which ethylene polymerization catalysts could be characterized. Finally, Matsiev et al. [25] have described the use of a tuning fork resonator for real-time monitoring of the viscoelastic properties of a medium in which polymerization is occurring. A variety of other approaches have been discussed at meetings, including high-throughput high-performance liquid chromatography and parallel differential scanning calorimetry devices.

III. CONCLUSION

The development of combinatorial catalysis has been remarkably active in both the academic and the commercial areas. There is every reason to anticipate that these methods will become widespread in industry and, to some degree, in academia. Many opportunities and challenges remain.

REFERENCES

1. Hanak, J.J., The multiple-sample concept in materials research: synthesis, compositional analysis and testing of entire multicomponent systems, *J. Mater. Sci. 5*, 964–971 (1970).
2. Dagani, R., A faster route to new materials, *Chem. Eng. News*, March 8, 51–60 (1999).
3. Xiang, X., Sun, X., Briceno, G., Lou, Y., Wang, K., Chang, H., Wallace-Freedman, W.G., Chen, S., Schultz, P.G., A combinatorial approach to materials discovery, *Science 268*, 1738–1740 (1995).
4. Bein, T., Efficient assays for combinatorial methods for the discovery of catalysts, *Angew. Chem. Int. Ed. 38*, 323–326 (1999).
5. Hoveyda, A.H., Catalyst discovery through combinatorial chemistry, *Chem. Biol. 5*, R187–R191 (1998).
6. Jandeleit, B., Weinberg, W.H., Putting catalysis on the fast track, *Chem. Ind. 19*, 795–798 (1998).
7. Jandeleit, B., Schaefer, D. J., Powers, T.S., Turner, H.W., Weinberg, W.H., Combinatorial materials science and catalysis, *Angew. Chem. Int. Ed. 38*, 4000–4038 (1999).

8. Schlogl, R., Combinatorial chemistry in heterogeneous catalysis: a new scientific approach or "The King's New Clothes?" *Angew. Chem. Int. Ed. 37*, 2333–2336 (1998).

9. Menger, F.M., Ding, J., Barragan, V., Combinatorial catalysis of an elimination reaction, *J. Org. Chem. 63*, 7578–7579 (1998).

10. Schwabacher, A.W., Shen, Y. and Johnson, C.W., Fourier transform combinatorial chemistry, *J. Am. Chem. Soc. 121*, 8669–8670 (1999).

11. Taylor, S. J., Morken, J. P., Thermographic selection of effective catalysts from an encoded polymer-bound library, *Science 280*, 267–270 (1998).

12. Reddington, E., Sapienza, A., Gurau, B., Viswanathan, R., Sarangapani, S., Smotkin, E.S., Mallouk, T.E., Combinatorial electrochemistry: a highly parallel, optical screening method for discovery of better electrocatalysts, *Science 280*, 1735–1737 (1998).

13. Akporiaye, D.E., Dahl, I.M., Karlsson, A., and Wendelbo, R., Combinatorial approach to the hydrothermal synthesis of zeolites, *Angew. Chem. Int. Ed. 37*, 609–611 (1999).

14. Moates, F.C., Somani, M., Annamalai, J., Richardson, J.T., Luss, D., Willson, R.C., Infrared thermographic screening of combinatorial libraries of heterogeneous catalysts, *Ind. Eng. Chem. Res. 35*, 4801–4803 (1996).

15. Holzwarth, A., Schmidt, H., Maier, W.F., Detection of catalytic activity in combinatorial libraries of heterogeneous catalysts by IR thermography, *Angew. Chem. Int. Ed. 37*, 2644–2647 (1998).

16. Reetz, M.T., Becker, M.H., Kuhling, K.M., and Holzwarth, A., Time-resolved IR thermographic detection and screening of enantioselectivity in catalytic reactions, *Angew. Chem. Int. Ed. 37*, 2647–2650 (1998).

17. Copeland, G.T., Miller, S.J., A chemosensor-based approach to catalyst discovery in solution and on solid support, *J. Am. Chem. Soc. 121*, 4306–4307 (1999).

18. Cooper, A.C., McAlexander, L.H., Lee, D. Torres, M.T., Crabtree, R.H., Reactive dyes as a method for rapid screening of homogeneous catalysts, *J. Am. Chem. Soc. 120*, 9971–9972 (1998).

19. Cong, P., Doolen, R.D., Fan, Q., Giaquinta, D.M., Guan, S., McFarland, E.W., Poojary, D.M., Self, K., Turner, H.W., Weinberg, W.H., High-throughput synthesis and screening of combinatorial heterogeneous catalyst libraries, *Angew. Chem. Int. Ed. 38*, 484–488 (1999).

20. Senkan, S.M., High-throughput screening of solid-state catalyst libraries, *Nature 394*, 350–353 (1998).

21. Senkan, S.M., Ozturk, S., Discovery and optimization of heterogeneous catalysts by using combinatorial chemistry, *Angew. Chem. Int. Ed. 38*, 791–795 (1999).

22. Willson, R.C., Catalyst testing process and apparatus, PCT Patent Application WO97/32208, 4 September, 1997.

23. McFarland, E.W., Archibald, W., Infrared spectroscopy and imaging of libraries, PCT Patent Application WO98/15813, 16 April, 1998.

24. Boussie, T.R., Coutard, C., Turner, H., Murphy, V., Powers, T.S., Solid-phase synthesis and encoding strategies for olefin polymerization catalyst libraries, *Angew. Chem. Int. Ed. 37*, 3273–3275 (1998).

25. Matsiev, L., Bennett, J., and McFarland, E., Method and apparatus for characterizing

materials by using a mechanical resonator, PCT Patent Application WO99/ 18431A1, 15 April, 1999.

26. Geysen, H.M., Meloen, R.H., Barteling, S.J., Use of peptide synthesis to probe viral antigens for epitopes to a resolution of a single amino acid, *Proc. Natl. Acad. Sci. U.S.A., 81*(13), 3998–4002 (1984).

27. Hill, C.L., Gall, R.D., The first combinatorially prepared and evaluated inorganic catalysts. Polyoxometalates for the aerobic oxidation of the mustard analog tetrahydrothiophene (THT). *J. Mol. Catal. A 114*, 103–111 (1996).

28. Cole, B.M., Shimizu, K.D., Krueger, C.A., Harrity, J.P.A., Snapper, M.L., and Hoveyda, A.H., Discovery of chiral catalysts through ligand diversity: Ti-catalyzed enantioselective addition of TMSCN to meso epoxides, *Angew. Chem. Int. Ed. 35*, 1668–1671 (1998).

Case Study 8-1
High-Throughput Screening of Titanium–Tripeptide Schiff Base Complexes for Catalytic Asymmetrical Addition of Trimethylsilylcyanide to Epoxides and Imines

Kevin W. Kuntz, Marc L. Snapper, and Amir H. Hoveyda
Boston College, Chestnut Hill, Massachusetts

ABSTRACT

Libraries of Ti-tripeptide ligands are synthesized and screened to identify catalysts for the enantioselective addition of cyanide to imines and epoxides.

Chemicals

1. *N*-Fmoc-L-glycine-OH (*N*-9-fluorenylmethoxycarbonyl-glycine-OH)
2. Fmoc-alanine-OH
3. Fmoc-valine-OH
4. Fmoc-leucine-OH
5. Fmoc-isoleucine-OH
6. Fmoc-*tert*-leucine-OH
7. Fmoc-norleucine-OH
8. Fmoc-aspartic acid (OtBu)-OH (*tert*-butyl ester)
9. Fmoc-glutamic acid (OtBu)-OH

10. Fmoc-threonine (tBu)-OH (*tert*-butyl ether)
11. Fmoc-threonine (Bn)-OH (benzyl ether)
12. Fmoc-threonine (Trt)-OH (trityl ether)
13. Fmoc-serine (tBu)-OH
14. Fmoc-serine (Bn)-OH
15. Fmoc-serine (Trt)-OH
16. Fmoc-aspargine (NHTrt)-OH (trityl amide)
17. Fmoc-glutamine (NHTrt)-OH
18. Fmoc-phenylalanine-OH
19. Fmoc-phenylglycine-OH
20. Fmoc-homophenylalanine-OH
21. Fmoc-cyclohexylglycine-OH
22. Fmoc-cyclohexylalanine-OH
23. Fmoc-tyrosine (tBu)-OH
24. Fmoc-histidine (NTrt)-OH
25. Fmoc-proline-OH
26. Fmoc-arginine (Tos)-OH (*p*-toluenesulfonamide)
27. Fmoc-α-methylalanine-OH
28. Fmoc-D-threonine (tBu)-OH
29. Fmoc-D-phenylalanine-OH
30. Fmoc-D-alanine
31. Salicylaldehyde
32. 3,5-Dibromosalicylaldehyde
33. 3,5-Dichlorosalicylaldehyde
34. 3,5-Diiodosalicylaldehyde
35. 3-Fluorosalicylaldehyde
36. 5-Methoxysalicylaldehyde
37. 5-Bromo-3-methoxysalicylaldehyde
38. 4-Hydroxysalicylaldehyde
39. 5-Bromosalicylaldehyde
40. 3-Methoxy-5-nitrosalicylaldehyde
41. 3,5-Dinitrosalicylaldehyde
42. 5-Hydroxysalicylaldehyde
43. 3,5-Di-*tert*-butylsalicylaldehyde
44. 5-Nitrosalicylaldehyde
45. 4-Methoxysalicylaldehyde
46. 3-Hydroxysalicylaldehyde
47. 5-Bromo-3-nitrosalicylaldehyde
48. 1-Hydroxy-2-naphthaldehyde
49. 2-Hydroxy-1-naphthaldehyde
50. 2-Hydroxy-3-naphthaldehyde
51. 2-Furaldehyde

52. 2-Methoxybenzaldehyde
53. 2,2-Dimethyl-3-hydroxypropionaldehyde
54. 3-Hydroxylbenzaldehyde
55. 4-Hydroxybenzaldehyde
56. Diisopropylcarbodiimide
57. Piperidine
58. *N,N*-Dimethylformamide
59. Triethylamine
60. Methanol
61. Toluene
62. Tetrahydrofuran
63. Wang glycine resin (Wang resin preloaded with Fmoc-glycine was obtained from Advanced Chemtech)
64. Titanium tetraisopropoxide
65. Trimethylsilyl cyanide (TMSCN)
66. Cyclohexene oxide
67. Benzaldehyde
68. Aminodiphenylmethane
69. Isopropanol
70. Silica gel
71. Ethyl acetate

Equipment and Supplies

1. Biospin chromatography columns (obtained from Bio-Rad)
2. Rotating device (Rotamix obtained from Appropriate Technical Resources, Inc.)
3. Gilson brand Pipetman (50–200 µL, adjustable)
4. Gilson brand Pipetman (0.5–5 mL, adjustable)
5. Finnpipette brand multipipeter (40–200 µL, adjustable)
6. Polypropylene 96-well plates (1 and 2 mL)
7. 96-Well filter plates (obtained from Whatman)
8. UniVac Vacuum Manifold (obtained from Whatman)
9. Jouan brand centrifugal vacuum concentrator
10. Platform shaker
11. Test tubes (10 × 75 mm)

I. INTRODUCTION

A compelling and important problem in modern chemistry involves the identification of small-molecule catalysts that emulate enzymes in their ability to facilitate selective chemical transformations. However, discovering such catalysts is not a simple matter. This difficulty arises partly because often only a small energy difference separates a nonselective catalyst from a selective one (typically about 1–2 kcal/mol variance in the transition state). Such differences arise unpredictably and from subtle variations in the catalyst structure. It is therefore difficult to predict a priori which catalyst is optimal, especially in the context of a multistep catalytic cycle. The above factors, and the fact that the most selective catalysts are also highly substrate-specific (the identity of the optimum catalyst varies), suggest that the catalyst "design" may be most effectively performed through high-throughput screening.

Accordingly, to find better ways to identify optimum catalysts we turned our attention to recent advances in combinatorial and high-throughput protocols. As stated above, this shift in strategy arose from the realization that if it is often the screening of large number of potential catalysts that leads to an important discovery, then why not do it systematically and more efficiently, and cover a broader range of catalyst candidates? The strategic decision, however, forced us to face various technological issues: the selection of effective catalysts requires an assay that can measure selectivity, kinetics, and turnover, and such assays are not efficient enough to be used to screen large (more than 100) libraries of catalysts (for recent reviews, see Refs. 1–3a).

To address the above concerns, we developed a positional optimization strategy to identify Ti-based catalysts for the enantioselective opening of meso-epoxides (see Ref. 4). This strategy involves screening one unit (amino acid or Schiff base) of a peptide-based ligand structure for selectivity while holding the other units constant. This approach is utilized in an iterative sense until the optimal catalyst is found. As an example, we utilized our ligand optimization protocol in solution to determine the most selective peptide-Schiff base ligand for the addition of cyanide to cyclohexene oxide (Fig. 1). Thus, the three structural modules of the chiral ligand were modified systematically. To determine the best AA1, 10 different ligand systems were prepared on the solid support (other two components were retained), cleaved, and screened in parallel fashion as catalysts in enantioselective formation of the cyanohydrin. The Tle residue emerged as the superior AA1 candidate. With Tle as AA1 and a 1,2-naphthyl Schiff base, the search for the most appropriate AA2 (second-generation ligands, 16 different amino acids) led us to Thr (t-Bu). With Tle as AA1 and Thr (t-Bu) as AA2, 13 aldehydes of varying steric and electronic properties were selected, synthesized, and screened; 3-fluorosalicaldehyde stood out as the best choice (third-generation). Through three rapid screenings of ligand classes consisting of 10–16 members, a ligand structure [AA1 = t-Leu, AA2 = Thr (t-Bu), aldehyde = 3-fluoro-salicylaldehyde] was identified that effects enantioselective formation of the cyanohydrin in 89% e.e.

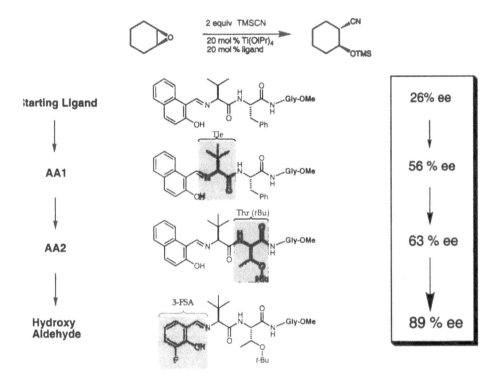

Figure 1 An iterative positional optimization approach was utilized to identify optimal ligand structures for the enantioselective addition of TMSCN (TMS = trimethylsilyl) to meso-epoxides.

The ligands can be screened in solution after cleavage from the resin by base-catalyzed transesterification and purification through a silica plug (as above), or while still attached to the resin (see Ref. 5). The number of components screened at each position is unrestricted, but this technique does not account for the possibility of cooperative effects between the different units. The approach proved highly fruitful, allowing us to quickly obtain ligand structures that give excellent selectivity for the addition of cyanide to several meso-epoxides.

Another advantage of our method is that it can lead to the discovery of the unexpected. For example, in our search for the most enantioselective Ti-ligand complex for the opening of cyclopentene oxide (see Ref. 5), it was observed that use of a ligand complex containing a particular amino acid results in the reversal of reaction selectivity without alteration of the absolute stereochemistry of the ligand (Fig. 2). A subtle modification in the constitution—not stereochemical identity—of one amino acid can, therefore, give rise to a reversal in selectivity.

With an effective ligand optimization strategy in hand, we turned our attention to the identification of a chiral catalyst for another important C-C bond forming process. Through the aforementioned catalyst screening protocol, we were able to identify specific Ti-peptide complexes that catalyze the addition of cyanide to imines (Strecker reaction; see Fig. 3 and Refs. 6 and 7) with outstanding enantioselectivity. Our catalyst screening approach led to the identification of catalysts that deliver amino nitriles with 90–97% e.e. and more than 90% yield for aromatic amino acids and 85% e.e. and 97% yield for the synthesis of *tert*-leucine. Many of the aminonitrile intermediates can be easily recrystallized to enantiopurity; importantly, a single hydrolysis/deprotection step provided the corresponding optically pure amino acids. One of the attractive features of this technique is that the simplicity of the library synthesis and screening allows for discovery of a unique metal–ligand complex for each substrate.

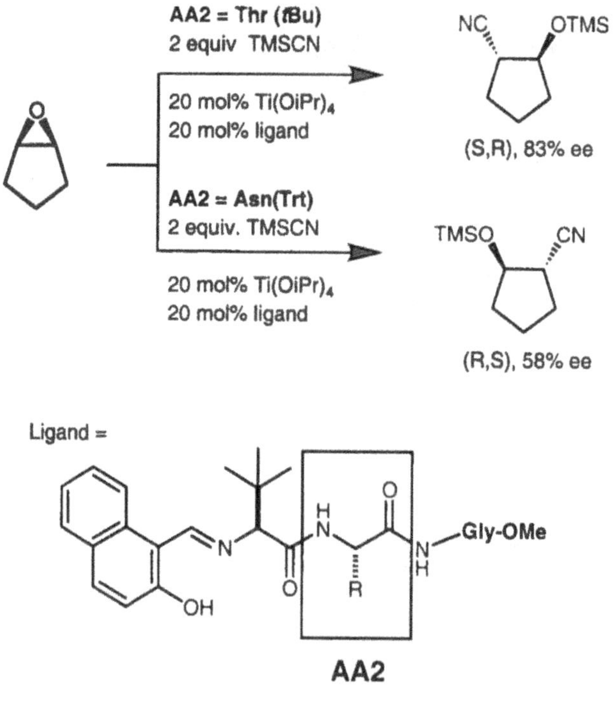

Figure 2 Subtle modifications of the catalyst structure can lead to significant variations in selectivity.

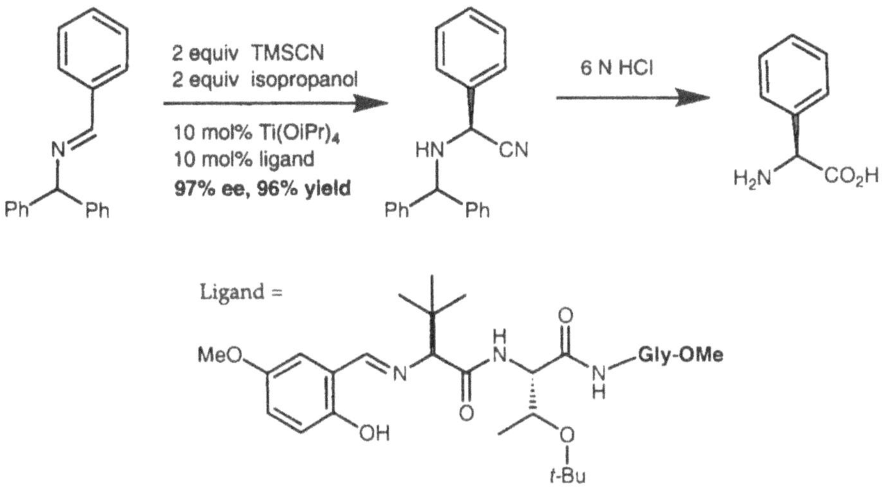

Ligand =

Figure 3 Outstanding catalysts for the synthesis of α-amino acids via the addition of cyanide to imines were discovered by employing the same screening strategy and modular ligand structure used for the opening of meso-epoxides.

II. GENERAL PROCEDURES

A. Preparation of Schiff Base Tripeptides on the Solid Phase

Reactions were performed in polypropylene Biospin chromatography tubes from Bio-Rad Laboratories with a fritted bottom and a cap on either end (Fig. 4). Wang-Gly-FMOC resin (100 mg, 0.04 mmol) was placed in the polypropylene reaction vessel and washed with DMF (3 × 1.5 mL). Rotation of a sample of the resin in DMF (1.5 mL) for 12 h led to swelling. At this point, the resin was washed with additional DMF (3 × 1.5 mL). Subsequent resin deprotection was effected by washing with 20% piperidine/DMF (1.5 mL), rotation of the sample for 1.5 h in 20% piperidine/DMF (1.5 mL), and then washing with DMF (10 × 1.5 mL). The FMOC-protected amino acid (0.17 mmol, AA2) was activated by treatment with an excess of DIC (49 μL, 0.25 mmol) in DMF (1.5 mL, 20 min). The resulting solution was added to H$_2$N-Gly-Wang, which was rotated for 1.5 h and then washed with DMF (10 × 1.5 mL). FMOC-AA2-Gly-Wang was deprotected by washing with 20% piperidine/DMF (1 × 1.5 mL), rotation for 1.5 h in 20% piperidine/DMF (1.5 mL), and then washing with DMF (10 × 1.5 mL). The second amino acid coupling and deprotection was performed in the same manner as before. Following coupling and deprotection, H$_2$N-AA1-AA2-Gly-

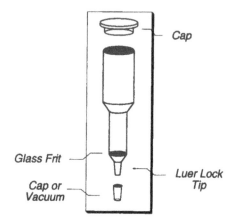

Figure 4 Biospin chromatography columns.

Wang was treated with 4.0 eq (0.17 mmol) of an *ortho*-hydroxyaldehyde in 1.5 mL DMF for 3 h and then washed with DMF (10 × 1.5 mL).

As mentioned previously, the ligand can be used either on or off the solid support. On the support, the SA-AA1-AA2-Gly-Wang resin was washed with toluene (3 × 1.5 mL), then dried under reduced pressure for 12 h. The resin was transferred to 96-well plates and stored in a desiccator until needed. Off the support the ligand was cleaved from the resin by rotation with 2.0 mL of triethylamine/DMF/MeOH (1:1:9) for 60 h. The solution was filtered and resin was washed thoroughly with distilled THF (3 × 2 mL). Solvent was removed in vacuo and the resulting yellow solid was taken up in EtOAc (1.0 mL) and loaded onto a pipet packed with a cotton plug and silica gel. The ligand was eluted with EtOAc (about 10 mL). The product was dissolved in toluene (5 mL) and concentrated on a rotary evaporator to azeotrope off water and DMF. On occasion, it proved necessary to azeotrope with toluene several times to remove all of the DMF. The product was dried in vacuo to give a pale to bright yellow solid. When the ligands were prepared on a large scale (about 5 g) the product was purified by recrystallization (methanol-water). Typical yields for the seven steps were 80–100%.

B. Procedure for Ligand Screening in Solution

The amounts of ligand and Ti salt shown below correspond to 10 mol % relative to the substrate and are the amounts used when we screened the catalysts for addition of TMCN to imines; 20 mol % catalyst was used when the libraries were screened for the addition of TMSCN to epoxides. The reactions were set up inside a glovebox under a nitrogen atmosphere. Chiral ligands (0.007 mmol) were placed in oven-dried test tubes (10 × 75 mm); the solid was subsequently dissolved in toluene (100 μL). The tubes were charged with Ti(O*i*-Pr)₄ (0.007 mmol, 100 μL of a 0.07 M solution in toluene) and the resulting yellow solutions stirred for 10 min at 22°C. Subsequently, the epoxide or imine (0.07 mmol, 100 μL of a 0.7 M solution in toluene) was added. The reaction vessels were capped with septa and sealed with Teflon tape. At this time, the reaction vessels were removed from the glovebox and placed into a test tube rack in a room maintained at 4°C and TMSCN (0.14 mmol, 100 μL of a 1.4 M solution in toluene) was added to each solution. The rack with the tubes was placed on a platform shaker and the reaction mixtures were agitated for 15 h at 4°C. The reactions were quenched by the addition of wet ether (0.5 mL; made by shaking ether with water and decanting), and the resulting mixture was passed through a plug of silica gel. All catalytic reactions were analyzed for conversion and enantioselectivity by chiral HPLC (Chiralpak AD for imines) or chiral GLC (Beta-dex 120 for epoxides).

C. Procedure for Ligand Screening on Solid Support

The ligands on resin (0.007 mmol) were weighed into individual wells of a 96-well plate. The plates were placed into a glovebox, and the reactions were set up inside of the glovebox under a nitrogen atmosphere. Toluene (100 μL) was added to each well, followed by Ti(O*i*-Pr)₄ (0.007 mmol, 100 μL of a 0.07 M solution in toluene). The plate was agitated on a shaker plate for 10 min at 22°C. Subsequently, the epoxide or imine (0.07 mmol, 100 μL of a 0.7 M solution in toluene) was added, followed by TMSCN (0.14 mmol, 100 μL of a 1.4 M solution in toluene). The plate was sealed with an adhesive aluminum liner and the reaction mixtures were agitated for 15 h at 22°C. Reactions were quenched by the addition of wet ether (0.5 mL) and filtered through a 96-well filter plate containing silica gel. The silica was washed with an additional 0.5 mL of ether. As before, reaction selectivity was determined by chiral HPLC (Chiralpak AD for imines) or chiral GLC (Beta-dex 120 for epoxides).

D. Commentary

Initially, all libraries were screened in solution with cleaved ligands. However, we subsequently illustrated that the selectivity trends were generally identical whether the ligands were screened in solution or still attached to the solid support. Our ability to examine chiral tripeptide Schiff bases on solid support allows us to increase the diversity of structures within a library. Specifically, in our initial studies, where ligand structures were evaluated after cleavage from support, diversity was limited as certain amino acids could not be inserted at the AA1 position; amino acids with aromatic and heteroatom-containing side chains proved particularly susceptible to epimerization. This restriction arises from the acidity of the AA1 α-proton that renders this site prone to epimerization during the cleavage process. Thus, elimination of the basic cleavage step permits access to a variety of previously unavailable tripeptide Schiff bases.

Several issues regarding the ligand optimization merit additional comment: (1) In general, the most effective ligands bear a Tle in the AA1 site (adjacent to the Schiff base) and Thr (t-Bu) in the AA2 position. The optimum Schiff base moiety of the ligand varies, depending on the substrate. (2) Specific ligands, when encountered in different libraries, may not promote reactions with identical levels of enantioselectivity. It is thus the *relative* levels of selectivity that are critical in the context of a library screening. Therefore, it is imperative that the screening of an entire library be carried out with identical reagents.

REFERENCES

1. AH Hoveyda, *Chem. Biol.* 1998; *5*:R187–R191.
2. KD Shimizu, ML Snapper, AH Hoveyda, *Chem. Eur. J.* 1998; *4*:1885–1889.
3. MB Francis, TF Jamison, EN Jacobsen, *Curr. Opin. Chem. Biol.* 1997; *2*:422–428.
3a. KW Kuntz, ML Snapper, AH Hoveyda, *Curr. Opin. Chem. Biol.* 1999; *3*:313–319.
4. BM Cole, KD Shimizu, CA Krueger, JP Harrity, ML Snapper, AH Hoveyda, *Angew Chem. Int. Ed. Engl.* 1996; *35*:1668–1671.
5. KD Shimizu, BM Cole, CA Krueger, KW Kuntz, ML Snapper, AH Hoveyda, *Angew. Chem. Int. Ed. Engl.* 1997; *36*:1704–1707.
6. CA Krueger, KW Kuntz, CD Dzierba, WG Wirschun, JD Gleason, ML Snapper, AH Hoveyda, *J. Am. Chem. Soc.* 1999, *121*:4284–4285.
7. JR Porter, WG Wirschun, KW Kuntz, ML Snapper, AH Hoveyda, *J. Am. Chem. Soc.* 2000; *122*:2657–2658.

Case Study 8-2
Selection of Effective Polymer-Bound Acyl Transfer Catalysts by Infrared Thermography

James P. Morken and Steven J. Taylor
University of North Carolina at Chapel Hill, Chapel Hill, North Carolina

ABSTRACT

This case study describes the preparation of an encoded polymer-bound chemical library by the split-and-pool synthesis technique. Infrared thermography is then used to detect the ability of each compound to catalyze the acylation of ethanol by acetic anhydride.

I. CHEMICALS

1. methylamine
2. 3-dimethylaminopropylamine
3. 1-(3-aminopropyl)imidazole
4. 4-(2-aminoethyl)morpholine
5. 3-(aminomethyl)pyridine
6. 4-amino-1-benzylpiperidine
7. 2-(aminomethyl)pyridine
8. 2-(2-aminoethyl)pyridine
9. aminodiphenylmethane
10. 1-adamantanamine
11. tritylamine
12. (*R*)-(+)-3-aminoquinuclidine dihydrochloride

Library Synthesis

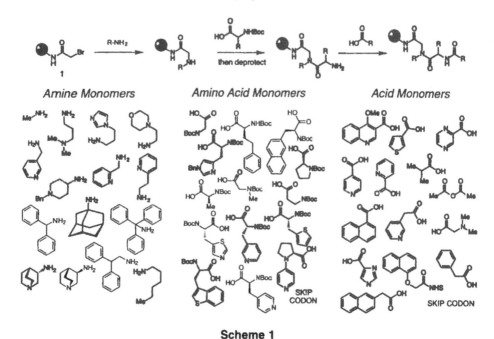

Scheme 1

13. (*S*)-(−)-3-aminoquinuclidine dihydrochloride
14. 2,2-diphenylethylamine
15. hexylamine
16. Boc-glycine-OH
17. Boc-im-benzyl-L-Histidine-OH
18. Boc-homophenylalanine-OH
19. Boc-3-(2-naphthyl)alanine-OH
20. Boc-proline-OH
21. Boc-alanine-OH
22. Boc-sarcosone-OH
23. Boc-β-alanine-OH
24. Boc-L-3-(4-thiazolyl)-alanine-OH
25. Boc-D-3-(3-pyridyl)-alanine
26. Boc-D-3-(2-thienyl)-alanine-OH
27. N-(4-pyridyl)-L-proline-OH
28. Boc-D-3-(3-benzothienyl)alanine-OH
29. Boc-L-3-(4-pyridyl)alanine

30. 4-methoxy-2-quinolinecarboxylic acid
31. 3-thiophenecarboxylic acid
32. 2-pyrazinecarboxylic acid
33. isonicotinic acid
34. picolinic acid
35. isobutyric acid
36. acetic anhydride
37. 1-naphthylacetic acid
38. 3-pyridylacetic acid HCl
39. *N,N*-dimethylglycine HCl
40. 4-imidazolecarboxylic acid
41. (2-paphthoxy)acetic acid *N*-hydroxysuccinimide ester
42. 2-naphthylacetic acid
43. phenylacetic acid
44. diisopropylethylamine (DIEA)
45. dimethylformamide (DMF)
46. 7-azahydroxybenzotriazolyltetramethyluronium hexafluorophosphate (HATU)
47. piperidine
48. trifluoroacetic acid (TFA)
49. aminomethylpolystyrene macrobeads (1 mmol/g, 500 μm, Rapp Polymere)

II. EQUIPMENT AND SUPPLIES

1. Biospin columns and stopcocks (Bio-Rad)
2. LabQuake rotator (Fisher Scientific)
3. Vacuum manifold (Bio-Rad)
4. Infrared thermographic camera (Cincinnati Electronics, IRIS 256)

III. INTRODUCTION

While current understanding of chemical reactivity often makes it possible to design an appropriate class of catalyst for a particular transformation, a large amount of trial and error is still involved in achieving the desired reactivity. Methods that allow for rapid synthesis and evaluation of many catalysts not only facilitate this trial-and-error process but allow for the evaluation of novel and potentially fruitful catalyst systems that, because of limited resources, might normally be overlooked [2]. In this case study we present a method for the rapid synthesis and evaluation of many catalyst systems through the combination of split/mix [3] combinatorial synthesis and infrared thermographic analysis [4].

According to the observation that most chemical reactions have a nonzero ΔH, reaction temperature may be used as an indicator of catalyst efficiency. For a given reaction, the temperature is a function of both the turnover rate and the ΔH for the reaction process. Advances in two-dimensional real-time infrared thermography make it possible to accurately monitor the temperature, and therefore the efficiency, of many polymer-bound catalysts at once [5]. By adding reagents for a reaction and measuring corresponding temperature changes in polymer bead–supported catalysts, it is possible to evaluate the catalyst's efficiency. Since all catalysts in a library screen are evaluated under the same reaction conditions, relative temperature differences between given beads are a direct result of differences in catalyst turnover rate. By simultaneous thermal imaging of all members of an encoded [6] polymer bead–supported library, in the presence of reagents for an acylation reaction (acetic anhydride, triethylamine, and ethanol) it is possible to select those beads that are become the hottest and hence those that are the most active catalysts for the acyl transfer reaction. Decoding of these catalyst beads reveals the chemical structure of the supported catalyst.

In this case study, we report the split pool–encoded synthesis and thermographic assay of a trimeric library of potential acyl transfer catalysts. The general synthesis plan is depicted in Scheme 1 and involves initial displacement of a bromide from polymer support with a collection of 15 amine monomers. Subsequent coupling and deprotection of 15 different Boc-protected amino acids to the resin then provides 225 different compounds (15 × 15). Lastly, coupling of 15 carboxylic acids completes the trimeric library. The library is encoded with the photocleavable electrophoric tags as described by Still [6]. Thermographic analysis of all library beads in the presence of ethanol, acetic anhydride, and triethylamine reveals effective catalysts for the acyl transfer reaction.

IV. GENERAL PROCEDURE

Except where noted, the following procedures are used for library synthesis and analysis. In general, all reactions are carried out in Biospin fritted columns (Bio-Rad) fitted with a polypropylene stopcock. Reaction solution is removed with the aid of a 20-vessel vacuum manifold available from Bio-Rad. Reactions are rotated on a LabQuake rotator available from Fisher Scientific.

A. HATU Activated Coupling Reaction

A 1.0 equivalent of polymer resin is swollen in DMF for 20 min and the solvent is removed by filtration. A DMF solution of diisopropylethylamine (5 eq) and carboxylic acid (4 eq) is added to the polymer resin (solution concentration is about 1.8 M in DIEA and about 1.5 M in acid, respectively). To the reaction mixture is then added a DMF solution of HATU (3 eq, concentration = 1.1 M) and the mixture allowed to rotate for the indicated time. The reagents are drained from the resin and the resin is washed successively with DMF, methylene chloride, methanol, DMF, methylene chloride, methanol, and finally DMF (resin is allowed to rotate for 10 min in each wash solvent).

B. Fmoc Deprotection

The resin is subjected to a 20% piperidine solution in DMF (1 mL per 100 mg resin) for 10 min. The solution is drained and the resin resubjected to fresh 20% piperidine solution for an additional 10 min. The resin is washed in triplicate with successive portions of DMF, CH_2Cl_2, and methanol (20 min per wash).

C. Boc Deprotection

Resin beads are washed twice with CH_2Cl_2 for 10 min. Vessel stopcocks are removed (stopcocks are unstable to TFA) and vessels capped with polypropylene caps that are supplied with columns. To the resin is added a freshly prepared solution of TFA in CH_2Cl_2 (1:1 vol/vol) and the resin allowed to rotate 20 min. The solution is drained and replaced with fresh reagents. After an additional 20 min of reaction, the solution is drained and washed three times with CH_2Cl_2 (10 min per wash), three times with methanol (10 min per wash), and three times with DMF (10 min per wash). The resin is then neutralized by washing for 10 min with a solution of 20% TEA in CH_2Cl_2. Finally, the resin is washed twice each with DMF and CH_2Cl_2 (5 min per wash).

D. Resin Splitting

Resin to be portioned is placed in a 50 mL polypropylene conical tube and diluted with DMF (total volume = 1 mL per portion). The bead solution is mixed by continuous bubbling with a gentle stream of nitrogen that is introduced through a glass pipette positioned with the tip 1 mm from the bottom of the vessel. To distribute beads among synthesis vessels, an Eppendorf Pipetteman is used to deliver 1 mL of the bead solution to each synthesis vessel.

E. Preparation of Bromoacetyl Resin

Fmoc-aminocaproic acid was coupled to 1.050 g of (aminomethyl)polystyrene resin (1.05 mmol/g, 1.10 mmol, 500 μm, Rapp Polymere) according to the general procedure and the N-terminal Fmoc protecting group removed also according to the general procedure. To the polymer resin was added 10 mL DMF, 1.86 g of bromoacetic acid (12 eq), and diisopropyl carbodiiimide (2.39 mL, 14 eq). After 1 h of reaction, the reagents were filtered away and the resin was washed in triplicate with successive 10-mL portions of DMF, CH_2Cl_2, and methanol (10 min per wash).

F. Coupling and Encoding First Monomer Position

Bromoacetyl resin (**1**) was divided among 15 synthesis vessels using the resin splitting technique. To each vessel was then added 0.5 mL DMF, 380 µL DIEA (2.19 mmol, 30 eq), and 15 eq amine monomer. Reaction was allowed to proceed at room temperature for 12 h. Reagents were removed by filtration and the resin was washed in triplicate with successive 0.8-mL portions of DMF, CH_2Cl_2, and methanol (10 min per wash). Electrophoric tags were then coupled (1.5 mol % tag) to polymer resin by the method of Still [6] and the resin pooled.

G. Encoding and Coupling of Second Monomer Position

Pooled resin obtained after coupling and encoding of the first position was redivided among 15 vessels using the general procedure. Electrophoric tags for the second position were then coupled (1.5 mol %) according to the procedure of Still. After encoding, the carboxylic acids for the second monomer position were coupled according to the general procedure for HATU-activated coupling (note that acids were coupled in 14 vessels, with the fifteenth vessel left blank). Resin was then pooled and subjected to TFA-mediated Boc deprotection according to the general procedure.

H. Encoding and Coupling of Third Monomer Position

Pooled resin obtained after encoding and coupling of the second position was redivided among 15 vessels using the general resin splitting procedure. Electrophoric tags for the third position were then coupled (1.5 mol %). After encoding, the carboxylic acids for the third monomer position were coupled according to the general procedure for HATU-activated coupling. Resin was then pooled and washed in triplicate with successive 0.8-mL portions of DMF, CH_2Cl_2, and methanol (10 min per wash).

I. Thermographic Selection of Polymer-Supported Catalysts

610 mg of encoded library beads were washed with three successive 20-mL portions of CH_2Cl_2 (20 min per wash). The polymer beads were added to a 120 ×

100 mm polypropylene tray containing 80 mL chloroform, 12 mL ethanol, 12 mL triethylamine, and 6 mL acetic anhydride. A Cincinnati Electronics IRIS 256 camera (25-mm lens, emissivity = 1.0), focused at the surface of the reaction solution, was used to image the temperature of polymer beads. Beads identified as among the hottest were retrieved with the aid of a sewing needle inserted into the eraser of a no. 2 pencil. The eye of the needle is used to ladle beads from solution. Selected beads were decoded according to the method of Still [6] to reveal the structure of active catalysts.

REFERENCES

1. S. J. Taylor and J. P. Morken, *Science*, 1998, *280*, 267
2. For recent reviews of combinatorial approaches to catalysis, see (a) M. B. Francis, T. F. Jamison, E. N. Jacobsen, *Curr. Opin. Chem. Biol.* 1998, *422*. (b) W. H. Weinberg, B. Iandeleit, K. Self, H. Turner, *Curr. Opin. Solid State Mater. Sci.* 1998, *3*, 104.
3. A. Furka, F. Sebestyen, M. Asgedom, G. Dibo, *Int. J. Pept. Protein Res.* 1991, *37*, 487.
4. For infrared thermal imaging of heterogeneous catalysts, see J. R. Brown, G. A. D'Netto, R. A. Schmitz. *Temporal Order*, L. Rensig and N. Jaeger, eds., p. 86, Springer-Verlag, Berlin (1985).
5. Infrared thermography has been used for the parallel imaging of 16 alumina-supported catalyst pellets used in the oxidation of H_2 gas with O_2 gas; see F. C. Moates, M. Somani, J. Annamalai, J. T. Richardson, D. Luss, R. C. Willson, *Ind. Eng. Chem. Res.* 1996, *35*, 4801.
6. M. H. J. Ohlmeyer, R. N. Swanson, L. W. Dillard, J. C. Reader, G. Asouline, R. Kobayashi, M. Wigler, W. C. Still, *Proc. Natl. Acad. Sci., USA* 1993, *90*, 10922.

Part IV
New Directions in
High-Throughput Synthesis

Chapter 9
Microfluidic Analysis, Screening, and Synthesis

C. Nicholas Hodge, Luc Bousse, and Michael R. Knapp
Caliper Technologies Corporation, Mountain View, California

I. MICROFLUIDIC ANALYSIS AND SCREENING

A. Introduction

Biological entities function through fine regulation of biochemical processes. This regulation is maintained through constant acquisition and interpretation of chemical and biochemical information. Interpretation leads to intervention in the case of perceived imbalance. The vast majority of chemical processes important to biological organisms are monitored unconsciously. Where possible and important, humans deliberately take chemical measurements, interpret their results in light of the potential implications of substance imbalance, and intervene by introducing chemicals that restore beneficial states.

Chemical and biochemical manipulation in the laboratory is generally a cumbersome process. The need for specialized equipment, laboratories, and highly trained personnel means that chemical synthesis and experimentation is done in only a fraction of situations where it would be useful. The importance of laboratory experimentation creates lively competition in the development of new techniques and equipment for producing results more simply or inexpensively. Nevertheless, except for the most commonly performed laboratory experimentation, such as clinical diagnostics for which expensive automated equipment is available, virtually all laboratory work is done with manual methods that are characterized by high labor costs, high reagent consumption, long turnaround times, relative imprecision, and poor reproducibility.

In fact, laboratory analysis is structured from the world of research where scientists endeavor to maximize their flexibility as experimentalists by designing equipment to perform one generically useful task, such as accurate liquid measurement, chemical partition, or optical spectroscopy. At the time a particular substance is synthesized or analyzed, a scientist uses his training and experience to combine the functions of the laboratory's single-function devices to create a characteristic sequence of processing steps that produce desired transformations, purifications, or detections of the substance in question. The principal complexity in the experimental process is therefore engaged once the sample is being processed.

1. Microtechnology

In the 1980s, scientists began to explore the feasibility of reversing the structure and building complexity into the instruments themselves. It was natural to try to emulate the technology trajectories that had proved so successful for information technology: miniaturization and integration. As everyone knows by now, the modern electronics industry was enabled by the techniques of repetitively modifying monocrystalline silicon wafers using lithography, etching, and deposition steps to create complex microscopic patterns of electrical connections, which when organized into functional processing systems are called *integrated circuits*. A central theme in the integrated circuit industry is the simultaneous fabrication of hundreds of such devices by reiterating the desired pattern on a large wafer of silicon or gallium arsenide and cutting the units apart after manufacture. The manufacturing process may require hundreds of steps and must be performed in specialized cleanrooms. However, since each batch of wafers contains many thousands of devices, the considerable costs of manufacture can be minimized.

Integrated circuits are built up from thin (1 nm to 2 μm) layers of conducting, insulating, and semiconducting films. Each film has a pattern etched into it, so that an exactly registered array of these layers forms components such as transistors, diodes, capacitors, etc. The patterns in the layers are created by lithography, which is the process of generating stencils in a radiation-sensitive material, with the stencil subsequently serving as a mask to permit etching only in selected areas of the film or substrate. The processing steps involved in the fabrication of an integrated circuit are a combination of thin-film deposition, oxidation, high-temperature annealing, ion implantation, and etching steps. At the risk of some oversimplification, the fabrication process can be summarized. First, a set of photomasks is created as follows:

1. A glass plate is coated with an opaque material, usually chrome.
2. The chrome is precisely removed from certain areas of the glass plate in a desired pattern, thus creating a mask.

Then, the active elements and interconnects are created with a series of deposition and etching steps:

1. A silicon or glass surface is coated with a photoresist, a material whose chemical properties change after exposure to radiation.

2. The photoresist-coated substrate is exposed to radiation through the chrome mask while being accurately aligned to any underlying pattern.
3. The coated, exposed substrate is treated with chemicals that accurately dissolve the photoresist where exposure has occurred (positive photoresist) or where the mask has protected the substrate from exposure (negative photoresist).
4. The underlying layer is selectively etched in the areas left open by the photoresist.
5. The photoresist is removed.
6. A new layer is deposited, or created by oxidation.
7. The sequence is repeated for the creation of multiple layers.

After completion of the integrated circuits, the completed wafer is cut or "diced" into chips, which are then mounted, at which point the input and output terminals can be connected to the outside.

Many different manufacturing processes exist for each of the above steps. Modest-complexity microchips contain many millions of connections capable of storing many millions of information bits. The most advanced integrated circuits of today contain several tens of millions of transistors, and a 100-million transistor circuit is expected in a few years [1].

2. Microfabrication for Purposes Other than Electronics

The combination of photolithography, deposition, and etching used to make three-dimensional devices is often called *micromachining*. In recent years, researchers have begun using the techniques of micromachining to create devices that perform functions other than purely electronic ones. An early example is silicon sensors for measuring quantities such as pressure and chemical concentrations [2]. In the late 1980s, a new development in this field was the creation of microdevices with moving parts fabricated with lithographic processes such as those described above. The earliest examples were microscopic motors [3,4], followed by resonating sensors such as accelerometers and gyroscopes, as well as other complex mechanical devices. Microfabricated pressure sensors [5] are now the basis of a major industry, accounting for $1.5 billion in sales in 1993. The automobile industry is the target of much development [6,7] (microfabricated accelerometers are an essential technology behind airbags). Microfabricated pumps and valves for microfluidic applications in the microliter range were another area of development [8–10]. The ultimate vision for these microelectromechanical systems (MEMSs) is the creation of complete sensing/intervention devices whereby multiple sensors convey complex information to higher level intelligence structures [4]. These, in turn, interpret the data and send instructions to appropriate actuators: elements capable of delivering mechanical or electrical stimuli to other systems or of intervening in an attempt to modify their environment according to the device's mission.

B. The Use of Microfabrication in Laboratory Science

1. Why Miniaturize?

There are many compelling reasons to recast laboratory methods in microscopic versions: (1) to achieve improved analytical performance; (2) to reduce costs in reagent or analyte consumption; (3) to reduce labor costs by integrating many automated processing steps into the same device; (4) to improve experimentation reproducibility through automation; and (5) to provide portable versions of experimentation systems that can be used at sites remote from traditional laboratory environments. While the first two items have interest in the short term, particularly in the research environment, microminiaturized instruments will deliver their truly revolutionary potential only when the latter three items are realized. Most work to date has produced components of systems; work on the integration of these components into complete microfabricated analytic systems is now progressing at a small number of sites.

2. Biosensors and Biochips

Though broadly used, the term *biosensor* has come to refer primarily to a device that can detect or quantify chemicals through specific interactions with a biomolecule in close proximity to a physical transducer capable of generating an electrical signal. The most frequently used biomolecules are antibodies and enzymes. The most frequent used transducers are electrochemical, optical, or thermal, but microbalance and acoustic wave methods have also been attempted. Despite great interest and research attention since the early 1980s, the biosensor idea has not proven robust commercially. Fewer than 20 commercialized products were available in 1993, accounting for revenues in the United States of approximately $100 million.

Biochips, on the other hand, have had robust commercial growth. Generally produced on glass, biochips are dense arrays of DNA fragments, up to 100,000 different molecules per square centimeter. These are used in nucleic acids hybridization experiments against complex solution phase labeled targets. In the most popular experiment, mRNA isolated from cells in a biological state of interest is amplified and fluorescently labeled in a series of biochemical reactions. Because the identity and position of each capture DNA molecule on the biochip is known, the pattern of fluorescence observed after the solution and solid-phase nucleic acids hybridize can be interpreted as an "RNA profile": a quantitative measurement of the expression of as many different genes as can be fit onto the array.

There are at least two ways to manufacture biochips. Affymetrix is still the most prominent company that makes these devices [11]. Their method of fabrication involves photochemically synthesizing the array one nucleotide at a time [12]. This requires a large series of masks, as many as four different masks per step, and well-controlled chemical environments. Most other methods use presynthesized DNA, either chemically or enzymatically synthesized (e.g., by the polymerase chain reaction, or PCR) and deposit them mechanically onto predetermined sites on a solid substrate, such as glass or nylon. Off-the-shelf robotics designed for this purpose can now be purchased. Nanogen has built chips of much more modest complexity (100–1000 molecules per chip) but with the interesting feature that each "pixel" is actively addressable with an electric field [13,14]. This permits hybridization that is much more rapid because target nucleic acids are concentrated at each site sequentially by electrophoresis. This effectively increases the concentration, permitting rapid experimentation.

Most conspicuous in the biochip domain is the fact that the chips perform only a small portion of the total laboratory manipulations necessary to make use of them. Processes such as nucleic acids isolation, amplification, labeling, and fragmentation are still done in the old way: in test tubes and beakers using pipets, electrophoresis, centrifugation, etc.

C. Microfluidic Laboratory Systems: Labs-on-a-Chip

A new approach to the miniaturized detection and manipulation of chemicals is currently being attempted. Unlike biosensors, which are based on qualitatively new ways of detecting interactions between molecules, the new approach is to create very small versions of traditional analytical methods and equipment. For example, the detection of antigen–antibody interactions is classically performed by preparing a labeled version of one or the other species, mixing this with a specimen, allowing binding to occur, then separating bound from free antigens and inferring the concentration of the molecular species in question from the label's partition profile. Interestingly, as typically performed in 96-well plates, these experiments are terribly inefficient. There are nearly one billion molecules contained in 100 μL of an analyte solution at the relatively low concentration of 1 nM. Even inexpensive optical systems are capable of detecting a million times less material than that. This inefficiency lies not in fundamental issues with the biochemical or chemical interactions taking place but rather with our inability to manipulate small volumes of fluid. The lab-on-a-chip trend is a recognition that if the basic processes of experimentation—reaction, partition, and product isolation—could be miniaturized, tremendous advantages in cost, data quantity, and data quality could be realized.

1. Integrated Analysis Systems

The group of Andreas Manz at Ciba-Geigy began publishing their ideas on this topic in the late 1980s, describing their concept of "miniaturized total analytic systems," or μ-TAS [15]. The idea was to integrate sampling, sample treatment, analysis, transport, and detection into single devices making use of micropumps and microvalves. In many cases in the early days, they advocated the utilization of electro-osmotic propulsion rather than pressure-driven flow as being more suited to the near-term requirements of rapid microfluidic transport [16]. Their published experimental work has focused largely on parameters of fluid control and electrophoretic separations. Jed Harrison in collaboration with Manz and later on his own at the University of Alberta brought many of these ideas to practice [17–23].

At about the same time, Mike Ramsey's group at the Oak Ridge National Laboratory published results showing various analytical methods on microchips, including electrochromatography, electrophoresis, and micellar electrokinetic chromatography [24–31]. They also provided the first glimpses into how truly integrated systems might work. Using electro-osmotic pumping with voltage-regulated valves, they created devices that accomplished reagent mixing, incubation, and reaction product separation and permitted sensitive, quantitative detection [32–34]. These devices could even be used to measure the kinetics of a chemical reaction, a difficult experiment to perform at the benchtop [27]. Interestingly, they also confirmed Manz' early prediction that integrated microdevices could provide enhancement in analytical performance. The reproducibility of their experimental determinations of reaction progress was greater than 99%, a level difficult to achieve with conventional equipment. This work has now been extended to many other areas of biochemistry, including enzymatic reactions and cell biology by several groups, including ours at Caliper Technologies.

In the majority of cases, workers using the lab-on-a-chip approach have employed fluorescent detection because it is a noncontact, highly sensitive, well-established technology. Especially with the advent of workable near-infrared dyes compatible with biochemistry, instrument development can take advantage of the tremendous amount of material available to serve the telecommunications industry. Also, given the nature of microfabrication techniques, some groups have begun to use electrochemical detection technology that integrates detection into the microchips themselves rather than as part of the processing instrument [35,36].

2. Existing Systems for Lab-on-a-Chip Experimentation

During the last 3 years, developers at Caliper Technologies have created platforms of LabChip experimentation. These are based on the idea that the LabChip® device itself can be a relatively passive and therefore inexpensive device, containing a particular pattern of interconnected channels having lateral dimensions of 10–100 μm and that are 10–50 μm deep (Fig. 1). The device is manufactured by the conventional processes of photolithography and wet-chemical etching described above [37,38]. It is constructed of different substances, primarily glass, quartz, or plastic (plastic microfluidic devices are manufactured from a microfabricated mold but can be replicated by embossing or injection molding methods). The channel structure is converted to a network of conduits by closing the top

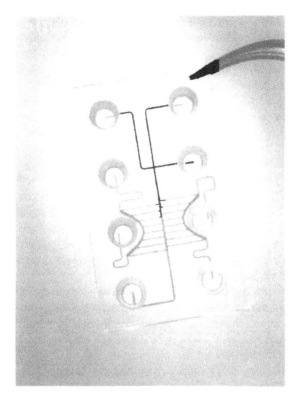

Figure 1 Microfabricated LabChip® fluidic processing device from Caliper Technologies. Thin lines are channels, which are filled with fluid during operations; circular structures are reservoirs for reagents in buffer or solvent.

with a second piece made of the same material, which can be flat. Typical volumes contained within such conduits are 1 nL per millimeter of channel. Channel intersections are used as valves, mixers, pipetors, etc. Portions of the channel structure perform incubations, separations, fractionation, reagent dilutions, and other aspects of the experiment. The channel structure is in this way a "hardwired" instruction set for nanovolume fluidic manipulations that produce the series of conversions and partitions necessary to acquire particular type of chemical or biochemical information.

A second level of control is created by introducing forces at the terminus of channels, thereby producing fluid and/or material movement within the channel network by electrophoresis or electro-osmosis. As described above, electrokinetic forces can be produced by placement of electrodes at the channel termini. Much in the way pinouts of a microchip in electronic systems allow for digital logic gates to be produced inside integrated circuits, electrokinetic forces can produce "fluidic logic gates" at intersections that are controlled actively by software and computer-controlled voltage supplies at run time. Pressure gradients produced with external sources or by clever arrangements of channel components can also engender movement of fluid within such networks.

With the combined control elements of the conduit pattern and active fluid or material actuation, remarkable biochemical or cell-based experimentation complexity can be achieved. At Caliper, we have used these elements to create assay systems typical in a pharmaceutical discovery environment, particularly for high-throughput screening of large libraries of organic compounds as potential drugs. These include assay development parametrics, such as the determination of kinetic constants and reagent optimization [39]. We use three basic strategies of biochemical assays: fluorogenesis, mobility shift, and fluorescence polarization spectroscopy. Fluorogenesis, the generation of a fluorescent signal upon interaction between reaction components, is a convenient methodology when feasible and is used routinely in our laboratory to perform enzyme assays and some binding assays (Fig. 2) [40]. Adaptations of fluorescence resonance energy transfer have extended usefulness of this strategy.

However, it is not always possible to find or generate chemicals that allow use of this convenient format. Mobility shift microchip assays are an alternative that take advantage of the fact that a labeled species usually migrates at a different rate in an electric field if it has been modified in a biochemical interaction. For example, fluorescently labeled peptides that are substrates for phosphorylation by a protein kinase gain a phosphate group that adds two negative charges to the molecule. In all cases we have tested, a simple and quick separation in a chip is enough to distinguish between the reaction substrate and the phosphorylated product [39]. This is a good example of how analytical separations integrated with reaction fluidics produce a valuable functionality not easily achieved by conventional assay modes.

Binding assays can also be done by mobility shift—a labeled molecule free in solution or bound to another molecule typically moves at different rates in an electric field. Even more conveniently, however, the rotational diffusion of a small, labeled molecule can be measured using fluorescence polarization spectroscopy. The fluorophore is excited with polarized light and the degree to which the molecule has rotated during its fluorescent lifetime is measured by comparing emitted light in both the parallel and perpendicular planes. A small molecule that is free in solution rotates more rapidly than one bound to another molecule, particularly a larger receptor; light from a small fluorescent molecule bound to a large, slowly rotating molecule remains polarized. Thus, the LabChip® microfluidic environment, coupled to this more sophisticated optical system, creates a useful assay methodology that has been used for applications as diverse as soluble receptor/ligand interactions, immunoassays, and nucleic acid hybridization analysis.

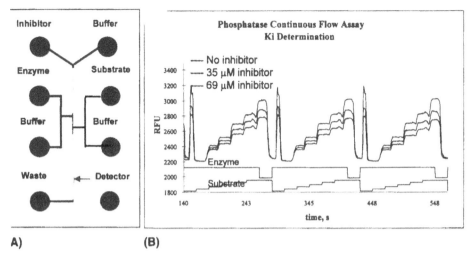

A)   (B)

Figure 2   (A) Schematic of an assay development chip for fluorogenic enzyme assays. Paired wells of reagents and diluent are used to vary concentration. An optical detector is positioned along the channel near the waste well at lower left to measure fluorescence and thus reaction rate. (B) Data from three replicates of an assay are shown, representing no inhibitor, low inhibitor concentration, and high inhibitor concentration, respectively. Each assay was run at six different substrate concentrations (stepped trace at bottom) and repeated three times in a continuous-flow format. The reproducibility between runs allows accurate kinetic constants to be extracted directly via software.

3.   Demonstrations of Integrated Assays on a Chip

As mentioned above, one of the most compelling reasons to perform assays on a microchip is the possibility of integrating several operations in a single device. This concept has by now been realized in several different types of assay.

For DNA assays, Jacobson et al. [32] combined the enzymatic cleavage of DNA into restriction fragments with electrophoretic separations. Another combination performed by several research groups is PCR amplification, followed by electrophoretic separation for sizing [34,41,42]. The research group of Richard Mathies at Berkeley has combined up to 96 identical DNA separations on a single chip to achieve the highest throughput in genotyping reported to date [43]. We have reported structures that are designed to achieve parallel separations without requiring hundreds of reagent wells [37].

Assays involving proteins are particularly useful to integrate because they often involve many steps. Hadd et al. [33] and Cohen et al. [40] report enzyme activity assays in which a chip is used to vary the concentration of the reagents

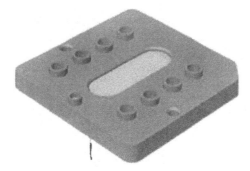

Figure 3 Caliper LabChip® sipping device with single-capillary inlet for compound library sampling.

by mixing and then incubating the mixture. This allows automated enzyme kinetic determinations. Note that these were the first electrokinetic microchip applications that did not involve separation. A step further in integration is the kinase assay discussed above [39], in which mixing, incubation, and separation must be combined to automatically measure the activity of enzymes in this important category. Integrated immunoassays, in which the mixing and incubation are also performed on-chip, were first reported by Chiem and Harrison [44].

Caliper automates the process of sample retrieval step in high-throughput screening applications by employing a Sipper™ chip, which contains a perpendicularly mounted fused silica capillary that joins with the main channel. A microtiter plate of any format containing compound dissolved in solvent is presented to the capillary, which sips 0.5–1 nL of dissolved compound per assay (Fig. 3). An example of a high-throughput kinase assay is shown in Figure 4. A single assay can be performed in seconds; ultrahigh throughput is accomplished by introducing multiple assay channels. Typically, only 1–2 nL of other reagents, such as enzyme and substrate, is also consumed per assay. The tiny quantities of re-

Figure 4 (A) Mobility shift assay. Protein kinase A and fluorescently labeled substrate are mixed in the main channel, and substrate and phosphorylated product are separated based on differential electrophoretic mobility. The resulting steady-state signal (baseline) is perturbed by the introduction of 1-nL pulses of inhibitor at the indicated concentrations. (B) High-throughput protein kinase A assay using mobility shift. The characteristic inhibitor signature indicates the location of six kinase inhibitors spiked randomly into two 96-well plates. Software determines the location and percent inhibition of each inhibitor with the aid of a reference dye shown in the lower trace. Ultrahigh throughput is achievable with chips employing multiple parallel sippers.

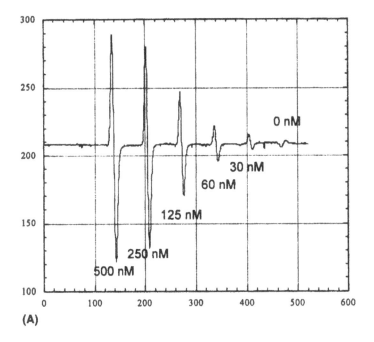

(A)

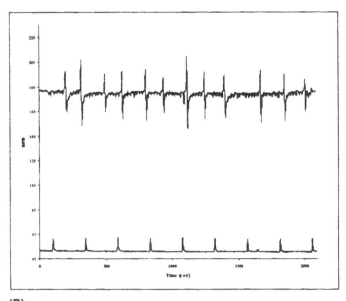

(B)

agents consumed under microfluidic conditions represent a major breakthrough for enabling screening of hard-to-obtain biological targets.

4. Cell-Based Assays

Increasingly, functional cell-based assays are being employed as a means to measure the effect of pharmaceutical agents on more complete biological systems. Cell-based assays offer the possibility to detect the coupling of a known inhibited target to unknown downstream effectors. Many think that this strategy can create a more relevant screen than that of an isolated biochemical activity.

We have deployed a general assay of calcium flux measurements to enable screening of G-protein-coupled receptor targets in LabChip® microdevices. Using both populational and single-cell assays, it is possible to use the sophisticated fluidic control of these devices to manipulate the cell's flow rate and control the concentration of agonists or antagonists, thereby creating assays that have complex kinetic features. High-throughput analysis of libraries can be achieved with very low consumption of the cell bioreagent; it is convenient to do assays on as few as 100 cells.

5. Closing the Loop

Perhaps the most ambitious aspiration for microfabricated instrumentation is to be part of a continuous monitor/intervention system whose behavior mimics that of normal homeostatic systems. Three steps needed to maintain the equilibrium of regulated systems are (1) determination of the level of a key analyte, (2) interpretation of the data, and (3) injection into the system of an appropriate modifier to bring it back to the desired state. As an example, many drugs (including insulin for diabetics) must be applied as a function of metabolic data. In most cases, because determining the level is difficult, the drug is applied in doses that cause significant peaks and valleys with respect to the desired concentration. A constant microchip-based monitor coupled to a drug delivery system could presumably do a much better job of maintaining the optimal dose. Such technology would not only reduce the costs of long-term care; it would presumably improve the quality of life for afflicted individuals. Other examples of feedback-controlled systems are discussed in the synthesis section of this chapter.

II. MICROFLUIDIC SYNTHESIS

A. Definitions

1. Microfluidics

As referred to in this chapter, microfluidics refers to continuous flow liquid-handling systems that move fluids through closed channels at low Reynolds numbers

velocity v

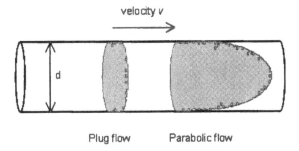

Plug flow   Parabolic flow

Figure 5   Examples of laminar flow.

such that fluid flow is laminar or viscous rather than turbulent [45] (Fig. 5). The
Reynolds number, Re, is the ratio of inertial force to viscous force experienced
by a particle in the liquid:

$$Re = vd\rho/\eta$$

where v is the velocity of the fluid in in cm/s, d is the characteristic distance
between the walls of the channel in cm (e.g., the diameter in a cylindrical channel), ρ is the density of the liquid in g/mL, and η is the coefficient of dynamic
viscosity in poise (g cm^{-1} s^{-1}). When the Reynolds number is less than about
2000, the flow can be assumed to be laminar. In most of the systems discussed
here, Re is close to 1, which is consistent with creeping flow [46]. Creeping flow
is not only nonturbulent but also minimizes areas of pooling and dead volume
as fluids pass by barriers and channel crossings. In pure water, where $\rho = 1$ and
$\eta = 0.01$ using the units defined above, creeping flow occurs when the product
of the diameter and velocity of the cylindrical channel is less than 0.01 as, for
example, when the diameter is 100 μm and the flow velocity is 1 mm/s. Under
these conditions, flow is time-independent and inertia-free. The absence of turbu-
lence and dead volume in microfluidic systems has profound consequences, as
is discussed in the section on pros and cons of microfluidics.

The inner diameter of the tubes used in capillary liquid and gas chromatography are also submillimeter, but the flow rates are much higher than those used
to date in microchips, and flow is not laminar. In addition, incorporating complex
liquid handling for integrated experiments requires cross-channels, which is difficult to realize in tubes without introducing pumps and valves and thus dead
volume and mechanical complexity. Therefore, the bulk of the work in microfluidic systems as defined here takes place in microchips that are fabricated by
methods described in the first section of this chapter.

2. Microfluidic Synthesis

Organic synthesis involves multiple steps: measuring and dispensing of reagents and solvents; mixing; applying reaction conditions such as heating, cooling, irradiation, and pressure; determining the extent of reaction; quenching the reaction; separating or purifying products; and identifying and quantitating products. This chapter focuses on progress toward the development of *complete* systems, i.e., systems in which all of the desired steps can proceed with very small amounts of material. Miniaturization of single steps in a full synthetic protocol is of interest but not within our scope. Because of the background of the authors and the focus of this volume, medicinal chemistry will be the focus of this chapter. However, similar trends are appearing in materials and polymer sciences as well.

3. Scale

As of the early 1980s, with the routine availability of Fourier Transform Nuclear Magnetic Resonance and sensitive mass spectrometers, synthetic chemists comfortably worked with amounts in the single milligrams (about 1–5 μmols). At the end of a long, difficult synthesis, as little as a few micrograms of product might be isolated for purposes of comparison with an authentic sample and spectral characterization. Effective miniaturization has not proceeded beyond this point, due to (1) inefficient techniques for handling very small amounts of liquid and solids; (2) inadequate sensitivity of common methods of spectral identification, and (3) the need for large amounts of the end product. In solid-phase synthesis, the product of a single 280-μm bead typically yields between 0.1 and 10 nmol of the final product of a three- or four-step sequence. However, solution phase synthesis remains in the macro domain, which we define as reaction volumes in the microliter range or greater. At "normal" concentrations used in organic synthesis—1–100 mM—about 1 nmol is the lower limit of solution phase synthesis, and even this is only achieved with special equipment and training. This chapter will focus on microfluidic solution phase synthesis.

Experience tells us that even with careful modeling and drug design techniques, only one in several hundred or several thousand compounds is of sufficient interest to warrant scale-up. In most fields, including medicinal and agricultural chemistry as well as in the search for materials with particular physical properties, only a few critical data points need to be obtained to rule out a compound as a candidate of interest. For example, in drug discovery, characterization of identity and purity and the activity of the molecule in primary screens is sufficient for the vast majority of compounds synthesized. Until recently, the demands of characterization and screening were such that macroscale synthesis was needed. However, developments in high-throughput screening and analytical instrumentation permit the use of much smaller quantities for these operations. Despite this, the current and foreseeable future paradigm of compound synthesis

and screening is to produce multiple milligrams of every compound to be tested
and to archive these (frequently unstable) compounds in perpetuity for future
testing on as yet unidentified targets. This approach adds a large operational overhead. Integration of analytical, screening, and synthetic functions into a truly
miniaturized format will obviate the need for large-scale synthesis and storage,
and encourage the scale-up, detailed characterization, and storage of only the
one-in-1000 molecules that are of significant interest. Plate-based, array-based,
and microfluidic integrated systems are all viable candidates for attaining this
vision; here we focus on advances in microfluidic synthetic chemistry. This is
an extremely young field. A brief summary of current work is followed by a
critical discussion of the benefits and disadvantages of pursuing synthesis in this
format. The chapter concludes with some thoughts on the progress and prospects
of integrated synthesis, screening, and analysis.

B. Summary of the Literature, with Examples

Despite increasing interest in the field, as evidenced by several well-attended
annual conferences, few detailed experimental accounts of organic synthesis in
a microfluidic format have been published. As of the writing of this review, citations from the primary, peer-reviewed literature are limited to a handful of papers
that are summarized below. This situation is not surprising because the current
barrier to progress in microfluidic synthesis is not due to a lack of challenging
synthetic problems but to the need for breakthroughs in instrumentation and microchip design that a synthetic or physical organic chemist can employ in his or
her research. As these tools become available, one would expect the number of
applied reports to increase substantially. Currently, however, the development of
chips and instruments is much more heavily focused on analytical systems for
diagnostics, genomics, and screening, as discussed in the first section. Earlier
reviews on microfluidic synthesis and microreactors include those by Ehrfeld
[47], Kamper [48], and DeWitt [49].

The chemical process industry began some years ago to develop microreactors for point-of-use production of materials that are unstable, explosive, or toxic,
so that small quantities could be produced on demand rather than synthesizing,
storing, and shipping large quantities [47,50]. For example [51,52], a gas phase
partial catalytic oxidation reactor chip was designed as follows: a 1.5 × 2.5 cm
silicon wafer was fabricated using the basic techniques described above to form
a sealed T-channel approximately 1 mm in diameter, as shown in Fig. 6. Platinum
heaters and sensors were patterned below the channel, and platinum was deposited in the main channel as the catalyst. The chip was epoxied to an aluminum
baseplate with gas inlet holes, which interfaced with oxygen and reactant gas
sources and flow controllers. The output gas was fed to a mass spectrometer.
Heating was controlled by applying a constant current to the platinum electrodes,

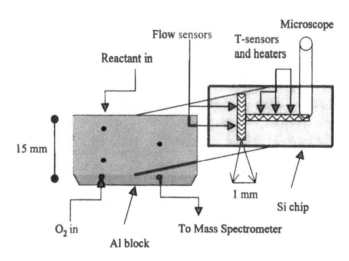

Figure 6   Reactor for catalytic oxidations, after [51].

and temperature was monitored by measuring electrical resistance and conversion based on a previously determined calibration curve. The effective heat dispersion by silicon allowed precise temperature control and prevented runaway chain-branching reactions. The rates of conversion, thermal behavior, and side product profile in the oxidation of ammonia to NO were determined. The estimated output of these prototypes was 10 g/day per chip, which is extremely efficient considering the size of the reactor. They are designed to operate in highly parallel format, allowing scale-up as needed at the point of use.

In this case, mixing, heating, and temperature measurement occurred on-chip, and the detection, although off-chip, was conducted in real time with the reaction using only a fraction of the efflux. Several laboratories have described pre- and postseparation derivitization on chip [19,27,28]. For example, Manz's laboratory describes the use of a fabricated silicon/glass hybrid chip as a microreactor for precolumn derivatization of amino acids [53]. The final chip measured 1.2 cm × 2.4 cm × 0.9 mm and contained a meandering reaction channel of width 1 mm and total depth 0.5 mm. Inlet and outlet holes were etched into the ends of the channel. On-chip heating and temperature-sensing elements were provided by sputtering thin film Cr-Pt resistors into grooves (adjacent to but isolated from the channels) that were formed by reactive ion etching. A solution of amino acids in buffer was premixed with the derivatizing agent 4-fluoro-7-nitrobenzofurazan and introduced onto the preheated reactor with a syringe pump. Percent conversion of each amino acid was monitored by directing the effluent into a cooling tube followed by an injector attached to an HPLC column and pump. The performance of the system was optimized by varying the preheating time and the temperature, and was compared with simply heating the mixture in a water bath before chromatography. The authors concluded that the reactor provided advantages in the areas of size of the unit, power consumption, and rapid rate of heating and cooling. The latter feature provided more reproducible conversions, with coefficients of variation in the peak areas in the 0.8–1.8% range (N = 7). This example only accomplishes heating and temperature sensing on-chip, with mixing and detection being done off-chip. The authors point out that introducing a feedback loop that changes the heating time based on the HPLC results is a simple extension of this work.

Harrison's group at the University of Alberta demonstrated that organic solvents can be pumped using electro-osmosis in chips that are similar in design to those used at Caliper for high-throughput screening and other biological applications [54] (Fig. 7). The chip measured 2 × 5.5 cm, and the trapezoidal channels were 90 μm deep and 190 μm across the top. Reservoirs filled with reagents made fluid contact with the channels. Electrodes in the reservoirs and at the waste

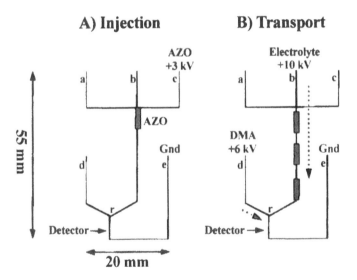

A) Injection   **B) Transport**

Figure 7 Schematic of PCRD2 microchip and the potential programs used. The channel lengths are approximately to scale. Reservoirs a, b, and e were filled with organic electrolyte solution. Reservoirs c and d contained 0.1% (m/v) *p*-nitrobenzenediazonium tetrafluoroborate and 2% (v/v) *N*, *N*-dimethylaniline in electrolyte, respectively. Loading step: a, b, and d floating; +3 kV and ground connected to c and e, respectively. Transport step: a and c floating; +10 kV, +6 kV, and ground connected to b, d, and e, respectively. (Used by permission of the American Chemical Society.)

well were used to inject and transport the fluids as shown. For demonstration purposes, a fast reaction that gave a strong absorbance signal on completion and proceeded under near-neutral conditions was chosen: the trapping of p-nitrobenzenediazonium ion by N,N-dimethylaniline to form a red dye. Absorbance was monitored at 488 nm using an argon ion laser source. Reactants were dissolved in either acetonitrile or methanol, and both solvents were successfully pumped by applying voltage to the electrodes, even in the absence of dissolved electrolytes. Addition of electrolyte prevented variation in flow rates over longer periods of time. Up to 10 h of repeated cycles of injection and reaction was demonstrated. To determine the reaction efficiency in the chip, the same solutions were reacted off-chip for 10 min. In acetonitrile, the reaction reached 22% of the conversion observed off-chip; in methanol, the reaction reached 37% of the off-chip value. The on-chip reaction times were 2.4 and 4.1 s for acetonitrile and methanol, respectively. The differences in these reaction time are attributable to the different viscosities of the solvents.

As mentioned in Section I, several groups have reported on-chip enzyme assays, including proteases, kinases, phosphatases, transferases, glycosidases, and esterases. These assays integrate dispensing, mixing, reaction, separation (in some cases), and detection (usually fluorescence). Ramsey's group at Oak Ridge National Laboratory recently demonstrated an acetylcholinesterase inhibition assay that included a derivatization step: enzyme-catalyzed acetylthiocholine hydrolysis is followed by introduction of coumarinylphenylmaleimide, which reacts quickly to form a fluorescent thioether [55]. Thus, two chemical transformations and a detection step occur in an integrated system in a matter of seconds. The efflux from a microchip has been demonstrated to be efficiently detected in an electrospray mass spectrometer, in some cases following on-chip separation (vide infra). The combination of these results with Harrison's demonstration of pumping organic solvents suggests that multistep organic transformations followed by detection in a nonaqueous environment is feasible.

These articles represent much of the work that has been published in the peer-reviewed literature with detailed experimental descriptions. Synthetic applications from other authors have been reported at the Micro Total Analysis Systems (μ-TAS) [56,57] and Microreaction Technology [58,59] conferences. A good general theoretical description of chip-based continuous flow reactor-separator systems can be found in a paper by Manz from μ-TAS '98 [60].

The most advanced microfluidic systems for synthesis are being commercialized by Orchid Biocomputer [49] based on technology developed with the David Sarnoff Institute [61–66]. The three-dimensional architecture in these chips allows precise delivery of reagents in solution to reaction wells that contain 650 nL volume (Fig. 8). The first chip contains 100 wells, with higher density formats under development. Hydantoins with two points of diversity were synthesized in a four-step process, with purities greater than or equal to 70%. The

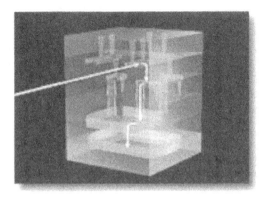

Figure 8 Orchid three-dimensional chemical synthesis chip. (Used by permission of Orchid Biocomputer, Inc.)

systems are designed for temperature control and inert atmospheres, and wash steps enable solid as well as solution phase synthesis. The chips are fabricated via multilayer planar processing methods. The systems have no moving parts: fluids are pumped by pressure, and valving is accomplished by stopping the flow at a brief widening of a channel and restarting flow by an applied electric field across the wider portion. This technology represents microfluidic liquid dispensing, although the mixing and chemistry takes place in traditional fixed reaction wells. Manz [67] has also filed a patent on a microchip for combinatorial chemistry applications.

C. Pros and Cons of Microfluidic Synthesis

Despite the dearth of synthetic experimental results in microfluidic formats, theoretical understanding and knowledge gained from analytical applications allow us to make some comments on the likely benefits and disadvantages of microfluidic synthesis. Many of these points are also discussed from different perspectives in the review articles cited above.

1. Advantages

Reaction time: When a reaction rate is diffusion-controlled and mixing is inefficient, a decrease in the reactor dimensions has an obvious advantage. The root-mean-square time (t) for the diffusion of a molecule across a one-dimensional vessel of length x is $t = x^2D$, where D is the diffusion coefficient in cm^2/s. A typical organic reactant has $D \sim 10^{-6}$. Thus, the diffusion-controlled reaction time increases with the square of the reactor dimension, and significant rate enhancements can be realized for these types of reaction in the microscale.

Cost and convenience: Both the gas phase microreactor example cited above and Caliper's high-throughput screening strategy illustrate the serial concept of microfluidic processing. In this paradigm, a reaction or series of reactions of very small volume takes place rapidly in each channel, and throughput is increased by parallelization. In the chemical process industry, where throughput is required to manufacture bulk substance, the same reaction occurs many times in each channel; in high-throughput synthesis and screening applications, a different reaction would take place in each channel. The advantages in microfluidic synthesis, if realized, would come from rapid manufacture of microreactors, efficient deployment and "scale-out" [48], increased automation, and decreased need for steps requiring the presence of technicians. The small capacity of individual channels is a disadvantage (see below).

Safety and temperature control: In the first example cited above, due to the high surface-to-volume ratio and the very efficient heat sink provided by the surrounding glass, the authors were unable to obtain explosive reactions of hydrogen and oxygen at concentrations that spontaneously detonate in normal-scale vessels. These effects also contribute to extremely sharp temperature control, as demonstrated by thermal cycling in PCR [34,41,42]. The use and production of nanoliters of toxic reagents on demand also has clear safety advantages over larger quantities that must be shipped and stored.

Waste stream and reagent usage reduction: This is one of the most important advantages of integrated miniaturization. In Caliper's high-throughput screening operations, a typical reaction occurs as follows: 1 nL of test compound is sipped onto the chip and joins the enzyme in the main channel for a few seconds of preincubation; substrate enters the main channel, and a few seconds later product formation is detected at the end of the channel. Total cycle times are between 6 and 15 s; at flow rates of 0.5–1 nL/s, this corresponds to 3–15 nL per reaction. Thus, a screen of 100,000 compounds would consume about 1 mL of solvent, as opposed to several liters for the same operations in 96-well plates. The effect of this miniaturization on the use of reagents is illustrated in Figure 9. In the parallel-synthesis Orchid Chemtel chips, even several solvent washes per four-step synthesis would only consume millilitres of solvent per 100 compounds synthesized. These numbers give some indication of the environmental benefits of decreasing scale to micrometer dimensions. It should be pointed out that a decrease in the waste stream is also a dramatic feature of pooled and split-mix synthesis.

Rapid optimization of conditions; direct coupling with assays or other downstream applications; feedback control: These are among the most exciting potential benefits of miniaturization and automation, and are the least likely to

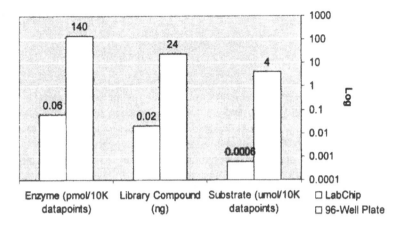

Figure 9 Log scale of comparison of reagent use in testing T-cell protein tyrosine phosphatase against 10,000 compounds. Substrate: DiFMUP. Consumption in 96-well plate on right bars; in LabChip® system on left bars. Note that use of 384 or more density plates still represents only small improvements vs. the orders of magnitude reduction with microfluidic chips.

be realized with current macroscale methods. See the discussion in the last section.

Mixing: As mentioned above, the mixing in a laminar flow system depends solely on diffusion, which can be both an advantage and a liability. Exquisite precision of contact times and concentrations can be achieved in microfluidic systems because the flow can be simulated mathematically with high accuracy and readily controlled [68]. However, a disadvantage is that turbulent flow systems provide increased rates of mixing that do not occur in laminar flow. Surface features and split channels (e.g., [69–72] have been designed to provide mixing that is more efficient than simple diffusion.

High surface-to-volume ratio: As the radius of a spherical reactor decreases, the surface area available to reactants decreases by the square of the radius, but the volume of the reactor decreases by the cube. In very small dimensions, the resulting S/V ratio is extremely high. This effect also offers advantages and challenges. As discussed earlier, very fast thermal cycling and pinpoint temperature control are possible due to the high surface area and the thermal properties of common chip materials, such as glass, quartz, silicon and many polymers. In addition, a large surface area is available for heterogeneous reactions, deposited catalysts, capture reagents, or solid phases for separation. However, problems due to interaction with the channel wall material are correspondingly magnified; side reactions or sticking that might normally not be observed on the milliliter or microliter scale can be magnified into more serious problems in these high-S/V domains.

2. Disadvantages

Detection: Some means of determining what has happened on-chip is obviously essential for fully integrated microfluidic systems, but the small sample amounts and volumes present challenges for sensitive detection. Several methods have been demonstrated, but none is yet robust enough for routine, high-throughput detection of small organic molecules on- or off-chip. Signal strength in absorption spectroscopy depends on path length, so it is a poor choice for channels narrower than 1 mm. Absorbance in the UV range has been measured across channels at high solute concentrations, and a fiberoptic cable was used to transmit and receive light lengthwise through a channel, increasing the effective pathlength and improving sensitivity [73]. Emission spectroscopy reduces the pathlength dependence and is very sensitive and useful in biological assays, but relatively few organic molecules are fluorescent. Thermal lens microscopy was applied to the detection of organic dye molecules at zeptomole sensitivity in a solvent/solvent extraction chip [57]. However, these methods are not useful for determining the identity of diverse organic species within mixtures, which is necessary for most synthetic applications. NMR spectroscopy has not been applied to microfluidic output, either on- or off-chip, to our knowledge; sensitivity would likely be a serious problem. Mass spectrometry (MS) has received the most attention due to its sensitivity and high information content. The Ramseys [74–76] and others [77–83] have reported direct coupling of a microchip to electrospray (ES) or MALDI-TOF mass spectrometers. Some practical problems still must be resolved, but early results indicate that ES-MS will be an extremely useful tool for microfluidic analysis and synthesis.

Amount of material: Clearly, if large amounts of material are needed, macroscale chemistry will remain the method of choice. Multiple parallel microreactors may also become an option for scale-out of reactions to produce laboratory quantities of simple synthetic targets.

Limitations in accurate liquid handling at low volumes: Synthetic chemists rarely work at volumes of under tens of microliters, not because their hands are inherently less steady than those of biologists but because of the need to dispense many different types of reagents and solvents under air- and moisture-free conditions. As syringes, pipets, and vessels become smaller, they also become less accurate and more likely to clog, and, as discussed above, the increased surface-to-volume ratios increase the chance of interactions with materials. Evaporation and absorption of ambient oxygen and water from air are also much more problematical for very small volumes. Affordable, accurate dispensing of submicroliter amounts is a recent development even for aqueous solutions and needs significant development for organic reagents. These topics are covered elsewhere in the context of parallel synthesis, but the implications for microfluidic synthesis are twofold: on the one hand, systems that dispense, react, and detect many different reagents and solvents in a true continuous flow format are unlikely to appear in the near future. On the other hand, early results suggest that, as was observed for combinatorial chemistry, a subset of chemistries that are compatible with the materials and dimensions of microchips may develop very quickly.

Dispersion: Even under creeping-flow conditions, a plug of solute in solvent in a flowing system will disperse over time, resulting in band broadening and dilution. One cause is normal diffusion, which occurs in all directions, including against the direction of flow, and can be controlled by increasing viscosity and by minimizing the time the plug resides in the channel. In addition, nonuniform dispersion occurs due to different pathlengths of parts of the plug when it navigates curves and corners, and due to different rates of fluid flow in different parts of the channel. Chip architectures have been designed that minimize these effects [84]. Because it is predictable mathematically under laminar flow conditions, dispersion can be useful in delivering controlled gradients of analyte concentration to a detection system.

High surface-to-volume ratios: See above.

Mixing: See above.

III. PROSPECTS FOR INTEGRATED AND FEEDBACK-CONTROLLED SYSTEMS

The promise of microfluidics lies in the potential for miniaturization and integration leading to automation. The most interesting and the most practical applications are those that are fully integrated because manual execution of intermediate steps becomes increasingly difficult at small scales. Microfluidic and other systems for delivering compounds to microfluidic screening platforms are already under development at Caliper and other laboratories. Successful examples of integrated microfluidic systems in other areas—DNA analysis, on-chip PCR, chromatography and MS detection, immunoassays, and simple chemical reaction followed by analysis—have been reported, as discussed above. The rate of progress over the past few years suggests that multistep microfluidic synthesis that is integrated with an analysis step will be a reality in the near term for certain chemistries that are repetitive and predictable. Extension of such systems to the synthesis and analysis of complex drug-like compound libraries will probably take considerably longer, unless current research efforts are significantly expanded.

Feedback control provides an impetus for investing more in research efforts, as the payoff is potentially very high. For example, a reaction parameter, such as the amount of desired product formed, can be monitored and the results sent back digitally to the software controlling reagent addition, temperature control, solvent choice, reagent choice, etc. These programs are standard fare in macroscale manufacturing and process development, but only for a few variables rather than the multivariate feedback and input control needed for optimization. The promise of microfluidics lies in multiple inputs, fast reaction and mixing times, and precise control of solvent and reagent movement. Why is this of such interest? Clearly, the limited menu of chemistries available to combinatorial chemists has been one of the major reasons for the recent resurgence of parallel synthesis, in addition to problems of purity, yield, and the quantity of sample needed for screening. The result is that true combinatorial expansion and testing has not been realized except in a few isolated cases. Rapid, feedback-controlled optimization protocols could provide a powerful research tool that could bring combinatorial chemistry into the mainstream because multiple experiments per minute could be carried out. Numerous improvements and technical advances will be needed to accomplish this goal, particularly in the area of rapid detection of small organic molecules in a microfluidic environment; but it is very hard to imagine this level of control and speed ever being achieved in macroscale systems.

In addition to reaction optimization, feedback control provides the opportunity for modifying synthesis based on a parameter such as binding affinity or inhibition of a biological target. This paradigm is the "holy grail" of drug discovery, since only active compounds would be selected and analyzed, reducing the need for characterization, storage, etc. It is clear that these applications will occur in the future; the time it will take to bring them to reality depends on the vision, ingenuity, and dedication of today's chemical researchers.

IV. SUMMARY

Combining traditional methods of biochemistry, cell biology, and chemistry with the wealth of experience in microfabrication from the microelectronics industry promises a revolutionary new approach to chemical and biological analysis and synthesis. Lab-on-a-chip microdevices allow for highly reproducible, automated experimentation, and the number and diversity of applications in these systems is growing rapidly.

REFERENCES

1. L Geppert. *IEEE Spectrum 36*:22–24, 1999.
2. P Bergveld and NF de Rooij. *Sensors and Actuators 1*:5–15, 1981.
3. H Fujita. *Proc. IEEE 86*:1721–1732, 1998.

4. KJ Gabriel. *Sci. Am. 273*:150–153, 1995.
5. M Esashi, S Sugiyama, K Ikeda, YL Wang, and H Miyashita. *Proc. IEEE 86*:1627–1639, 1998.
6. DS Eddy and DR Sparks. *Proc. IEEE 86*:1747–1755, 1998.
7. N Yazdi, F Ayazi and K Najafi. *Proc. IEEE 86*:1640–1659, 1998.
8. HTG van Lintel, FCM van de Pol, and S Bouwstra. *Sensors and Actuators 15*:153–167, 1998.
9. BH van der Schoot, A van den Berg, S Jeanneret, and NF de Rooij. A miniaturized chemical analysis system using two silicon micro pumps. *Transducers 91*, San Francisco, 1991, 789–791.
10. P Gravesen, J Branebjerg, and OS Jensen. *J. Micromech. Microeng. 3*:168–182, 1993.
11. M Chee, R Yang, E Hubbell, A Berno, XC Huang, D Stern, J Winkler, DJ Lockhart, MS Morris, and SP Fodor. *Science 274*:610–614, 1996.
12. S Fodor, JL Read, MC Pirrung, L Stryer, AT Lu, and D Solas. *Science 251*:767–773, 1991.
13. RG Sosnowski, E Tu, WF Butler, JP O'Connell, and MJ Heller. *Proc Natl Acad Sci USA 94*:1119–1123, 1997.
14. CF Edman, DE Raymond, DJ Wu, E Tu, RG Sosnowski, WF Butler, M Nerenberg, and MJ Heller. *Nucleic Acids Res 25*:4907–4914, 1997.
15. A Manz, N Graber, and HM Widmer. *Sensors and Actuators B 1*:244–248, 1990.
16. A Manz. *Trends Anal. Chem. 10*:144, 1991.
17. DJ Harrison, A Manz, Z Fan, H Ludi, and HM Widmer. *Anal. Chem. 64*:1926–1932, 1992.
18. A Manz, DJ Harrison, EMJ Verpoorte, JC Fettinger, A Paulus, H Lüdi, and HM Widmer. *J. Chromatogr. 593*:253–258, 1992.
19. DJ Harrison, K Fluri, K Seiler, Z Fan, CS Effenhauser, and A Manz. *Science 261*:895–897, 1993.
20. K Seiler, DJ Harrison, and A Manz. *Anal. Chem. 65*:1481–1488, 1993.
21. K Seiler, ZH Fan, K Fluri, and DJ Harrison. *Anal. Chem. 66*:3485–3491, 1994.
22. ZH Fan and DJ Harrison. *Anal. Chem. 66*:177–184, 1994.
23. DJ Harrison, K Fluri, N Chiem, T Tang, and Z Fan. Micromachining chemical and biochemical analysis and reaction systems on glass substrates. *Transducers '95*, Stockholm, Sweden, 1995, 752–755.
24. SC Jacobson, R Hergenröder, LB Koutny, RJ Warmack, and JM Ramsey. *Anal. Chem. 66*:1107–1113, 1994.
25. SC Jacobson, R Hergenröder, LB Koutny, and JM Ramsey. *Anal. Chem. 66*:1114–1118, 1994.
26. SC Jacobson, R Hergenröder, LB Koutny, and JM Ramsey. *Anal. Chem. 66*:2369–2373, 1994.
27. SC Jacobson, LB Koutny, R Hergenroder, AW Moore, and JM Ramsey. *Anal. Chem. 66*:3472–3476, 1994.
28. SC Jacobson, R Hergenroder, LB Koutny, and JM Ramsey. *Anal. Chem. 66*:4127–4132, 1994.
29. JM Ramsey, SC Jacobson, and MR Knapp. *Nature Med. 1*:1093–1096, 1995.
30. SC Jacobson and JM Ramsey. *Electrophoresis 16*:481–486, 1995.

31. SC Jacobson, AW Moore, and JM Ramsey. *Anal. Chem. 67*:2059–2063, 1995.
32. SC Jacobson and JM Ramsey. *Anal. Chem. 68*:720, 1996.
33. AG Hadd, DE Raymond, JW Halliwell, SC Jacobson, and JM Ramsey. *Anal. Chem. 69*:3407–3412, 1997.
34. LC Waters, S Jacobson, N Kroutchinina, J Khandurina, RS Foote, and JM Ramsey. *Anal. Chem. 70*:158–162, 1998.
35. AT Woolley, K Lao, AN Glazer, and RA Mathies. *Anal. Chem. 70*:684–688, 1998.
36. JS Rossier, A Schwarz, F Reymond, R Ferrigno, F Bianchi, and HH Girault. *Electrophoresis 20*:727–731, 1999.
37. L Bousse, AR Kopf-Sill, and JW Parce. Parallelism in integrated fluidic circuits. *Systems and Technologies for Clinical Diagnostics and Drug Discovery*, San Jose, 1998, 179:186.
38. A Kopf-Sill, T Nikiforov, L Bousse, R Nagle, and JW Parce. Complexity and performance of on-chip biochemical assays. *Micro- and nanofabricated electro-optical Mechanical Systems for Biomedical and Environmental Applications*, San Jose, 1997, 172–179.
39. CB Cohen, E Chin-Dixon, S Jeong, and TT Nikiforov. *Anal. Biochem. 273*:1999.
40. CB Cohen, TT Nikiforov, and JW Parce. Microchip assays for enzyme analyses. *Pittsburgh Conference on Analytical Chemistry and Applied Spectroscopy*, New Orleans, 1998, Abstract 456.
41. AT Woolley, D Hadley, P Landre, AJ deMello, RA Mathies, and MA Northrup. *Anal. Chem. 68*:4081–4086, 1996.
42. LC Waters, SC Jacobson, N Kroutchinina, J Khandurina, RS Foote, and JM Ramsey. *Anal. Chem. 70*:5172–5176, 1998.
43. PC Simpson, D Roach, AT Woolley, T Thorsen, R Johnston, GF Sensabaugh, and RA Mathies. *Proc Natl Acad Sci U S A 95*:2256–2261, 1998.
44. NH Chiem and DJ Harrison. *Clin Chem 44*:591–598, 1998.
45. JF Douglas, JM Gasiorek, and JA Swaffileld. *Fluid Mechanics*. New York: John Wiley and Sons, 1995.
46. M Van Dyke. *An Album of Fluid Motion*. Stanford, CA: Parabolic Press, 1982.
47. W Ehrfeld, V Hessel, and H Lehr. *Microreactors for Chemical Synthesis and Biotechnology—Current Developments and Future Applications*. Berlin: Springer-Verlag, 1998, 233–252.
48. K Kamper, W Ehrfeld, J. Dopper, V. Hessel, H Lehr, H Lowe, T Richter, and A Wolf. Microfluidic Components for Biological and Chemical Reactors. *Proc. IEEE Annu. Int. Workshop Micro Electro Mech. Syst.*, New York, 1997, 338–343.
49. S DeWitt. *Curr. Opin. Chem. Biol. 3*:350–356, 1999.
50. RS Benson and JW Ponton. *Chem. Eng. Res. Des. 71*:160–168, 1993.
51. R Srinivasan, IM Hsing, PE Berger, KF Jensen, SL Firebaugh, MA Schmidt, MP Harold, JJ Lerou, and JF Riley. *AIChE J. 43*:3059–3069, 1997.
52. KF Jensen, SL Firebaugh, AJ Franz, D Quiram, R Srinivasan, and MA Schmidt. Integrated gas phase microreactors. *Micro Total Systems Analysis '98*, Banff, 1998, 463–468.
53. JCT Eijkel, A Prak, S Cowen, DH Craston, and A Manz. *J. Chromatogr. A 815*: 265–271, 1998.

54. H Salimi-Moosavi, T Tang, and DJ Harrison. *J. Am. Chem. Soc. 119*:8716–8717, 1997.

55. AG Hadd, SC Jacobson, and JM Ramsey. *Anal. Chem. 71*:5206–5212, 1999.

56. H Mensinger, T Richter, V Hessel, J Dopper, and W Ehrfeld. Microreactor with integrated static mixer and analysis system. Micro Total Analysis Systems, *Proc. uTAS 94*, Mainz, 1995, 237–243.

57. T Kitamori, M Fujinama, T Odake, M Tokeshi and T Sawada. Photothermal ultrasensitive detection and microchemistry in the integrated chemistry lab. *Micro Total Systems Analysis '98*, Banff, 1998, 295–298.

58. JR Burns, C Ramshaw, AJ Bull and P Harston. Development of a microreactor for chemical production. *Proceedings of the First International Conference on Microreaction Technology*, 1998, 127–133.

59. B Gruber, M Almstetter, and M Heilingbruner. Microreactors as processors for chemical computers. *Proceedings of the First International Conference on Microreaction Technology*, 1998, 195–203.

60. A Manz, F Bessoth, and MU Kopp. Continuous flow versus batch process—a few examples. *Micro Total Systems Analysis '98*, Banff, 1998, 235–240.

61. Z Fan, P York, S Cherukuri. Chip fabrication for combinatorial chemistry. *Proc. Electrochem. Soc. (Microstructures and Microfabricated Systems)*, 1997, 86–93.

62. PJ Zanzucchi, SC Cherukuri, and SE McBride. U.S. Patent No. 5,575,069, 1996.

63. PJ Zanzucchi, SC Cherukuri, SE McBride, and AK Judd. U.S. Patent No. 5,643,738, 1997.

64. PJ Zanzucchi, SE McBride, CA Burton, and SC Cherukuri. U.S. Patent No. 5,632,876, 1997.

65. SC Cherukuri, RR Demers, ZH Fan, AW Levine, SE McBride, and PJ Zanzucchi. U.S. Patent No. 5,603,351, 1997.

66. PJ Zanzucchi, SE McBride, and SC Cherukuri. U.S. Patent No. 5,681,484, 1997.

67. AMB Manz, M. Smith, PCT Application WO 98/23368, 1998.

68. SV Ermakov, SC Jacobson, and JM Ramsey. Computer simulations for microchip electrophoresis. *Micro Total Analysis Systems '98*, Banff, 1998, 149–152.

69. T Fujii, K Hosokawa, S Shoji, A Yotsumoto, T Nojima, and I Endo. Development of a microfabricated biochemical workbench-Improving the mixing efficiency. *Micro Total Analysis Systems '98*, Banff, 1998, 173–176.

70. A Yotsumoto, R Nakamura, S Shoji, and T Wada. Fabrication of an integrated mixing/reaction micro flow cell for u-TAS. *Micro Total Analysis Systems '98*, Banff, 1998, 185–188.

71. Z Yang, H Goto, M Matsumoto, and T Yada. Micro mixer incorporated with piezoelectrically driven valveless micropump. *Micro Total Analysis Systems '98*, Banff, 1998, 177–180.

72. J Voldman, ML Gray, and MA Schmidt. Liquid mixing studies with an integrated mixer/valve. *Micro Total Analysis Systems '98*, Banff, 1998, 181–184.

73. Z Liang, N Chiem, G Ocvirk, T Tang, K Fluri, and DJ Harrison. *Anal. Chem. 68*: 1040–1046, 1996.

74. RS Ramsey and JS Ramsey. *Anal. Chem. 69*:1174–1178, 1997.

75. IM Lazar, RS Ramsey, S Sundberg, and JM Ramsey. *Anal. Chem. 71*:3627–3631, 1999.

76. MS Kriger, KD Cook, and RS Ramsey. *Anal. Chem.* 67:385–389, 1995.
77. Q Xue, F Foret, YM Dunayevskiy, PM Zavracky, NE McGruer, and BL Karger. *Anal. Chem.* 69:426–430, 1997.
78. B Zhang, H Liu, BL Karger, and F Foret. *Anal. Chem.* 71:3258–3264, 1999.
79. PL Ross, PA Davis, and P Belgrader. *Anal. Chem.* 70:2067–2073, 1998.
80. D Figeys, Y Ning, and R Aebersold. *Anal. Chem.* 69:3153–3160, 1997.
81. D Figeys, SP Gygi, G McKinnon, and R Aebersold. *Anal. Chem.* 70:3728–3734, 1998.
82. F Foret, DP Kirby, P Vouros, and BL Karger. *Electrophoresis* 17:1829–1832, 1996.
83. NH Bings, CD Skinner, C Wang, CL Colyer, DJ Harrison, J Li, and P Thibault. Coupling electrospray mass spectrometry to microfluidic devices with low dead volume connections. *Micro Total Analysis Systems '98*, Banff, 1998, 141–144.
84. CT Culbertson, SC Jacobson, and JM Ramsey. Minimizing dispersion introduced by turns on microchips. *Micro Total Analysis Systems '98*, Banff, 1998, 161–164.

Chapter 10
Sensors

David R. Walt
Tufts University, Medford, Massachusetts

I. INTRODUCTION

A. Sensors

Sensors are entities that detect the presence of a particular physical or chemical parameter. The former types of sensors are referred to as physical sensors, whereas the latter are referred to as chemical sensors. Sensors come in many forms: a thermometer is a temperature sensor; a barometer is a pressure sensor; a pH meter is an acid sensor. Biological cells have receptors for peptides such as insulin, hormones such as estrogen, ions such as sodium, and so forth, that can also be considered sensors.

Chemical sensors combine a molecular recognition element with a transducer (Fig. 1). Recognition elements can be either small molecules or macromolecules. Small molecules include such chemical species as crown ethers, cyclodextrins, calixarenes, and other specialized molecules. Macromolecules include antibodies, proteins, nucleic acids, zeolites, protein pores, and polymers. The area of molecular recognition is one of considerable activity both from the perspective of drug design and by providing new sensor recognition elements [1].

Binding to any of these receptors is insufficient to create a sensor. It is essential that the binding event become coupled to a transduction event. Binding to the recognition element elicits a response that is coupled to the transduction element, such as electrochemical-, optical-, thermal-, or mass-sensitive responses (Table 1). Within each of these major categories, there are numerous transduction mechanisms that can be employed. For example, within the electrochemical sensor category there are potentiometric, conductometric, and amperometric mecha-

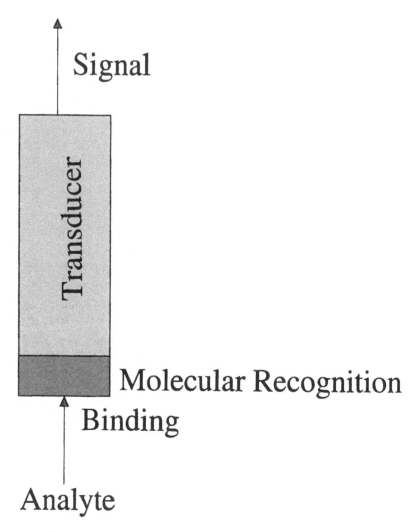

Figure 1 Sensors combine recognition and transduction elements.

nisms. Even within each subcategory there are specific techniques employed that distinguish one modality from another. In the optical sensor area, subcategories include fluorescence, absorbance, polarization, surface plasmon resonance, as well as a variety of different spectroscopic regions such as ultraviolet-visible (UV-Vis), near-infrared (near-IR), and IR.

Table 1   Transduction Mechanisms for Sensors

Electrochemical	Optical	Thermal	Mass
Potentiometric	Fluorescence	Thermometer	Quartz crystal micro-balance
Amperometric	Absorbance	Thermistor	Surface acoustic wave
Conductometric	Polarization	Thermocouple	
	Chemi/Bioluminescence	Pyroelectric	Shear acoustic plate mode
	Interferometry		
	Surface plasmon resonance		

B. Biosensors

Sensors can also be prepared using biological recognition elements [2,3]. For example, enzymes, antibodies, and cellular receptors can be incorporated into the sensor. Such sensors are referred to as biosensors (Fig. 2). The best known example of biosensors is the glucose biosensor in which the enzyme glucose oxidase is immobilized to a transducing element. Glucose oxidase catalyzes the conversion of glucose and oxygen to gluconic acid and hydrogen peroxide. Glucose biosensors are created by coupling this enzyme to a transducer that can measure the disappearance of oxygen, the appearance of acid or hydrogen peroxide, or the extent of electron flow from the enzyme.

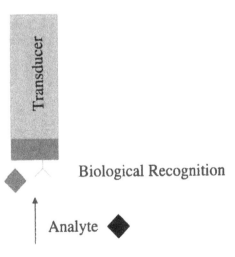

Figure 2   Biosensors employ a biological recognition element to generate selectivity.

A second type of biosensor employs biological cells. In this type of biosensor, living or intact cells are immobilized on the sensing surface. Cellular metabolism generates certain products. These products can be measured by the transducers described above. In addition, molecular biological approaches can be employed to cause the cells to generate a particular signal (i.e., fluorescence). For example, numerous groups are engineering different cell lines to express either green fluorescent protein (GFP) or luciferase when they are activated by some exogenous substance [4–6]. Cells can be measured for activation or toxicity upon exposure to different analytes. In this way, functional assays can be created in which cellular responses rather than simple binding events are measured.

II. INFORMATION REQUIRED

This last statement underscores one of the most important aspects of sensing: What information is required (Table 2)? In many cases, it is sufficient simply to know whether a particular substance is present or absent. For example, a smoke detector contains a sensor that is sensitive to smoke and alerts the user when smoke is present. In other cases it is insufficient to know simply whether a material is present; one needs to know how much of the material is present, i.e., quantification. For example, with glucose sensing, diabetics need to know what their glucose level is when they make a measurement so that they can adjust their diet or insulin dose accordingly. Simple binding sensors can accomplish these types of detection. In other cases, the nature of the desired information is more complex. For example, one may wish to know whether a particular substance is fresh or stale, sweet or bitter, floral, herbal, healthful, dangerous, and so forth. These attributes may describe qualities or functions rather than specific molecules. The information required is more complex and may involve many compo-

Table 2 Information Required from Sensors Varies in Complexity

Simple	Complex/smarter
Is it present?	Is it bad?
Alarm Yes/No	What does it resemble?
What? i.e., qualitative	Surrogate
How much? i.e., quantitative	What will it do?
	Binding, function

nents. For high-throughput screening, such as for drug discovery, one may want to know what type of activity a particular molecule exhibits, what type of structure a molecule resembles, what its pharmacokinetic or pharmacodynamic properties are, etc. For such information, smart sensors are required that can evaluate numerous aspects of the sample being tested.

III. SENSOR FORMATS

A. Single Sensors

The most common sensors are capable of measuring a single parameter. A wide variety of such sensors exist for the determination of various fermentation and clinical parameters, including glucose, pH, oxygen, CO_2, temperature, pressure, etc. [7]. Such sensors tend to be limited in terms of the amount of information they can provide.

B. Arrays

1. Selective Arrays

An alternative sensor format is that of arrays. In this format, a multitude of sensing elements are present that can measure either the same parameter in many regions of a sample or many parameters in a single sample. A number of different array formats have been devised, including chip-based systems that may operate on the basis of surface acoustic wave [8–10], microelectrode arrays [11,12], and optical arrays [13–18]. Such arrays or chips can be prepared by ink-jet printing [19–21], photolithography [22–28], and photodeposition [14–17]. A simple way to visualize such an array is to think of a microtiter plate, a format that enables typically one assay per well to be conducted, with the ability to run 96, 384, or 1536 assays simultaneously. An example of a high density array is the gene chip in which a multitude of different oligonucleotide sequences are positioned on the array surface, enabling the user to interrogate the presence or absence of all the represented sequences simultaneously. The majority of such arrays employ the same chemistry and transduction mechanisms used for single sensors. They are simply technical expedients for creating multianalyte sensitivity.

2. Cross-Reactive Arrays

In contrast to specific arrays, there is an increasing movement toward cross-reactive arrays [29–34]. In this format, sensors do not possess the exquisite selectivity and specificity exhibited by the types of sensors described above. Sensors are either broadly sensitive or semiselective. The specificity arises from the signals generated from the pattern of response across the entire array rather than from any single sensing element (Fig. 3). In such arrays, the response patterns for particular analytes, mixtures, or properties are collected by exposing the arrays to samples with known compositions or properties. The response pattern obtained is used to train a computational network, such as a neural network. During the training phase, the network learns to correlate certain patterns or features of response with particular analytes or sample compositions. Once the network is trained, it is capable of evaluating unknown samples and ascertaining whether or not those analytes are present. Recent studies have suggested that it may be possible to use such arrays as smart sensors in the manner described above. In this way, features or properties of new materials may be evaluated in terms of how closely they correlate to existing materials with known properties.

For high-throughput screening purposes, arrays offer many advantages over single sensors. They provide the ability to test a multitude of reactivities or properties simultaneously. Such arrays may even be capable of providing information that is inaccessible using single selective sensors. On the other hand, arrays tend to be larger than single sensors and consequently require more sample. Thus, the additional information provided by arrays must be balanced by the need to prepare more of each test material.

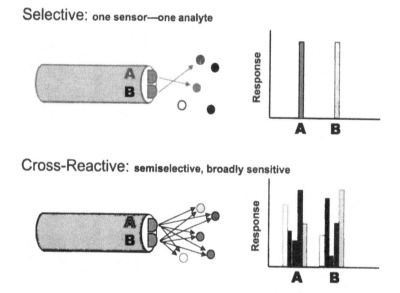

Figure 3 Cross-reactive arrays use pattern recognition.

3. Sample Preparation

It must also be mentioned that any sensor, whether single or array-formatted, requires sample preparation and delivery steps. In the ideal case, no preparation is required and the sensor is simply put in contact with the sample to be analyzed. For example, by placing the end of the sensor into blood, urine, a combinatorial compound library, etc., the sample is defined. In other cases, sample preparation may require cell lysis, DNA amplification, particulate removal, filtration, precipitation, or other partial purification scheme before exposing the sample to the sensor. If such a preparation scheme is required, the integration of sensors into microfluidic systems, as discussed in a previous chapter, may be necessary.

IV. BIOLOGY AS A MODEL

One of the best ways to create sensors for screening is to use biology for conceptual inspiration. Often biological systems can provide the sensitivity and specificity needed in a given application.

V. OPTICAL SENSORS

A. Operation Principles

To illustrate some of the above concepts, I will present some of the array-based sensing work presently being carried out in my laboratory. We employ optical fiber sensors in which we use fluorescence as the transduction mechanism [35]. Optical fibers transmit light via total internal reflection all the way to the end of the fiber. A fluorescent material is immobilized on the fiber tip that ideally responds reversibly to the presence of analyte. For the genosensing applications discussed below, "reversible" means that one must expose the sensor to a dissociation medium to regenerate the sensor. Monochromatic light is used to excite the fluorescent material bound to the fiber tip. Fluorescent light emerging from the material is emitted isotropically (Fig. 4), and some of it gets captured by the same fiber. Therefore, one color of light is introduced into the fiber and a detector captures the returning fluorescence.

B. Optical Arrays

1. Architecture

The fibers we use for sensors are not the typical single-core, single-clad fibers employed in the telecommunications industry. We employ arrays of fibers as

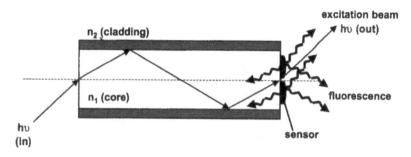

Figure 4   Structure of an optical fiber showing how fluorescent light is captured.

shown in Figure 5. The sensor is typically 350 μm in diameter and contains 6000 individual optical fibers that are arranged and fused. Arrays may be larger; for example, 0.5-mm-diameter arrays contain between 10,000 and 20,000 individual fiber elements. Such arrays have the capability of transmitting images through the fibers.

In Figure 5, we see a region of the optical fiber array. The individual cores of the optical fibers are approximately 3.5 μm in diameter. The cladding does not transmit light and confines light to these cores, as seen in the darker material surrounding the cores.

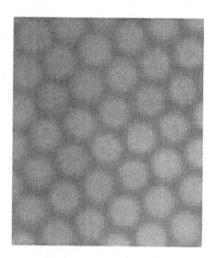

Figure 5   SEM of an optical imaging array.

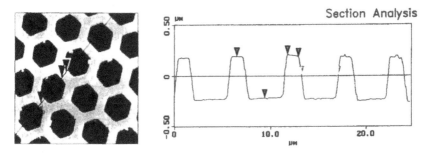

Figure 6 Wells can be etched into the array. (left) AFM image. (right) AFM profile of etched wells.

2. Beads in Wells

(a) Preparation.   These optical fiber arrays can be etched differentially such that the core etches selectively and creates an array of microwells [36] (Fig. 6). Fibers can be control-etched to any desired depth. At the bottom of each of these microwells is an optical fiber creating an optically wired array of microwells.

When a drop of a solution containing latex microspheres is dripped onto the fiber, the latex beads spontaneously fall into the tiny microwells [36] (Fig. 7). The beads are held via dispersion forces. The well size is about 3.5 µm in diameter, the latex bead about 3.1 µm in diameter, and a single exposure to a drop of beads provides a bead in 96% of the wells. Fibers can also be tapered, etched, and filled with nanobeads to provide even smaller features.

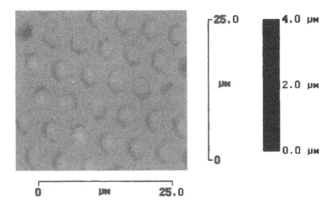

Figure 7 AFM image of beads distributed in the array.

(b) Encoding. To prepare sensors, we first prepare a solution of latex spheres and attach a desired sensing chemistry to the surface. Next, each bead type is encoded with a unique fluorescent dye combination. The encoding dyes absorb and/or emit in a different region of the spectrum from the analytical signal. Each type of sensing bead is labeled with a different combination of encoding dyes that distinguishes it from the other sensing beads.

To create a sensing array, one starts with a polished imaging fiber with etched microwells; beads are then distributed randomly into the wells. Because each bead carries an optical bar code, the position of all of the different sensors in the array can be mapped [37].

3. Specific Sensors

(a) Genosensors. One major application for such arrays is in the area of geno-sensors. We have immobilized 25 different oligonucleotide probes on encoded microspheres. Each oligonucleotide probe sequence is attached to beads by reaction with a solution containing approximately 20% solids (i.e., the beads occupy 20% of the volume). A 1-mL solution of 3.1-μm beads contains 5.8 billion beads that are virtually identical in their response properties. A library of encoded beads containing different oligonucleotide sequences is prepared and mixed to form a sensor bead pool. The pool is then distributed randomly on the end of the optical array.

After assembly, the array is put into a solution containing labeled targets and only those sensors that are perfect complements will hybridize and give a fluorescent signal. Determining the probe sequence on each of the microspheres is a simple matter; the optical bar codes allow each bead to be decoded. Such decoding can be accomplished for the entire array in a few seconds using image processing techniques. Figure 8 shows images of different patterns for consecutive hybridizations with different labeled targets. With this format, each probe sequence is represented as multiple members of each bead type, thus providing the capability of eliminating both false positives and false negatives.

(b) Cell-Based Sensors. Another approach to arrays involves distributing different living cells in each well. Different cell types with different sensitivities are first encoded by attaching a combination of nontoxic dyes to different cell populations. These dyes are membrane probes with a hydrophobic tail sticking into the cell membrane. Cells are then mixed to afford a population of cells. The cell suspension is put into contact with the fiber and the cells gravitationally settle into the wells. Because each cell is encoded, one has an array of cells with multiple sensitivities represented. Such an array can be used for functional screening on a cell-by-cell basis. Multiple copies of each of these cell types in the array provide redundancy so as to avoid basing the measurement on only one cell.

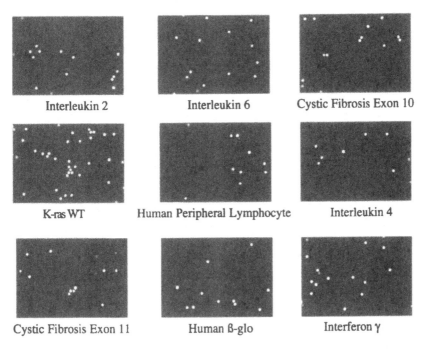

Figure 8   Images of sequential hybridizations with an encoded oligonucleotide array.

4.   Cross-Reactive Sensors

(a)   Vapor Sensors.   Let us return to a discussion of smart sensors based on pattern recognition. These sensors are biologically inspired by the mammalian olfactory system [38]. In the mammalian olfactory system, approximately 1000 receptors are distributed over several million receptor cells in the olfactory epithelium in the nasal cavity. On average, we have about 10,000 cells expressing each type of receptor. When we smell a pure odor we stimulate about 50% of the cells; individual receptors possess broad responses to a multitude of vapors. They are not selective for only one vapor. Each odor generates a different pattern of activity.

This pattern is sent to the olfactory bulb, which acts as a preprocessor and then sends the information to the cortical region of the brain, which registers the pattern. Animals, including humans, are trained to recognize odors. We learn by experience to correlate a particular pattern in the brain with the name or quality an odor elicits. This cross-reactive sensing paradigm can be used with artificial sensors to recognize patterns and to train computational networks to recognize the presence of particular odors or mixtures of odors.

We designed the artificial system with the following biologically inspired features: (1) A delivery system providing vapor pulses generates a temporal response. This response is more information-rich than a simple equilibrium measurement. (2) Semiselective sensors that cover a wide dynamic range of concentrations. (3) A built-in amplification system with multidimensional sensing capabilities. (4) A computational pattern recognition program to process these complex signals.

We created an array of semiselective sensors on a bundle of optical fibers [39–45]. The fibers contain different dye/polymer combinations as sensing materials. Each sensor generates a different response upon exposure to a particular vapor. The response is due to a variety of phenomena, such as fluorescence, polymer swelling, and vapor partitioning. A vapor delivery system is used to apply a vapor pulse to the sensor array.

As can be seen in Figure 9, different sensors respond by giving different signal intensities and respond at different rates to a vapor pulse. It is perhaps most important to recognize the information richness in the responses of the individual sensors in the array. The digitized data are used to train a computational network [12]. After training, the array is shelved for some time. Then the array is exposed to the same vapors and asked to identify the substance being tested based on prior training.

(b)   Sensor Summing.   One final aspect of the present approach that relates directly to biological systems is that of having multiple redundant sensors present in the array. Figure 10 shows the response of individual beads and the sum of

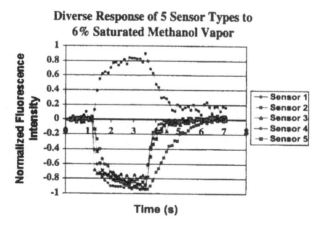

Figure 9   Cross-reactive sensor array responds with diverse signals from each sensing element.

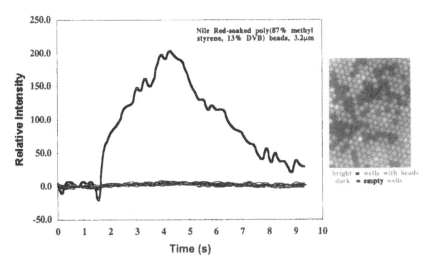

Figure 10   The signal-to-noise ratio improves by summing the signals from multiple beads. Baseline signals are from individual beads. Bold trace is due to summed response.

all the individual bends. As the number of beads increases, the signal improves and the noise cancels resulting in a much higher signal-to-noise ratio.

VI. SUMMARY AND FUTURE PROSPECTS

For high-throughput screening, there are a variety of different analytes that can be measured: small and large organic molecules, peptides and proteins, and oligonucleotides. Arrays composed of a multitude of sensing elements will undoubtedly provide more information than single sensors. Protein arrays composed of immobilized receptors or antibodies may provide additional information for examining how new structures interact with known targets. In this context, the utility of cross-reactive sensor arrays cannot be overemphasized. As listed in Table 2, by training such cross-reactive arrays to recognize patterns corresponding to particular properties (e.g., solubility, toxicity, reactivity, etc.), it may be possible to employ such arrays as surrogates in lieu of explicit testing of such properties.

Finally, it is clear that a variety of transduction mechanisms will be useful for screening. Advances in miniaturization, microfluidics, new materials and sample preparation will ultimately contribute to robust, user-friendly, miniature systems with ultrahigh-throughput capability.

REFERENCES

1. For a review of molecular recognition, see *Chem. Rev. 97(5)*, 1997.
2. A. P. Turner, I Karube, G. S. Wilson, eds., *Biosensors: Fundamentals and Applications*, Oxford University Press, 1987.
3. E. Kress-Rogers, ed., *Handbook of Biosensors and Electronic Noses*, CRC Press, 1997.
4. J. C. Lewis, A. Feltus, M. C. Ensor, S. Ramanathan, and S. Daunert, Applications of reporter genes in analytical chemistry, *Anal. Chem., 70*, 579A–585A (1998).
5. S. Ramanathan, R. Shetty, J. Mercer, and S. Daunert, The green fluorescent protein in the design of a living biosensor for arabinose, *Anal. Chem., 71*, 763–768 (1999).
6. S. Ramanathan, M. C. Ensor, and S. Daunert, Metal-resistance bacterial systems in sensors, *Trends Biotechnol., 15*, 500–506 (1997).
7. M. Uttamlal and D. Walt, Optical sensors, in *Encyclopedia of Bioprocess Technology: Fermentation, Biocatalysis and Bioseparation*, John Wiley and Sons, 1999.
8. K. Chen, D. Liu, L. Nie, and S. Yao, *Biosens. Bioelectron. 11*, 515 (1996).
9. E. T. Zellers, and M. Han, *Anal. Chem., 68*, 2409–2418 (1996).
10. E. T. Zellers, S. A. Batterman, M. Han, and S. J. Patrash, *Anal. Chem., 67*, 1092–1106 (1995).
11. R. M. Wightman and D. O. Wipf, In *Electroanalytical Chemistry*, A. J. Bard, ed., Marcel Dekker, 1989, Vol. 15, pp. 267–353.
12. C. A. Amatore, In *Physical Electrochemistry: Principles, Methods, and Applications*; I. Rubenstein, e.d., Marcel Dekker, 1995; pp. 131–208.
13. A. E. Bruno, S. Barnard, M. Rouilly, A. Waldner, J. Berger, and M. Ehrat, *Anal. Chem., 69*, 507–513 (1997).
14. L. Li, and D. R. Walt, *Anal. Chem., 67*, 3746–3752 (1995).
15. B. G. Healey, and D. R. Walt, *Anal. Chem., 67*, 4471–4476 (1995).
16. J. A. Ferguson, B. G. Healey, K. S. Bronk, S. M. Barnard, and D. R. Walt, *Anal. Chem. Acta., 340*, 123 (1997).
17. K. L. Michael, J. A. Ferguson, B. G. Healey, A. A. Panova, P. Pantano, and D. R. Walt, In *Polymers in Sensors: Theory and Practice*, N. Akmal and A. M. Usmani, eds., ACS Symposium Series 690, American Chemical Society, Washington, DC 273–289 (1998).
18. C. A. Browne, D. H. Tarrant, M. S. Olteanu, J. W. Mullens, and E. L. Chronister, *Anal. Chem., 68*, 2289 (1996).
19. J. Kimura, Y. Kawana, and T. Kuriyama, *Biosensors, 4*, 41 (1988).
20. J. D. Newman, A. P. F. Turner, and G. Marrazza, *Anal. Chim. Acta., 262*, 13–17 (1992).
21. A. V. Lemmo, J. T. Fisher, H. M. Geysen, and D. J. Rose, *Anal. Chem., 69*, 543 (1997).
22. S. P. A. Fodor, J. L. Read, M. C. Pirrung, L. Stryer, A. T. Lu, and D. Solas, *Science, 251*, 767–773 (1991).
23. S. P. A. Fodor, R. P. Rava, X. C. Huang, A. C. Pease, C. P. Holmes, and C. L. Adams, *Nature, 364*, 555–556 (1993).
24. A. C. Pease, D. Solas, E. J. Sullivan, M. T. Cronin, C. P. Homes, and S. P. A. Fodor, *Proc. Natl. Acad. Sci. U.S.A., 91*, 5022–5026 (1994).

25. R. S. Matson, J. Rampal, S. L. Pentoney, Jr., P. D. Anderson, and P. Coassin, *Anal. Biochem.*, *224*, 110–116 (1995).
26. M. Chee, R. Yang, E. Hubbell, A. Berno, X. C. Huang, D. Stern, J. Winkler, D. J. Lockheart, M. S. Morris, and S. P. A. Fodor, *Science*, *274*, 610–614 (1996).
27. G. McGall, J. Labadie, P. Brock, G. Wallraff, T. Nguyen, and W. Hinsberg, *Proc. Natl. Acad. Sci. U.S.A.*, *93*, 13555–13560 (1996).
28. M. Eggers and D. Ehrlich, *Hematol. Pathol.*, *9*, 1–15 (1995).
29. H. V. Shurmer, The electronic nose, *Anal. Proc. Int. Anal. Commun.*, *31*, 39–40 (1994).
30. J. W. Gardner and P. N. Bartlett, A brief history of electronic noses, *Sens. Actuat. B*, 18–19, 211–220 (1994).
31. P. Mielle, Electronic noses: towards the objective instrumental characterization of food aroma, *Trends Food Sci.*, *7*, 432–438 (1996).
32. M. A. Craven, J. W. Gardner, and P. N. Bartlett, Electronic noses—development and future prospects, *Trends Anal. Chem.*, *15*, 486–493 (1996).
33. J. Gardner and P. N. Bartlett, *Electronic Noses: Principles and Applications*, Oxford University Press New York, 1999.
34. T. A. Dickinson, D. R. Walt, J. White, and J. Kauer, Monitoring systems based on "artificial Noses" *Trends Biotechnol.*, *16*(6), 250–258 (1998).
35. D. R. Walt, Fiber-optic imaging sensors, *Acc. Chem. Res.*, *31*, 267–268 (1998).
36. P. Pantano and D. R. Walt, Ordered nanowell arrays, *Chem. Mater.*, *8*, 2832–2835 (1996).
37. K. L. Michael, L. C. Taylor, S. L. Schultz, and D. R. Walt, Randomly-ordered addressable high-density optical sensor arrays, *Anal. Chem.*, *70*(7), 1242–1248 (1998).
38. J. Kauer, *Trends Neurosci.*, *14*, 79–85 (1991).
39. J. White, J. S. Kauer, T. A. Dickinson, and D. R. Walt, Rapid analyte recognition in a device based on optical sensors and the olfactory system, *Anal. Chem. 68*(13), 2191–2202 (1996).
40. T. A. Dickinson, J. S. White, J. S. Kauer, and D. R. Walt, A chemical-detecting system based on a cross-reactive optical sensor array, *Nature*, *382*, 697–700 (1996).
41. T. A. Dickinson, D. R. Walt, J. White, and J. S. Kauer, Generating sensor diversity through combinatorial polymer synthesis, *Anal. Chem.*, *69*(17), 3413–3418 (1997).
42. J. White, J. S. Kauer, T. A. Dickinson, and D. R. Walt, An olfactory neuronal network for vapor recognition in an artificial nose, *Biol. Cyber. 78*, 245–251 (1998).
43. S. R. Johnson, J. M. Sutter, H. L. Engelhardt, P. C. Jurs, J. White, J. S. Kauer, T. A. Dickinson, and D. R. Walt, Identification of multiple analytes using an optical sensor array and pattern recognition neural networks.
44. D. R. Walt, T. Dickinson, J. White, J. Kauer, S. Johnson, H. Engelhardt, J. Sutter, and P. Jurs, Optical sensor arrays for odor recognition. *Biosens. Bioelectron.*, *13*, 697–699 (1998).
45. T. A. Dickinson, K. L. Michael, J. S. Kauer, and D. R. Walt, Convergent, self-encoded bead sensor arrays in the design of an artificial nose, *Anal. Chem. 71*, 2192–2198 (1999).

Chapter 11
Use of Magnetic Separation Coupled with Power Ultrasound for High-Throughput Solid-Phase Organic Synthesis

Irving Sucholeiki, J. Manuel Perez, and Patrick D. Owens
Solid Phase Sciences Corporation, Medford, Massachusetts

I. THE SEPARATION PROBLEM

In Chapter 1 we mentioned various advantages to the use of a solid support in organic chemistry. It was also mentioned that the most common way to separate a solid-phase support from the soluble components of a reaction mixture is filtration. Although filtration has been the method of choice, it has limitations when applied to an automated system that warrant the development of other approaches. One such limitation is the difficulty in automating the simultaneous running of thousands of solid-phase reactions in an economical manner. This is due to the fact that for each reaction vessel that is added to an array, separate valves must be used to allow a user to selectively add or remove liquid from that specific vessel. This makes the cost of expanding the number of fully automated reactions that a machine can run rise in proportion with the increasing number of parallel reactions.

Another limitation with the use of filtration in separating solids from liquids in an automated system is the periodic clogging of the pores of the various filters used. Clogging of the filter pores comes from three primary sources: (1) fines produced through the fragmentation of the solid support itself during the mixing process; (2) precipitation of reagents and/or byproducts produced during the reaction; and (3) precipitation of reagents and/or byproducts produced during washing of the support. Whether one uses a polymer, glass, or paper filter and whether one uses standard filtration or reverse filtration, the periodic clogging of the pores of these filters makes the concept of separating and washing large numbers of supports by computer control a risky proposition. This is because in most automated solid-phase synthesizers there is no way to determine if the pores of a filter contained in a reaction vessel have become clogged. It is possible for a sensor to be placed in a vessel to monitor the presence or absence of a liquid. However, current sensor technology relies primarily on conductance to differentiate between liquid and air, a property that is easy to measure in an aqueous-based system but difficult in the non-aqueous-based environment of most solid-phase reactions.

There are other limitations to the use of filtration in automated solid-phase organic synthesis, such as in the area of microchemistry. When the volume of a solid-phase reaction is sufficiently reduced, low enough, say to a few microliters, the use of filtration as a means of separating a solid from a liquid can become very problematical. A field that has had some success in translating some of its techniques into automation is immunodiagnostics. Exposing of antibody-bound paramagnetic beads to a magnetic field can be used to separate antibody-bound antigen from unbound antigen in very small volumes [1,2]. Magnetic separation methods have also been applied successfully in cell sorting [3–5]. There are several advantages to the use of magnetic separation over simple filtration. One can separate out particles in small reaction volumes. With magnetic separation, clogged filters are no longer a problem, one simply aspirates the liquid and non-magnetic material (precipitates) using a large-diameter needle (Fig. 1). If aspiration is accomplished using a liquid-handling robot, expanding the number of vessels that can be magnetically separated and aspirated will only require extending the time for the robot to cover the larger number of vessels, but not the addition of expensive hardware.

The use of magnetic separation in the field of solid-supported organic chemistry has been slow in coming due to the instability exhibited by many of the currently available paramagnetic supports in organic solvents such as dimethylformamide (DMF) and methylene chloride. Upon exposure to these solvents, the typical polymer-coated magnetic beads dissolve or fragment. Some paramagnetic supports utilize high levels of polymer crosslinking to stabilize the magnetite encased within, whereas others utilize a silica-based coating. Unfortunately, these more stable types of paramagnetic supports exhibit relatively poor loading capacities (typically less than 0.2 mmol reactive groups/g) making them not very economical to use for practical solid-phase organic synthesis [6,7].

One way to make paramagnetic polymer beads stable in organic solvent is to enhance the levels of cross-linking around the magnetite core. As was mentioned in Chapter 1, a major problem with this approach is that higher levels of cross-linking (more than 1–2%) reduce the extent to which the support expands

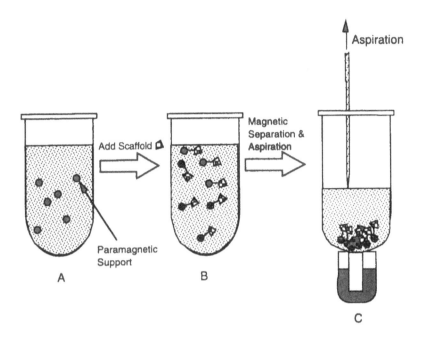

Figure 1 (A) Addition of paramagnetic support to reaction mixture. (B) Attachment of scaffold to support. (C) Magnetic separation of scaffold-bound support followed by aspiration of liquid and nonmagnetic precipitates.

and contracts and decreases the level of reactivity. A solution to this problem involved synthesizing a composite paramagnetic bead incorporating zones of high cross-linking only in those areas where the magnetite is located. The simplest way of creating such a composite was to first create primary beads **1** (5–10 μm in diameter) composed of magnetite crystals encased in highly cross-linked polystyrene, then to enmesh these primary beads **1** into a larger bead composed of 1–2% cross-linked polystyrene (100–200 μm in diameter) (Scheme 1) [8a,8b]. In order to incorporate functionality into the new composite bead, a functionalized styrene monomer such as chloromethylstyrene was used to enmesh the primary beads **1** to give the resulting chloromethylated, polystyrene composite bead **2** (1 mmol chloromethyl groups/g of paramagnetic bead). Treating the chloromethylated support **2** with potassium phthalimide in refluxing DMF followed by deprotection with hydrazine gave the resulting aminomethylated composite support **3** (0.7–0.9 mmol aminomethyl groups/g). Alternatively, the primary beads can be enmeshed with unfunctionalized polystyrene and subsequently functionalized by treating the support with butyllithium and an electrophile such as CO_2 to give the resulting carboxyl-functionalized support **5** (0.5 mmol carboxy

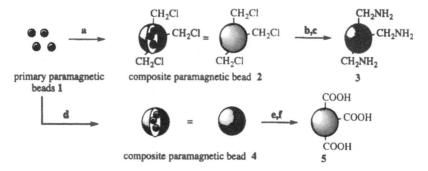

Scheme 1 *Reagents*: (a) chloromethylstyrene; (b) potassium phthalimide; (c) N_2H_4, ethanol; (d) styrene; (e) butyllithium, CO_2; (f) H^+.

groups/g, Scheme 1). Composite beads **2–5** are stable, expand in organic solvents, and exhibit loading capacities comparable to those of standard solid-phase synthesis supports.

These functionalized, high-loading composite supports **2**, **3**, and **5** can be used in solid-phase organic synthesis and for scavenging impurities in solution (Fig. 2) [8a,8b]. For example, treating the aminomethylated composite support (**3**) with iodoacetic acid and diisopropylcarbodiimide (DIC) produces an iodoacetamide-functionalized support **6** (0.64 mmol iodoacetamides/g of support; Scheme 2). This derivatized support can be used as a high-loading scavenger resin for free thiols in solution.

As a demonstration of this application, iodoacetamide support **6** (300 mg, 0.19 mol of iodoacetamide) was added to a mixture composed of equal molar quantities (0.06 mmol) of 2-mercapto-3-nitropyridine (**7**) and 4-nitroaniline (**8**) and diisopropylethylamine (DIEA) in (DMF) as solvent. After stirring the mixture for 26 min, 98% of the mercaptan (**7**) was found to bind to the resin.

The high stability of the composite paramagnetic supports **2**, **3**, and **5** and the ease to which one can magnetically separate them makes them ideal for combinatorial library production. For example, to the aminomethyl support **3** was attached the hydroxymethylbenzoic acid linker **10** using DIC to give the hydroxymethyl support **11** (Scheme 3). 4-Nitrophenylacetic acid (**12**) was then attached to the hydroxymethyl support **11**, followed by nitro reduction using $SnCl_2$ to give the aniline-bound support **14** [9]. The loading capacity of this derivatized support **14** was measured by elemental analysis to be around 0.5 mmol aminophenylacetate/g of support. The resin-bound aniline is then treated with an excess of 2-bromo-5-nitropyridine at 75°C to give the corresponding resin-bound pyridine amide **15** (Scheme 3). A small portion of the resulting 2-anilino-5-nitro-

Figure 2 (A) Flask containing high-loading paramagnetic beads suspended in methylene chloride. (B) Flask placed over magnetic separator causes paramagnetic beads to
adhere to the bottom edges of the flask, making the center free for aspiration.

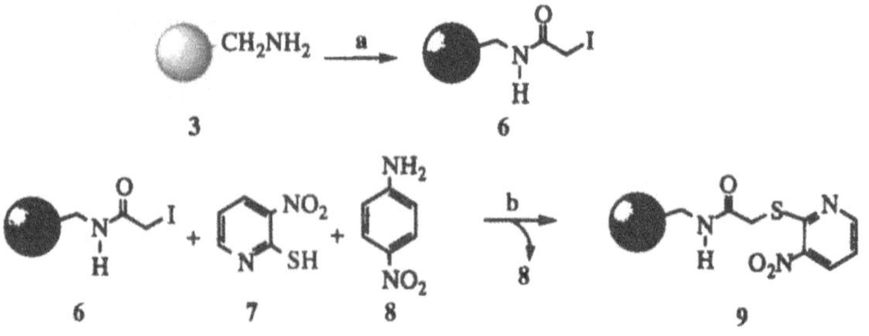

Scheme 2 *Reagents*: (a) iodoacetic acid, DIC, methylene chloride; (b) DIEA, DMF.

pyridine-bound resin **15** was cleaved with sodium methoxide in methanol-tetrahy-drofuran to give the resulting nitro-coupled product **16** in 60% overall yield [9]. The resin-bound nitro compound **15** was reduced to the corresponding amine and treated with various alkylating agents to generate a library of highly function-alized pyridylbenzenebenzamides and pyridylbenzenesulfonamides, an example of which (**19**) is shown in Scheme 4.

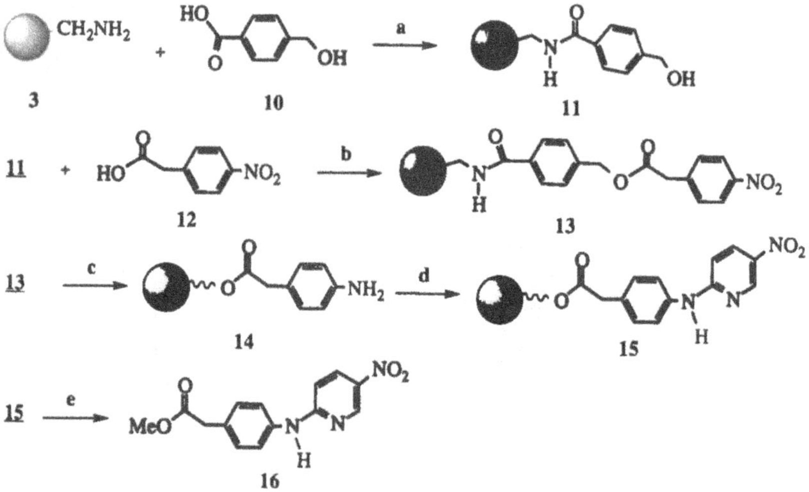

Scheme 3 *Reagents*: (a) DIC, CH$_2$Cl$_2$; (b) DIC, CH$_2$Cl$_2$; (c) SnCl$_2$·2H$_2$O, DMF; (d) 2-bromo-5-nitropyridine, DIEA, DMF; (e) NaOme-MeOH, THF.

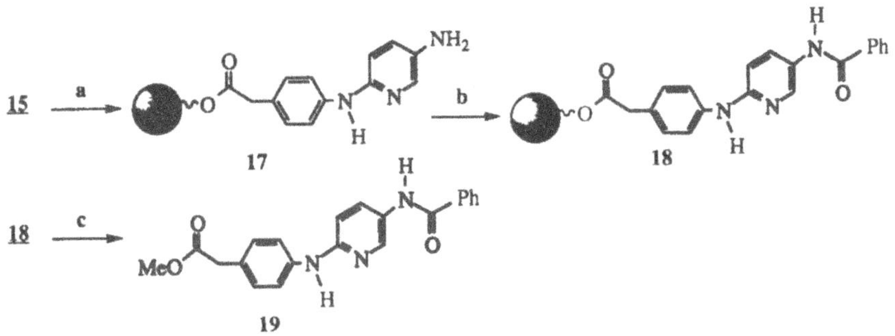

Scheme 4 *Reagents*: (a) SnCl$_2$·2H$_2$O, DMF; (b) Acyl chloride or sultonyl chloride, DIEA, DMF; (c) NaOMe-MeOH, THF.

II. THE MIXING PROBLEM

In Chapter 1 we asked what method of mixing was best for a solid-phase reaction. The answer was not clear because the few studies that have been conducted in this area have not yielded enough data to conclusively point to one method as being better at affecting the solid-phase reaction rate. Most automated solid-phase synthesizers utilize stirring, gas bubbling, or orbital shaking as the primary methods of mixing the heterogeneous reactions. Of these three methods, orbital shaking seems to predominate, even though very little data have been published demonstrating that orbital shaking is superior to the other two methods. This is puzzling because implementing orbital shaking into a fully automated system for the synthesis of large compound arrays (several hundred reactions) is difficult and expensive. One form of agitation that has been found to enhance the solid-phase reaction rate is high-energy ultrasound or power ultrasound [10]. Application of power ultrasound in solid-phase peptide synthesis has been found to accelerate the observed peptide coupling and cleavage reaction rates [11–13]. The observed rate enhancements may be attributed to several phenomena. However, they are attributable not to the sound waves directly impacting the resin-bound molecules but on the effect of the sound waves on the propagating liquid. To better comprehend how ultrasound impacts a solid-phase reaction one must understand how ultrasound impacts a liquid medium. As a sound wave propagates through a liquid medium it exerts pressure on that liquid that causes both rarefaction and compression of the solvent molecules parallel to the propagating axis. The first result of this rarefaction and compression of the solvent molecules is what is termed *acoustic streaming*, which sets the liquid medium in motion [14]. If there exists in the liquid medium tiny air bubbles, another phenomenon occurs that is termed *cavitation*, which is the expansion and contraction of these bubbles or air pockets [15]. In an expansion cycle, the bubble grows until it can no longer absorb further energy; then the bubble collapses. If the bubble does not collapse it enters a contraction cycle. In a contraction cycle the bubble contracts while generating heat. At some point, the bubble may become unstable and instead of expanding again, implodes during the compression stage. There are three direct results of a cavitation bubble implosion, the first being generation of a great deal of localized heat that is quickly dissipated in the surrounding medium. The second is the formation of high-speed liquid "jets" due to the pressure differential surrounding the bubble implosion. The third is the emission of a pressure pulse or shock wave immediately following the implosion of the cavitation bubble [16]. How do these events affect a heterogeneous reaction? The forces derived from cavitational bubble collapsing can be intense enough to, for example, shatter metal particles into much smaller fragments [17]. Under milder conditions, these forces can enhance the mixing process. Specifically, the creation of pressure differentials around the imploding cavitation bubble can create high-speed solvent jets that accelerate the movement of solute particles to and from the matrix of the solid-phase support (Fig. 3).

The obvious question one should ask is why ultrasound has not been quickly embraced by the solid-phase synthesis community as a mixing method

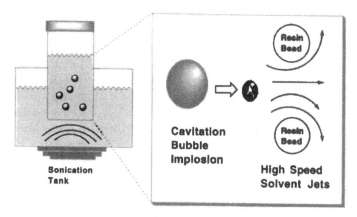

Figure 3 (left) Application of bath ultrasound to a heterogeneous reaction. (right) Within the reaction medium, cavitation bubble collapse results in pressure differentials that cause the formation of high-speed solvent jets that circulate throughout the reaction medium.

for high-throughput solid-phase organic syntheses? There are many possible explanations, but one main obstacle to the use of ultrasound for mixing solid-phase
reactions is the difficulty in applying it in the presence of filtration-type manifolds. However, incorporation of ultrasound into a multiple, parallel solid-phase
synthesis format is relatively easy when it is combined with the use of a paramagnetic support because only an external magnetic field is required to separate the
support from the soluble components of the reaction mixture (Fig. 4). Those
nonmagnetic components of the reaction mixture can then be removed by aspiration (Fig. 4) [18]. Other benefits of using power ultrasound in solid-phase organic
synthesis include solvent degassing while the reaction is running, as well as the
ease in controlling the reaction temperature via the bath temperature of the sonicating tank.

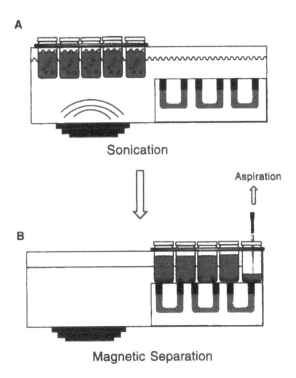

Figure 4 (A) Application of high-energy, bath ultrasound in a multiple solid-phase reaction format. (B) Application of a magnetic field separates paramagnetic particles from the
nonmagnetic components of the reaction mixture, which are then removed by aspiration.

III. USE OF ULTRASONIC MIXING TO ENHANCE SOLID PHASE REACTION RATES

As a specific example of the use of ultrasound in solid-phase organic chemistry, we began with the hydroxymethyl-derivatized, paramagnetic support **11** (0.51 mmol hydroxymethyl/g of support) and added an excess of fluorenylmethoxy-carbonyl (Fmoc) protected L-proline, benzotriazole-1-yloxytrispyrrolidinophosphonium-hexafluorophosphate (PyBop) coupling reagent and DIEA in DMF under the presence of bath ultrasound to give the resulting resin bound Fmoc protected proline (0.32 mmol Fmoc/g of support) (Scheme 5) [19]. Ultrasound was applied to the coupling reaction via a computer controlled, immersible 1200-W, 25- or 40-kHz ultrasonic transducer at 30% power for 24 h (Fig. 5). Temperature control was achieved via a computer-controlled, positive displacement heater-chiller unit that was directly connected to the sonicating bath (Fig. 5). Reactions were also run using 20 kHz probe-type cup-horn transducers with similar results. The percent power level used depended on the reactivity of the overall solid-phase reaction. When the reactions were run at higher power levels (>50%) and for longer periods (>10 h) the paramagnetic beads were found to fracture into smaller fragments that were easily separated using a magnetic field. Application of power ultrasound to non-paramagnetic-based supports could be problematical because such fragmentation phenomenon would produce fines small enough to clog the pores of most filtration-type systems.

Deprotection of the Fmoc protecting group was accomplished using 50% piperidine in DMF under similar ultrasonic conditions to give the resin-bound L-proline **20** (Scheme 5). Alkylation of the resin bound L-proline with 2-nitrobenzyl bromide **5** and DIEA in the presence of power ultrasound gave the resulting

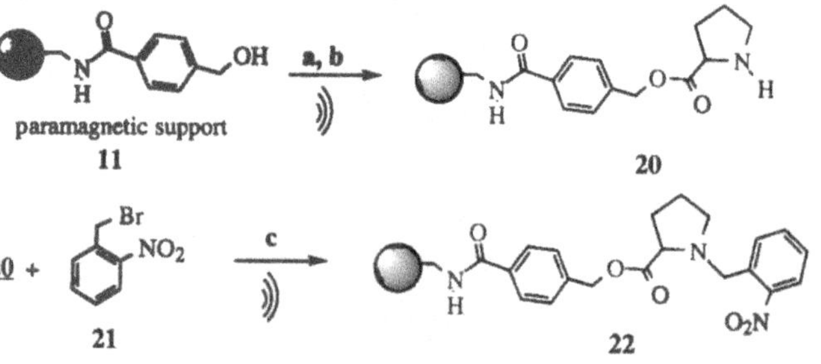

Scheme 5 *Reagents*: (a) Fmoc-L-proline, PyBop, DIEA, 30% power; (b) 50% piperidine-DMF, 30% power; (c) DIEA, DMF, 100% power.

Figure 5   Computer-controlled sonicating synthesizer incorporating an open-loop, positive-displacement heater-chiller unit and a two-arm x,y,z liquid-handling robot.

N-benzyl adduct **22**. It was possible to monitor by UV spectrophotometry the disappearance of the 2-nitrobenzyl bromide **21** in solution under varying agitation conditions. Figure 6A shows the course of the reaction as a function of time in the presence of ultrasound, stirring, and neither ultrasound nor stirring. The results show that at constant temperature ($t = 25°C$) an enhancement in the reaction rate was observed when ultrasound was applied ($t_{1/2} = 12.4$ min) to the solid-phase reaction as compared to stirring ($t_{1/2} = 20.1$ min).

As was mentioned earlier, the creation of pressure differentials around the imploding cavitation bubbles during ultrasound can create high-speed solvent jets that can enhance solvent-solute mixing in and around the resin bead and may explain the observed rate enhancements seen in Figure 6A. Further proof of this mechanism was provided by monitoring the release of the non-covalently bound 2-nitrobenzyl bromide from the support as a function of time. Figure 6B shows that beads exposed to ultrasound release 2-nitrobenzyl bromide faster ($t_{1/2} = 30$ s) than when stirred ($t_{1/2} = 45$ s) [19].

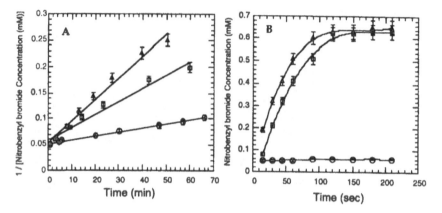

Figure 6 (A) Plot of 1/[2-nitrobenzyl bromide concentration] as a function of time and in the presence of **21** and DIEA while exposed to ultrasound, stirring, and no agitation (unstirred). (B) Plot of the release of 2-nitrobenzyl bromide in DMF as a function of time in the presence of ultrasound, stirring, and no agitation. $\triangle$, 25 kHz ultrasound; $\square$, stirring; $\bigcirc$, no agitation (unstirred).

IV. USE OF ULTRASONIC MIXING TO IMPROVE RESIN WASHING

The data shown in Figure 6B suggest that through the use of ultrasonic mixing one can enhance the diffusion rate of molecules traveling into and out of the solid-phase support. From the data presented in Figure 6B, one can see that use of ultrasound to enhance the flow of solute molecules into and out of a support can also be used to improve the efficiency of resin washing. For example, there are many instances when the starting material and/or reagents used in a solid-phase reaction are not cleanly washed off a solid support. Sometimes a reagent or a reagent's byproduct will precipitate throughout the reaction medium. In cases where precipitation occurs within the support itself, repeated washings may not be enough to remove the contaminant. This can pose a serious problem if the contaminant finds its way into the final cleaved product. A case in point is the $SnCl_2$ reduction of a resin-bound arylnitro group to an amine. Although this reaction has been commonly used in combinatorial library production, little comment has been made on the possibility of further tin contamination after resin washing. However, experiments in our laboratory have shown that repeated washing of a resin after undergoing $SnCl_2$ reduction does not entirely remove all of the resin-bound tin [20]. In fact, we have found that after a $SnCl_2$ mediated reduction, washing a support using the standard plethora of solvents followed by filtration can leave as much as 1.5% tin by weight on the support (Table 1). We have also found that application of low-energy ultrasonic mixing during resin washing can dramatically reduce this residual resin-bound tin to levels far below that found using standard shaking (Table 1).

Table 1 Comparison of Washing Procedure for Removal of Resin-Bound Tin

Resin agitation method[a]	Resin separation method	% Sn in resin after washing[b]
Shaking	Filtration	1.5
Ultrasound	Mag. separation/aspiration	0.3
No agitation	Filtration	1.5
No agitation	Mag. separation/aspiration	1.2

[a] Resins were washed equally with 5×1 mL DMF, 5×1 mL MeOH, and 5×1 mL CH_2Cl_2.
[b] Obtained by Sn elemental analysis.

V. SUMMARY AND CONCLUSION

As solid-phase organic chemistry continues to have a greater role in drug discovery, new techniques will need to be developed that can better translate the process into automation. There is little doubt that many of the tasks done manually by bench chemists today will be automated in the future, and that will include every aspect of solid-phase organic chemistry. However, it is clear that today's robotic technology is still somewhat crude and expensive to apply within a laboratory setting. In fact, many of the electromechanical components (i.e., solenoid valves, stepper motors) that make up many of today's automated synthesizers are not very different from those produced in the mid-1970's. Unlike software development, which has evolved at a spectacular pace, the lag in the development of robotic hardware has made the classic use of, say, filtration very cost-ineffective when applied to full automation. Magnetic separation, on the other hand, has several advantages that make it perfectly suited for today's automation technology. For example, when magnetic separation is coupled with a standard x,y,z liquid-handling robot, the cost of washing and separating individual solid-phase reactions drops with an increasing number of reactions run. The fact that one no longer has to worry about clogged filters makes the use of magnetic separation a more reliable means for computer-controlled separation. In addition, use of magnetic separation can also be a great advantage when applied to microchemistry because it allows one an easier way to work with small reaction volumes. Lastly, the use of magnetic separation allows for the application of ultrasonic mixing in a multiple, parallel, solid-phase synthesis format; a technique that has been shown to enhance both the solid phase reaction rate and the efficiency of resin washing. Until robotic and sensor technology catches up with the dramatic strides in software development, new methods, such as magnetic separation and ultrasonic mixing, will have to be introduced if the dream of a fully automated laboratory is to become a practical reality.

ACKNOWLEDGMENT

The authors would like to thank the National Institutes of Health through its SBIR program for partial funding of this work.

REFERENCES

1. J. Ugelstad, T. Ellingsen, A. Berge, and B. Helgee, PCT Int. Appl. WO83/03920 (1983).
2. J. G. Fjeld, H. B. Benestad, T. Stigbrand, and K. Nustad, *J. Immunol. Meth. 109*, 1 (1988).
3. J. G. Treleaven, J. Gibson, J. Ugelstad, A. Rembaum, T. Philip, G. C. Caine, and J. Kemshead, *Lancet, 14*, 70 (1984).
4. S. Miltenyi, W. Muller, W. Weichel, and A. Radbruch, *Cytometry, 11*, 231 (1990).
5. R. Padmanabhan, C. D. Corsico, T. H. Howard, W. Holter, C. M. Fordis, M. Willingham, and N. Bruce, *Anal. Biochem., 170*, 341 (1988).
6. M. J. Szymonifka, and K. T. Chapman, *Tetrahedron Lett.*, 36, 1597 (1995).
7. Benner, S. A. US 4,638,032 (1987).
8a. I. Sucholeiki and J. M. Perez, *Tetrahedron Lett.*, 40, 3531 (1999).
8b. A polyethylene glycol (PEG) grafted version of paramagnetic support 3 has also been developed which exhibits improved swelling characteristics. I. Sucholeiki, US 5,858,534 (1999).
9. J. M. Perez and I. Sucholeiki, In *Innovation and Perspectives in Solid Phase Synthesis & Combinatorial Libraries—Proceedings of the 6th International Symposium*, (R. Epton, ed.), York, Mayflower Scientific, Birmingham (in press).
10. For an introduction to power ultrasound see T. J. Mason. In *Practical Sonochemistry: User's Guide to Applications in Chemistry and Chemical Engineering*, Ellis Horwood Ltd., 1991.
11. S. Takahashi, and Y. Shimonishi, *Chemistry Lett.*, 51–56 (1974).
12. A. M. Bray, L. M. Lagniton, R. M. Valerio, and N. J. Maeji, *Tetrahedron Lett.*, 35, 9079–9082 (1994).
13. V. Krchnák, and J. Vágner, *Pept. Res.*, 3, 182–193 (1990).
14. J. L. Luche, In *Synthetic Organic Sonochemistry*, Plenum Press, New York, 8–11 (1998).
15. K. S. Suslick, *Science*, 247, 1439–1445 (1990).
16. Ref. 14, pp. 32–36.
17. K. S. Suslick, M. Fang, and T. Hyeon, *J. Am. Chem. Soc., 118*, 11960–11961 (1996).
18. I. Sucholeiki, US 5,779,985 (1998); I. Sucholeiki, US 5,962,338 (1999), and other patents pending.
19. J. M. Perez, E. J. Wilhelm, and I. Sucholeiki, *Bioorg. Med. Chem. Lett.*, 10, 171–174 (2000).
20. I. Sucholeiki, and J. M. Perez, unpublished results.

Author Index

Subject Index

Milton Keynes UK
Ingram Content Group UK Ltd.
UKHW020319111024
449327UK00040B/1401

Contents

Preface

Welcome to *Tree-based methods for statistical learning (with examples in R)*. Tree-based methods, as viewed in this book, refer to a broad family of algorithms that rely on *decision trees*, of which this book attempts to provide a thorough treatment. This is not a general statistical or machine learning book, nor is it an R book. Consequently, some familiarity with both would be useful, but I've tried to keep the core material as accessible and practical as possible to a broad audience (even if you're not an R programmer or master of statistical and machine learning). That being said, I'm a firm believer in learning by doing, and in understanding concept through code examples. To that end, almost every major section in this book is followed-up by general programming examples to help further drive the material home. Therefore, this book necessarily involves a lot of code snippets.

Who is this book for?

This book is primarily aimed at researchers and practitioners who want to go beyond a fundamental understanding of tree-based methods, such as decision trees and tree-based ensembles. It could also serve as a useful supplementary text for a graduate level course on statistical and machine learning. Some parts of the book necessarily involve more math and notation than others, but where possible, I try to use code to make the concepts more comprehensible. For example, Chapter 3 involves a bit of linear algebra and intimidating matrix notation, but the math-oriented sections can often be skipped without sacrificing too much in the way of understanding the core concepts; the adjacent code examples should also help drive the main concepts home by connecting the math to simple coding logic.

Nonetheless, this book does assume some familiarity with the basics of statistical and machine learning, as well as the R programming language. Useful references and resources are provided in the introductory material in Chapter 1. While I try to provide sufficient detail and background where possible, some topics could only be given cursory treatment, though, whenever possible, I try to point the more ambitious reader in the right direction in terms of references.

Companion website

There is a companion website for this book located at

https://bgreenwell.github.io/treebook/.

This is where I plan to include chapter exercises, code to reproduce most of the examples and figures in the book, errata, and various supplementary material.

Contributions from the community are more than welcome! If you notice something is missing from the website (e.g., the code to reproduce one of the figures or examples) or notice an issue in the book (e.g., typos or problems with the material), please don't hesitate to reach out. A good place to report such problems is the companion website's GitHub issues tab located at

https://github.com/bgreenwell/treebook/issues.

Even if it's a section of the material you found confusing or hard to understand, I want to hear about it!

The treemisc package

Along with the companion website, there's also a companion R package, called **treemisc** [Greenwell, 2021c], that houses a number of the data sets and functions used throughout this book. Installation instructions and documentation can be found in the package's GitHub repository at

https://github.com/bgreenwell/treemisc.

Colorblindess

This book contains many visuals in color. I have tried as much as possible to keep every figure colorblind friendly. For the most part, I use the Okabe-Ito color palette, designed by Masataka Okabe and Kei Ito (`https://jfly.uni-koeln.de/color/`), which is available in R (`>=4.0.0`); see `?grDevices::palette.colors` for details. If you find any of the visuals hard to read (whether due to color blindness or not) please consider reporting it so that it can be corrected in the next available printing/version.

Acknowledgments

I'm extremely grateful to Bradley Boehmke, who back in 2016 asked me to help him write "Hands-On Machine Learning with R" [Boehmke and Greenwell, 2020]. Without that experience, I would not have had the confidence (nor the skill or patience) to prepare this book on my own. Thank you, Brad.

Also, a huge thanks to Alex Gutman and Jay Cunningham, who both agreed to provide feedback on an earlier draft of this book. Their reviews and attention to detail have ultimately led to a much improved presentation of the material. Thank you both.

Lastly, I cannot express how much I owe to my wonderful wife Jennifer, and our three kids: Julia, Lillian, and Oliver. You help inspire all I do and keep me sane, and I truly appreciate you putting up with me while I worked on this book.

Brandon M. Greenwell

1

Introduction

Ever play a game called *twenty questions*? If you have, then you should have no trouble understanding the basics of how decision trees work. A decision tree is essentially a set of sequential yes or no questions regarding the available features in an attempt to make an accurate prediction (or classification). For example, "Is systolic blood pressure less than 120 mm Hg?" or "Does it have three leaves?". The answer to each question determines the next follow-up question. For example, in trying to determine whether or not a particular plant in your backyard is poison ivy, if you answered yes to whether or not it has three leaves[a], the next question you (or the decision tree) might ask is "Does it have notched leaves?", and so forth. The overall idea is to ask a series of good and intelligent questions using the available data that will hopefully lead to an accurate prediction. What mostly differs between the various decision tree algorithms I'll discuss in this book is the nature of the questions asked and how they're determined given a set of training observations.

Compared to other nonparametric algorithms, there's also a bit more transparency in how decision trees make predictions. For example, in classifying poison ivy, a decision tree might use a simple rule such as, "If it has three leaves AND and the leaves are notched, then it's likely poison ivy." Being able to interpret the output from a model is crucial in understanding how a model makes predictions and conveying the results to others.

But we don't often just ask random questions, whether we're playing twenty questions or trying to determine if a particular plant is poison ivy. Our inquiries tend to have a hierarchy, in that we often start with the most general questions that we think will narrow down the possibilities the most, then follow up with more refined questions to home in on the answer. Decision trees work in a similar way in that the first handful of questions tends to be the most important, while the questions further down the tree are just smaller refinements to further improve accuracy.

I'll return to talking about trees in Section 1.2. Next, I'll turn the discussion to some basic (but important) ideas in statistical and machine learning.

1.1   Select topics in statistical and machine learning

This section is intended to provide a (very) brief overview of a handful of topics from statistical and machine learning that will be useful to know for some of the material to come. Select topics, like the *bias-variance tradeoff* and *right censoring*, are important to several areas and examples in this book, and so I'll spend the next few sections highlighting some of these important ideas. This is by no means intended to act as a primer, or even just a basic introduction to statistical and machine learning in general, but rather to highlight key topics

---

[a]"Leaves of three, let it be." is a common rule of thumb for avoiding contact with poison ivy.

that will help introduce more advanced topics later in the book. This book does assume, however, that readers have at least some general background or exposure to common topics in statistics and machine learning (like hypothesis testing, cross-validation, and hyperparameter tuning). If you're looking for a more thorough overview of statistical and machine learning, I'd suggest starting with James et al. [2021]. For a deeper dive, go with Hastie et al. [2009]. Both books are freely available for download, if you choose not to purchase a hard copy. Harrell [2015] and Matloff [2017], while more statistical in nature, offer valuable takes on several concepts fundamental to statistical and machine learning, and I highly recommend each.

1.1.1 Statistical jargon and conventions

To start, let's introduce ourselves to some of the notation used throughout the book; additional mathematical notation will be introduced when necessary in the chapters that follow. Since this book is primarily concerned with fitting tree-based models for prediction and description, I'll often be talking about independent variables (what we use to predict) and dependent variables (what we want to predict). The independent variables are referred to as either features or predictors (maybe even as covariates[b] or regressors at one point or another). The dependent variable is referred to as the response, target, or outcome variable. Generic features are denoted by x or x_1, x_2, and so on, and the response using y. In most cases, bold symbols typically refer to matrices (usually uppercase Latin letters) or column vectors (usually lowercase Latin letters). For example, $\boldsymbol{X}$ typically represents an $N \times p$ matrix of p features from a data set with N rows (or observations/records); $\boldsymbol{x}_i$ denotes the i-th row of $\boldsymbol{X}$, whereas $\boldsymbol{x}$ (or sometimes $\boldsymbol{x}_0$) refers to the p predictor values of a (single) generic observation.

As far as variable types go, this book is mainly concerned with three:

- Nominal categorical (i.e., categorical where the order of categories doesn't matter). Examples include gender, eye color, zip code, or blood type.

- Ordered categorical (i.e., categorical where the order of categories matters). Examples include socioeconomic status (e.g., low < middle < high), age range (e.g., [0-10yrs.] < [11-20yrs.] < ...), or satisfaction rating (e.g., not satisfied < somewhat satisfied < very satisfied). Ordered categorical variables are sometimes referred to as ordinal.

- Ordered numeric. Examples include age or temperature measured on a continuum, height, weight, or concentration.

---

[b]Technically, a covariate refers to a predictor that we think is related to the response, but whose effect is not of direct interest (e.g., we may not care to interpret its effect, but we include it to improve the overall model; think *analysis of covariance*).

I'll often refer to ordered numeric variables as either numeric or continuous, and both ordered categorical and numeric variables collectively as ordered variables; note that ordered categorical variables can arbitrarily be mapped to integers as long as the original ordering is preserved (for example, low < med < high → 1 < 2 < 3).

Many tree-based algorithms only make the distinction between ordered and nominal variables. Categorical variables, whether ordered or nominal, will often be referred to as *factors* (e.g., temperature was recorded as an ordered factor with levels: freezing < cold < warm < hot). There's also the concept of *censored variables* (usually the response), but I'll defer discussion of censored outcomes to the example in Section 1.4.9.

The *learning sample*, also called the *training data*, is often denoted by $d_{trn} = \{(\boldsymbol{x}_i, y_i)\}_{i=1}^{N}$, where N is the sample size, and $\boldsymbol{x}_i$ is the i-th row of training features (e.g., $x_{1i}, x_{2i}, \dots$).

On the rare occasion where I'm referring to a *random variable*, I'll typically use an upper case Latin letter (e.g., Y or X) or the lowercase Greek letter ϵ; other Greek letters, like β or θ, will generally represent the fixed, but unknown parameters of a model. The operators E and V will denote the expected value (i.e., mean) and variance of a random variable, respectively. Several probability distributions are also used throughout this book and are denoted using a mix of capital letters (sometimes in a calligraphic font) and Greek letters, for example:

- $\mathcal{U}(0, 1)$ represents a continuous uniform distribution over the interval $[0, 1]$.

- $N(\mu, \sigma)$ represents a normal (or Gaussian) distribution with mean μ and standard deviation σ.

- χ^2_ν represents a chi-squared distribution with ν degrees of freedom.

To say that $X_1, X_2, \dots, X_p$ are a random sample from some arbitrary distribution $\mathcal{D}$, parameterized by $\boldsymbol{\theta}$, I'll write $\{X_i\}_{i=1}^{N} \overset{iid}{\sim} \mathcal{D}(\boldsymbol{\theta})$; the *iid* stands for *independently and identically distributed*.

Whenever possible, I try to emphasize words and terms you may be unfamiliar with using *italicized text*, and I encourage you to "google" them for more details, but not knowing them should not distract you from the fundamental ideas presented throughout this book.

1.1.2   Supervised learning

Supervised learning can often be thought of as an exercise in *function approximation*. For simplicity, we often assume that the response variable, Y, is related to a set of predictors, $\boldsymbol{x}$, through a model with additive error:

$$Y = f(\boldsymbol{x}) + \epsilon, \tag{1.1}$$

where ϵ is a random variable with mean zero (i.e., $\mathrm{E}(\epsilon) = 0$) and is assumed to be independent of $\boldsymbol{x}$. Note that the response is also a random variable here since it is a function of ϵ. The function $f(\boldsymbol{x})$ is fixed and represents the *systematic* part of the relationship between Y and $\boldsymbol{x}$. As is almost always the case, the true relationship between Y and $\boldsymbol{x}$ is often statistical in nature (i.e., not deterministic) and the additive error helps to capture the non-deterministic aspect of this relationship (e.g., unobserved predictors, measurement error, etc.).

Since we assume $\mathrm{E}(\epsilon) = 0$, it turns out that $f(\boldsymbol{x})$ can be viewed as a *conditional expectation*:

$$\mathrm{E}(Y|\boldsymbol{x}) = f(\boldsymbol{x}),$$

where we can interpret $f(\boldsymbol{x})$ as the mean response for all observations with predictor values equal to $\boldsymbol{x}$. In the case of *J-class classification* (Section 1.1.2.3), we can still view $f(\boldsymbol{x})$ as a conditional mean; in this case it's the conditional proportion corresponding to a particular class: $\mathrm{E}(Y = j|\boldsymbol{x})$, which can be interpreted as an estimate of $\mathrm{Pr}(Y = j|\boldsymbol{x})$—the probability that Y belongs to class j given a particular set of predictor values $\boldsymbol{x}$. In this sense, class probability estimation is really a regression problem.

The term "supervised" in supervised learning refers to the fact that we use labeled training data[c] $\{y_i, \boldsymbol{x}_i\}_{i=1}^N$ and an algorithm to learn a reasonable mapping between the observed response values, y_i, and a set of predictor values, $\boldsymbol{x}_i$. Without a labeled response column, the task would be *unsupervised* (Section 1.1.3), and the analytic goal would be different.

An estimate $\hat{f}(\boldsymbol{x})$ of $f(\boldsymbol{x})$ can be used for either *description* or *prediction* (or both). I'll briefly discuss the meaning of each in turn next.

1.1.2.1 Description

In supervised learning, descriptive tasks are often concerned with determining which features have the most impact on $\hat{f}(\boldsymbol{x})$ and how. For example, in supervised learning problems, we are often interested in determining

- which predictors are the most "important" for prediction (feature importance);

- the marginal impact of each predictor (or a subset of the important ones) on the predicted outcome (feature effects).

---

[c]The labels here are provided by the response values $\{y_i\}_{i=1}^N$.

For example, in the Ames housing data (Section 1.4.7), we may be interested in determining which predictors are most influential on the predicted sale price in a fitted model. We may also be interested in how a particular feature (e.g., overall house size) functionally relates to the predicted sale price from a fitted model.

Questions like these are relatively straightforward to glean from simpler models, like an *additive linear model* or a simple decision tree. However, this type of information is often hidden in more complicated nonparametric models—like *neural networks* (NNs) and *support vector machines* (SVMs)—which unfortunately has given rise to the term "black box" models. In Chapter 6, I'll look at several model-agnostic techniques that can be helpful in extracting relevant descriptive information from any supervised learning model.

1.1.2.2   Prediction

As the name implies, prediction tasks are concerned with predicting future or unobserved outcomes. For example, we may be interested in predicting the sale price for a new home given a set of relevant features. This could, in theory, be a useful starting point in setting the listing price for a home, or trying to help infer whether or not a particular house is under- or over-valued. Great care must be taken in such problems, however, as the outcome variable (the sale price of homes, in this case) can be complex in nature and the available data may not be enough to adequately capture sudden changes in the distribution of future response values; a bit more on this in Section 1.4.7.

It should be stressed that prediction and description often go hand in hand. Description helps provide transparency in how a model's predictions are generated. Transparency helps reveal potential issues and biases and therefore can help increase trust or distrust in a model's predictions. Would you feel comfortable putting a model into production if you did not have some understanding as to how different subsets of features contribute to the model's predictions?

Single decision trees, while often great descriptors, seldom make for good predictors, at least when compared to more contemporary techniques. Nonetheless, as we'll see in Part I of this book, single decision trees are sometimes the right tool for the job, but it just so happens that more accurate decision trees tend to be harder to interpret. This is especially true for the decision trees discussed in Chapter 4, which are flexible enough to achieve good performance, but often pay a price in interpretability.

1.1.2.3 Classification vs. regression

Supervised learning tasks generally fall into one of two categories: *classification* or *regression*. Regression is used in a very general sense here, and often refers to any supervised learning task with an ordered outcome. Examples of ordered outcomes might be sale price or wine quality on a scale of 0–10 (essentially, an ordered category).

In classification, the response is categorical and the objective is to "classify" new observations into one of J possible categories. In the mushroom classification example (Section 1.4.4), for instance, the goal is to classify new mushrooms as either edible or poisonous on the basis of simple observational attributes about each (like the color and odor of each mushroom). In this example, $J = 2$ (edible or poisonous) and the task is one of *binary classification*. When $J > 2$, the task is referred to as *multiclass classification*.

1.1.2.4 Discrimination vs. prediction

Pure classification is almost never the goal, as we are usually not interested in directly classifying observations into one of J categories. Instead, interest often lies in estimating the conditional probability of class membership. That is to say, it is often far more informative to estimate $\{\Pr(Y = j|\boldsymbol{x}_0)\}_{j=1}^{J}$ as opposed to predicting the class membership of some observation $\boldsymbol{x}_0$. Even when the term "classification" is used, the underlying goal is usually that of estimating class membership probabilities conditional on the feature values[d].

Frank Harrell, a prominent biostatistician, couldn't have said it better:

It is important to distinguish prediction and classification. In many decision making contexts, classification represents a premature decision, because classification combines prediction and decision making and usurps the decision maker in specifying costs of wrong decisions. The classification rule must be reformulated if costs/utilities or sampling criteria change. Predictions are separate from decisions and can be used by any decision maker.

Classification is best used with non-stochastic/deterministic outcomes that occur frequently, and not when two individuals with identical inputs can easily have different outcomes. For the latter, modeling tendencies (i.e., probabilities) is key.

---

[d]The term "classification" is actually abused quite often in practice (and in this book), as it is often used in situations where the true goal is class probability estimation.

Classification should be used when outcomes are distinct and predictors are strong enough to provide, for all subjects, a probability near 1.0 for one of the outcomes.

Frank Harrell

https://www.fharrell.com/post/classification/

A related issue, known as *class imbalance* is discussed in Section 7.9.4.

1.1.2.5   The bias-variance tradeoff

The terms *overfitting* and *underfitting* are used throughout this book (more so the former), but what do they mean? Overfitting occurs when your model is too complex, and has gone past any signal in the data and is starting to fit the noise. Underfitting, on the other hand, refers to when a model is too simple and does not adequately capture any of the signal in the data. In both cases, the model will not generalize well to new data.

A model that is overfitting the learning sample often exhibits lower bias but has higher variance when compared to a model that is underfitting, which often exhibits higher bias but lower variance. This tradeoff is more specifically referred to as the bias-variance tradeoff. Excellent discussions of this topic can be found in Matloff [2017, Sec. 1.11] and Hastie et al. [2009, Sec. 7.2–7.3]; the latter provides more of a theoretical view.

For the additive error model (1.1) with constant variance σ^2, Hastie et al. [2009] show that the mean square prediction error for an arbitrary observation x_0 can be decomposed into

$$\mathrm{E}\left[\left(Y - \hat{f}(x_0)\right)^2 | x = x_0\right] = \sigma^2 + \mathrm{Bias}^2\left[\hat{f}(x_0)\right] + \mathrm{V}\left[\hat{f}(x_0)\right],$$

where σ^2 represents irreducible error that we cannot do anything about, regardless of how well we estimate $f(x_0)$. Normally, increasing the complexity of $\hat{f}$ will cause the squared bias (middle term) to decrease and the variance (last term) to increase, and vice versa.

For illustration, consider the data displayed in Figure 1.1, which consists of a random sample of $N = 100$ observations from a simple quadratic model with additive Gaussian noise:

$$Y = 1 + 0.5X^2 + \epsilon,$$

where $X \sim \mathcal{U}(0, 1)$, and $\epsilon \sim \mathcal{N}(0, \sigma = 0.1)$. Figure 1.1 also shows the fitted mean response from three linear models: a simple linear regression model (left), a polynomial modelpolynomial regression of degree two (middle), and a polynomial modelpolynomial regression of degree 20. Clearly, the simple linear and 20-th degree polynomial modelspolynomial regression are underfitting and overfitting, respectively, while the quadratic model (the correct fit) provides the best tradeoff.

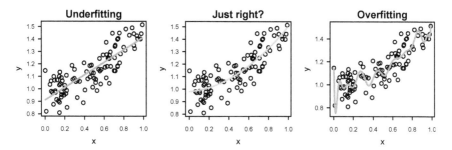

FIGURE 1.1: Fitted mean response from three linear models applied to the quadratic example. Left: a simple linear model (i.e., degree one polynomial). Middle: a quadratic model (i.e., the correct model). Right: a 20-th degree polynomial model.

The same tradeoff applies equally to classification as well. In probabilistic classification, for example, the MSE between the true and predicted class probabilities can also be decomposed into parts due to irreducible error, squared bias, and variance; see Manning et al. [2008, pp. 308–314] for details.

To visually illustrate the bias-variance tradeoff for classification, Figure 1.2 shows $N = 500$ observations generated from the simple "twonorm" benchmark problem; see `?mlbench::mlbench.twonorm` in R for details and references. Here, the classes correspond to two bivariate normal distributions with mean vectors $\boldsymbol{\mu}_1 = \left(\sqrt{2}, \sqrt{2}\right)^\top$ (yellow points) and $\boldsymbol{\mu}_2 = -\boldsymbol{\mu}_1$ (blue points) and unit covariance matrix (i.e., $\boldsymbol{\Sigma} = \boldsymbol{I}_2$). Figure 1.2 also shows two different *decision boundaries*[e]. The dashed line corresponds to the optimal or *Bayes decision boundary* (i.e., the best we can do in this problem), which in this case is linear. The second decision boundary (solid line) corresponds to a simple *k*-*nearest neighbor* (*k*-NN) classifier with $k = 1$[f]. Clearly, the 1-NN model is overfitting and produces an overly complex decision boundary (e.g., the three little islands) compared to the optimal linear boundary.

---

[e]In two dimensions, the decision boundary from a classifier is the boundary that seperates the predictor space into disjoint sets, one for each class. They can be useful for understanding and comparing the flexibility and performance of different classifiers.

[f]A 1-NN model classifies a new observation according to the class of its nearest neighbor in the learning sample; in this case, "nearest neighbor" is defined as the closest observation in the training set as measured by the Euclidean distance.

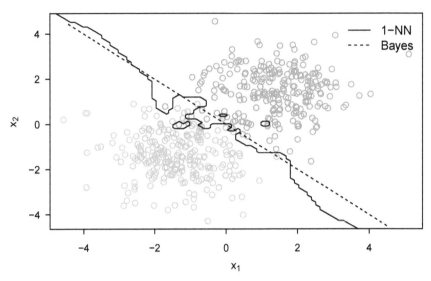

FIGURE 1.2: Decision boundaries for the twonorm benchmark example. In this case, the Bayes decision boundary is linear (dashed line). Also shown is the decision boundary from a 1-NN model (solid line), which is clearly overfitting; notice the three little islands.

1.1.3 Unsupervised learning

In *unsupervised learning*, there is no working response y; hence, the learning sample comprises just features $\{x_i\}_{i=1}^N$. The general goal is to determine whether not the data form any "interesting groups;" for example, whether or not the data form natural *clusters*. A useful application is in detecting *outliers* or *anomalies*. An example of using decision trees for the purpose of anomaly detection is given in Section 7.8.5. Another example, related to the identification of potentially mislabeled response values, is given in Section 7.6.1.

1.2 Why trees?

There are a number of great modeling tools available to data scientists. But what makes a modeling tool good? Is being able to achieve competitive accuracy all that matters? Of course not. According to the late Leo Breiman, a good modeling tool, at a minimum, should:

- be applicable to both classification (binary and multiclass) and regression;
- be competitive in terms of accuracy;

- be capable of handling large data sets;

- handle missing values effectively.

Additionally, Leo also believed that good modeling tools should be able to:

- tell you which predictors are important;

- tell you how the predictors functionally relate to the response;

- tell you if and how the predictors interact;

- tell you how the data cluster;

- tell you if there are any novel cases or outliers.

As we'll see throughout this book, tree-based methods naturally check a lot of these boxes. While individual trees are particularly capable of handling a wide range of problems, their main disadvantage is that they don't often perform as well compared to more complex models, like NNs or SVMs; however, in Part II of this book, I'll discuss powerful ways to combine decision trees into *ensembles* that are quite competitive with current state-of-the-art algorithms, while still adhering to several of the features of a good modeling tool outlined above.

Table 1.1 compares several characteristics of different statistical learning algorithms; this is a modified recreation of Table 10.1 from Hastie et al. [2009, p. 351]. MARS, which stands for *multivariate adaptive regression splines* [Friedman, 1991], can be thought of as an extension of *linear models* that automatically handles variable selection, nonlinear relationships, and interactions between predictors[g]; MARSis discussed in the online supplementary material on the book website. Note how tree-based methods tend to have a lot of desirable properties. The idea of tree-based ensembles is hopefully to retain many of these useful properties while improving predictive performance.

Table 1.1 is not in full agreement with Table 10.1 from Hastie et al. [2009, p. 351], on which it is based. For example, individual trees receive a   for predictive performance in Table 1.1, indicating that they can be fairly accurate in certain problems, especially modern decision tree algorithms, like GUIDE (Chapter 4).

Trees also make great *off-the-shelf* modeling tools. The term off-the-shelf, at least as it's used here, implies a procedure that can be usefully applied to a wide range of data sets without requiring much in the way of pre-processing

---

[g]The terms "MARS" is actually trademarked and licensed to Salford Systems (which was acquired by Minitab in 2017), hence, open source implementations often go by different names. For example, a popular open source implementation of MARSis the fantastic **earth** package [Milborrow, 2021a] in R.

TABLE 1.1: Comparison of several characteristics between different statistical learning algorithms. This is a modified recreation of Table 10.1 from Hastie et al. [2009, p. 351]; here, TBEs stands for "tree-based ensembles", ✖ means "bad", ✔ means "good", and ◉ means "so-so".

Characteristic	NNs	SVMs	MARS	Trees	TBEs
Can naturally handle data of mixed type	✖	✖	✔	✔	✔
Can naturally handle missing values	✖	✖	✔	✔	✔
Robust to outliers in the predictors	✖	✖	✖	✔	✔
Insensitive to monotone transformations of the predictors	✖	✖	✖	✔	✔
Scales well with large N	✖	✖	✔	✔	✔
Ability to deal with irrelevant features	✖	✖	✔	✔	✔
Ability to extract linear combinations of the predictors	✔	✔	✖	✖	✖
Descriptive power (interpretability)	✖	✖	◉	◉	◉
Predictive power	✔	✔	◉	◉	✔

or careful tuning. You can generally just hit "Run" and get something useful. THAT IS NOT TO SAY THAT YOU SHOULD NOT PUT TIME AND EFFORT INTO CLEANING UP YOUR DATA AND CAREFULLY TUNING THESE MODELS. Rather, trees can work seamlessly in rather messy data situations (e.g., outliers, missing values, skewness, mixed data types, etc.) without requiring the level of pre-processing necessary for other algorithms to "just work" (e.g., neural networks). For example, even if I'm not using a decision tree for the final model, I will often use it as a first pass as it can give me a quick and dirty picture of the data, and any serious issues (which can easily be missed in the exploratory phase) will often be highlighted (e.g., extreme *target leakage* or accidentally leaving in an ID column). In other words, trees make great exploratory tools, especially when dealing with potentially messy data.

1.2.1   A brief history of decision trees

Decision trees have a long and rich history in both statistics and computer science, and have been around for many decades. However, decision trees arguably got their true start in the social sciences. Motivated by the need for finding interaction effects in complex survey data, Morgan and Sonquist [1963] developed and published the first decision tree algorithm for regression called *automatic interaction and detection* (AID). Starting at the root node, AID recursively partitions the data into two homogeneous subgroups, called *child nodes*, by maximizing the between-node sum of squares, similar to the process described in Section 2.3. AID continues successively bisecting each resulting child node until the reduction in the within-node sum of squares is less than some prespecified threshold.

Messenger and Mandell [1972] extended AID to classification in their *theta automatic interaction detection* (THAID) algorithm. The *theta criterion* used in THAID to choose splits maximizes the sum of the number of observations in each modal category.

The *chi-square automatic interaction detection* (CHAID) algorithm, introduced in Kass [1980], improved upon AID by countering some of its initial criticisms; CHAID was original developed for classification and later extended to also handle regression problems. Similar to the decision tree algorithms discussed in Chapters 3–4, CHAID employs statistical tests and stopping rules to select the splitting variables and split points. In particular, CHAID relies on chi-squared tests, which require discretizing ordered variables into bins. Compared to AID and THAID, CHAID was unique in that it allowed *multiway splits* (which typically require larger sample sizes, otherwise the child nodes can become too small rather quickly) and included a separate category for missing values.

Despite the novelty of AID, THAID, and CHAID, it wasn't until Breiman et al. [1984] introduced the more general *classification and regression tree* (CART) algorithm[h], that tree-based algorithms started to catch on in the statistical community. CART-like decision trees are the topic of Chapter 2. A similar tree-based algorithm, called C4.5 [Quinlan, 1993], which evolved into the current C5.0 algorithm[i], has become very similar to CART in many regards; hence, I focus on CART in this book and discuss the details of C4.5/C5.0 in the online supplementary material.

CART helped generate renewed interest in partitioning methods, and we've seen that evolution unfold over the last several decades. While the history is rife with advancements, the first part of this book will focus on three of the most important tree-based algorithms:

Chapter 2: Classification and regression trees (CART).

Chapter 3: Conditional inference trees (CTree).

---

[h]Like MARS, the term "CART" is also trademarked and licensed to Salford Systems, which is now part of Minitab.

[i]C50 is now open source and available in the R package **C50** [Kuhn and Quinlan, 2021].

Chapter 4: Generalized, unbiased, interaction detection, and estimation (GUIDE).

GUIDE itself has a rather long and interesting history, and evolved out of several earlier well-known tree algorithms—like QUEST [Loh and Shih, 1997] and CRUISE [Kim and Loh, 2001][j]. If you're interested in a more thorough overview of the history of tree-based algorithms, I highly encourage you to read Loh [2014].

1.2.2   The anatomy of a simple decision tree

In this section, I'll look at the basic parts of a typical decision tree (perhaps tree topology would've been a cooler section header).

A typical (binary) decision tree is displayed in Figure 1.3. The tree is made up of *nodes* and *branches*; the path between two consecutive branches is called an *edge*. The nodes are the points at which a branch occurs. Here, we have three *internal nodes*, labeled $\mathcal{I}_1$, $\mathcal{I}_2$, and $\mathcal{I}_3$, and five terminal (or leaf) nodes, labeled $\mathcal{L}_1$, $\mathcal{L}_2$, $\mathcal{L}_3$, $\mathcal{L}_4$, and $\mathcal{L}_5$. The tree is binary because it only uses two-way splits; that is, at each node, a split results in only two branches, labeled "Yes" and "No"[k]. The path taken at each internal node depends on whether or not the corresponding split condition is satisfied. For example, an observation with $x_1 = 0.33$ and $x_5 = 1.19$ would find itself in terminal node $\mathcal{L}_2$, regardless of the values of the other predictors.

The split conditions (or just splits) for an ordered predictor x have the form $x < c$ vs. $x \geq c$, where c is in the domain of x (typically the midpoint between two consecutive x values in the learning sample); note that the same type of splits are used for ordered factors since we just need to preserve the natural ordering of the categories (e.g., $x <$ medium vs. $x \geq$ medium)[l]. Splits on (unordered) categorical variables have the form $x \in S_1$ vs. $x \in S_2$, where $S = S_1 \cup S_2$ is the full set of unique categories in x.

Each tree begins with a *root node* containing the entire learning sample. Starting with the root node, the training data are split into two non-overlapping groups, one going left, and the other right, depending on whether or not the first split condition is satisfied by each observation. The process is repeated on each subgroup (or child node) until each observation reaches a terminal node.

---

[j]For those interested, QUEST stands for *quick, unbiased, and efficient statistical tree* and CRUISE stands for *classification rule with unbiased interaction selection and estimation*.

[k]Some decision tree algorithms allow multiway (i.e., > 2) splits, but none are really discussed in this book.

[l]For splits on ordered variables, it generally doesn't matter whether the left branches in a tree correspond to $x < c$ or $x \geq c$, as long as you're consistent.

At each node in a tree, we can compute several quantities that may be of interest, for example:

- the number (or fraction) of observations from the learning sample in that node;

- for classification models, the number of observations classified correctly (or incorrectly) in that node (as determined by the majority class in that node).

- for probabilistic models, the proportion of observations in that node belonging to each class;

- the current fitted or predicted value if this were a leaf or terminal node (e.g., the mean response or class proportions).

The terminal nodes themselves are usually characterized by a statistical summary of the training response values in each, like the sample mean for regression, a frequency table for classification, or the Kaplan-Meier estimator/curve for censored outcomes (see the example in Section 1.4.9). These summaries can be used to produce fitted values and predictions for new observations[m]. For example, if the response is continuous (i.e., a regression tree), then new observations that occupy a terminal node might be assigned a predicted value equal to the mean response of the training observations defining that node.

1.2.2.1 Example: survival on the Titanic

To further illustrate, let's look at the tree diagram in Figure 1.4. This CART-like decision tree was constructed using the well-known Titanic data set and is trying to separate passengers who survived from those who didn't using readily available information about each. The data, which I'll revisit in Section 7.9.3, contain $N = 1,309$ observations (i.e., passengers) on the following six variables:

- `survived`: binary indicator of passenger survival (the response);

- `pclass`: integer specifying passenger class (i.e., 1–3);

- `age`: passenger age in years;

- `sex`: factor giving the sex of each passenger (i.e., male/female);

- `sibsp`: integer specifying the number of siblings/spouses aboard;

- `parch`: integer specifying the number of parents/children aboard.

The variable `pclass` is commonly treated as nominal categorical, but here the natural ordering has been taken into account. The tree split the passengers into six relatively homogeneous groups (terminal nodes) based on four of the above

---

[m]Fitted values are just the predicted values for each observation in the learning sample.

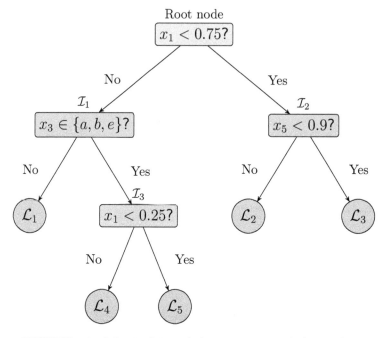

FIGURE 1.3: A basic (binary) decision tree with four splits.

five available features; `parch` is the only variable not selected to partition the data. The terminal nodes (nodes 6–11) each contain a node summary giving the proportion of surviving passengers in that node. As we'll see in later chapters, these proportions can be used as class probability estimates.

For example, the tree diagram estimates that first and second class female passengers had a 93% chance of survival; the percentage displayed in the bottom of each node corresponds to the fraction of training observations used to define that node. Given what you know about the ill-fated Titanic, does the tree diagram make sense to you? Does it appear that women and children were given priority and had a higher chance of survival (i.e., "women and children first")? Perhaps, unless you were a third-class passenger.

In Part I of this book, I'll look at how several popular decision tree algorithms choose which variables to split on (splitters) and how each split condition is determined (e.g., `age < 9.5`). Part II of this book will then look at how to improve the accuracy and generalization performance of a single tree by combining several hundred or thousand individual trees together.

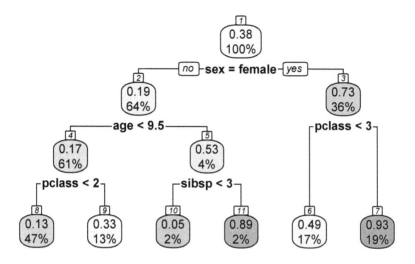

FIGURE 1.4: Decision tree diagram summarizing survival probability aboard the ill-fated Titanic.

1.3 Why R?

Why not?

1.3.1 No really, why R?

I grew up on R (and SAS), but I've chosen it for this book primarily for one reason: it currently provides the best support and access to a wide range of tree-based methods (both classic and modern-day)[n]. For example, I'm not aware of any non-R open source implementation of tree-based methods that provides full support for nominal categorical variables without requiring them to be encoded numerically[o]. Another good example is *conditional inference trees* [Hothorn et al., 2006c], the topic of Chapter 3, which are only implemented in R (as far as I'm aware). Nonetheless, there is at least one Python example included in this book!

---

[n]This book was also written in R (and LaTeX) using the wonderful **knitr** package Xie [2021].

[o]There's been some progress in areas of scikit-learn and other open source software that will be discussed in Chapter 8.

While I have a strong appreciation for the power of base R (i.e., the core language), I'm extremely appreciative of, and often rely on, the amazing ecosystem of contributed packages. Nonetheless, I've decided to keep the use of external packages to a minimum (aside from those packages related to the core tree-based methods discussed in this book), and instead rely on vanilla R programming as much as possible. This choice was made for several (highly opinionated) reasons:

- it will make the book easier to maintain going forward, as the code examples will (hopefully) continue to work for many years to come without much modification;

- using standard R programming constructs (e.g., `for` loops and their apply-style functional replacements, like `lapply()`) will make the material easier to comprehend for non-R programmers, and easier to translate to other open source languages, like Julia and Python;

- it emphasizes the basic concepts of the methods being introduced, rather than focusing on current best coding practices and cool packages, of which there are many. (Please don't send me hate mail for using `sapply()` instead of `vapply()` or the family of map functions available in **purrr** [Henry and Wickham, 2020].)

Note that I've tried to be as aggressive as possible in terms of commenting upon the various code snippets scattered throughout this book and the online supplementary material. You should pay careful attention to these comments as they often link a particular line or section of code to a specific step in an algorithm, try to explain a hacky approach I'm using, and so on.

Each chapter includes one or more software sections, which highlight both R and non-R implementations of the relevant algorithms under discussion. Additionally, each chapter also contains R-specific software example sections (usually at the end of each chapter), which demonstrate use of relevant tree-specific software on actual data (either simulated or real).

There are many great resources for learning R, but I would argue that the online manual, "An Introduction to R", which can be found at

`https://cran.r-project.org/doc/manuals/r-release/R-intro.html`,

is a great place to start. Book-wise, I think that Matloff [2011] and Wickham [2019] are some of the best resources for learning R (note that the latter is freely available to read online). If you're interested in a more hands-on introduction to statistical and machine learning with R, I'll happily self-promote Boehmke and Greenwell [2020].

1.3.2   Software information and conventions

Each chapter contains at least one software section, which points to relevant implementations of the ideas discussed in the corresponding chapter. And while this book focuses on R, these sections will also allude to additional implementations in other open source programming languages, like Python and Julia. Furthermore, several code snippets are contained throughout the book to help solidify certain concepts (mostly in R).

Be warned, I `occasionally.use.dots` in variable and function names—old R programming habits die hard. Package names are in bold text (e.g., **rpart**), inline code and function names are in typewriter font (e.g., `sapply()`), and file names are in sans serif font (e.g., path/to/filename.txt). In situations where it may not be obvious which package a function belongs to, I'll use the notation `foo::bar()`, where `bar()` is the name of a function in package **foo**.

I often allude to the documentation and help pages for specific R functions. For example, you can view the documentation for function `foo()` in package **bar** by typing `?foo::bar` or `help("foo", package = "bar")` at the R console. It's a good idea to read these help pages as they will often provide more useful details, further references, and example usage. For base R functions— that is, functions available in R's **base** package—I omit the package name (e.g., `?kronecker`). I also make heavy use of R's `apply()`-family of functions throughout the book, often for brevity and to avoid longer code snippets based on `for` loops. If you're unfamiliar with these, I encourage you to start with the help pages for both `apply()` and `lapply()`.

R package vignettes (when available) often provide more in-depth details on specific functionality available in a particular package. You can browse any available vignettes for a CRAN package, say **foo**, by visiting the package's homepage on CRAN at

$$\texttt{https://cran.r-project.org/package=foo.}$$

You can also use the **utils** package to view package vignettes during an active R session. For example, the vignettes accompanying the R package **rpart** [Therneau and Atkinson, 2019], which is heavily used in Chapter 2, can be found at `https://CRAN.R-project.org/package=rpart` or by typing `utils::vignette("bar", package = "foo")` at the R console.

There's a myriad of R packages available for fitting tree-based models, and this book only covers a handful. If you're not familiar with CRAN's task views, you should be. They provide useful guidance on which packages on CRAN are relevant to a certain topic (e.g., machine learning). The task view on statistical and machine learning, for example, which can be found at

$$\texttt{https://cran.r-project.org/web/views/MachineLearning.html,}$$

lists several R packages useful for fitting tree-based models across a wide variety of situations. For instance, it lists **RWeka** [Hornik, 2021] as providing an open source interface to the J4.8-variant of C4.5 and M5 (see the online supplementary material on the book website). A brief description of all available task views can be found at `https://cran.r-project.org/web/views/`.

Keep in mind that the focus of this book is to help you build a deeper understanding of tree-based methods, it is not a programming book. Nonetheless, writing, running, and experimenting with code is one of the best ways to learn this subject, in my opinion.

This book uses a couple of graphical parameters and themes for plotting that are set behind the scene. So don't fret if your plots don't look exactly the same when running the code. This book uses a mix of base R and **ggplot2** [Wickham et al., 2021a] graphics, though, I think there's a **lattice** [Sarkar, 2021] graphic or two floating around somewhere. For **ggplot2**-based graphics, I use the `theme_bw()` theme, which can be set at the top level (i.e., for all plots) using `theme_set(theme_bw())`. Most of the base R graphics in this book use the following `par()` settings (see `?graphics::par` for details on each argument):

```
par(
  mar = c(4, 4, 0.1, 0.1),   # may be different for a handful of figures
  cex.lab = 0.95,
  cex.axis = 0.8,
  mgp = c(2, 0.7, 0),
  tcl = -0.3,
  las = 1
)
```

Some of the base R graphics in this book use a slightly different setting for the `mar` argument (e.g., to make room for plots that also have a top axis, like Figure 8.12 on page 349).

1.4   Some example data sets

The examples in this book make use of several data sets, both real and simulated, and both small and large. Many of the data sets are available in the **treemisc** package that accompanies this book (or another R package), but many are also available for download from the book's website:

`https://bgreenwell.github.io/treebook/datasets.html`.

In this section, I'll introduce a handful of the data sets used in the examples throughout this book. Some of these data sets are pretty common, and are often used in other texts or articles to illustrate concepts or compare and benchmark performance.

1.4.1 Swiss banknotes

The Swiss banknote data [Flury and Riedwyl, 1988] contain measurements from 200 Swiss 1000-franc banknotes: 100 genuine and 100 counterfeit. There are six available predictors, each giving the length (in mm) of a different dimension for each bill (e.g., the length of the diagonal). The response variable is a 0/1 indicator for whether or not the bill was genuine/counterfeit. This is a small data set that will be useful when exploring how some classification trees are constructed. The code snippet below generates a simple scatterplot matrix of the data, which is displayed in Figure 1.5:

```
bn <- treemisc::banknote
cols <- palette.colors(3, palette = "Okabe-Ito")
pairs(bn[, 1L:6L], col = adjustcolor(cols[bn$y + 2], alpha.f = 0.5),
      pch = c(1, 2)[bn$y + 1], cex = 0.7)
```

Note how good some of the features are at discriminating between the two classes (e.g., `top` and `diagonal`). This is a small data set that will be used to illustrate fundamental concepts in decision tree building in Chapters 2–3.

1.4.2 New York air quality measurements

The New York air quality data contain daily air quality measurements in New York from May through September of 1973 (153 days). The data are conveniently available in R's built-in **datasets** package; see `?datasets::airquality` for details and the original source. The main variables include:

- `Ozone`: the mean ozone (in parts per billion) from 1300 to 1500 hours at Roosevelt Island;

- `Solar.R`: the solar radiation (in Langleys) in the frequency band 4000–7700 Angstroms from 0800 to 1200 hours at Central Park;

- `Wind`: the average wind speed (in miles per hour) at 0700 and 1000 hours at LaGuardia Airport;

- `Temp`: the maximum daily temperature (in degrees Fahrenheit) at La Guardia Airport.

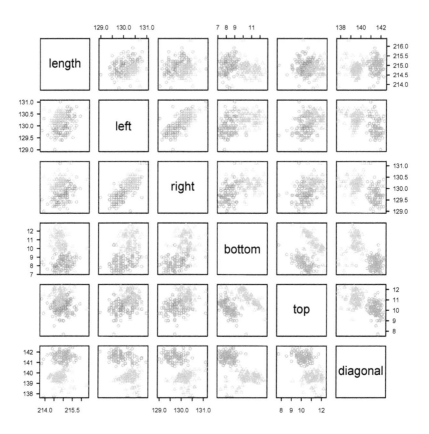

FIGURE 1.5: Scatterplot matrix of the Swiss banknote data. The black circles and orange triangles correspond to genuine and counterfeit banknotes, respectively.

The month (1–12) and day of the month (1–31) are also available in the columns Month and Day, respectively. In these data, Ozone is treated as a response variable.

This is another small data set that will be useful when exploring how some regression trees are constructed. A simple scatterplot matrix of the data is constructed below; see Figure 1.6. The upper diagonal scatterplots each contain a *LOWESS smooth*[p] of the data (red curve). Note that there's a relatively strong nonlinear relationship between Ozone and both Temp and Wind, compared to the others.

---

[p]A LOWESS smoother is a nonparametric smooth based on locally-weighted polynomial regression; see [Cleveland, 1979] for details.

```
aq <- datasets::airquality
color <- adjustcolor("forestgreen", alpha.f = 0.5)
ps <- function(x, y, ...) {  # custom panel function
  panel.smooth(x, y, col = color, col.smooth = "black",
               cex = 0.7, lwd = 2)
}
pairs(aq, cex = 0.7, upper.panel = ps, col = color)
```

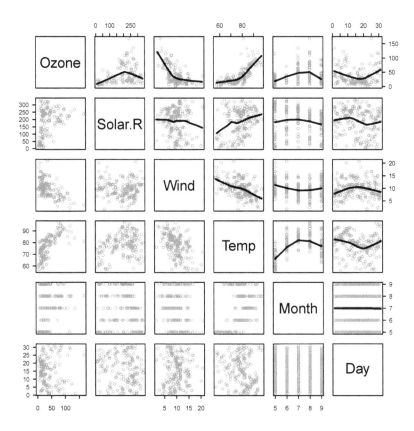

FIGURE 1.6: Scatterplot matrix of the New York air quality data. Each black curve in the upper panel represents a LOWESS smoother.

1.4.3 The Friedman 1 benchmark problem

The Friedman 1 benchmark problem [Breiman, 1996a, Friedman, 1991] uses simulated regression data with 10 input features according to:

$$Y = 10\sin\left(\pi X_1 X_2\right) + 20\left(X_3 - 0.5\right)^2 + 10X_4 + 5X_5 + \epsilon, \qquad (1.2)$$

where $\epsilon \sim \mathcal{N}\left(0, \sigma\right)$ and the input features are all independent uniform random variables on the interval $[0, 1]$: $\{X_j\}_{j=1}^{1} 0 \overset{iid}{\sim} \mathcal{U}\left(0, 1\right)$. Notice how X_6–X_{10} are unrelated to the response Y.

These data can be generated in R using the `mlbench.friedman1` function from package **mlbench** [Leisch and Dimitriadou., 2021]. Here, I'll use the `gen_friedman1` function from package **treemisc**, which allows you to generate any number of features $\geq$ 5; similar to the `make_friedman1` function in scikit-learn's **sklearn.datasets** module for Python. See `?treemisc::gen_friedman1` for details. Below, I generate a sample of $N = 5$ observations from (1.2) with only seven features (so it prints nicely):

```
set.seed(943)  # for reproducibility
treemisc::gen_friedman1(5, nx = 7, sigma = 0.1)
```

```
#>       y    x1    x2    x3    x4    x5     x6     x7
#> 1 18.5 0.346 0.853 0.655 0.839 0.293 0.3408 0.0573
#> 2 13.7 0.442 0.691 0.214 0.108 0.543 0.1616 0.5055
#> 3 10.8 0.223 0.789 0.807 0.252 0.257 0.8595 0.7248
#> 4 18.9 0.859 0.520 0.891 0.129 0.936 0.0348 0.7105
#> 5 14.5 0.181 0.590 0.893 0.611 0.415 0.4104 0.2636
```

From (1.2), it should be clear that features X_1–X_5 are the most important! (The others don't influence Y at all.) Also, based on the form of the model, we'd expect X_4 to be the most important feature, probably followed by X_1 and X_2 (both comparably important), with X_5 probably being less important. The influence of X_3 is harder to determine due to its quadratic nature, but it seems likely that this nonlinearity will suppress the variable's influence over its observed range (i.e., $[0, 1]$). Since the true nature of $\mathrm{E}\left(Y|\boldsymbol{x}\right)$ is known, albeit somewhat complex (e.g., nonlinear relationships and an explicit interaction effect), these data are useful in testing out different model interpretability techniques (at least on numeric features), like those discussed in Section 6. Since these data are convenient to generate, I'll use them in a couple of small-scale simulations throughout this book.

1.4.4   Mushroom edibility

The mushroom edibility data is one of my favorite data sets. It contains 8124 mushrooms described in terms of 22 different physical characteristics, like odor and spore print color. The response variable (`Edibility`) is a binary indicator for whether or not each mushroom is `Edible` or `Poisonous`. The data are available from the UCI Machine Learning repository at `https://archive.ics.`

uci.edu/ml/datasets/mushroom, but can also be obtained from **treemisc**; see ?treemisc::mushroom for details and the original source.

What's interesting about these data (at least to me) is that every single variable, both predictor and response, is categorical. These data will be helpful in illustrating how certain decision tree algorithms deal with categorical predictors when choosing splits. A *mosaic plot* showing the relationship between mushroom edibility and odor (one of the better discriminators between edible and poisonous mushrooms in this sample) is constructed below; see Figure 1.7.

The area of each tile is proportional to the number of observations in the particular category. The mosaic plot indicates that the poisonous group is dominated by mushrooms with a strong or unpleasant odor. Hence, we might surmise that poisonous mushrooms tend to be associated with strong or unpleasant odors.

```
mushroom <- treemisc::mushroom
mosaicplot(~ Edibility + odor, data = mushroom, color = TRUE,
           las = 1, main = "", cex.axis = 0.6)
```

1.4.5   Spam or ham?

These data refer to $N = 4,601$ emails classified as either spam (i.e., junk email) of non-spam (i.e. "ham") that were collected at Hewlett-Packard (HP) Labs. In addition to the class label, there are 57 predictors giving the relative frequency of certain words and characters in each email. For example, the column charDollar gives the relative frequency of dollar signs ($) appearing in each email. The data are available from the UCI Machine Learning repository at

https://archive.ics.uci.edu/ml/datasets/spambase.

In R, the data can be loaded from the **kernlab** package [Karatzoglou et al., 2019]; see ?kernlab::spam for further details.

Below, I load the data into R, check the frequency of spam and non-spam emails, then look at the average relative frequency of several different words and characters between each:

```
data(spam, package = "kernlab")

# Distribution of ham and spam
table(spam$type)

#>
#> nonspam    spam
#>    2788    1813
```

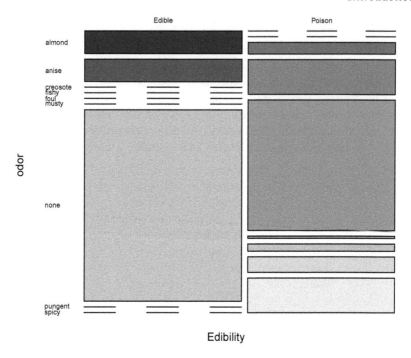

FIGURE 1.7: Mosaic plot visualizing the relationship between mushroom edibility and odor. The area of each tile is proportional to the number of observations in the particular category.

```
# Compute average relative frequency of different words and characters
aggregate(cbind(remove, charDollar, hp, parts, direct) ~ type,
          data = spam, FUN = mean)

#>      type  remove charDollar      hp   parts direct
#> 1 nonspam 0.00938     0.0116  0.8955 0.01872 0.0831
#> 2    spam 0.27541     0.1745  0.0175 0.00471 0.0367
```

Notice how the first three variables show a much larger difference between spam and non-spam emails; we might expect these to be important predictors (at least compared to the other two) in classifying new HP emails as spam vs. non-spam. For example, given that these emails all came from Hewlett-Packard Labs, the fact that the non-spam emails contain a much higher relative frequency of the word **hp** makes sense (email spam was not as clever back in 1998).

As a preview of what's to come, the code chunk below fits a basic decision tree with three splits (i.e., it asks three yes or no questions) to a 70% random sample of the data. It also takes into account the specified assumption that classifying a non-spam email as spam is five times more costly than classifying a spam email as non-spam. We'll learn all about **rpart** and the steps taken below in Chapter 2.

```r
library(rpart)
library(treemisc)

# Split into train/test sets using a 70/30 split
set.seed(852)  # for reproducibility
id <- sample.int(nrow(spam), size = floor(0.7 * nrow(spam)))
spam.trn <- spam[id, ]   # training data
spam.tst <- spam[-id, ]   # test data

# Fit a simple classification tree
loss <- matrix(c(0, 1, 5, 0), nrow = 2)  # misclassification costs
spam.cart <- rpart(type ~ ., data = spam.trn, cp = 0,
                   parms = list("loss" = loss))
cp <- spam.cart$cptable
cp <- cp[cp[, "nsplit"] == 3, "CP"]  # CP associated with 3 splits
spam.cart.pruned <- prune(spam.cart, cp = cp)  # grab smaller subtree

# Display tree diagram
tree_diagram(spam.cart.pruned)
```

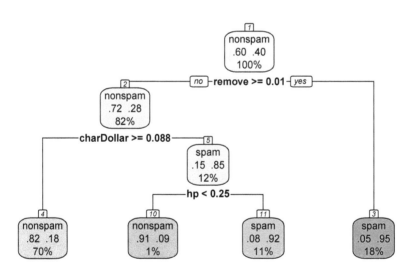

FIGURE 1.8: Decision tree diagram for a simple classification tree applied to the email spam learning sample.

The associated tree diagram is displayed in Figure 1.8. This tree is too simple and underfits the training data (I'll re-analyze these data using an ensemble in Chapter 5). Nonetheless, simple decision trees can often be displayed as a small set of simple *rules*. As a set of mutually exclusive and exhaustive rules, the tree diagram in Figure 1.8 translates to:

```
Rule 1 (path to terminal node 3)
IF   remove >= 0.01
THEN classification = spam (probability = 0.95)


Rule 2 (path to terminal node 11)
IF   remove < 0.01 AND
|    charDollar >= 0.088 AND
|    hp < 0.25
THEN classification = spam (probability = 0.92)


Rule 3 (path to terminal node 10)
IF   remove < 0.01 AND
|    charDollar >= 0.088 AND
|    hp >= 0.25
THEN classification = non-spam (probability = 0.91)


Rule 4 (path to terminal node 4)
IF   remove < 0.01 AND
|    charDollar < 0.088 AND
THEN classification = non-spam (probability = 0.82)
```

The first rule, for instance, states that if the relative frequency of the word "remove" is 0.01 or larger, then we would classify the email as spam with probability 0.95.

1.4.6 Employee attrition

The employee attrition data contain (simulated) human resources analytics data of employees that stay and leave a particular company. The main objective with these data, according to the original source, is to "Uncover the factors that lead to employee attrition..." Such factors include age, job satisfaction, and commute distance. The response variable is `Attrition`, which is a binary indicator for whether or not the employee left (`Attrition = Yes`) or stayed (`Attrition = No`). The data are conveniently available via the R package **modeldata** [Kuhn, 2021]; they can also be obtained from the following IBM GitHub repository: `https://github.com/IBM/employee-attrition-aif360`. To load the data in R, use:

```
data(attrition, package = "modeldata")

# Distribution of class outcomes
```

```
table(attrition$Attrition)

#>
#>   No  Yes
#> 1233  237
```

1.4.7 Predicting home prices in Ames, Iowa

The Ames housing data [De Cock, 2011], which are available in the R package **AmesHousing** [Kuhn, 2020], contain information from the Ames As-
sessor's Office used in computing assessed values for individual residential
properties sold in Ames, Iowa from 2006–2010; online documentation describing the data set can be found at `http://jse.amstat.org/v19n3/decock/`
`DataDocumentation.txt`. These data are often used as a more contemporary
replacement to the often cited—and ethically challenged [Carlisle, 2019]—
Boston housing data [Harrison and Rubinfeld, 1978].

The data set contains $N = 2,930$ observations on 81 variables. The response
variable here is the final sale price of the home (`Sale_Price`). The remaining
80 variables, which I'll treat as predictors, are a mix of both ordered and
categorical features.

In the code chunk below, I'll load the data into R and split it into train/test
sets using a 70/30 split, which I'll use in several examples throughout this
book (note that for plotting purposes, mostly to avoid large numbers on the
y-axis, I'll rescale the response by dividing by 1,000):

```
ames <- as.data.frame(AmesHousing::make_ames())
ames$Sale_Price <- ames$Sale_Price / 1000  # rescale response
set.seed(2101)  # for reproducibility
trn.id <- sample.int(nrow(ames), size = floor(0.7 * nrow(ames)))
ames.trn <- ames[trn.id, ]  # training data/learning sample
ames.tst <- ames[-trn.id, ]  # test data
```

Figure 1.9 shows a scatterplot of sale price vs. above grade (ground) living
area in square feet (`Gr_Liv_Area`) from the 70% learning sample. Above grade
living area, as we'll see in later chapters, is arguably one of the more important predictors in this data set (as you might expect). It is evident from this
plot that *heteroscedasticity* is present, with variation in sale price increasing
with home size. Linear models assume constant variance whenever relying on
the usual normal theory standard errors and confidence intervals for interpretation. Outliers are another potential problem.

```
plot(Sale_Price ~ Gr_Liv_Area, data = ames.trn,
     col = adjustcolor(1, alpha.f = 0.5),
     xlab = "Above grade square footage",
     ylab = "Sale price / 1000")
```

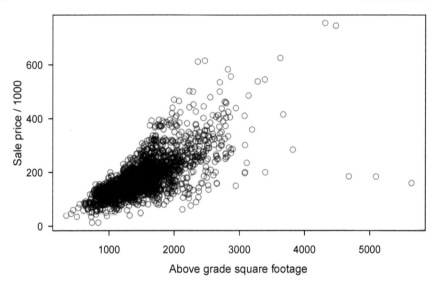

FIGURE 1.9: Scatterplot of sale price vs. above grade (ground) living area in square feet from the Ames housing training sample; here you can see five or so potential outliers.

Note that predictions based solely on these data should not be used alone in setting the sale price of a home. I mean, they could, but they would likely not perform well over time. There are many complexities involved in valuing a home, and housing markets change over time. With the data at hand, it can be hard to predict such changes, especially during the initial Covid-19 outbreak during which the majority of this book was written (many things became rather hard to predict and forecast). However, such a model could be a useful place to start, especially for descriptive purposes.

1.4.8   Wine quality ratings

These data are related to red and white variants of the Portuguese "Vinho Verde" wine; for details, see Cortez et al. [2009]. Due to privacy and logistic issues, only physicochemical and sensory variables are available (e.g., there is no data about grape types, wine brand, wine selling price, etc.). The response variable here is the wine quality score (`quality`), which is an ordered integer in the range 0–10.

The data are available in the R package **treemisc** and can be used for classification or regression, but given the ordinal nature of the response, the latter is more appropriate; see `?treemisc::wine`. The data can also be downloaded from the UCI Machine Learning repository at `https://archive.ics.uci.`

`edu/ml/datasets/wine+quality`. Outlier detection algorithms could be used to detect the few excellent or poor wines. Also, it is not known if all the available predictors are relevant.

Below, I load the data into R and look at the distribution of quality scores by wine type (e.g., red or white):

```
wine <- treemisc::wine
xtabs(~ type + quality, data = wine)

#>        quality
#> type      3    4    5    6    7    8    9
#>    red    10   53  681  638  199   18    0
#>    white  20  163 1457 2198  880  175    5
```

Note that most wines (red or white) are mediocre and relatively few have very high or low scores. The response here, while truly an integer in the range 0–10, is often treated as binary by arbitrarily discretizing the ordered response into "low quality" and "high quality" wines. A more appropriate analysis, which utilizes the fact that the response is ordered, is given in Section 3.5.2.

1.4.9   Mayo Clinic primary biliary cholangitis study

This example concerns data from a study by the Mayo Clinic on *primary biliary cholangitis* (PBC) of the liver conducted between January 1974 and May 1984; follow-up continued through July 1986. PBC is an autoimmune disease leading to destruction of the small bile ducts in the liver. There were a total of $N = 418$ patients whose *survival time* and *censoring indicator* were known (I'll discuss what these mean briefly). The goal was to compare the drug *D-penicillamine* with a placebo in terms of survival probability. The drug was ultimately found to be ineffective; see, for example, Fleming and Harrington [1991, p. 2] and Ahn and Loh [1994] (the latter employs a tree-based analysis). An additional 16 potential covariates are included which I'll investigate further as predictors in Section 3.5.3.

Below, I load the data from the **survival** package [Therneau, 2021] and do some prep work. For starters, I'll only consider the subset of patients who were randomized into the D-penicillamine and placebo groups; see `?survival::pbc` for details. Second, I'll consider the small number of subjects who underwent liver transplant to be censored at the day of transplant[q]:

```
library(survival)

pbc2 <- pbc[!is.na(pbc$trt), ]   # omit non-randomized subjects
pbc2$id <- NULL   # remove ID column
```

---

[q]As mentioned in Harrell [2015, Sec. 8.9], liver transplantation was rather uncommon at the time the data were collected, so it still constitutes a natural history study for PBC.

```
# Consider transplant patients to be censored at day of transplant
pbc2$status <- ifelse(pbc2$status == 2, 1, 0)

# Look at frequency of death and censored observations
table(pbc2$status)

#>
#>   0   1
#> 187 125
```

In this sample, 125 subjects died (i.e., experienced the event of interest) and the remaining 187 were considered censored (i.e., we only know they did not die before dropping out, receiving a transplant, or reaching the end of the study period).

In survival studies (like this one), the dependent variable of interest is often *time until some event occurs*; in this example, the event of interest is death. However, medical studies cannot go on forever, and sometimes subjects drop out or are otherwise lost to follow-up. In these situations, we may not have observed the event time, but we at least have some partial information. For example, some of the subjects may have survived beyond the study period, or perhaps some dropped out due to other circumstances. Regardless of the specific reason, we at least have some partial information on these subjects, which survival analysis (also referred to as *time-to-event* or *reliability analysis*) takes into account.

The scatterplot in Figure 1.10 shows the survival times for the first ten subjects in the PBC data, with an indicator for whether or not each observation was censored. The first subject, for example, was recorded dead at $t = 400$ days, while subject two was censored at $t = 4,500$ days.

In survival analysis, the response variable typically has the form

$$Y = \min\left(T, C\right),$$

where T is the survival time and C is the *censoring time*. In this book, I'll only consider right censoring (the most common form of censoring), where $T \geq Y$. In this case, all we know is that the true event time is at least as large as the observed time[r]. For example, if we were studying the failure time of some motor in a machine, we might have observed a failure at time $t = 56$ hours, or perhaps the study ended at $t = 100$ hours, so all we know is that the true failure time would have occurred some time after that.

To indicate that a particular observation is censored, we can use a censoring indicator:

---

[r] *Left censoring* and *interval censoring* are other common forms of censoring.

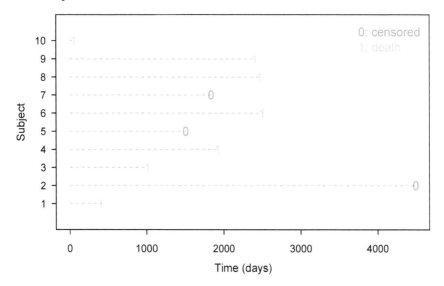

FIGURE 1.10: Survival times for the first ten (randomized) subjects in the Mayo Clinic PBC data.

$$\delta = \begin{cases} 1 & \text{if } T \leq C \\ 0 & \text{if } T > C \text{ (i.e., censored)} \end{cases},$$

where $\delta = 1$ implies that we observed the true survival time and $\delta = 0$ indicates a right censored observation (i.e., we only know the subject survived past time C). A common cause for right censoring in medical studies is that the study ended before the event of interest (e.g., death) occurred or perhaps some of the individuals dropped out or were lost to follow-up; in either case, we only have partial information. As we'll see, several classes of decision tree algorithms can be extended to handle right censored outcomes. Examples are provided in Sections 3.5.3 (single decision tree) and 8.9.1 (ensemble of decision trees).

A common summary of interest in survival studies is the *survival function*:

$$S(t) = \Pr(T > t), \tag{1.3}$$

which describes the probability of surviving longer than time t. The Kaplan-Meier (or product limit) estimator is a nonparametric statistic used for estimating the survival function in the presence of censoring (if there isn't any censoring, then we could just use the ordinary *empirical distribution function*).

The details are beyond the scope of this book, but the `survfit` function from package **survival** can do the heavy lifting for us.

In the code snippet below, I call `survfit` to estimate and plot the survival curves for both the drug and placebo groups; see Figure 1.11. Here, you can see that the estimated survival curves between the treatment and control group are similar, indicating that D-penicillamine is rather ineffective. The *log-rank test* can be used to test for differences between the survival distributions of two groups. Some decision tree algorithms for the analysis of survival data use the log-rank test to help partition the data; see, for example, Segal [1988] and Leblanc and Crowley [1993].

```
palette("Okabe-Ito")
plot(survfit(Surv(time, status) ~ trt, data = pbc2), col = 2:3,
     conf.int = FALSE, las = 1, xlab = "Days until death",
     ylab = "Estimated survival probability")
legend("bottomleft", legend = c("Penicillmain", "Placebo"),
        lty = 1, col = 2:3, text.col = 2:3, inset = 0.01, bty = "n")
palette("default")
```

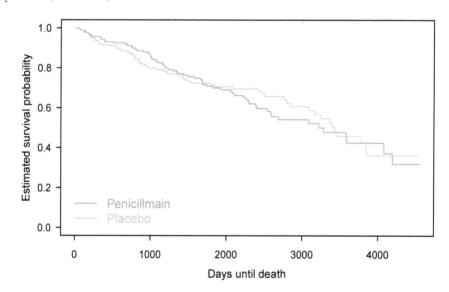

FIGURE 1.11: Kaplan-Meier estimate of the survival function for the randomized subjects in the Mayo Clinic PBC data by treatment group (i.e., drug vs. placebo). The median survival times are 3282 days (drug) and 3428 days (placebo).

In Section 3.5.3, we'll see how a simple tree-based analysis can estimate the survival function conditional on a set of predictors, denoted $\hat{S}(t|\boldsymbol{x})$, by partitioning the learning sample into non-overlapping groups with similar survival rates; here, we'll see further evidence that D-penicillamine was not effective in improving survival. For a thorough overview of survival analysis, my gold standard has always been Klein and Moeschberger [2003].

1.5 There ain't no such thing as a free lunch

Too often, we see papers or hear arguments claiming that some cool new algorithm A is better than some existing algorithms B and C at doing D. This is mostly baloney, as any experienced statistician or modeler would tell you that no one procedure or algorithm is uniformly superior across all situations. That being said, you should not walk away from this book with the impression that tree-based methods are superior to any other algorithm or modeling tool. They are powerful and flexible tools for sure, but that doesn't always mean they're the right tool for the job. Consider them as simply another tool to include in your modeling and analysis toolbox.

1.6 Outline of this book

This book is about decision trees, both individual trees (Part I) and ensembles thereof (Part II). There are a large number of decision tree algorithms in existence, and entire books have even been dedicated to some. Consequently, I had to be quite selective in choosing the topics to present in detail in this book, which has mostly been guided by my experiences with tree-based methods over the years in both academics and industry. As mentioned in Loh [2014], "There are so many recursive partitioning algorithms in the literature that it is nowadays very hard to see the wood for the trees."

I'll discuss some of the major, and most important tree-based algorithms in current use today. However, due to time and page constraints, several important algorithms and extensions didn't make the final cut, and are instead discussed in the (free) online supplementary material that can be found on the book website. These methods include:

- C5.0 [Kuhn and Johnson, 2013, Sec. 14.6], the successor to C4.5 [Quinlan, 1993], which is similar enough to CART that including it in a separate chapter would be largely redundant with Chapter 2;

- MARS, which was briefly mentioned in Section 1.2 (see Table 1.1), is essentially an extension of linear models (and CART) that automatically handles variable selection, nonlinear relationships, and interactions between predictors;

- *rule-based models*, like Certifiable Optimal RulE ListS [Angelino et al., 2018], or CORELS for short, which are very much like decision trees, but with an emphasis on producing a small number of simple rules (i.e., short sequences of yes or no questions).

Decision trees remain one of the most flexible and practical tools in the data science toolbox, whether for description or prediction. While they are most commonly used for prediction problems in an ensemble (see Chapters 5–8), individual decision trees are still one of the most useful off-the-shelf analytic tools available (e.g., they can be used for missing value imputation, description, and variable ranking and selection, to name a few).

The rest of this book is split into two parts:

Part I: Individual decision trees. Common decision tree algorithms, like CART (Chapter 2), CTree (Chapter 3), and GUIDE (Chapter 4), are brought into focus. I'll discuss both the basics and the nitty-gritty details which are often glossed over in other texts, or buried in the literature. These algorithms form the building blocks upon which many current state-of-the-art prediction algorithms are built. Such algorithms are the focus of Part II.

Part II: Decision tree ensembles. While Part I will highlight several useful decision tree algorithms, it will become apparent that individual trees rarely make good predictors, at least when compared to other popular algorithms, like neural networks and *random forests* (Chapter 7). Fortunately, we can often improve their performance by combining the predictions from several hundred or thousand individual trees together. There are several ways this can be accomplished, and Chapter 5 presents two popular and general strategies: *bagging* and *boosting*. Chapters 7–8 then dive deeper into specialized versions of bagging and boosting, respectively.

Each chapter contains numerous software examples that help solidify the main concepts, typically, only involving minimal package use and developing ideas from scratch. Tree-specific software and longer examples, however, are typically reserved for the end of each chapter, after the main ideas have been presented.

Part I

Decision trees

2

Binary recursive partitioning with CART

I'm always thinking one step ahead, like a carpenter that makes stairs.

Andy Bernard

The Office

This is arguably the most important chapter in the book. It is long, and rather involved, but serves as the foundation to more contemporary partitioning algorithms, like *conditional inference trees* CTree (Chapter 3), *generalized, unbiased, interaction detection, and estimation* (Chapter 4), and tree-based ensembles, such as *random forests* (Chapter 7) and *gradient boosting machines* (Chapter 8).

2.1 Introduction

In this chapter, I'll discuss one of the most general (and powerful) tree-based algorithms in current practice: *binary recursive partitioning*. This treatment of the subject follows closely with the open source routines available in the **rpart** package [Therneau and Atkinson, 2019], the details of which can be found in the corresponding package vignettes which can be accessed directly from R using `browseVignettes("rpart")` (they can also be found on the package's CRAN landing page at `https://cran.r-project.org/package=rpart`). The **rpart** package, which is discussed in depth in Section 2.9, is a modern

implementation of the *classification and regression tree* (CART)[a] procedures
proposed in Breiman et al. [1984]. But don't let the words "classification" and
"regression" in the name CART fool you; the procedure is general enough
to be applied to many different types of data (e.g., categorical, continuous,
multivariate, count, and censored outcomes). However, the primary focus of
this chapter will be on standard classification and regression.

Figure 2.1 shows two separate scatterplots, each of which has been divided
into three non-overlapping rectangular regions. The left plot contains $N = 200$
Swiss banknotes (Section 1.4.1) that have been identified as either genuine
(purple circles) or counterfeit (yellow triangles). The x-axis and y-axis correspond to the length (in mm) of the top and bottom edges of each bill,
respectively. Clearly there's some separation between the classes using just
these two features. We could use these three regions to classify new banknotes
as either genuine or counterfeit according to the majority class in whichever
region they belong to. For example, any banknote that lands in Region 3 will
be classified as counterfeit, since the majority of training observations that occupy it are counterfeit. In this way, the top and right edges of Region 2 form a
decision boundary that can be used for classifying new Swiss banknotes.

Similarly, the right plot shows the relationship between temperature (degrees
F), wind speed (mph), and ozone level (the response, measured in ppb) for
the New York air quality data (Section 1.4.2); brighter points indicate higher
ozone readings. The regions were selected in a way that tries to minimize
the response variance within each, subject to some additional constraints. To
predict the ozone level for a new data point x, we could use the average
response rate from whichever region x falls in (i.e., the prediction surface is a
step function).

This is the overall goal of CART, that is, to divide the feature space into
non-overlapping rectangular regions that have similar response rates in each,
which can then be used for description or prediction. For example, from a
description standpoint, we can see that counterfeit Swiss banknotes tend to
have abnormally longer top and bottom edges.

In more than two dimensions (i.e., more than two predictors), the disjoint
regions are formed by *hyperrectangles*. Why rectangular regions? Rectangular regions are simpler and more computationally feasible to find; they also
tend to yield a more interpretable model that can be represented using a convenient tree diagram. In particular, we want the resulting regions to be as
homogeneous as possible with respect to the response variable. The challenge
is in defining the regions. For example, how many regions should we use and
where should we draw the lines? Obviously we could continue refining each

[a] As mentioned in Section 1.2, the term "CART" is trademarked; hence, all the open
source implementations go by other names. For brevity, I'll use the acronym CART to refer
to the broad class of implementations that follow the original ideas in Breiman et al. [1984],
which includes **rpart** and scikit-learn's **sklearn.tree** module scikit-learn.

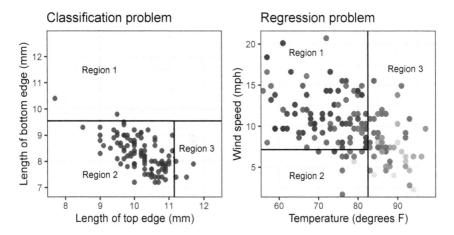

FIGURE 2.1: Scatterplots of two data sets split into three non-overlapping rectangular regions. The regions were selected so that the response values within each were as homogenous as possible. Left: a binary classification problem concerning 200 Swiss banknotes that have been identified as either genuine (purple circles) or counterfeit (yellow triangles). Right: a regression problem (brighter spots indicate higher average response rates within each bin).

region in the left side of Figure 2.1 by making more partitions, but this would eventually lead to overfitting.

The term "binary recursive partitioning" is quite descriptive of the general CART procedure, which I'll discuss in detail in the next section for the classification case. The word binary refers to the binary (or two-way) nature of the splits used to construct the trees (i.e., each split partitions a set of observations into two non-overlapping subsets). The word recursive refers to the *greedy* nature of the algorithm in choosing splits sequentially (i.e., the algorithm does not look ahead to find splits that are globally optimal in any sense; it only tries to find the next best split). And of course, partitioning refers to the way splits attempt to partition a set of observations into non-overlapping subgroups with homogeneous response values.

2.2   Classification trees

The construction of *classification trees* (categorical outcome) and *regression trees* (continuous outcome) is very similar. However, classification trees involve some subtle nuances that are easy to overlook, so I'll consider them in detail first. To begin, let's go back to the Swiss banknote data from Figure 2.1. As discussed in Section 1.4.1, these data contain six continuous measurements on 200 Swiss 1000-franc banknotes: 100 genuine and 100 counterfeit. The goal is to use the six available features to classify new Swiss banknotes as either genuine or counterfeit.

The code chunk below loads the data into R and prints the first few observations:

```
head(bn <- treemisc::banknote)   # load and peek at data

#>    length left right bottom  top diagonal y
#> 1     215  131   131    9.0  9.7      141 0
#> 2     215  130   130    8.1  9.5      142 0
#> 3     215  130   130    8.7  9.6      142 0
#> 4     215  130   130    7.5 10.4      142 0
#> 5     215  130   130   10.4  7.7      142 0
#> 6     216  131   130    9.0 10.1      141 0
```

A tree diagram representation of the Swiss banknote regions from Figure 2.1 is displayed in Figure 2.2. The bottom number in each node gives the fraction of observations that pass through that node (hence, the root node displays 100%). The values in the middle give the proportion of counterfeit and genuine banknotes, respectively, and the class printed at the top corresponds to the larger fraction (i.e., whichever class holds the majority in the node). The number above each node gives the corresponding node number. This is an example classification tree that can be used to classify new Swiss banknotes. For example, any Swiss banknote with **bottom >= 9.55** would be classified as counterfeit ($y = 1$); note that the split points are rounded for display purposes in Figure 2.2. The proportion of counterfeit bills in this node is 0.977 and can be used as an estimate of $\Pr(Y = 1|\boldsymbol{x})$; but more on this later.

From this tree, we can construct three simple rules for classifying new Swiss banknotes using just the bottom and top length of each bill:

```
Rule 1 (path to terminal node 2)
IF    bottom >= 9.55 (mm)
THEN classification = Counterfeit (probability = 0.977)

Rule 2 (path to terminal node 6)
IF    bottom < 9.55 (mm) AND top >= 11.15 (mm)
THEN classification = Counterfeit (probability = 0.765)

Rule 3 (path to terminal node 7)
IF    bottom < 9.55 (mm) AND top < 11.15 (mm)
THEN classification = Genuine (probability = 0.989)
```

This tree was found using the CART algorithm as implemented in **rpart**; the corresponding R code is used in Section 2.9.1. But how did CART determine which features to split on and which split point to use for each? Since this is a binary classification problem, CART searched for the predictor/split combinations that "best" separated the genuine banknotes from the counterfeit ones (I'll discuss how "best" is determined in the next section).

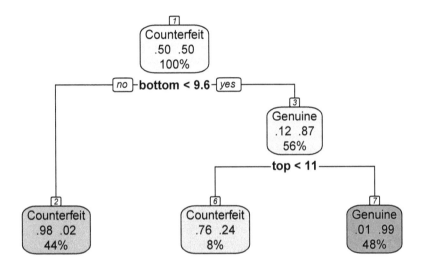

FIGURE 2.2: Example decision tree diagram for classifying Swiss banknotes as counterfeit or genuine.

2.2.1 Splits on ordered variables

Let's first discuss in general how CART finds the "best" split for an ordered variable. A hypothetical split S of an arbitrary node A into left and right child nodes, denoted A_L and A_R, respectively, is shown in Figure 2.3. If A contains N observations, then S partitions A into subsets A_L and A_R with node sizes N_L and N_R, respectively; note that $N_L + N_R = N$. Since the splitting process we're about to describe applies to any node in a tree, we can assume without loss of generality that A is the root node, which contains the entire learning sample (that is, all of the training data that will be used in constructing the tree). For now, I'll assume that all of the features are ordered, which includes both continuous and ordered categorical variables (I'll discuss splits for nominal categorical features in Section 2.4). The first step is to partition the root node in a way that "best separates" the individual class labels into two child nodes; I'll discuss ways to measure how well a particular split separates the class labels momentarily.

The split S depicted in Figure 2.3 can be summarized via a 2-by-2 *contingency table* giving the number of observations from each class that go to the left or

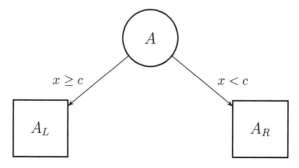

FIGURE 2.3: Hypothetical split for some parent node A into two child nodes using a continuous feature x with split point c.

TABLE 2.1: Confusion table summarizing the split S depicted in Figure 2.3.

	$y = 0$	$y = 1$	
$A_L : x \geq c$	N_{0,A_L}	N_{1,A_L}	N_{A_L}
$A_R : x < c$	N_{0,A_R}	N_{1,A_R}	N_{A_R}
	$N_{0,A}$	$N_{1,A}$	

right child node. Table 2.1 gives such a summary for a binary 0/1 outcome. For example, N_{0,A_L} is the number of observations belonging to class $y = 0$ that went to the left child node. The row and column margins are also displayed.

CART takes a *greedy* approach to tree construction. At each step in the splitting process, CART uses an exhaustive search to look for the next best (i.e., locally optimal) split, which does not necessarily lead to a globally optimal tree structure. This offers a reasonable trade-off between simplicity and complexity—otherwise the algorithm would have to consider all future potential splits at each step, which would lead to a combinatorial explosion. Let's turn our attention now to how CART chooses to split a node.

Let's assume the outcome is binary with $J = 2$ classes that are arbitrarily coded as 0/1 (e.g., for failure/success). For a continuous feature x with k

distinct values, CART will consider $k-1$ potential splits.[b] Typically, the midpoints of any two consecutive unique values are used as potential split points; for example, if x has unique values $\{1, 3, 7\}$ in the learning sample, then CART will consider $\{2, 5\}$ for potential split points. With $k-1$ potential splits to consider, which one does CART choose to partition the data? Ideally, it'd be the split that gives the "best separation" of the class labels (e.g., genuine and counterfeit banknotes, or edible and poisonous mushrooms). So how do we define the goodness of a particular split? Enter *node impurity* measures.

Ideally, we want the two resulting child nodes, A_L and A_R, to be as homogeneous as possible with respect to the class labels (e.g., all 0s or all 1s, if possible). To that end, we'd like to construct some function $i(A)$ that measures the *impurity* of a particular node A. At one extreme, A could be a *pure node*, that is, contain either all 0s or all 1s, in which case $i(A) = 0$. At the other extreme, the class labels in A are uniformly distributed (i.e., a 50/50 mix of 0s and 1s)—this is a worst-case scenario and the worst split possible. In this situation, the impurity function, $i(A)$, should be at a maximum.

Two common measures of node impurity used in CART are the *Gini index* and *cross-entropy* (or just *entropy* for short). For a response with J classes, these are defined as:

$$i(A) = \begin{cases} \sum_{j=1}^{J} p_j(A)(1 - p_j(A)) & \text{Gini index} \\ -\sum_{j=1}^{J} p_j(A) \log(p_j(A)) & \text{Cross-entropy} \end{cases}, \qquad (2.1)$$

where $p_j(A)$ is the expected proportion of observations in A that belong to class j; note that $i(A)$ is a function of the $p_j(A)$ $(j = 1, 2, \ldots, J)$. To avoid problems with $\log(0)$ in (2.1), we define $0 \log(0) \equiv 0$.

Another splitting measure, called the *twoing splitting rule* [Breiman et al., 1984, pp. 104–106], is only implemented in proprietary software (at least I'm not aware of any open source implementations of CART that support it). The twoing method tends to generate more balanced splits than the Gini or cross-entropy methods. For a binary response, the twoing criterion is equivalent to the Gini index. See Breiman [1996c] for additional details.

Before continuing, we need to introduce some more notation. Let N be the number of observations in the learning sample and N_j be the number of observations in the learning sample that belong to class j (i.e., $\sum_{j=1}^{J} N_j = N$). Similarly, let N_A be the number of observations in node A, and $N_{j,A}$ be the

---

[b]For large data sets, k may be too large, and approximate solutions can be used for scalability; for example, binning x by constructing a histogram on GPUs (Graphical Processing Units) [Zhang et al., 2017], which can then be used to quickly find a nearly optimal split.

number of observations in A that belong to class j. We can estimate $p_j(A)$ with $N_{j,A}/N_A$, the proportion of observations in A that belong to class j.[c]

For binary 0/1 outcomes, if we let $p = p_1(A)$ be the expected proportion of 1s in A, then (2.1) simplifies to

$$i(A) = \begin{cases} 2p(1-p) & \text{Gini index} \\ -p\log(p) - (1-p)\log(1-p) & \text{Cross-entropy} \end{cases}. \qquad (2.2)$$

These are plotted in Figure 2.4 below.

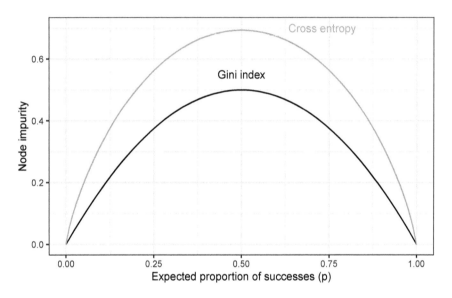

FIGURE 2.4: Common impurity measures for two-class classification problems, as a function of the the expected proportion of successes (p).

You may wonder why I'm not considering *misclassification error* as a measure of impurity. As it turns out, misclassification error is not a useful impurity measure for deciding splits; see, for example, Hastie et al. [2009, Section 9.2.3]. However, misclassification error can be a useful measure of the *risk* associated with a tree and is used in *decision tree pruning* (Section 2.5).

---

[c]Technically, we should use $p_j(A) \propto \pi_j\left(N_{j,A}/N_j\right)$, where π_j represents the true proportion of class j in the population of interest (called the prior for class j), but I'll come back to this in Section 2.2.4. For now, let's take $\pi_j = N_j/N$, the observed proportion of observations in the learning sample that belong to class j—this assumption is not always valid (e.g., when the data have been *downsampled*), but simplifies the formulas in this section, so I'll leave the complexities to Section 2.2.4.

Now that we have some notion of node impurity, we can define a measure for the quality of a particular split. In essence, the quality of an ordered split $S = \{x, c\}$ (see Figure 2.3), often called the *gain* of S, denoted $\Delta \mathcal{I}(S, A)$, is defined as the degree to which the two resulting child nodes, A_L and A_R, reduce the impurity of the parent node A:

$$
\begin{aligned}
\Delta \mathcal{I}(S, A) &= p(A) i(A) - [p(A_L) i(A_L) + p(A_R) i(A_R)] \\
&= p(A) i(A) - p(A_L) i(A_L) - p(A_R) i(A_R)
\end{aligned}
\tag{2.3}
$$

Here, $p(A)$, $p(A_L)$, and $p(A_R)$ correspond to the expected proportion of new observations in nodes A, A_L, and A_R, respectively. For example, we can interpret $p(A_L)$ as the probability of a case falling in the left child node A_L. If A is the root node, then $p(A) = 1$; otherwise, we can estimate it with N_A/N. For the two-class problem (i.e., $J = 2$), we can estimate $p(A_L)$ and $p(A_R)$ with the corresponding proportion of training cases in A_L and A_R, respectively. For instance, we can estimate $p(A_L)$ with N_{A_L}/N.

In essence, we want to find the split S (i.e., $x < c$ vs. $x \geq c$) associated with the maximum gain: $S_{best} = \arg\max_S \Delta \mathcal{I}(S, A)$. To this end, CART performs an exhaustive search through all features and potential splits therein, and chooses the split with maximal gain to partition A into two child nodes. This process is then repeated recursively in each resulting child node until a *saturated tree* has been constructed (i.e., no more splits are possible) or some suitable stopping criteria have been met (e.g., the specified maximum depth of the tree has been reached). While CART's approach to choosing the best split seems complicated, we'll implement it in R from scratch and apply it to the Swiss banknote data set in Section 2.2.2.

2.2.1.1 So which is it in practice, Gini or entropy?

For binary trees, Breiman [1996c] noted that the Gini index tends to prefer splits that put the most frequent class into one pure node, and the remaining classes into the other. Both entropy and the twoing splitting rules, on the other hand, put their emphasis on balancing the class sizes in the two child nodes. In problems with a small number of classes (i.e., $J = 2$), the Gini and entropy criteria tend to produce similar results.

Géron [2019, pp.183–184] echoes similar thoughts to Breiman's: "So should you use Gini impurity or entropy? The truth is, most of the time it does not make a big difference: they lead to similar trees. Gini impurity is slightly faster to compute, so it is a good default. However, when they differ, Gini impurity tends to isolate the most frequent class in its own branch of the tree, while entropy tends to produce slightly more balanced trees."

2.2.2   Example: Swiss banknotes

Returning to the Swiss banknote example, our goal is to find the first split
condition that "best" separates the genuine banknotes from the counterfeit
ones. Here, we'll restrict our attention to just two features: `top` and `bottom`,
which give the length (in mm) of the top and bottom edge, respectively. (We're
restricting attention to these two features because, as we'll see later, `diagonal`
is too good a predictor and leads to a less interesting illustration of finding
splits.) Since this is a classification problem, we can use cross-entropy or the
Gini index to measure the goodness of each split; here, we'll use the Gini index
and leave implementing cross-entropy as an exercise for the reader.

A simple R function for computing the Gini index in the two-class case is
given below. This function takes the binary target values as input, which are
assumed to be coded as 0/1 (which corresponds to genuine/counterfeit, in this
example); compare the function below to (2.2).

```
gini <- function(y) {  # y should be coded as 0/1
  p <- mean(y)  # proportion of successes (or 1s)
  2 * p * (1 - p)  # Gini index
}
```

To find the optimal split $S = \{x, c\}$, where x is an ordered (but numeric)
feature and c is in its domain, we need to search through every possible value
of c. This can be done, for example, by searching through the midpoints of
the sorted, unique values of x. For each split, we then need to compute the
weighted impurity of the current (or parent) node, as well as the weighted
impurities of the resulting left and right child nodes; then we find which split
point resulted in the largest gain (2.3).

A simple R function, called `splits()`, for carrying out these steps is given
below. Here, `node` is a data frame containing the observations in a particular
node (i.e., a subset of the learning sample), while `x` and `y` give the column
names in `node` corresponding to the (ordered numeric) feature of interest
and the (binary or 0/1) target, respectively. The argument `n` specifies the
number of observations in the learning sample; this is needed to compute
the probabilities $p(A)$, $p(A_L)$, and $p(A_R)$ used in (2.3). The use of `drop =
TRUE` in the definitions of the variables `left` and `right` ensures the results are
coerced to the lowest possible dimension. The `drop` argument in subsetting
arrays and matrices is used a lot in this book; for details, see ¿[' and `?drop`
for additional details.

```
splits <- function(node, x, y, n) {  # y should be coded as 0/1
  xvals <- sort(unique(node[[x]]))  # sorted, unique values
  xvals <- xvals[-length(xvals)] + diff(xvals) / 2  # midpoints
  res <- matrix(nrow = length(xvals), ncol = 2)  # to store results
  colnames(res) <- c("cutpoint", "gain")
  for (i in seq_along(xvals)) {  # loop through each midpoint
```

```
    left <- node[node[[x]] >= xvals[i], y, drop = TRUE]   # left child
    right <- node[node[[x]] < xvals[i], y, drop = TRUE]   # right child
    p <- c(nrow(node), length(left), length(right)) / n   # proportions
    gain <- p[1L] * gini(node[[y]]) -    # Equation (2.3)
      p[2L] * gini(left) - p[3L] * gini(right)
    res[i, ] <- c(xvals[i], gain)   # store split point and gain
  }
  res   # return matrix of results
}
```

Let's test this function out on the full data set (i.e., A is the root node) and find the optimal split point for **bottom**. To start, we'll find the gain that is associated with each possible split point and plot the results:

```
res <- splits(bn, x = "bottom", y = "y", n = nrow(bn))
head(res, n = 5)   # peek at first five rows

#>        cutpoint     gain
#> [1,]      7.25  0.00761
#> [2,]      7.35  0.01020
#> [3,]      7.45  0.01948
#> [4,]      7.55  0.03045
#> [5,]      7.65  0.03616

plot(res, type = "b", col = 2, las = 1,
     xlab = "Split value for bottom edge length (mm)",
     ylab = "Gain")   # Figure 2.5
```

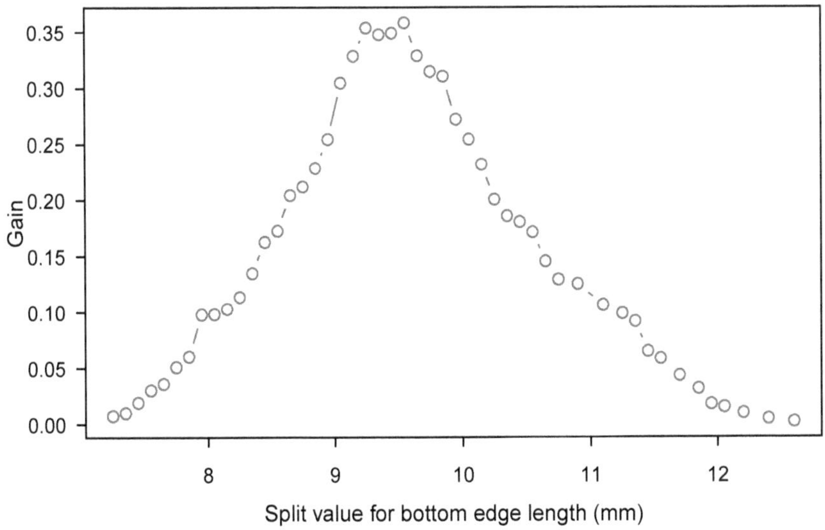

FIGURE 2.5: Reduction to the root node impurity as a function of the split value c for the bottom edge length (mm).

Figure 2.5 shows the split value c as a function of gain (or goodness of split). We can extract the exact cutpoint associated with the largest gain using

```
res[which.max(res[, "gain"]), ]   # extract row with maximum gain
```

```
#> cutpoint      gain
#>    9.550     0.358
```

Here, we can see that the optimal split point for `bottom` is 9.55 mm. A typical tree algorithm based on an exhaustive search would do this for each feature and pick one feature with the largest overall gain. Since all the features in `banknote` are continuous, we can just apply `splits()` to each feature to see which predictor would be used to first split the training data (i.e., the root node). To make things easier, let's write a wrapper function that calls `splits()` for any number of features, finds the split point associated with the largest gain for each, and then returns the best predictor/cutpoint pair. This is accomplished by the `find_best_split()` function below:

```
find_best_split <- function(node, x, y, n) {
  res <- matrix(nrow = length(x), ncol = 2)  # to store output
  rownames(res) <- x  # set row names to feature names
  colnames(res) <- c("cutpoint", "gain")  # column names
  for (xname in x) {  # loop through each feature
    # Compute optimal split
    cutpoints <- splits(node, x = xname, y = y, n = n)
    res[xname, ] <- cutpoints[which.max(cutpoints[, "gain"]), ]
  }
  res[which.max(res[, "gain"]), , drop = FALSE]
}
```

Now we're ready to start recursively partitioning the `banknote` data set. The code chunk below uses `find_best_split()` on the root node (i.e., the full learning sample) to find the best split between the features `top` and `bottom`:

```
features <- c("top", "bottom")  # feature names
find_best_split(bn, x = features, y = "y", n = nrow(bn))
```

```
#>          cutpoint  gain
#> bottom       9.55 0.358
```

Using the Gini index, the best way to separate genuine bills from counterfeit ones, using only the lengths of the top and bottom edges, is to separate the banknotes according to whether or not `bottom >= 9.55` (mm), which partitions the root node (i.e., full learning sample) into two relatively homogeneous subgroups (or child nodes):

```
left <- bn[bn$bottom >= 9.55, ]   # left child node
right <- bn[bn$bottom < 9.55, ]   # right child node
```

```
table(left$y)  # class distribution in left child node
```

```
#>
#>   0  1
#>   2 86
```

```
table(right$y)  # class distribution in right child node
```

```
#>
#>   0  1
#>  98 14
```

It makes no difference which node we consider the left or right child node; here I chose them for consistency with the tree diagram from Figure 2.2. Notice how the left child node is nearly pure, since 86 of the 88 observations (98%) in that node are counterfeit. While we could try to further partition this node using another split, it will likely lead to overfitting. The right node, on the other hand, is less homogeneous, with 14 of the 112 observations being counterfeit, and could potentially benefit from further splitting, as shown below:

```
find_best_split(right, x = features, y = "y", n = nrow(bn))
```

```
#>     cutpoint  gain
#> top    11.1 0.082
```

The next best split used `top` with a split value of $c = 11.15$ (mm) and a corresponding gain of 0.082. The resulting child nodes from this split are more homogenous but still not pure.

These two splits match the tree structure from Figure 2.2, which was obtained using actual tree fitting software, but more on that later. Without any stopping criteria defined, the partitioning algorithm could continue splitting until all terminal nodes are pure (a saturated tree). In Section 2.5, we'll discuss how to select an optimal number of splits (e.g., based on cross-validation). Saturated (or nearly full grown) trees are not generally useful on their own; however, in Chapter 5, we'll discuss a simple ensemble technique for improving the performance of individual trees by aggregating the results from several hundred (or even thousand) saturated trees.

2.2.3   Fitted values and predictions

Fitted values and predictions for new observations are obtained by passing records down the tree and seeing which terminal nodes they fall in. Recall that every terminal node in a fitted tree comprises some subset of the original training instances. If A is a terminal node, then any observation x (training or new) that lands in A would be assigned to the majority class in A: $\arg\max_{j \in \{1,2,...,J\}} N_{j,A}$; tie breaking can be handled in a number of ways (e.g., drawing straws). The predicted probability of x belonging to class j, which is often of more interest (and more useful) than the classification of x, is given by the proportion of training observations in A that belong to class j:

$$\widehat{\Pr}\left(Y = j|\boldsymbol{x}\right) = p_j\left(A\right) = N_{j,A}/N_A, \quad j = 1, 2, \ldots, J.$$

In the Swiss banknote tree (Figure 2.2; p. 43), any Swiss banknote with `bottom` `>= 9.55` (mm) would be classified as counterfeit (since the majority of observations in the corresponding terminal node are counterfeit) with a predicted probability of $86/(86 + 2) = 0.977$; note that the fitted probabilities in Figure 2.2 have been rounded to two decimal places, which is why they are not identical to the results we computed by hand in the previous section.

In summary, terminal nodes in a CART-like tree are summarized by a single statistic (or sometime multiple statistics, like the individual class proportions for J-class classification), which is then used to obtain fitted value and predictions—all observations that are predicted to be in the same terminal node also receive the same prediction. In classification trees, terminal nodes can be summarized by the majority class or the individual class proportions which are then used to generate classifications or predicted class probabilities for each of the J classes, respectively. Similarly, the terminal nodes in a CART-like regression tree (Section 2.3) can be summarized by the mean or median response, typically the former.

2.2.4   Class priors and misclassification costs

In Section 2.2.3, I mentioned that classifying new observations is done via a majority vote.[d] Similarly, predicted class probabilities can be obtained using the observed class proportions in the terminal nodes. This is a reasonable thing to do if the data are a random sample from some population of interest and the observed frequencies of each target class reflect the true balance in the population. If the observed class frequencies are off (e.g., the data have been *downsampled, upsampled,* or the design used to collect the data intentionally over-sampled the minority class to get a representative sample), then it may be beneficial to reweight the observations in a way that reflects the true class proportions, especially when searching for the best splits.

The common but often misguided practice of artificially rebalancing the class labels is especially interesting. Frank Harrell, who we briefly met in Section 1.1.2.4, once wrote

---

[d] For more than two classes (i.e., $J > 2$), a *plurality* vote is used.

A special problem with classifiers illustrates an important issue. Users of machine classifiers know that a highly imbalanced sample with regard to a binary outcome variable y results in a strange classifier. For example, if the sample has 1,000 diseased patients and 1,000,000 non-diseased patients, the best classifier may classify everyone as non-diseased; you will be correct 0.999 of the time. For this reason the odd practice of subsampling the controls is used in an attempt to balance the frequencies and get some variation that will lead to sensible looking classifiers (users of regression models would never exclude good data to get an answer). Then they have to, in some ill-defined way, construct the classifier to make up for biasing the sample. It is simply the case that a classifier trained to a 1/2 prevalence situation will not be applicable to a population with a 1/1,000 prevalence. The classifier would have to be re-trained on the new sample, and the patterns detected may change greatly.

Frank Harrell

https://www.fharrell.com/post/classification/

Fortunately, CART can flexibly handle imbalanced class labels without changing the learning sample. At a high level, we can assign specific unequal losses or penalties on a one-by-one basis to each type of misclassification error; in binary classification, there are two types of misclassification errors we can make: misclassify a 0 as a 1 (a *false positive*) or misclassify a 1 as a 0 (a false negative). The CART algorithm can account for these unequal losses or misclassification costs when deciding on splits and making predictions. Unfortunately, it seems that many practitioners are either unaware, or fail to take advantage of this feature.

Our discussion of splitting nodes in Section 2.2.1 implicitly made several assumptions about the available data. For instance, estimating $p_j(A)$ with $N_{j,A}/N_A$, the proportion of observations in node A that belong to class j, **assumes the training data are a random sample from some population of interest**. In particular, it assumes that the true prior probability of observing class j, denoted π_j, can be estimated with the observed proportion of class j observations in the training data; that is, $\pi_j \approx N_j/N$. If the observed class proportions are off (e.g., the data have been downsampled or the minority class has intentionally been over-sampled to over-represent rare cases), then $N_{j,A}/N_A$ is no longer a reasonable estimate of $p_j(A)$. Instead, we should be using $p_j(A) \propto \pi_j N_{j,A}/N_j$, where we scale the $\{p_j(A)\}_{j=1}^J$ to sum to one. Note that if we take π_j to be the observed class proportions, then $\pi_j = N_j/N$ and

$p_j(A)$ reduces to the observed proportion of observations in A that belong to class j. Similarly, when determining the "best" split in Section 2.2.1, we weighted the impurity of the two resulting child nodes, A_L and A_R, by the expected proportions of new observations going to each. If the data are not a random sample, then we should estimate $p(A)$ with $p(A) \approx \sum_{j=1}^{J} \pi_j N_{j,A}/N_j$; and similarly for $p(A_L)$ and $p(A_R)$.

Again, if we take π_j to be the observed class proportions in the learning sample, like we assumed in Section 2.2.1, then we can estimate $p(A)$ with N_A/N, the proportion of observations in node A. However, this is not always realistic. Think about the Swiss banknote data. These data consist of a 50/50 split of both counterfeit and genuine banknotes, which is not likely to be representative of the true class distributions. Nonetheless, I can't find any background information on how these data were collected. So, without additional information about the true class distributions, there's not much we can do. The example given in Section 2.9.5 demonstrates the use of CART with updated class prior information from historical data.

What's important to remember is that the prior class probabilities, $\{\pi_j\}_{j=1}^{J}$, affect the choice of splits in a tree and how the terminal nodes are summarized (e.g., how fitted values and new predictions are computed).

Increasing/decreasing the prior probabilities for certain classes essentially tricks CART into attaching more/less importance to those classes. In other words, it will try harder to correctly predict the classes associated with higher priors at the expense of less accurately predicting the other ones; in this sense, the prior probabilities can be seen as a *tuning parameter* in decision tree construction, especially if you want to attach more importance to correctly classifying certain classes. However, in some cases, it may be more natural to think about the specific costs associated with certain misclassifications. For example, with binary outcomes, it is often the case that false positives are more severe than false negatives, or vice versa. In the mushroom classification example (Section 1.4.4), it would be far worst to misclassify a poisonous mushroom as edible (a false negative, assuming poisonous represents the positive class, or class of interest) than to misclassify an edible mushroom as poisonous (a false positive). The next section introduces a general strategy for incorporating unequal losses, called *altered priors*; a second strategy, called the *generalized Gini index*, is discussed in the "Introduction to Rpart" vignette; see `vignette("longintro", package = "rpart")` for details.

2.2.4.1 Altered priors

Let $\boldsymbol{L}$ be a $J \times J$ *loss matrix* with entries $L_{i,j}$ representing the loss (or cost) associated with misclassifying an i as a j. We can define the *risk* of a node A as

$$r(A) = \sum_{j=1}^{J} p_j(A) \times L_{j,\tau_A}, \qquad (2.4)$$

where τ_A is the class assigned to A, if A were a terminal node, such that this risk is minimized. Since $p_j(A)$ depends on the prior class probabilities, risk is a function of both misclassification costs and class priors.

As a consequence, we can take misclassifcation costs into account by absorbing them into the priors for each class; this is referred to as the *altered priors* method. In particular, if

$$L_{i,j} = \begin{cases} L_i & i \neq j \\ 0 & i = j \end{cases}$$

then we can use the prior approach discussed above with the priors altered according to

$$\tilde{\pi}_i = \pi_j L_i / \sum_{j=1}^{J} \pi_j L_j, \qquad (2.5)$$

where π_j is the prior (observed or specified) associated with class j ($j = 1, 2, \ldots, J$). This is always possible for binary classification (i.e., $J = 2$). For multiclass problems (i.e., $J \geq 3$), we can use (2.5) with $L_i = \sum_{j=1}^{J} L_{i,j}$.

For details and further discussion, see Berk [2008, pp. 122–128] or the "Introduction to Rpart" vignette in package **rpart** (use `vignette("longintro", package = "rpart")` at the R console).

2.2.4.2 Example: employee attrition

To illustrate, let's walk through a detailed example using the employee attrition data set (Section 1.4.6). Figure 2.6 displays two classification trees fit to the employee attrition data, each with a max depth of two.[e] The only difference is that the tree on the left used the observed class priors $\pi_{no} = 1233/1470 = 0.839$ and $\pi_{yes} = 237/1470 = 0.161$ (i.e., it treats both types of misclassifications as equal). The tree on the right used altered priors based on the following loss (or misclassification cost) matrix:

---

[e]The depth of a decision tree is the maximum of the number of edges from the root node to each terminal node and is a common tuning parameter; see Section 8.3.2.

$$
\begin{array}{cc}
 & \begin{array}{cc} \text{No} & \text{Yes} \end{array} \\
\boldsymbol{L} = \begin{array}{c} \text{No} \\ \text{Yes} \end{array} & \begin{pmatrix} 0 & 1 \\ 8 & 0 \end{pmatrix},
\end{array}
$$

where the rows represent the true class and the columns represent the predicted class. For example, we're saying that it is 8 times more costly to misclassify a Yes (employee will leave due to attrition) as a No (employee will not leave due to attrition) than it is to misclassify a No as a Yes. Using this loss matrix, we can compute the altered priors as follows:

$$
\tilde{\pi}_{no} \propto (0+1)\,\pi_{no} = 1233/1470 = 0.839
$$
$$
\tilde{\pi}_{yes} \propto (8+0)\,\pi_{yes} = 8\,(237/1470) = 1.290
$$

Rescaling so that $\tilde{\pi}_{no} + \tilde{\pi}_{yes} = 1$ gives $\tilde{\pi}_{no} = 0.394$ and $\tilde{\pi}_{yes} = 0.606$. Notice how altering the priors resulted in a tree with different splits and node summaries.

The confusion matrix from each tree applied to the learning sample is shown in Table 2.2. Altering the priors by specifying a higher cost for misclassifying the Yeses increased the number of true negatives (assuming No represents the positive class) from 48 to 233, albeit at the expense of decreasing the number of true negatives from 1212 to 163. Finding the right balance is application-specific and requires a lot of thought and collaboration with subject matter experts.

TABLE 2.2: Confusion matrix from the trees in Figure 2.6.

		\multicolumn{4}{c}{Observed class}		
		\multicolumn{2}{c}{Default priors}	\multicolumn{2}{c}{Altered priors}
		No	Yes	No	Yes
Predicted class	No	1212	189	163	4
	Yes	21	48	1070	233

The tree structure on the left of Figure 2.6 uses the same calculations we worked through for the Swiss banknote example, so let's walk through some of the calculations for the tree on the right.

In any particular node A, we estimate $p_{no}(A) \propto \tilde{\pi}_{no} \times N_{no,A}/N_{no}$ and $p_{yes}(A) \propto \tilde{\pi}_{yes} \times N_{yes,A}/N_{yes}$, which are rescaled to sum to one. For instance, if A is the root node, we have $p_{no}(A) = \tilde{\pi}_{no} = 0.394$ since $N_{no,A}/N_{no} = 1233/1233 = 1$. Similarly, $p_{yes}(A) = 0.606$. We can then calculate the impurity of the root node using the Gini index:

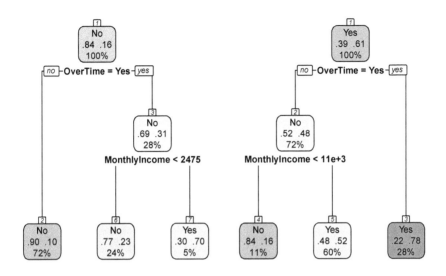

FIGURE 2.6: Decision trees for the employee attrition example. Left: default (i.e., observed) class priors. Right: altered class priors.

$$i(A) = 2 \times p_{no}(A) \times (1 - p_{no}(A)) = 0.478.$$

If we split the data according to `Overtime = Yes` (right branch) vs. `Overtime = No` (left branch), we have the following:

```
left <- attrition[attrition$OverTime == "No", ]   # left child
right <- attrition[attrition$OverTime == "Yes", ]  # right child
table(attrition$Attrition)   # class frequencies

#>
#>   No  Yes
#> 1233  237

table(left$Attrition)   # class frequencies (left node)

#>
#>  No Yes
#> 944 110

table(right$Attrition)   # class frequencies (right node)

#>
#>  No Yes
#> 289 127
```

For the left child node, we have

$$p\left(A_{L}\right) = \tilde{\pi}_{no} \times \left(944/1233\right) + \tilde{\pi}_{yes} \times \left(110/237\right) = 0.583,$$
$$p_{no}\left(A_{L}\right) = \tilde{\pi}_{no} \times \left(944/1233\right) / p\left(A_{L}\right) = 0.518,$$
$$p_{yes}\left(A_{L}\right) = 1 - p_{no}\left(A_{L}\right) = 0.482,$$
$$i\left(A_{L}\right) = 2 \times p_{no}\left(A_{L}\right) \times \left(1 - p_{no}\left(A_{L}\right)\right) = 0.499.$$

Similarly, for the right child node, we have:

$$p\left(A_{R}\right) = \tilde{\pi}_{no} \times \left(289/1233\right) + \tilde{\pi}_{yes} \times \left(127/237\right) = 0.417,$$
$$p_{no}\left(A_{R}\right) = \tilde{\pi}_{no} \times \left(289/1233\right) / p\left(A_{R}\right) = 0.221,$$
$$p_{yes}\left(A_{R}\right) = 1 - p_{no}\left(A_{R}\right) = 0.779,$$
$$i\left(A_{L}\right) = 2 \times p_{no}\left(A_{R}\right) \times \left(1 - p_{no}\left(A_{R}\right)\right) = 0.345.$$

And the gain for this split is

$$p\left(A\right) \times i\left(A\right) - p\left(A_{L}\right) \times i\left(A_{L}\right) - p\left(A_{R}\right) \times i\left(A_{R}\right) = 0.043.$$

In Section 2.9.4, we'll verify these calculations using open source tree software that follows the same CART-like procedure for altered priors.

This wraps our discussion of CART's search for the best split for an ordered variable in classification trees. Before discussing the search for splits on categorical features, I'll introduce the concept of a *regression tree*; that is, a decision tree with a continuous outcome.

2.3   Regression trees

Up to this point, our discussion of splitting nodes applies primarily to the case of CART-like classification trees. In CART, regression trees are constructed in nearly the same way as classification trees. The only real difference is that rather than finding the predictor/split combination that gives the greatest reduction in the within-node impurity, we look for the predictor/split combination that gives the greatest reduction in node *sum of squared errors (SSE)*:

$$\Delta \mathcal{I}\left(S, A\right) = SSE_{A} - \left(SSE_{A_{L}} + SSE_{A_{R}}\right), \qquad (2.6)$$

where, for example, $SSE_A = \sum_{i=1}^{N_A} (y_i - \bar{y})$ is the SSE within node A; recall that N_A is the number of training records in node A. This is equivalent to choosing the split that maximizes the between-groups sum-of-squares in an *analysis of variance* (ANOVA); in fact, in **rpart**, this split rule is referred to as the `"anova"` method (see `?rpart::rpart`). Note the similarities and differences between Equations (2.3) and (2.6).

To speed up the search for the best split, open source implementations, like **rpart** and scikit-learn's **sklearn.tree** module, do not directly search for splits that maximize (2.6) directly, but rather an equivalent proxy that's more efficient to compute. For example, it can be shown that

$$SSE_A = SSE_{A_L} + SSE_{A_R} + \frac{N_{A_L} N_{A_R}}{N_A} (\bar{y}_L - \bar{y}_R)^2, \qquad (2.7)$$

where $\bar{y}_L$ and $\bar{y}_R$ give the sample mean for the left and right child nodes of A, respectively. This implies that maximizing (2.6) is equivalent to maximizing the last term in (2.7), which makes sense, since we want the child nodes to be as different as possible (i.e., a greater difference in the mean responses).

In the regression case, we don't have to worry about priors or node probabilities. The terminal nodes are summarized by the mean response in each (the sample median is another possibility), and these are used for producing fitted values and predictions. For example, if a new observation x were to occupy some node terminal node A, then $\hat{f}(x) = \sum_{i=1}^{N_A} y_{i,A}/N_A$, where $y_{i,A}$ denotes the i-th response value from the learning sample that resides in terminal node A.

Aside from being useful in their own right, regression trees, as presented here, serve as the basic building blocks for *gradient tree boosting* (Chapter 8), one of the most powerful tree-based ensemble algorithms available.

2.3.1 Example: New York air quality measurements

Consider, for example, the `airquality` data frame introduced in Section 1.4.2, which contains daily air quality measurements in New York from May to September of 1973. A regression tree with a single split was fit to the data and the corresponding tree diagram is displayed in the left side of Figure 2.7. Here, the chosen splitter was temperature (in degrees Fahrenheit). Each node displays the predicted ozone concentration for all observations that fall in that node (top number) as well as the proportion of training observations in each (bottom number). According to this tree, the predicted ozone concentration is given by the simple rule:

$$\widehat{\text{Ozone}} = \begin{cases} 26.544 & \text{if Temp < 82.5} \\ 75.405 & \text{if Temp >= 82.5} \end{cases}.$$

The estimated regression surface is plotted in the right side of Figure 2.7. Note that the estimated prediction surface from a regression tree is essentially a step function, which makes it hard for decision trees to capture arbitrarily smooth or linear response surfaces.

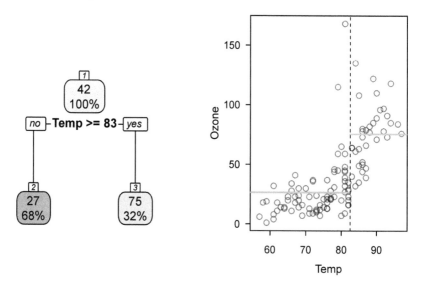

FIGURE 2.7: Decision stump predicting ozone concentration as a function of temperature. Left: tree diagram. Right: estimated regression function; a vertical dashed line is drawn at the split point $c = 82.5$ (the tree diagram on the left rounded up to the nearest integer).

To manually find the first partition and reconstruct the tree in Figure 2.7, we'll start by creating a simple function to calculate the within-node SSE. Note that these data contain a few missing values[f] (or NAs in R), so I set **na.rm = TRUE** in order to remove them before computing the results.

```
sse <- function(y, na.rm = TRUE) {
  sum((y - mean(y, na.rm = na.rm)) ^ 2, na.rm = na.rm)
}
```

Next, I'll modify the **splits()** function from Section 2.2.2 to work for the regression case:

---

[f]CART is actually pretty clever in how it handles missing values in the predictors, but more on this in Section 2.7.

```
splits.sse <- function(node, x, y) {
  xvals <- sort(unique(node[[x]]))  # sorted, unique values
  xvals<- xvals[-length(xvals)] + diff(xvals) / 2  # midpoints
  res <- matrix(nrow = length(xvals), ncol = 2)
  colnames(res) <- c("cutpoint", "gain")
  for (i in seq_along(xvals)) {  # loop through each feature
    left <- node[node[[x]] >= xvals[i], y, drop = TRUE]  # left
    right <- node[node[[x]] < xvals[i], y, drop = TRUE]  # right
    gain <- sse(node[[y]]) - sse(left) - sse(right)  # Equation (2.6)
    res[i, ] <- c(xvals[i], gain)  # store cutpoint and associated gain
  }
  res  # return matrix of results
}
```

Before applying this function to the air quality data, I'll remove the 37 rows that have a missing response value. The possible split points for `Temp`, along with their associated gains, are displayed in Figure 2.8. (To make the y-axis look nicer on the plot, the gain values were divided by 1,000.)

```
# Find optimal split for `Temp`
aq <- airquality[!is.na(airquality$Ozone), ]
res <- splits.sse(aq, x = "Temp", y = "Ozone")
res[which.max(res[, "gain"]), ]

#> cutpoint     gain
#>     82.5  60158.5

# Plot results
res[, "gain"] <- res[, "gain"] / 1000  # rescale for plotting
plot(res, type = "b", col = 2, las = 1,
     xlab = "Temperature split value (degrees Fahrenheit)",
     ylab = "Gain/1000")
abline(v = 82.5, lty = 2, col = 2)
```

To show that temperature is the best primary splitter for the root node, we can use **sapply()** to find the optimal cutpoint for all five features.:

```
features <- c("Solar.R", "Wind", "Temp", "Month", "Day")
sapply(features, FUN = function(xname) {
  res <- splits.sse(aq, x = xname, y = "Ozone")
  res[which.max(res[, "gain"]), ]
})
```

#>	Solar.R	Wind	Temp	Month	Day
#> cutpoint	153	6.6	82.5	6.5	24.5
#> gain	29721	50591.2	60158.5	14511.3	10282.8

Clearly, the split associated with the largest gain is `Temp`, followed by `Wind`, `Solar.R`, `Month`, and `Day`.

A regression tree in one predictor produces a step function, as was seen in the right side of Figure 2.7. The same idea extends to higher dimensions as well.

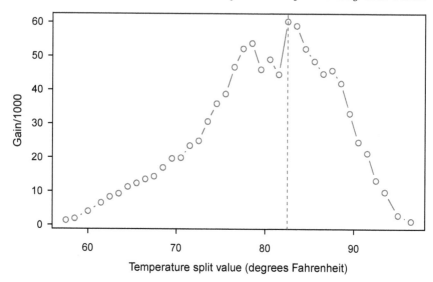

FIGURE 2.8: Potential split points for temperature as a function of gain. The maximum gain occurs at a temperature of 82.5 °F (the dashed vertical line).

For example, suppose we considered splitting on Wind next. Using the same procedures previously described, we would find that the next best partition occurs in the left child node using Wind with a cutpoint of 7.15 (mph). The corresponding tree diagram is displayed on the left side of Figure 2.9. If we stop splitting here, the result is a regression tree in two features. The corresponding prediction function, displayed on the right side of Figure 2.9, is a surface that's constant over each terminal node.

2.4   Categorical splits

Up to this point, we've only considered splits for ordered predictors, which have the form $x < c$ vs. $x \geq c$, where c is in the domain of x. But what about splits involving nominal categorical features? If x is ordinal (i.e., an ordered category, like low < medium < high), then we can map its ordered categories to the integers $1, 2, \ldots, J$, where J is the number of unique categories, and split as if x were originally numeric. If x is nominal (i.e., the order of the categories has no meaning), then we have to consider all possible ways to split

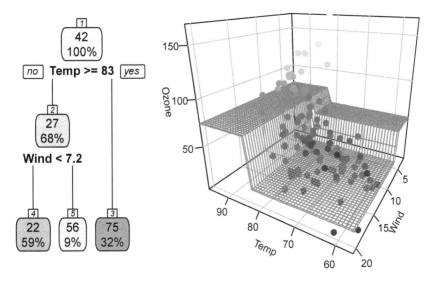

FIGURE 2.9: Regression tree diagram (left) and corresponding regression surface (right) for the air quality data. These are the same splits shown in Figure 2.1.

x into two mutually disjoint groups. For example, if x took on the categories $\{a, b, c\}$, then we could form a total three splits:

- $x \in \{a\}$ vs. $x \in \{b, c\}$;

- $x \in \{b\}$ vs. $x \in \{a, c\}$;

- $x \in \{c\}$ vs. $x \in \{a, b\}$.

For a nominal predictor with J categories, there are a total of $2^{J-1} - 1$ potential splits to search through, which can be computationally prohibitive for large J; for $J \geq 21$, we'd have to search more than a million splits! Fortunately, for ordered or binary outcomes, there is a computational shortcut that can be exploited for the splitting rules discussed in this chapter (i.e., Gini index, entropy, and SSE). This is discussed, for example, in Hastie et al. [2009, Sec. 9.2.4] and the "User Written Split Functions" vignette in package **rpart** (use `vignette("usercode", package = "rpart")` at the R console).

In short, the optimal split for a nominal predictor x at some node A can be found by first ordering the individual categories of x by their average response value—for example, the proportion of successes in the binary outcome case—and then finding the best split using this new ordinal variable.[g] This reduces

---

[g]This is equivalent to performing *mean/target encoding* [Micci-Barreca, 2001] prior to searching for the best split at each node; see Section 2.4.3.

the total number of possible splits from $2^{J-1} - 1$ to $J - 1$, an appreciable reduction in the total number of splits that must be searched. It will also still result in the optimal split when using the Gini index, cross-entropy, or SSE splitting rules discussed earlier. A proof for the Gini and entropy measures is provided in Ripley [1996, p. 218], with Chou [1991] providing a proof for a more general family of impurity measures. For multiclass problems (i.e., $J > 2$), no such computational shortcut exists, although efficient search methods have been proposed in Sleumer [1969] and Loh and Vanichsetakul [1988].

2.4.1   Example: mushroom edibility

To illustrate, let's return to the mushroom edibility example, which contains all categorical features and a binary response. A simple classification tree diagram for the data is shown in Figure 2.10. The tree contains two splits on the features `odor` and `spore.print.color`. Since the response (`Edibility`) is binary, we can use the shortcut approach to build the tree using the same process for ordered splits, as long as we apply the Gini or entropy splitting criterion; here, I'll use the Gini index since it's already built into our previously defined `find_best_split()` function.

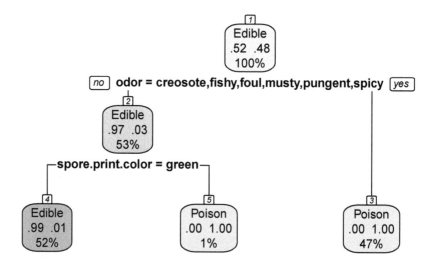

FIGURE 2.10: Example classification tree for determining the edibility of mushrooms.

For each mushroom attribute, the individual categories need to be mapped to the proportion of successes within each. For this example, I'll refer to the outcome class `Poison` as a success and re-encode the target as 0/1 for

Edible/Poison. I'll also remove the `veil.type` feature because it only takes on a single value (i.e., it has zero variance) and can contribute nothing to the partitioning:

```
m <- treemisc::mushroom  # load mushroom data
m$veil.type <- NULL  # remove useless feature
m$Edibility <- ifelse(m$Edibility == "Poison", 1, 0)
m2 <- m  # make a copy of the original data
```

To illustrate the main idea, let's look at a frequency table for the `veil.color` predictor, which has four unique categories:

```
table(m2$veil.color)
```

```
#>
#>  brown orange  white yellow
#>     96     96   7924      8
```

We need to find the mean response within each category—in this case, the proportion of poisonous mushrooms—and then map those back to the original feature values. For instance we would re-encode all the values of `"white"` in `veil.color` as 0.493 because $3908/7924 \approx 0.493$ of the mushrooms with `veil.color = "white"` are poisonous. This can be done in any number of ways, and here I'll write a simple function, called `ordinalize()`, that returns a list with two components: `map`, which contains the numeric value each category gets mapped to, and `encoded`, which contains the re-encoded feature values.

```
ordinalize <- function(x, y) {  # convert nominal to ordered
  map <- tapply(y, INDEX = x, FUN = mean)
  list("mapping" = map, "encoded" = map[x])
}
```

```
# Check which numeric values `veil.color` gets mapped to
ordinalize(m2$veil.color, m2$Edibility)$map
```

```
#>  brown orange  white yellow
#>  0.000  0.000  0.493  1.000
```

Next, I'll write a simple `for` loop that uses `ordinalize()` to numerically re-encode each feature column in the `m2` data frame:

```
xnames <- setdiff(names(m2), "Edibility")
for (xname in xnames) {  # mean/target encode each feature
  m2[[xname]] <- ordinalize(m2[[xname]], y = m2[["Edibility"]])$encoded
}
```

```
# Take a peek at the re-encoded data
m2[1L:8L, 1L:5L]
```

```
#>    Edibility cap.shape cap.surface cap.color bruises
#> 1          1     0.467       0.552     0.447   0.185
```

```
#> 2          0      0.467      0.552      0.627   0.185
#> 3          0      0.106      0.552      0.308   0.185
#> 4          1      0.467      0.536      0.308   0.185
#> 5          0      0.467      0.552      0.439   0.693
#> 6          0      0.467      0.536      0.627   0.185
#> 7          0      0.106      0.552      0.308   0.185
#> 8          0      0.106      0.536      0.308   0.185
```

Since all the categorical features have been re-encoded numerically, we can use our previously defined `find_best_split()` function to partition the data. Starting with the root node (i.e., the full learning sample), we obtain:

```
find_best_split(m2, x = xnames, y = "Edibility", n = nrow(m2))

#>         cutpoint  gain
#> odor      0.517 0.471

# Summarize split
left <- m[m2$odor >= 0.5170068, ]
right <- m[m2$odor < 0.5170068, ]

table(left$Edibility)  # non-pure node

#>
#>    1
#> 3796

table(right$Edibility)  # pure node

#>
#>    0    1
#> 4208  120
```

The first split uses **odor**, with a mean/target encoded split point of $c = 0.517$ and a corresponding gain of 0.471. Since the resulting right child node is pure (in this case, all poisonous), let's continue partitioning with the left one:

```
# Ordinalize left child node and find next best split
right.ord <- right
for (xname in xnames) {  # mean/target encode each feature
  right.ord[[xname]] <-
    ordinalize(right.ord[[xname]],
               y = right.ord[["Edibility"]])$encoded
}

# Find best split in newly "ordinalized" predictors
find_best_split(right.ord, x = xnames, y = "Edibility", n = nrow(m2))

#>                      cutpoint  gain
#> spore.print.color      0.538 0.017
```

The next split is based on **spore.print.color**, with a mean/target encoded split point $c = 0.538$ and a corresponding gain of 0.017, which is equivalent to separating mushrooms based on whether or not they have a green spore print.

To map these splits back to their corresponding categories, we can look at the $map component from the output of `ordinalize()` on each split variable:

```
sort(ordinalize(m$odor, m$Edibility)$map)
#>    almond     anise     none creosote    fishy      foul
#>     0.000     0.000    0.034    1.000    1.000     1.000
#>     musty   pungent    spicy
#>     1.000     1.000    1.000

sort(ordinalize(right[["spore.print.color"]],
                y = right[["Edibility"]])$map)
#>     black     brown     buff chocolate   orange
#>    0.0000    0.0000   0.0000    0.0000   0.0000
#>    purple    yellow    white     green
#>    0.0000    0.0000   0.0769    1.0000
```

For example, the split point for `odor` was 0.517 (the midpoint between 0.034 1.00), and every feature mapped to a re-encoded `odor` value ≥ 0.517 is used to construct the first partition; see the first split in Figure 2.10. In Section 2.9.2, we'll verify these results (e.g., the computed gain for both splits) using CART-like software in R.

2.4.2   Be wary of categoricals with high cardinality

One drawback of CART-like decision trees is that they tend to favor categorical features with high cardinality (i.e., large J), even if they are mostly irrelevant.[h] For categorical features with large J, for example, there are so many potential splits that the tree is more likely to find a good split just by chance. Think about the extreme case where a nominal feature x is different and unique in every row of the learning sample, like a row ID column. The split variable selection bias in CART-like decision trees has been discussed plenty in the literature; see, for example, Breiman et al. [1984, p. 42], Segal [1988], and Hothorn et al. [2006c] (and the additional references therein)

To illustrate the issue, I added ten random categorical features (`cat1`–`cat10`) to the `airquality` data set from Section 2.3.1, each with a cardinality of $J = 26$ (they're just random letters from the alphabet). A default regression tree was fit to the data using **rpart**, and the resulting tree diagram is displayed in Figure 2.11. Notice that all of the splits, aside from the first, use the completely irrelevant categorical features that were added! In Section 2.5 we'll look at a

---

[h]This bias actually extends to any predictor with lots of potential split points, whether ordered or nominal.

general *pruning* technique that can be helpful in screening out pure noise variables.

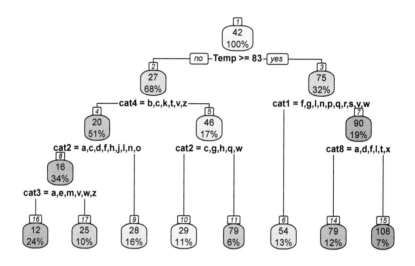

FIGURE 2.11: A decision tree fit to a copy of the air quality data set that includes ten completely random categorical features, each with cardinality 26.

In some cases, it's possible to reduce the number of potential categories to something more manageable—like lumping rare categories together, or combining categories into a smaller set of meaningful subgroups (e.g., combining zip or area codes into a smaller set of larger geographic areas).

The partitioning algorithms discussed in Chapters 3–4 address the split selection bias issue more directly by separating the exhaustive search over all possible splits for each feature into two sequential steps, where the optimal split point is found only after a splitting variable has been selected.

2.4.3   To encode, or not to encode?

When dealing with categorical data, we are often concerned with how to encode such features. In linear models, for example, we often employ *dummy encoding* or *effect encoding*, depending on the task at hand. Similarly, *one-hot-encoding* (OHE), closely related to dummy encoding, is often used in general machine learning problems outside of (generalized) linear models. And there are plenty of other ways to encode categorical variables, depending on the algorithm and task at hand.

As you've already seen, decision trees can naturally handle variables of any type without special encoding, although we did see that a local form of mean/target encoding can be used to reduce the computational burden imposed by nominal categorical splits. Nonetheless, using an encoding strategy, like OHE, can sometimes improve the predictive performance or interpretability of a tree-based model; see Kuhn and Johnson [2013, Sec. 14.7] for a brief discussion on the use of OHE in tree-based methods. Further, some tree-based software, like Scikit-learn's **sklearn.tree** module, require all features to be numeric—forcing users to employ different encoding schemes for categorical features. See Boehmke and Greenwell [2020, Chap. 3] for details on different encoding strategies (with examples in R), and further references.

2.5   Building a decision tree

In the previous sections, we talked about the basics of splitting a node (i.e., partitioning some subset of the learning sample). Building a CART-like decision tree starts by splitting the root node, and then recursively applying the same splitting procedure to every resulting child node until a saturated tree is obtained (i.e., all terminal nodes are pure) or other stopping criteria are met. In essence, the partitioning stops when at least one of the following conditions are met:

- all the terminal nodes are pure;
- the specified maximum tree depth has been reached;
- the minimum number of observations that must exist in a node in order for a split to be attempted has been reached;
- no further splits are able to decrease the overall lack of fit by a specified factor;
- and so forth.

This often results in an overly complex tree structure that overfits the learning sample; that is, it has low bias, but high variance.

To illustrate, consider a random sample of size $N = 500$, generated from the following sine wave with Gaussian noise:

$$Y = \sin(X) + \epsilon,$$

where $X \sim \mathcal{U}(0, 2\pi)$ and $\epsilon \sim \mathcal{N}(0, \sigma = 0.3)$. A scatterplot of the data, along with the true response function, is shown in Figure 2.12.

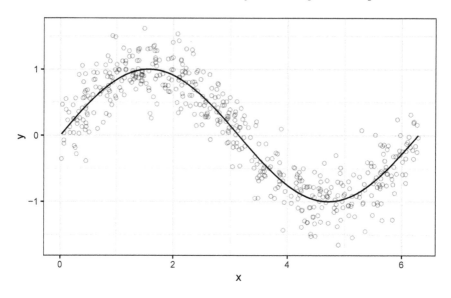

FIGURE 2.12: Data generated from a simple sine wave with Gaussian noise. The black curve shows the true mean response $E(Y|X = x) = \sin(x)$.

Figure 2.13 shows the prediction function from two regression trees fit to the same data.[i] The tree on the left is too complex and has too many splits, and exhibits high variance, but low bias (i.e., it fits the current sample well, but the tree structure will vary wildly from one sample to the next because it's mostly fitting the noise here); unstable models, like this one are often referred to as *unstable learners* (more on this in Section 5.1). The tree on the right, which is a simple decision stump (i.e., a tree with only a single split), is too simple, and will also not be useful for prediction because it has extremely high bias, but low variance (i.e., it doesn't fit the data too well, but the tree structure will be more stable from sample to sample); such a weak performing model is often referred to as a *weak learner* (more on this in Section 5.2).

Neither tree is likely to be accurate when applied to a different sample from the same model; the ensemble methods discussed in Part II of this book can improve the performance of both weak and unstable learners. When using a single decision tree, however, the question we need to answer is, How complex should we make the tree? Ideally, we should have stopped splitting nodes at some *subtree* along the way, but where?

A rather careless approach is to build a tree by only splitting nodes that meet some threshold on prediction error. However, this is shortsighted because a low-quality split early on may lead to a very good split later in the tree. The standard approach to finding an optimal subtree—basically, determining when

---

[i]The associated tree diagrams are shown in the top left and bottom right of Figure 2.14 (p. 73), respectively.

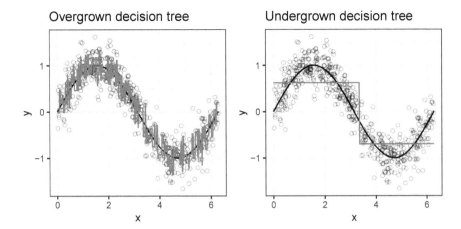

FIGURE 2.13: Regression trees applied to the sine wave example. Left: this tree is too complex (i.e., low bias and high variance). Right: this tree is too simple (i.e., high bias and low variance).

we should have stopped splitting nodes—is called *cost-complexity pruning*, or *weakest link pruning* [Breiman et al., 1984], or just pruning for short. Other pruning procedures are discussed in Ripley [1996, pp. 226–231] and Zhang and Singer [2010, pp. 44–49]. Pruning a decision tree is quite analogous to the process of *backward elimination* in multiple linear regression—start with a complex tree with too many splits, and *prune off* leaves whose contributions aren't enough to offset the added complexity. The details are covered in the next section.

2.5.1   Cost-complexity pruning

The idea of pruning a decision tree is similar to the process of backward elimination in multiple linear regression. In essence, we build a large tree with too many splits, denoted $\mathcal{T}_0$, and then prune it back by collapsing internal nodes until we find some optimal subtree, denoted $\mathcal{T}_{opt}$, that meets a certain criterion, like having the smallest cross-validation error.

Let $\{A_k\}_{k=1}^K$ be the terminal nodes of some tree $\mathcal{T}$, where $|\mathcal{T}| = K$ is the number of terminal nodes, or size of $\mathcal{T}$. Recall that the overall goal of CART is to extract homogenous subgroups (i.e., terminal nodes). In this sense, the overall quality (or *risk*) of the tree depends on the quality of its terminal nodes. We define the risk of the tree to be $R\left(\mathcal{T}\right) = \sum_{k=1}^K p\left(A_k\right) \times r\left(A_k\right)$, where $r\left(A_k\right)$ is some measure of the quality of the k-th terminal node; see (2.4) on page 55. For regression trees, $R\left(\mathcal{T}\right)$ is the error sum of squares (SSE). For classification trees based on the observed class priors and equal misclassification costs (i.e., $L_{i,j} = 1$ for all $i \neq j$), $R(\mathcal{T})$ is simply the proportion of observations misclassified in the learning sample.

Building a tree to minimize $R(\mathcal{T})$ will always lead to a saturated tree, resulting in a model with little or no bias but often high variance (i.e., overfitting the learning sample). Instead, we penalize the complexity (or size) of the tree by minimizing

$$R_\alpha(\mathcal{T}) = R(\mathcal{T}) + \alpha|\mathcal{T}|,$$

where $\alpha \geq 0$ is a tuning parameter controlling the trade-off between the complexity of the tree, $|\mathcal{T}|$, and how well it fits the training data, $R(\mathcal{T})$. In this sense, $R_\alpha(\mathcal{T})$ can be viewed as a penalized objective function similar to what's used in *regularized regression*; see, for example, Hastie et al. [2009, Chap. 3] or Boehmke and Greenwell [2020, Chap. 6]. When $\alpha = 0$, no penalty is incurred, resulting in the most complex tree $\mathcal{T}_0$. On the other extreme, we can always find a large enough value of α that results in a decision tree with no splits (i.e., the root node). Choosing the right value of α is important and can be done using cross-validation or other methods; a specific cross-validation approach is covered in Section 2.5.2.

Breiman et al. [1984, Chap. 10] showed that for each α, there exists a unique smallest subtree, denoted $\mathcal{T}_\alpha$, that minimizes $R_\alpha(\mathcal{T})$. This result is important because it guarantees that no two equally sized subtrees of $\mathcal{T}_0$ will have the same value of $R_\alpha(\mathcal{T})$. To obtain $\mathcal{T}_\alpha$, start pruning $\mathcal{T}_0$ by successively collapsing the internal node that produces the smallest per-node increase to $R(\mathcal{T})$, and continue until reaching the root node. This process results in a (finite) sequence of nested subtrees (see Figure 2.14 on page 73 for an example) that contains $\mathcal{T}_\alpha$; for details, see Breiman et al. [1984, Chap. 10] or Ripley [1996, Sec. 7.2].

To illustrate, take $\mathcal{T}_0$ to be the left tree in Figure 2.12, which has a total of 154 splits. The corresponding tree diagram is displayed in the top left of Figure 2.14. The rest of the tree diagrams in Figure 2.14 correspond to the last 15 trees in the pruning sequence (minus the root node), ending with a decision stump. The optimal subtree, $\mathcal{T}_\alpha$, which has a total of 20 splits (or 21 terminal nodes), was found using 10-fold cross-validation and is highlighted in green.

For comparison, I compared how each subtree performed on an independent test set of 500 new observations. For each subtree in the pruned sequence, the prediction error on the test set, measured as $1 - R^2$, where R^2 is the squared Pearson correlation between the observed and fitted values, was computed. Both the test and cross-validation errors are displayed in Figure 2.15. Here, the results are similar, but the test error suggests a slightly simpler tree with only 18 splits.

FIGURE 2.14: Nested subtrees for the sine wave example. The optimal subtree, chosen via 10-fold cross-validation, is highlighted in green.

So how is the sequence of α values determined? For any internal node A, we can find α using

$$\alpha = \frac{R(A) - R(\mathcal{T}_A)}{|\mathcal{T}_A| - 1},$$

where $\mathcal{T}_A$ is the subtree rooted at node A. To start pruning, we need to find the first threshold value α_1, which is just the smallest α value among the $|\mathcal{T}| - 1$ internal nodes of the tree $\mathcal{T}$. Once α_1 is obtained, we prune the tree by collapsing one of the $|\mathcal{T}| - 1$ internal nodes and making it a terminal node whenever

$$\alpha_1 \geq \frac{R(A) - R(\mathcal{T}_A)}{|\mathcal{T}_A| - 1}.$$

This results in the optimal subtree, $\mathcal{T}_{\alpha_1}$, associated with $\alpha = \alpha_1$. Starting with $\mathcal{T}_{\alpha_1}$, we then continue this process by finding α_2 in the same way we found α_1 for the full tree $\mathcal{T}$. The process is continued until reaching the root node. It might sound confusing, but we'll walk through the calculations using the mushroom example in the next section.

The **rpart** package, which is used extensively throughout this chapter, employs a slightly friendlier, and rescaled, version of the cost-complexity parameter α, which they denote as *cp*. Specifically, **rpart** uses

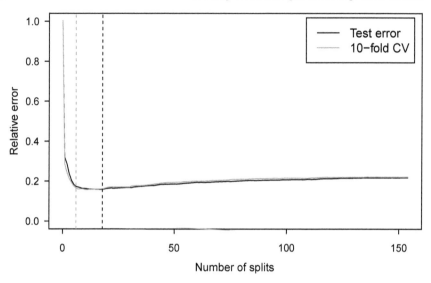

FIGURE 2.15: Relative error based on the test set (black curve) and 10-fold cross-validation (yellow curve) vs. the number of splits for the sine wave example. The vertical yellow line shows the optimal number of splits based on 10-fold cross-validation, while the vertical black line shows the optimal number of splits based on the independent test set.

$$R_{cp}\left(\mathcal{T}\right) \equiv R\left(\mathcal{T}\right) + cp \times |\mathcal{T}| \times R\left(\mathcal{T}_1\right),$$

where $\mathcal{T}_1$ is the tree with zero splits (i.e., the root node). Compared to α, cp is unitless, and a value of $cp = 1$ will always result in a tree with zero splits. The complexity parameter, cp, can also be used as a stopping rule during tree construction. In many open source implementations of CART, whenever $cp > 0$, any split that does not decrease the overall lack of fit by a factor of cp is not attempted. In a regression tree, for instance, this means that the overall R^2 must increase by cp at each step for a split to occur. The main idea is to reduce computation time by avoiding potentially unworthy splits. However, this runs the risk of not finding potentially much better splits further down the tree.

2.5.1.1   Example: mushroom edibility

Let's drive the main ideas home by calculating a few α values to prune a simple tree for the mushroom edibility data. Consider again a simple decision tree for the mushroom edibility data which is displayed in Figure 2.16. This is a simple tree with only three splits, but we'll use it to illustrate how pruning works and how the sequence of α values is computed. For clarity, the number of observations in each class is displayed within each node, and the node numbers appear at the top of each node. For example, node 8 contains 4208 edible mushrooms and 24 poisonous ones. The assigned classification, or majority class, is printed above the class frequencies in each node. This tree was also built using the observed class priors and equal misclassification costs; hence, $R\left(\mathcal{T}\right)$ is just the proportion of misclassifications in the learning sample: $24/8124 \approx 0.003$.

Let A_i, $i \in \{1, 2, 3, 4, 5, 8, 9\}$ denote the seven nodes of the tree in Figure 2.16; in **rpart**, the left and right child nodes for any node numbered x are always numbered $2x$ and $2x+1$, respectively (the root node always corresponds to $x = 1$). We can compute the risk of any terminal node using $R\left(A_i\right) = N_{j,A}/N_A$. For example, nodes A_5–A_7 all have a risk of zero (since they are pure nodes).

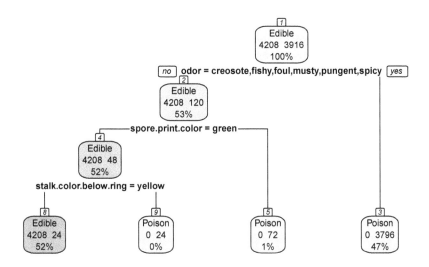

FIGURE 2.16: Classification tree with three splits for the mushroom edibility data. The overall risk of the tree is $24/8124 \approx 0.003$.

To find α_1, we need to first compute α for each of the $|\mathcal{T}_0| - 1 = 3$ internal nodes of the tree, and find which one is the smallest; use the tree diagram in Figure 2.16 to follow along. The α values for the three internal nodes are computed as follows:

$$\alpha_{A_1} = \left(3916/8124 - 24/8124\right) / \left(4 - 1\right) \approx 0.160$$
$$\alpha_{A_2} = \left(120/8124 - 24/8124\right) / \left(3 - 1\right) \approx 0.006 \;.$$
$$\alpha_{A_4} = \left(48/8124 - 24/8124\right) / \left(2 - 1\right) \approx 0.003$$

Since α_{A_4} is the smallest, we collapse node A_4, resulting in the next optimal subtree in the sequence, $\mathcal{T}_{\alpha_1}$, which is displayed in the left side of Figure 2.17. The cost-complexity of this tree is $R_{\alpha_1}(\mathcal{T}_{\alpha_1}) = 0.015$. To find α_2, we start with $\mathcal{T}_{\alpha_1}$ and repeat the process by first finding the smallest α value associated with the $|\mathcal{T}_{\alpha_1}| - 1 = 2$ internal nodes of $\mathcal{T}_{\alpha_1}$. These are given by

$$\alpha_{A_1} = (3916/8124 - 48/8124) / (3 - 1) \approx 0.238$$
$$\alpha_{A_2} = (120/8124 - 48/8124) / (2 - 1) \approx 0.009 \text{ ,}$$

making $\alpha_2 = 0.009$. We would then prune the current subtree, $\mathcal{T}_{\alpha_2}$, by collapsing A_2 into a terminal node, resulting in the decision stump displayed in the right side of Figure 2.17. This makes only one possibility for $\alpha_3 = (3916/8124 - 120/8124) / (2 - 1) \approx 0.467$, which results in the root node after pruning the decision stump, $\mathcal{T}_{\alpha_3}$. In the end, we have the following sequence of α values: $(\alpha_1 = 0.003, \alpha_2 = 0.009, \alpha_3 = 0.467)$. In practice, we would use cross-validation, or some other validation procedure, to select a reasonable value of the complexity parameter α from this sequence. The next two sections discuss choosing α using k-fold cross-validation.

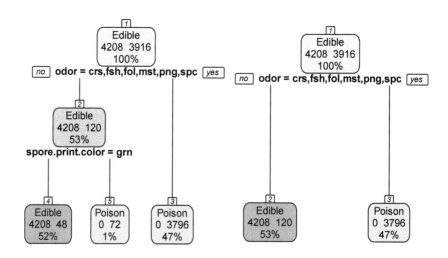

FIGURE 2.17: Optimal subtrees in the sequence with minimum cost-complexity. Since the original tree contains only three splits, there are only two possible subtrees, not counting the tree with zero splits. Here the category names have been truncated to three letters to fit more compactly in the display.

2.5.2 Cross-validation

Once the sequence $\alpha_1, \alpha_2, ..., \alpha_{k-1}$ has been found, we still need to estimate the overall risk/quality of the corresponding sequence of nested subtrees, $R_{\alpha_i}(\mathcal{T})$, for $i = 1, 2, \dots, k - 1$. Breiman et al. [1984, Chap. 11] suggested picking α using a separate validation set or k-fold cross-validation. The latter is more computational, but tends to be preferred since it makes use of all available data, and both tend to lead to similar results. The procedure described in Algorithm 2.1 below follows the implementation in the **rpart** package in R (see the "Introduction to Rpart" vignette):

Algorithm 2.1 K-fold cross-validation for cost-complexity pruning.

---

1) Fit the full model to the learning sample to obtain $\alpha_1, \alpha_2, ..., \alpha_{k-1}$.

2) Define β_i according to

$$\beta_i = \begin{cases} 0 & i = 1 \\ \sqrt{\alpha_{i-1}\alpha_i} & i = 2, 3, \dots, m - 1 \\ \infty & i = m \end{cases}.$$

Since any value of α in the interval $(\alpha_i, \alpha_{i+1}]$ results in the same subtree, we instead consider the sequence of β_i's, which represent typical values within each range using the geometric midpoint.

3) Divide the data into k groups (or folds), $D_1, D_2, \dots, D_k$, with approximately k/N observations in each (N being the number of rows in the learning sample). For $i = 1, 2, \dots, k$, do the following:

   a) Fit the full model to the learning sample, but omit the subset D_i, and find the sequence of optimal subtrees $\mathcal{T}_{\beta_1}, \mathcal{T}_{\beta_2}, \dots, \mathcal{T}_{\beta_k}$.

   b) Compute the prediction error from each tree on the validation set D_i.

4) For each subtree, aggregate the results by averaging the k out-of-sample prediction errors.

5) Return $\mathcal{T}_\beta$ from the initial sequence of trees based on the full learning sample, where β corresponds to the β_i associated with the smallest prediction error in step 4).

2.5.2.1   The 1-SE rule

When choosing α with k-fold cross-validation, Breiman et al. [1984, Sec. 3.4.3] recommend using the *1-SE rule*, and argue that it is useful in screening out irrelevant features. The 1-SE rule suggests using the most parsimonious tree (i.e., the one with fewest splits) whose cross-validation error is no more than one standard error above the cross-validation error of the best model. This of course requires an estimate of the standard error during cross-validation. A heuristic estimate of the standard error can be found in Breiman et al. [1984, pp. 306–309] or Zhang and Singer [2010, pp. 42–43], but the formula isn't pretty! Applying cost-complexity pruning using cross-validation, with or without the 1-SE rule, would almost surely remove all of the nonsensical splits seen in Figure 2.11. (In fact, this was the case after applying 10-fold cross-validation using the 1-SE rule.)

2.6   Hyperparameters and tuning

There are essentially three hyperparameters associated with CART-like decision trees:

1) the maximum depth or number of splits;
2) the maximum size of any terminal node;
3) the cost-complexity parameter *cp*.

Different software will have different names for these parameters and different default values. Arguably, *cp* is the most flexible and important tuning parameter in CART, and a good strategy is to relax the maximum depth and size of the terminal nodes as much as possible, and use cost-complexity pruning to find an optimal subtree using k-fold cross-validation, or some other validation procedure. In some cases, Chapter 7, for example, trees are intentionally grown to maximal or near maximal depth (in some cases, leaving only a single observation in each terminal node).

2.7   Missing data and surrogate splits

One of the best features of CART is the flexibility with which missing values can be handled. More traditional statistical models, like linear or logistic regression, will often discard any observations with missing values. CART, through the use of *surrogate splits*, can utilize all observations that have non-missing response values and at least one non-missing value for the predictors. Surrogate splits are essentially splits using other available features with non-missing values. The basic idea, which is fully described in Breiman et al. [1984, Sec. 5.3], is to estimate (or *impute*) the missing data point using the other available features.

Consider the decision stump in Figure 2.18, which corresponds to the optimal tree for the Swiss banknote data when using all available features.

What if we wanted to classify a new observation which had a missing value for diagonal? The surrogate approach finds *surrogate variables* for the missing splitter by building decision stumps, one for each of the other features (in this case, length, left, right, bottom, and top), to predict the binary response, denoted below by $y^\star$, formed by the original split:

$$y^\star = \begin{cases} 0 & \text{if diagonal} \geq 140.65 \\ 1 & \text{if diagonal} < 140.65 \end{cases}.$$

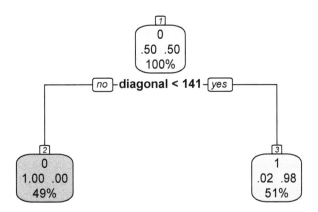

FIGURE 2.18: Decision stump for the Swiss banknote example.

For each feature, the optimal split is chosen using the procedure described in Section 2.2.1. (Note that when looking for surrogates, we do not bother to incorporate priors or losses since none are defined for $y^\star$.) In addition to the optimal split for each feature, we also consider the *majority rule*, which just uses the majority class. Once the surrogates have been determined, they're ranked in terms of misclassification error, and any surrogate that does worse than the majority class is discarded. Some implementations, like R's **rpart** package, further require surrogate splits to send at least two observations to each of the left and right child nodes.

Returning to the Swiss banknote example, let's find the surrogate splits for the primary split on `diagonal` depicted in Figure 2.18. We can find the surrogate splits using the same splitting process as before, albeit with our new target variable y^*:

```
bn2 <- treemisc::banknote  # load Swiss banknote data
bn2$y <- ifelse(bn2$diagonal >= 140.65, 1, 0)  # new target
bn2$diagonal <- NULL  # remove column
features <- c("length", "left", "right", "bottom", "top")
res <- sapply(features, FUN = function(feature) {
  find_best_split(bn2, x = feature, y = "y", n = nrow(bn2))
})
rownames(res) <- c("cutpoint", "gain")
res[, order(res["gain", ], decreasing = TRUE)]

#>              bottom    right     top     left    length
#> cutpoint     9.550  129.850  10.950  130.050  215.1500
#> gain         0.343    0.169   0.157    0.137    0.0344
```

In this case, the ranked surrogate splits—in descending order of importance—are `bottom`, `right`, `top`, `left`, and `length`; the corresponding split points for each are also shown in the output (we'll verify these results in Section 2.9 using real tree software). If we were to use the decision stump in Figure 2.18 to classify a new banknote with a missing value for `diagonal`, the tree would use the next best surrogate split instead (in this case, whether or not `bottom >= 9.550`).

For each surrogate, we could also compute its *agreement* and *adjusted agreement* with the primary split, which are used in **rpart**'s definition of *variable importance* (Section 2.8). The agreement between a primary split and its surrogate is just the proportion of observations they send in the same direction. The adjusted agreement adjusts this proportion by subtracting the number of observations sent one way or another using the majority rule. An example is given in Section 2.9.

2.7.1   Other missing value strategies

Aside from being able to handle missing predictor values directly, classification trees can be extremely useful in examining patterns of missing data [Harrell, 2015, Sec. 3.2]. For example, CART can be used to describe observations that tend to have missing values (a description problem). This can be done by growing a classification tree using a target variable that's just a binary indicator for whether or not a variable of interest is missing; see Harrell [2015, pp. 302–304] for an example using real data in R.

It can also be informative to construct missing value indicators for each predictor under consideration. Imagine, for example, that you work for a bank and that part of your job is to help determine who should be denied for a loan and who should not. A missing credit score on a particular loan application might be an obvious red flag, and indicative of somebody with a bad credit history, hence, an important indicator in determining whether or not to approve them for a loan. A similar strategy for categorical variables is to treat missing values as an actual category. As noted in van Buuren [2018, Sec. 1.3.7], the missing value indicator method may have its uses in particular situations but fails as a generic method to handle missing data (e.g., it does not allow for missing data in the response and can lead to biased regression estimates across a wide range of scenarios).

Imputation—filling in missing values with a reasonable guess—is another common strategy, and trees make great candidates for imputation models (e.g., they're fully nonparametric and naturally support both classification and regression).

Using CART for the purpose of missing value imputation has been suggested by several authors; see van Buuren [2018, Sec. 3.5] for details and several references. A generally useful approach is to use CART to generate *multiple imputations* [van Buuren, 2018, Sec. 3.5] via the *bootstrap* method[j] (see Davison and Hinkley [1997] for an overview of different bootstrap methods); multiple imputation is now widely accepted as one of the best general methods for dealing with incomplete data [van Buuren, 2018, Sec. 2.1.2].

The basic steps are outlined in Algorithm 2.2; also see `?mice::cart` for details on its implementation in the **mice** package [van Buuren and Groothuis-Oudshoorn, 2021]. Here, it is assumed that the response y corresponds to the predictor with incomplete observations (i.e., contains missing values) and that the predictors correspond to the original predictors with complete information (i.e., no missing values).

As described in Doove et al. [2014] and van Buuren [2018, Sec. 3.5], this process can be repeated m times using the bootstrap to produce m imputed data sets. As noted in van Buuren [2018, Sec. 3.5], Algorithm 2.2 is a form of *predictive mean matching* [van Buuren, 2018, Sec. 3.4], where the "predictive mean" is instead calculated by CART, as opposed to a regression model. An example using CART for multiple imputation is provided in Section 7.9.3.

But what if you're using a decision tree as the model, and not just as a means for imputation: should you rely on surrogate splits or a different strategy,

---

[j]Unless stated otherwise, a bootstrap sample refers to a random sample of size N with replacement from a set of N observations; hence, some of the original obervations will be sampled more than once and some not at all.

Algorithm 2.2 Tree-based missing value imputation.

---

1) fit a decision tree (e.g., CART) to the complete observations;

2) find the terminal assigned to each observation with a missing y value;

3) for each missing y value, randomly draw an observed response value from the terminal node to which it's assigned (i.e., the complete response values from the learning sample that summarize the node) to use for the imputed value.

---

like imputation? Feelders [2000] suggests that imputation (especially multiple imputation), if done properly, tends to outperform trees based on surrogate splits. However, one should still consider whether or not the potential for improved performance outweighs the additional effort required in specifying an appropriate imputation scheme. Feelders further notes that with "...moderate amounts of missing data (say 10% or less) one can avoid generating imputations and just use surrogate splits."

2.8   Variable importance

In practice, it may be useful, or even necessary, to reduce the number of features in a model. One way to accomplish this is to rank them in some order of importance and use a subset of the top features. Loh [2012] showed that using a subset of only the important variables can lead to increased prediction accuracy. Reducing the number of features can also decrease model training time and increase interpretability. However, lack of a proper definition of "importance" has led to many variable importance measures being proposed; see Greenwell and Boehmke [2020] for some discussion and further references.

Decision trees probably offer the most natural model-specific approach to quantifying the importance of predictors. In a binary decision tree, at each node A, a single predictor is used to partition the data into two homogeneous groups. The chosen predictor is the one that maximizes (2.3). The relative importance of predictor x is the sum of the squared improvements over all internal nodes of the tree for which x was chosen as the primary splitter; see Breiman et al. [1984] for details. This idea also extends to regression trees and ensembles of decision trees, such as those discussed in Chapters 5–8.

When surrogate splits are enabled, they can be accounted for in the quantification of variable importance. In particular, a variable may appear in the tree more than once, either as a primary or surrogate splitter. The variable importance measure for a feature is the sum of the gains associated with each split for which it was the primary variable, plus the gains (adjusted for agreement) associated with each split for which it was a surrogate. The notation is a bit involved, but the interested reader is pointed to Loh and Zhou [2021, Sec. 3].

Including surrogate information can help improve interpretation when you have strongly correlated or redundant features. For instance, imagine two features x_1 and x_2 that are essentially redundant. If we only counted gains where each variable was a primary splitter, these two features would likely split the importance, with neither showing up as important as they should. Comparing the variable importance rankings with and without competing splitters (i.e., non-primary splitters) can also be informative. Variables that appear to be important, but rarely split nodes, are probably highly correlated with the primary splitters and contain very similar information.

The relative variable importance standardizes the raw importance values for ease of interpretation. The relative importance is typically defined as the percent improvement with respect to the most important predictor, and is often reported in statistical software. The relative importance of the most important feature is always 100%. So, if x_3 is the most important feature, and the relative importance of another feature, say x_5, is 83%, you can say that x_5 is roughly 83% as important as x_3.

It is well known, however, that CART-like variable importance scores are biased; see Loh and Zhou [2021] for a thorough (and more recent) review. According to Loh and Zhou, a variable importance procedure is said to be unbiased if all predictors have the same mean importance score when they are independent of the response. Solutions to CART's variable importance bias, which really stems from CART's split selection bias (Section 2.4.2), are discussed in several places throughout this book; see, for example, the discussion in Section 7.5.1.

2.9   Software and examples

Packages **rpart** and **tree** [Ripley, 2021] provide modern implementations of the CART algorithm in R, although **rpart** is recommended over **tree**, and so we won't be discussing the latter. The name **rpart** comes from the acronym for (binary) Recursive PARTitioning. Beyond simple classification and regression trees, **rpart** can also be used to model Poisson counts (e.g., the number of occurrences of some event per unit of time), and censored outcomes. Note

that **rpart** is extendable[k] and several R packages on CRAN extend **rpart**
in various ways. For example, **rpartScore** [Galimberti et al., 2012] can be
used to build classification trees for ordinal responses within the same CART-
like framework, and **rpart.LAD** [Dlugosz, 2020] can be used to fit regression
trees based on *least absolute deviation* [Breiman et al., 1984, Sec. 8.11]. The
treemisc package provides some utility functions to support **rpart**, for example, to implement pruning based on the 1-SE rule (Section 2.5.2.1). The
R package **treevalues** [Neufeld, 2022] can be used to construct confidence
intervals and p-values for the mean response within a node or the difference in
mean response between two nodes in a CART-like regression tree (built using
the package **rpart**); see [Neufeld et al., 2021] for details.

CART-like decision trees are implemented in many other open source languages as well. Scikit-learn's **sklearn.tree** module offers extensive decision
tree functionality, but doesn't support categorical features, unless they've been
numerically re-encoded. The **DecisionTree.jl** package[l] for Julia provides an
implementation of CART and random forest (Chapter 7), but is rather limited in terms of features, especially when compared to R and Python's tree
libraries. Decision trees are also implemented in Spark MLlib [Meng et al.,
2016], Spark's open-source distributed machine learning library.[m]

The following examples illustrate the basic use of **rpart** for building decision
trees. We'll confirm the results we computed manually in previous sections as
well as construct decision trees for new data sets.

An excellent case study using decision trees in R to identify email spam is
provided in Nolan and Lang [2015, Chapter 3].

2.9.1 Example: Swiss banknotes

In Section 2.2.2, I restricted my attention to just two predictors, `top` and
`bottom`, and walked through the steps of constructing a two-split tree by hand
(i.e., a tree with three terminal nodes). Here, I'll use the **rpart** package to
reconstruct the same tree and to confirm my previous split calculations.

By default, **rpart** uses the Gini splitting rule, equal misclassification costs,
and the observed class priors[n] when building a classification tree; hence, we do
not need to set any additional arguments (we'll do that in the next section).

---

[k]As described in the "User Written Split Functions" vignette; see `vignette("usercode", package = "rpart")` for details.
[l]`https://github.com/bensadeghi/DecisionTree.jl`.
[m]Spark has various interfaces, including R and Python; an example using R will be given in Section 7.9.5.
[n]Note that the balanced nature of these data is not very realistic, unless roughly half the Swiss banknotes truly are counterfeit. However, without any additional information about the true class priors, there's not much that can be done here.

However, for ease of interpretation, I'll re-encode the outcome y from 0/1 to Genuine/Counterfeit°:

```
library(rpart)

# Load the Swiss banknote data and re-encode the response
bn <- banknote
bn$y <- ifelse(bn$y == 0, "Genuine", "Counterfeit")

# Fit a CART-like tree using top and bottom as the only features
(bn.tree <- rpart(y ~ top + bottom, data = bn))
#> n= 200
#>
#> node), split, n, loss, yval, (yprob)
#>        * denotes terminal node
#>
#> 1) root 200 100 Counterfeit (0.5000 0.5000)
#>   2) bottom>=9.55 88    2 Counterfeit (0.9773 0.0227) *
#>   3) bottom< 9.55 112  14 Genuine (0.1250 0.8750)
#>     6) top>=11.1 17     4 Counterfeit (0.7647 0.2353) *
#>     7) top< 11.1 95     1 Genuine (0.0105 0.9895) *
```

Note that this is the same tree that was displayed in Figure 2.2 (p. 43). The output from printing an **"rpart"** object can seem intimidating at first, especially for large trees, so let's take a closer look. The output is split into three sections. The first section gives N, the number of rows in the learning sample (or root node). The middle section, starting with **node)**, indicates the format of the tree structure that follows. The last section, starting at **1)**, provides a a brief summary of the tree structure. All the nodes of the tree are numbered, with **1)** indicating the root node and lines ending with a * indicating the terminal nodes. The topology of the tree is conveyed through indented lines; for example, nodes **2)** and **3)** are nested within **1)**; the left and right child nodes for any node numbered x are always numbered $2x$ and $2x + 1$, respectively.

For each node we can also see the split that was used, the number of observations it captured, the deviance or loss (in this case, the number of observations misclassified in that node), the fitted value (in this case, the classification given to observations in that node), and the proportion of each class in the node. Take node **2)**, for example. This is a terminal node, the left child of node **1)**, and contains 88 of the $N = 200$ observations (two of which are genuine banknotes). Any observation landing in node **2)** will be classified as counterfeit with a predicted probability of 0.977.

°I could leave the response numerically encoded as 0/1, but then I would need to tell **rpart** to treat this as a classification problem by setting **method = "class"** in the call to rpart().

If you want even more verbose output, with details about each split, you can
use the **summary()** method:

```
summary(bn.tree)   # print more verbose tree summary

#> Call:
#> rpart(formula = y ~ top + bottom, data = bn)
#>   n= 200
#>
#>      CP nsplit rel error xerror    xstd
#> 1 0.84      0      1.00   1.14 0.0700
#> 2 0.09      1      0.16   0.19 0.0415
#> 3 0.01      2      0.07   0.12 0.0336
#>
#> Variable importance
#> bottom     top
#>     66      34
#>
#> Node number 1: 200 observations,     complexity param=0.84
#>   predicted class=Counterfeit  expected loss=0.5  P(node) =1
#>      class counts:   100    100
#>     probabilities: 0.500 0.500
#>   left son=2 (88 obs) right son=3 (112 obs)
#>   Primary splits:
#>       bottom < 9.55 to the right, improve=71.6, (0 missing)
#>       top    < 11   to the right, improve=30.7, (0 missing)
#>   Surrogate splits:
#>       top < 11   to the right, agree=0.685, adj=0.284, (0 split)
#>
#> Node number 2: 88 observations
#>   predicted class=Counterfeit  expected loss=0.0227  P(node) =0.44
#>      class counts:    86     2
#>     probabilities: 0.977 0.023
#>
#> Node number 3: 112 observations,     complexity param=0.09
#>   predicted class=Genuine      expected loss=0.125  P(node) =0.56
#>      class counts:    14    98
#>     probabilities: 0.125 0.875
#>   left son=6 (17 obs) right son=7 (95 obs)
#>   Primary splits:
#>       top    < 11.1 to the right, improve=16.40, (0 missing)
#>       bottom < 9.25 to the right, improve= 2.42, (0 missing)
#>
#> Node number 6: 17 observations
#>   predicted class=Counterfeit  expected loss=0.235  P(node) =0.085
#>      class counts:    13     4
#>     probabilities: 0.765 0.235
#>
#> Node nmsber 7: 95 observations
#>   predicted class=Genuine      expected loss=0.0105  P(node) =0.475
```

```
#>      class counts:    1    94
#>      probabilities: 0.011 0.989
```

Here, we can see each primary splitter, along with its corresponding split point and gain (i.e., a measure of the quality of the split). For example, using `bottom` `< 9.55` yielded the greatest improvement and was selected as the first primary split. The reported improvement (`improve=71.59091`) is $N \times \Delta\mathcal{I}(s, A)$, hence why it differs from the output of our previously defined `splits()` function, which just uses $\Delta\mathcal{I}(s, A)$; but you can check the math: $71.59091/200 = 0.358$, which is the same value we obtained by hand back in Section 2.2.2. Woot!

Before continuing, let's refit the tree using all available features:

```
summary(rpart(y ~ ., data = bn, method = "class"))

#> Call:
#> rpart(formula = y ~ ., data = bn, method = "class")
#>   n= 200
#>
#>      CP nsplit rel error xerror   xstd
#> 1 0.98      0      1.00   1.12 0.0702
#> 2 0.01      1      0.02   0.03 0.0172
#>
#> Variable importance
#> diagonal    bottom    right      top    left   length
#>       28        22       15       14      14        6
#>
#> Node number 1: 200 observations,    complexity param=0.98
#>    predicted class=Counterfeit  expected loss=0.5  P(node) =1
#>      class counts:    100    100
#>      probabilities: 0.500 0.500
#>    left son=2 (102 obs) right son=3 (98 obs)
#>    Primary splits:
#>        diagonal < 141   to the left,  improve=96.1, (0 missing)
#>        bottom   < 9.55 to the right, improve=71.6, (0 missing)
#>        right    < 130   to the right, improve=34.3, (0 missing)
#>        top      < 11    to the right, improve=30.7, (0 missing)
#>        left     < 130   to the right, improve=27.8, (0 missing)
#>    Surrogate splits:
#>        bottom < 9.25 to the right, agree=0.910, adj=0.816, (0 split)
#>        right  < 130   to the right, agree=0.785, adj=0.561, (0 split)
#>        top    < 11    to the right, agree=0.765, adj=0.520, (0 split)
#>        left   < 130   to the right, agree=0.760, adj=0.510, (0 split)
#>        length < 215   to the left,  agree=0.620, adj=0.224, (0 split)
#>
#> Node number 2: 102 observations
#>    predicted class=Counterfeit  expected loss=0.0196  P(node) =0.51
#>      class counts:    100      2
```

```
#>      probabilities: 0.980 0.020
#>
#> Node number 3: 98 observations
#>      predicted class=Genuine      expected loss=0  P(node) =0.49
#>        class counts:     0    98
#>      probabilities: 0.000 1.000
```

Using all the predictors results in the same decision stump that was displayed in Figure 2.18. As it turns out, the best tree uses a single split on the length of the diagonal (in mm) and only misclassifies two of the genuine banknotes in the learning sample. In addition to the chosen splitter, `diagonal`, we also see a description of the competing splits (four by default) and surrogate splits (five by default); note that these match the surrogate splits I found manually back in Section 2.7. For example, if I didn't include `diagonal` as a potential feature, then `bottom` would've been selected as the primary splitter because it gave the next best reduction to weighted impurity (`improve=71.59091`).

While the **rpart** package provides `plot()` and `text()` methods for plotting and labeling tree diagrams, respectively, the resulting figures are not as polished as those produced by other packages; for example, **rpart.plot** [Milborrow, 2021b] and **partykit** [Hothorn and Zeileis, 2021]. All the tree diagrams in this chapter were constructed using a simple wrapper function around `rpart.plot()` called `tree_diagram()`, which is part of **treemisc**; see `?rpart.plot::rpart.plot` and `?treemisc::tree_diagram` for details. For example, the tree diagram from Figure 2.2 (p. 43) can be constructed using:

```
treemisc::tree_diagram(bn.tree)
```

Figure 2.19 shows a tree diagram depicting the primary split (left) as well as the second best surrogate split (right). In the printout from `summary()`, we also see the computed agreement and adjusted agreement for each surrogate. From Figure 2.19, we can see that the surrogate sends $(66 + 91))/200 \approx 0.785$ of the observations in the same direction as the primary split (agreement). The majority rule gets 102 correct, giving an adjusted agreement of $(66 + 91 - 102))/(200 - 102) \approx 0.561$.

2.9.2   Example: mushroom edibility

In this section, we'll use **rpart** to fit a classification tree to the mushroom data, and explore a bit more of the output and fitting process. Recall from Section 1.4.4, that the overall objective is to find a simple rule of thumb (if possible) for avoiding potentially poisonous mushrooms. For now, I'll stick with **rpart**'s defaults (e.g., the splitting rule is the Gini index), but set com-

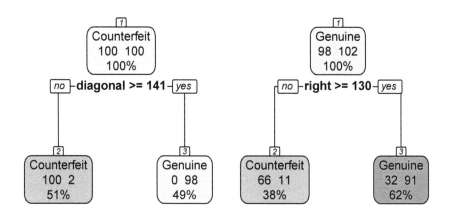

FIGURE 2.19: Decision stump for the Swiss banknote example (left) and one of the surrogate splits (right).

plexity parameter, *cp*, to zero (`cp = 0`) for a more complex tree.[P] Although the tree construction itself is not random, the internal cross-validation results are, so I'll also set the random number seed before calling **rpart()**:

```
mushroom <- treemisc::mushroom
```

```
# Fit a default tree with zero penalty on tree size
set.seed(1054)  # for reproducibility
(mushroom.tree <- rpart(Edibility ~ ., data = mushroom, cp = 0))
```

```
#> n= 8124
#>
#> node), split, n, loss, yval, (yprob)
#>       * denotes terminal node
#>
#>  1) root 8124 3920 Edible (0.51797 0.48203)
#>    2) odor=almond,anise,none 4328   120 Edible (0.97227 0.02773)
#>      4) spore.print.color=black,brown,buff,chocolate,orange,purple...
#>        8) stalk.color.below.ring=brown,gray,orange,pink,red,white ...
#>         16) stalk.color.below.ring=gray,orange,pink,red,white 4152...
#>           32) habitat=grasses,meadows,paths,urban,waste,woods 3952...
#>           33) habitat=leaves 200     8 Edible (0.96000 0.04000)
#>             66) cap.surface=smooth 192     0 Edible (1.00000 0.0000...
#>             67) cap.surface=grooves,scaly 8     0 Poison (0.00000 1...
```

[P]The default setting in **rpart** is cp = 0.01.

```
#>           17) stalk.color.below.ring=brown 80    16 Edible (0.80000 0...
#>             34) stalk.root=bulbous 64    0 Edible (1.00000 0.00000) *
#>             35) stalk.root=missing 16    0 Poison (0.00000 1.00000) *
#>         9) stalk.color.below.ring=yellow 24    0 Poison (0.00000 1....
#>       5) spore.print.color=green 72    0 Poison (0.00000 1.00000) *
#>     3) odor=creosote,fishy,foul,musty,pungent,spicy 3796    0 Poiso...
```

This is a complex tree with many splits, so let's use **treemisc**'s
`tree_diagram()` function to plot it (see Figure 2.20).

```
tree_diagram(mushroom.tree)  # Figure 2.20
```

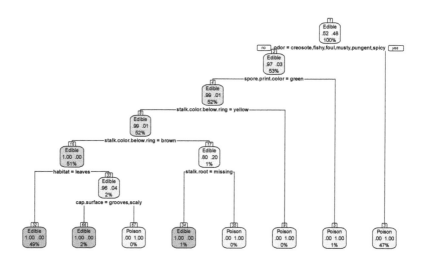

FIGURE 2.20: Decision tree diagram for classifying the edibility of mushrooms.

Setting `cp = 0` won't necessarily result in the most complex or saturated tree.
This is because `rpart()` sets a number of additional parameters by default,
many of which help control the maximum size of the tree; these are printed
below for the current `mushroom.tree` object. For instance, `minsplit`, which
defaults to 20, controls the number of observations that must exist in a node
before a split can be attempted.

```
unlist(mushroom.tree$control)
```

```
#>        minsplit        minbucket              cp
#>              20                7               0
#>      maxcompete      maxsurrogate     usesurrogate
#>               4                5               2
#> surrogatestyle        maxdepth            xval
```

```
#>                   0              30              10
```

You can change any of these parameters via `rpart()`'s `control` argument, or by passing them directly to `rpart()` via the ... (pronounced dot-dot-dot) argument.[q] For example, the two calls to `rpart()` below are equivalent. Each one fits a classification tree but changes the default complexity parameter from 0.01 to 0 (`cp = 0`) and the number of internal cross-validations from ten to five (`xval = 5`); see `?rpart::rpart.control` for further details about all the configurable parameters.

```
ctrl <- rpart.control(cp = 0, xval = 5)  # can also be a names list
tree <- rpart(Edibility ~ ., data = mushroom, control = ctrl)
tree <- rpart(Edibility ~ ., data = mushroom, cp = 0, xval = 5)
```

Another useful option in `rpart()` is the `parms` argument, which controls how nodes are split in the tree[r]; it must be a named list whenever supplied. Below we print the `tree$parms` component, which in this case is a list containing the class priors, loss matrix, and impurity function used in constructing the tree.

```
mushroom.tree$parms
```

```
#> $prior
#>     1     2
#> 0.518 0.482
#>
#> $loss
#>      [,1] [,2]
#> [1,]    0    1
#> [2,]    1    0
#>
#> $split
#> [1] 1
```

The `$prior` component defaults to the class frequencies in the root node, which can easily be verified:

```
proportions(table(mushroom$Edibility))  # observed class proportions
```

```
#>
#> Edible Poison
#>  0.518  0.482
```

The loss matrix, given by component `$loss`, defaults to equal losses for false positives and false negatives (the off diagonals); there's no loss associated with a correct classification (i.e., the diagonal entries are always zero).

---

[q]In R, functions can have a special ... argument that allows them to take any number of additional arguments; see Wickham [2019, Sec. 6.6] for details.

[r]The `parms` argument only applies to response variables that are categorical (classification trees), counts (Poisson regression), or censored (survival analysis)

The `$split` component displays either a 1 (`split = "gini"`) or 2 (`split = "information"`)[s] (partial matching is allowed). All of these can be changed from their respective defaults by passing a named list to the **parms** argument in the call to `rpart()`. For example, to use the entropy splitting rule[t], run the following:

```
parms <- list("split" = "information")  # use cross-entropy split rule
rpart(Edibility ~ ., data = mushroom, parms = parms)
```

Specifying a loss matrix in **rpart** isn't well-documented, unfortunately. For binary outcomes, the matrix has the following structure:

$$L = \begin{bmatrix} TP & FP \\ FN & TN \end{bmatrix},$$

where rows represent the observed classes and columns represent the assigned classes. Here, TP, FP, FN, and TN stand for *true positive*, *false positive*, *false negative*, and *true negative*, respectively; for example, a false negative is the case in which the tree misclassifies a 1 as a 0. The order of the rows/columns correspond to the same order as the categories when sorted alphabetically or numerically.

Since there is no cost for correct classification, we take $TP = TN = 0$. Setting $FP = FN = c$, for some constant c (i.e., treat FPs and FNs equally), will always result in the same splits (although, the internal statistics used in selecting the splits will be scaled differently). When misclassification costs are not equal, specify the appropriate values in the loss matrix. For example, the following tree would treat false negatives (i.e., misclassifying poisonous mushrooms as edible) as five times more costly than false positives (i.e., misclassifying edible mushrooms as poisonous). We could also obtain the same tree by computing the altered priors based on this loss matrix and supplying them via the **parms** argument, but this is left as an exercise to the reader.

```
levels(mushroom$Edibility)  # inspect order of levels
(loss <- matrix(c(0, 5, 1, 0), nrow = 2))  # loss matrix
rpart(Edibility ~ ., data = mushroom, parms = list("loss" = loss))
```

The variable importance scores (Section 2.8) are contained in the `$variable.importance` component of the `mushroom.tree` object; they're also printed at the top of the output from `summary()`, but rescaled to sum to 100.

---

[s]In **rpart**, setting `split = "information"` corresponds to using the cross-entropy split rule discussed in Section 2.2.1.

[t]Users can also supply their own custom splitting rules. The steps for doing so are well documented in **rpart**'s vignette on "User Written Split Functions": `utils::vignette("usercode", package = "rpart")`.

```
mushroom.tree$variable.importance
```

```
#>                   odor        spore.print.color
#>               3823.407                 2834.187
#>             gill.color stalk.surface.above.ring
#>               2322.460                 2035.816
#> stalk.surface.below.ring                ring.type
#>               2030.555                 2026.526
#>    stalk.color.below.ring               stalk.root
#>                 53.933                   25.600
#>    stalk.color.above.ring               veil.color
#>                 17.546                   16.315
#>            cap.surface                cap.color
#>                 15.360                   14.032
#>                habitat                cap.shape
#>                 13.409                    3.840
#>         gill.attachment
#>                  0.585
```

In many cases, predictors that weren't used in the tree will have a non-zero importance score. The reason is that surrogate splits are also incorporated into the calculation. In particular, a variable may effectively appear in the tree more than once, either as a primary or surrogate splitter. The variable importance measure for a particular feature is the sum of the gains associated with each split for which it was the primary variable, plus the gains (adjusted for agreement) associated with each split for which it was a surrogate. You can turn off surrogates by setting `maxsurrogate = 0` in `rpart.control()`.

How does k-fold cross-validation (Section 2.5.2) in **rpart** work? The `rpart()` function does internal 10-fold cross-validation by default. According to **rpart**'s documentation, 10-fold cross-validation is a reasonable default, and has been shown to be very reliable for screening out "pure noise" features. The number of folds (k) can be changed, however, using the `xval` argument in `rpart.control()`.

You can visualize the cross-validation results of an `"rpart"` object using `plotcp()`, as illustrated in Figure 2.21 for the `mushroom.tree` object. A good rule of thumb in choosing `cp` for pruning is to use the leftmost value for which the average cross-validation score lies below the horizontal line; this coincides with the 1-SE rule discussed in Section 2.5.2.1. The columns labeled `"xerror"` and `"xstd"` provide the cross-validated risk and its corresponding standard error, respectively (Section 2.5).

```
plotcp(mushroom.tree, upper = "splits", las = 1)  # Figure 2.21
mushroom.tree$cptable  # print cross-validation results
```

```
#>        CP nsplit rel error    xerror      xstd
#> 1 0.96936      0   1.00000  1.000000  0.011501
#> 2 0.01839      1   0.03064  0.030644  0.002777
```

```
#> 3 0.00613        2     0.01226 0.012257 0.001764
#> 4 0.00204        3     0.00613 0.006129 0.001249
#> 5 0.00102        5     0.00204 0.002043 0.000722
#> 6 0.00000        7     0.00000 0.000511 0.000361
```

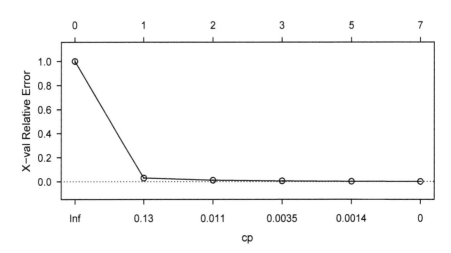

FIGURE 2.21: Cross-validation results from the fitted `mushroom.tree` object. A good choice of *cp* for pruning is often the leftmost value for which the mean lies below the horizontal line; this corresponds to the optimum value based on the 1-SE rule.

Don't be confused by the fact that the *cp* values between `printcp()` (and hence the `$cptable` component of an `"rpart"` object) and `plotcp()` don't match. The latter just plots the geometric means of the CP column listed in `printcp()` (these relate to the β_i values used in the *k*-fold cross-validation procedure described in Section 2.5). Any *cp* value between two consecutive rows will produce the same tree. For instance, any *cp* value between 0.002 and 0.001 will produce a tree with five splits. Also, these correspond to a scaled version of the complexity values α_i from Section 2.5. Note that **rpart** scales the CP column, as well as the error columns, by a factor inversely proportional to the risk of the root node, so that the associated training error (`"rel error"`) for the root node is always one (i.e., the first row in the table); which in this case is $1/(3916/8124) \approx 2.075$. Dividing through by this scaling factor should return the raw α_i values; the first three correspond to the values I computed by hand back in Section 2.5:

```
mushroom.tree$cptable[1L:3L, "CP"] / (8124 / 3916)
```

```
#>        1        2        3
```

```
#> 0.46726 0.00886 0.00295
```

Consequently, setting `cp = 1` will always result in a tree with no splits. The default, `cp = 0.01`, has been shown to be useful at "pre-pruning" the trees in a way such that the cross-validation step results in only the removal of 1–2 layers, although it can also occasionally over-prune. In practice, it seems best to set `cp = 0`, or some other number smaller than the default, and use the cross-validation results to choose an optimal subtree.

Using the 1-SE rule would suggest a tree with 5, or possibly 7, splits. However, since our main objective is to construct a simple rule-of-thumb for classifying the edibility of mushrooms, it seems like the simpler model with only a single split (i.e., a decision stump) will suffice; it only misclassifies 3% of the poisonous mushrooms as edible. To prune an **rpart** tree, use the `prune()` function with a specified value of the complexity parameter:

```
tree_diagram(prune(mushroom.tree, cp = 0.1))  # Figure 2.22
```

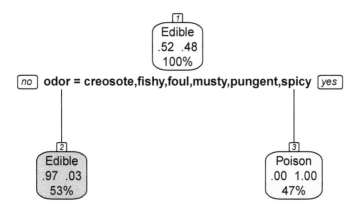

FIGURE 2.22: Decision tree diagram for classifying the edibility of mushrooms; in this case, the result is a decision stump.

The tree diagram displayed in Figure 2.22 provides us with a handy rule of thumb for classifying mushrooms as either edible or poisonous. If the mushroom smells fishy, foul, musty, pungent, spicy, or like creosote, it's likely poisonous. In other words, if it smells bad, don't eat it!

2.9.3   Example: predicting home prices

In this section, I'll use **rpart** to build a regression to the Ames housing data (Section 1.4.7). I'll also show how to easily prune an **rpart** tree using the 1-SE rule via **treemisc**'s `prune_se()` function. The code chunk below loads in the data before splitting it into train/test sets using a 70/30 split:

```
ames <- as.data.frame(AmesHousing::make_ames())
ames$Sale_Price <- ames$Sale_Price / 1000  # rescale response
set.seed(2101)  # for reproducibility
trn.id <- sample.int(nrow(ames), size = floor(0.7 * nrow(ames)))
ames.trn <- ames[trn.id, ]  # training data/learning sample
ames.tst <- ames[-trn.id, ]  # test data
```

Next, I'll intentionally fit an overgrown regression tree by setting `cp` = 0; in a regression tree, **rpart** will not attempt a split unless it increases the overall R^2 by `cp`, so setting `cp` = 0 will cause the tree to continue splitting until some other stopping criterion is met, such as minimum node size (in **rpart**, the default minimum number of observations that must exist in a node in order for a split to be attempted is 20). I'll also compare the RMSE between the train and test sets:

```
library(rpart)
library(treemisc)  # for prune_se() function

# Fit a regression tree with no penalty on complexity
set.seed(1547)  # for reproducibility
ames.tree <- rpart(Sale_Price ~ ., data = ames.trn, cp = 0)

rmse <- function(pred, obs) {  # computes RMSE
  sqrt(mean((pred - obs) ^ 2))
}

# Compute train RMSE
rmse(predict(ames.tree, newdata = ames.trn), obs = ames.trn$Sale_Price)
```
```
#> [1] 23.4
```
```
# Compute test RMSE
rmse(predict(ames.tree, newdata = ames.tst), obs = ames.tst$Sale_Price)
```
```
#> [1] 31.4
```

The tree is likely overfitting, as indicated by the relatively large discrepancy between the train and test RMSE. Let's see if pruning the tree can help. The `prune_se()` function from **treemisc** can be used to prune **rpart** trees using the 1-SE rule, as illustrated below:

```
ames.tree.1se <- prune_se(ames.tree, se = 1)  # prune using 1-SE rule

# Train RMSE on pruned tree
```

```
rmse(predict(ames.tree.1se, newdata = ames.trn),
     obs = ames.trn$Sale_Price)
```

```
#> [1] 29.5
```

```
# Test RMSE on pruned tree
rmse(predict(ames.tree.1se, newdata = ames.tst),
     obs = ames.tst$Sale_Price)
```

```
#> [1] 34.1
```

A smaller discrepancy, but the pruned tree is slightly less accurate than the unpruned tree on the test set. So did pruning really help here? It depends on how you look at it. Both trees are displayed in Figure 2.23 without text or labels. The unpruned tree has 169 splits while pruning with the 1-SE rule and cross-validation resulted in a subtree with only 33 splits—a much more parsimonious tree

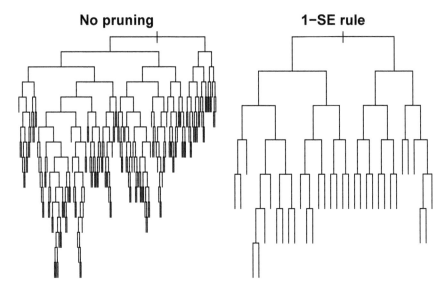

FIGURE 2.23: Regression trees for the Ames housing example. Left: unpruned regression tree. Right: pruned regression tree using 10-fold cross-validation and the 1-SE rule.

Even with pruning, we still ended up with a subtree that is too complex to easily interpret. In situations like this, it can be helpful to use different post hoc interpretability techniques to help the end user interpret the model in a way more understandable for humans. For instance, it can be quite informative to look at a plot of variable importance scores, like the Cleveland dot plot displayed in Figure ??; here, the importance scores are scaled to sum to 1 (see the code chunk below). From the results, we can see that the overall quality of the home (`Overall_Qual`), neighborhood (`Neighborhood`), and basement quality (`Bsmt_Qual`) were some of the key features used to partition the data into groups of homes with similar sale prices.

```
vi <- sort(ames.tree.1se$variable.importance, decreasing = TRUE)
vi <- vi / sum(vi)  # scale to sum to 1
dotchart(vi[1:10], xlab = "Variable importance", pch = 19)
```

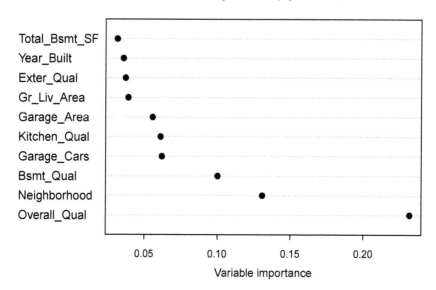

FIGURE 2.24: Variable importance plot for the top ten predictors in the pruned decision tree for the Ames housing data.

While variable importance plots can be useful, they don't tell us anything about the nature of the relationship between each feature and the predicted outcome. For instance, how does the above grade square footage (`Gr_Liv_Area`) impact the predicted sale price on average? This is precisely what *partial dependence plots* (PDPs) can tell us; see Section 6.2.1. In a nutshell, PDPs are low-dimensional graphical renderings of the prediction function so that the relationship between the outcome and predictors of interest can be more easily understood. PDPs, along with other interpretability techniques, are discussed in more detail in Chapter 6. For now, I'll just introduce the **pdp** package [Greenwell, 2021b], and show how it can be used to help visualize the relationship between above grade square footage and the predicted sale price:

```
library(ggplot2)
library(pdp)

# Compute partial dependence of predicted Sale_Price on Gr_Liv_Area
pd <- partial(ames.tree.1se, pred.var = "Gr_Liv_Area")
```

```
autoplot(pd, rug = TRUE, train = ames.trn) +   # Figure 2.25
  ylab("Partial dependence")
```

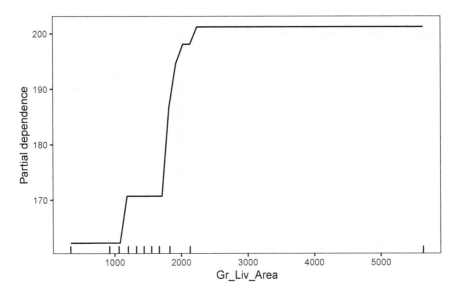

FIGURE 2.25: Partial dependence of `Sale_Price` on `Gr_Liv_Area` from the pruned regression tree. There's evidence of a monotonic increasing relationship between above grade square footage (`Gr_Liv_Area`) and the predicted sale price, while accounting for the average effect of all the other predictors.

Note that the *y*-axis is on the same scale as the response, and in this case, represents the averaged predicted sale price across the entire learning sample for a range of values of `Gr_Liv_Area`. The rug display (or 1-dimensional plot) on the *x*-axis shows the distribution of `Gr_Liv_Area` in the training data, with a tick mark at the min/max and each *decile*. As you would expect, larger size homes are associated with higher average predicted sales. Full details on the **pdp** package are given in Greenwell [2017].

Decision trees, especially smaller ones, can be rather self-explanatory. However, it is often the case that a usefully discriminating tree is too large to interpret by inspection. Variable importance scores, PDPs, and other interpretibility techniques, can be used to help understand any tree, regardless of size or complexity; these techniques are even more critical for understanding the output from more complex models, like the tree-based ensembles discussed in Chapters 5–8.

2.9.4   Example: employee attrition

In this example, I'll revisit the employee attrition data (Section 1.4.6) and build a classification tree using **rpart** with altered priors to help understand drivers of employee attrition and confirm my previous calculations from Section 2.2.4.1.

Figure 2.6 showed two classification trees for the employee attrition data, one using the default priors and one with altered priors based on a specific loss matrix with unequal misclassification costs. In **rpart**, you can specify the loss matrix, priors, or both—it's quite flexible!

The next code chunk fits three depth-two classification trees to the employee attrition data. The first tree (**tree1**) assumes equal misclassification costs and uses the default (i.e., observed) class priors. The other two trees use different, but equivalent, approaches: **tree2** uses the previously defined loss matrix from Section 2.2.4.1, while **tree3** uses the associated altered priors I computed by hand back in Section 2.2.4.1. Although the internal statistics used in constructing each tree differ slightly, both trees are equivalent in terms of splits and will make the same classifications. The resulting tree diagrams are displayed in Figure 2.26.

```
data(attrition, package = "modeldata")

# Fit classification trees with default priors and costs
set.seed(904)  # for reproducibility
tree1 <- rpart(Attrition ~ OverTime + MonthlyIncome, data = attrition,
               maxdepth = 2, cp = 0)

# Specify unequal misclassification costs
loss <- matrix(c(0, 8, 1, 0), nrow = 2)
tree2 <- rpart(Attrition ~ OverTime + MonthlyIncome, data = attrition,
               maxdepth = 2, cp = 0, parms = list("loss" = loss))

# Equivalent approach using altered priors
tree3 <- rpart(Attrition ~ OverTime + MonthlyIncome, data = attrition,
               maxdepth = 2, cp = 0,
               parms = list("prior" = c(1 - 0.6059444, 0.6059444)))

# Display trees side by side (Figure 2.26)
par(mfrow = c(1, 3))
tree_diagram(tree1)  # default costs and priors
tree_diagram(tree2)  # unequal costs
tree_diagram(tree3)  # altered priors
```

The subtle difference between **tree2** and **tree3** is that the within-node class proportions for **tree2** are not adjusted for cost/loss; hence, the predicted class probabilities will not match between the two trees. In essence, the tree based on the loss matrix (**tree2**) makes classifications using a predicted probability

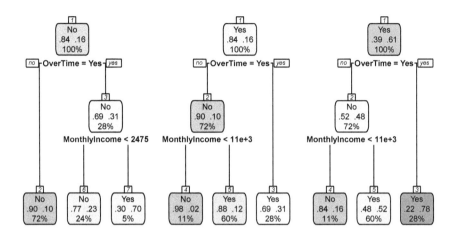

FIGURE 2.26: Decision trees fit to the employee attrition data set. Left: default priors and equal costs. *Middle:* Unequal costs specified via a loss matrix. Right: altered priors equivalent to the costs associated with the middle tree.

threshold different from 0.5^{u} (i.e., classification is no longer based on a majority vote). On the other hand, in the altered priors tree (**tree3**), the expected class proportions are adjusted so that classification is still done via a majority vote (i.e., if A is the far right terminal node, then any observation in A would be classified as **Yes** since $p_{yes}(A) = 0.779 > 0.5$.

To summarize this particular example, specifying a loss matrix produced a tree and adjusted the probability threshold used for classification, whereas using altered priors produced a tree and adjusted the predicted probabilities to keep the threshold at 0.5; but both are equivalent in terms of the classifications they'll assign to new observations.

Note that the order of the rows and columns of the loss matrix, and the order of the priors, is the same as the sorted order for the response categories. For instance, in the matrix below, the first row/column corresponds to class **No** because it appears first in alphabetical order:

```
levels(attrition$Attrition)  # can be changed with relevel()

#> [1] "No"  "Yes"

matrix(c(0, 8, 1, 0), nrow = 2)
```

---

[u]In general, the default probability threshold for classification is $1/J$, where J is the number of classes.

```
#>       [,1] [,2]
#> [1,]    0    1
#> [2,]    8    0
```

Although I restricted each tree to a max depth of two, the tree on the far left side of Figure 2.26 actually coincides with the optimal tree I would've obtained using 10-fold cross-validation and the 1-SE rule (assuming unequal misclassification costs, of course), and shows that having to work overtime, as well as having a lower monthly income, was associated with the highest predicted probability of attrition ($p = 0.70$)

To further illustrate pruning using the 1-SE rule in **rpart**, let's look at a tree based on altered priors (**tree3**) with the full set of features. Below, I refit the same altered priors tree using all available features to maximum depth (i.e., intentionally overgrow the tree); the cross-validation results and pruned tree diagram are displayed in Figure 2.27 (p. 109). The top plot in Figure 2.27 shows the cost-complexity pruning results as a function of the number of splits (top axis). If I were to prune using the 1-SE rule, I would select the tree corresponding to the point farthest to the left that's below the horizontal line (the horizontal line corresponds to 1-SE above the minimum error). In this case, we would end up with a tree containing just four splits, as seen in the bottom plot in Figure 2.27.

```
library(rpart)

# Saturated tree with altered priors using all predictors
att.cart <-
  rpart(Attrition ~ ., data = attrition, cp = 0, minsplit = 2,
        parms = list("prior" = c(1 - 0.6059444, 0.6059444)))

# Plot pruning results (Figure 2.27)
par(mfrow = c(2, 1))
plotcp(att.cart, upper = "splits")
(att.cart.1se <- prune_se(att.cart, se = 1))

#> n= 1470
#>
#> node), split, n, loss, yval, (yprob)
#>       * denotes terminal node
#>
#>  1) root 1470 579.0 Yes (0.394 0.606)
#>    2) OverTime=No 1054 413.0 No (0.518 0.482)
#>      4) TotalWorkingYears>=2.5 966 304.0 No (0.577 0.423)
#>        8) StockOptionLevel>=0.5 568 117.0 No (0.684 0.316) *
#>        9) StockOptionLevel< 0.5 398 163.0 Yes (0.465 0.535)
#>         18) JobRole=Healthcare_Representative,Human_Resources,Mana...
#>         19) JobRole=Laboratory_Technician,Research_Scientist,Sales...
#>      5) TotalWorkingYears< 2.5 88  27.7 Yes (0.203 0.797) *
```

```
#>    3) OverTime=Yes 416 136.0 Yes (0.221 0.779) *
tree_diagram(att.cart.1se, tweak = 0.8)
```

2.9.5   Example: letter image recognition

The goal of this example is to build an image classifier using a simple decision tree that incorporates additional information about the true class priors
for a multiclass outcome with $J = 26$ classes. The letter image recognition
data, which are available in R via the **mlbench** package [Leisch and Dimitriadou., 2021][v], contains 20,000 observations on 17 variables. Each observation
corresponds to a distorted black-and-white rectangular pixel image of a capital letter from the English alphabet (i.e., A–Z). A total of 16 ordered features
was derived from each image (e.g., statistical moments and edge counts) which
were then scaled to to be integers in the range 0–15. The objective is to identify
each letter using the 16 ordered features. Frey and Slate [1991] first analyzed
the data using *Holland-style adaptive classifiers*, and reported an accuracy of
just over 80%.

To start, I'll load the data and split it into train/test sets using a 70/30
split:

```
library(treemisc)  # for prune_se()

data(LetterRecognition, package = "mlbench")
lr <- LetterRecognition  # shorter name for brevity
set.seed(1051)  # for reproducibility
trn.ids <- sample(nrow(lr), size = 14000, replace = FALSE)
lr.trn <- lr[trn.ids, ]  # training data
lr.tst <- lr[-trn.ids, ]  # test data
```

Next, I'll use `rpart()` to fit a classification tree that's been pruned using the
1-SE rule with 10-fold cross-validation, and see how accurate it is on the test
sample:

```
set.seed(1703)  # for reproducibility
lr.cart <- rpart(lettr ~ ., data = lr.trn, cp = 0, xval = 10)
lr.cart <- prune_se(lr.cart, se = 1)  # prune using 1-SE rule

# Compute accuracy on test set
pred <- predict(lr.cart, newdata = lr.tst, type = "class")
sum(diag(table(pred, lr.tst$lettr))) / length(pred)
#> [1] 0.813
```

---

[v]The data are also available from the UCI Machine Learning Repository: `https:`
`//archive.ics.uci.edu/ml/datasets/Letter+Recognition`.

Nice, an overall accuracy of 81%—similar to the best accuracy reported in Frey and Slate [1991]. But don't be fooled, the data have been artificially balanced:

```
table(lr$lettr)
```

```
#>
#>    A   B   C   D   E   F   G   H   I   J   K   L   M
#> 789 766 736 805 768 775 773 734 755 747 739 761 792
#>    N   O   P   Q   R   S   T   U   V   W   X   Y   Z
#> 783 753 803 783 758 748 796 813 764 752 787 786 734
```

As noted in Matloff [2017, Sec. 5.8], this is unrealistic since some letters tend to appear much more frequently than others in English text. For example, assuming balanced classes, `rpart()` uses a prior probability for the letter "A" of $\pi_A = 1/26 \approx 0.0384$, when in fact π_A is closer to 0.0855.[w]

A proper analysis should take these frequencies into account, which is easy enough to do with classification trees. Fortunately, the correct letter frequencies are conveniently available in the **regtools** package [Matloff, 2019]. Below, I'll refit the model including the updated (and more realistic) class priors. I'll then sample the test data so that the resulting class frequencies more accurately reflect the true prior probabilities of each letter:

```
data(ltrfreqs, package = "regtools")

# Compute correct class priors
priors <- ltrfreqs$percent
priors <- priors / sum(priors)  # class priors should sum to 1
names(priors) <- ltrfreqs$ltr
priors <- priors[order(ltrfreqs$ltr)]

# Refit tree using correct priors
set.seed(1718)  # for reproducibility
lr.cart.priors <- rpart(lettr ~ ., data = lr.trn, cp = 0,
                        parms = list(prior = priors))
lr.cart.priors <- prune_se(lr.cart.priors, se = 1)

# Sample test set to reflect correct class frequencies
ltrfreqs2 <- ltrfreqs
names(ltrfreqs2) <- c("lettr", "prior")
ltrfreqs2$prior <- ltrfreqs2$prior / sum(ltrfreqs2$prior)
temp <- merge(lr.tst, ltrfreqs2)  # merge the two data sets
set.seed(1107)  # for reproducibility
lr.tst2 <- temp[sample(nrow(temp), replace = TRUE,
                       prob = temp$prior), ]
```

Finally, let's compare the two CART fits on the modified test set that reflects the more correct class frequencies:

---

[w]Based on the English letter frequencies reported at http://practicalcryptography.com.

```
pred2 <- predict(lr.cart, newdata = lr.tst2, type = "class")
pred3 <- predict(lr.cart.priors, newdata = lr.tst2, type = "class")
sum(diag(table(pred2, lr.tst2$lettr))) / length(pred2)
```

```
#> [1] 0.803
```

```
sum(diag(table(pred3, lr.tst2$lettr))) / length(pred3)
```

```
#> [1] 0.858
```

While the error of the original model based on equal priors decreased (albeit, not by much), the tree incorporating true prior information did much better. Woot!

2.10   Discussion

CART is one of the best off-the-shelf machine learning algorithms in existence, but it's not without its drawbacks. In closing, let's summarize many of the advantages and disadvantages of the CART algorithm for decision tree induction. While the emphasis here is on CART, much of this discussion equally applies to other decision tree algorithms as well (e.g., C4.5/C5.0).

2.10.1   Advantages of CART

Small trees are easy to interpret. Decision trees are often hailed as being simple and interpretable, relative to more complex algorithms. However, this is really only true for small trees with relatively few splits, like the one from Figure 2.22 (p. 95).

Trees scale well to large N. Individual decision trees scale incredibly well to large data sets, especially if most of the features are ordered, or categorical with relatively few categories. Even in the extreme cases, various shortcuts and approximations can be used to reduce computational burden (see, for example, Section 2.4).

The leaves form a natural clustering of the data. All observations that coinhabit a terminal node necessarily satisfied all the same conditions when traversing the tree; in this sense, the records within a terminal node should be similar with respect to the feature values and can be considered *nearest neighbors*. We'll revisit this idea in Section 7.6.

Trees can handle data of all types. Trees can naturally handle data of mixed types, and categorical features do not necessarily have to be numerically re-encoded, like in linear regression or neural networks. Trees are also invariant to monotone transformations of the predictors; that is, they only care about the rank order of the values of each ordered feature. For example, there's no need to apply logarithmic or square root transformations to any of the features like you might in a linear model.

Automatic variable selection. CART selects variables and splits one step at a time ("...like a carpenter that makes stairs"); hence the quote at the beginning of the chapter. If a variable cannot meaningfully partition the data into homogeneous subgroups, it will not likely be selected as a primary splitter. If it does, it'll likely get snipped off during the pruning phase.

Trees can naturally handle missing data. As discussed in Section 2.7, CART can naturally avoid many of the problems caused by missing data by using surrogate splits (i.e., back up splitters that can be used whenever missing values are encountered during prediction).

Trees are completely nonparametric. CART is fully nonparametric. It does not require any distributional assumptions, and the user does not have to specify any parametric form for the model, like in linear regression. It can also automatically handle nonlinear relationships (although it tends be quite biased since it uses step function to approximate potentially smooth surfaces) and interactions.

2.10.2   Disadvantages of CART

Large trees are difficult to interpret. Large tree diagrams, like the ones in Figure 2.23 (p. 97), can be difficult to interpret and are probably not very useful to the end user. Fortunately, various interpretbility techniques, like variable importance plots and PDPs, can help alleviate this problem. Such techniques are the topic of Chapter 6.

CART's splitting algorithm is quite greedy. CART makes splits that are locally optimal. That is, the algorithm does not look through all possible tree structures to globally optimize some performance metric; that would be unfeasible, even for a small number of features. Instead, the algorithm recursively partitions the data by looking for the next best split at each stage. This is analogous to the difference between *forward-stepwise selection* and *best-subset selection*. Greedy algorithms use a more constrained search and tend to have lower variance but often pay the price in bias. Chapters 5–8 discuss several strategies for breaking the bias-variance-tradeoff by combining many decision trees together.

Splits lower on the tree are less accurate. Data is essentially taken away after each split, making splits further down the tree less accurate (and noisy) compared to splits near the root node. This is part of the reason why binary splits are used in the first place. While some decision tree algorithms allow multiway splits (e.g., CHAID and C4.5/C5.0), this is not a good strategy in general as the data would be fragmented too quickly, and the search for locally optimal splits becomes more challenging.

Trees contain complex interactions. CART finds splits in a sequential manner, and all splits in the tree depend on any that came before it. Once a final tree structure has been identified, the resulting prediction equation can be written using a linear model in *indicator functions*. For example, the prediction equation for the tree diagram in Figure 2.9 (p. 63) can be written as follows:

$$\widehat{\texttt{Ozone}} = 75.41 \times I\left(\texttt{Temp} \geq 82.50\right) + 55.60 \times I\left(\texttt{Temp} < 82.50 \,\&\, \texttt{Wind} < 7.15\right)$$
$$+ 22.33 \times I\left(\texttt{Temp} < 82.50 \,\&\, \texttt{Wind} \geq 7.15\right),$$

where $I\left(\cdot\right)$ is the *indicator function* that evaluates to one whenever its argument is true, and zero otherwise. The right-hand side can be re-written more generally as $f\left(\texttt{Temp}\right) + f\left(\texttt{Temp}, \texttt{Wind}\right)$, where the second term explicitly models an interaction effect between `Temp` and `Wind`. As you can imagine, a more complex tree with a larger number of splits easily leads to a model with high-order interaction effects. The presence of high-order interaction effects can make interpreting the main effects (i.e., the effect of individual predictors) more challenging.

Biased variable selection. As briefly discussed in Section 2.4.2, CART's split selection strategy is biased towards features with many potential split points, such as categorical predictors with high cardinality. More contemporary decision tree algorithms, like those discussed in Chapters 3–4, are unbiased in this sense.

Trees are essentially step functions. Trees can have a hard time adapting to smooth and/or linear response surfaces. Recall the twonorm problem from Section 1.1.2.5, where the optimal decision boundary is linear. I fit an pruned **rpart** tree to the same sample using 10-fold cross-validation and the 1-SE rule; the resulting decision boundary (along with the optimal Bayes rule) is displayed in Figure 2.28 (p. 110). Of course, I could increase the number of splits resulting in smaller steps, but in practice this often leads to overfitting and poor generalizability. This lack of smoothness causes more problems in the regression setting.

Trees are noisy. A common criticism of decision trees is that they are considered *unstable predictors*; this was also noted in the original CART monograph; see Breiman et al. [1984, Section 5.5.2]. By unstable, I mean high variance or, in other words, the tree structure (and therefore predictions) can vary, often wildly, from one sample to the next. For example, at any node in a particular tree, there may be several competing splits that result in nearly the same decrease in node impurity and different samples may lead to different choices among these similar performing split contenders.

To illustrate, let's look at six independent samples of size $N = 3,220$ from the email spam data described in Section 1.4.5 ($\approx 70\%$ training sample), and fit a CART-like tree to each using a maximum of four splits. The results are displayed in Figure 2.29. Note the difference in structure and split variables across the six trees.

Fortunately, tree stability isn't always as problematic as it sounds (or looks). According to Zhang and Singer [2010, p. 57], "...the real cause for concern [in practice] regarding tree stability is the psychological effect of the appearance of a tree." Even though the structure of the tree can vary from sample to sample, Breiman et al. [1984, pp. 156–159] argued that competitive trees, while differing in appearance, can give fairly stable and consistent predictions. Strategies for improving the stability and performance of decision trees are discussed in Chapters 5–8.

Instability is not a feature specific to trees, though. For example, traditional model selection techniques in linear regression—like forward selection, backward elimination, and hybrid variations thereof—all suffer from the same problem. However, averaging can improve the accuracy of unstable predictors, like overgrown decision trees, through variance reduction [Breiman, 1996a]; more on this in Chapter 5.

2.11 Recommended reading

First and foremost, I highly recommend reading the original CART monograph [Breiman et al., 1984]. For a more approachable and thorough discussion of CART, I'd recommend Berk [2008, Chap. 3] (note that there's now a second edition of this book). I also recommend reading the vignettes accompanying the **rpart** package; R users can launch these from an active R session, as mentioned throughout this chapter, but they're also available from **rpart**'s CRAN landing page: https://cran.r-project.org/package=rpart. Scikit-learn's **sklearn.tree** module documentation is also pretty solid: https://scikit-learn.org/stable/modules/tree.html. There's also a fantastic talk by Dan Steinberg about CART, called "Data Science Tricks With the Single Decision Tree," that can be found on YouTube:

https://www.youtube.com/watch?v=JVbU_tS6zKo&feature=youtu.be

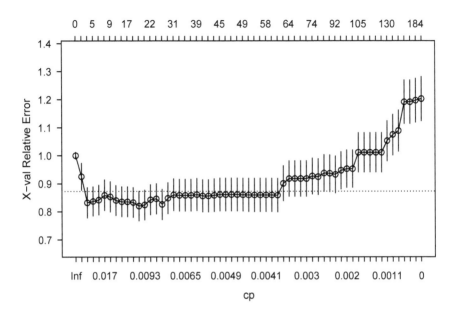

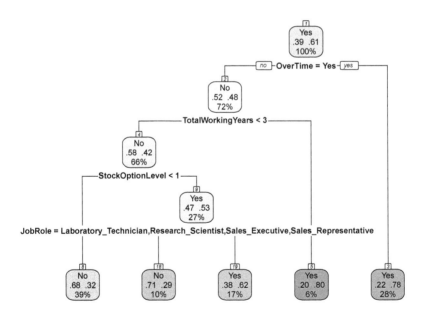

FIGURE 2.27: Employee attrition decision tree with altered priors. Top: 10-fold cross-validation results. Bottom: pruned tree using the 1-SE rule (which corresponds to the left-most point in the 10-fold cross-validation results that lies beneath the horizontal dotted line).

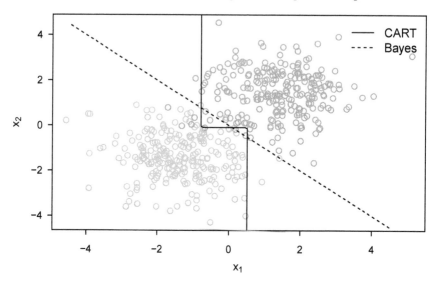

FIGURE 2.28: Decision boundaries for the twonorm benchmark example. In this case, the Bayes decision boundary is linear (dashed line). Also shown is the decision boundary from the pruned classification tree (solid line). The axis-oriented nature of decision tree splits makes it difficult to adapt to linear or smooth decision surfaces.

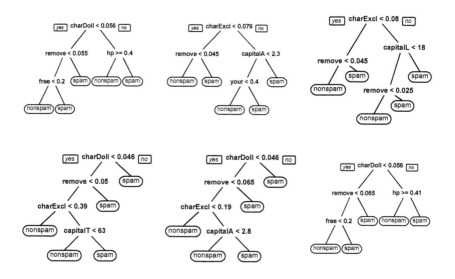

FIGURE 2.29: CART-like trees applied to six independent samples from the email spam data; for plotting purposes, each tree is restricted to just four splits.

3

Conditional inference trees

The trees that are slow to grow bear the best fruit.

Moliere

The abundance of different tree algorithms makes it rather difficult to keep up with current developments. While CART-like decision trees (Chapter 2) are still popular for classification and regression problems, there have been numerous developments and improvements over exhaustive search procedures, like CART, especially from the year 2000 onward. In this chapter, I'll discuss one of the more important developments: unbiased recursive partitioning via *conditional inference*.

3.1 Introduction

As discussed in Section 2.4.2, CART has a bias towards selecting variables with many potential split points. These issues also extend to C4.5 and C5.0 (which are discussed in the online complements), or any other recursive partitioning procedure that performs an exhaustive search over all possible splits to maximize some measure of node impurity.

To illustrate, consider a regression scenario where none of the predictors have any association to the response. Following Loh [2014] (see his rejoiner to the discussions), I'll look at 5,000 simulated data sets, each of which consisted of 100 observations on the following eight variables:

- y: a $N(0,1)$ random variable (the response variable);
- ch2: a $\chi^2(2)$ random variable;

- m2: a random factor with 2 equiprobable categories;

- m4: a random factor with 4 equiprobable categories;

- m10: a random factor with 10 equiprobable categories;

- m20: a random factor with 20 equiprobable categories;

- nor: an $N(0,1)$ random variable;

- uni: a $\mathcal{U}(0,1)$ random variable.

Keep in mind that none of the seven predictors have any relationship to the response. For each simulated data set, two types of regression trees were fit, each with a single split (i.e., a decision stump):

- a CART-like decision tree, as discussed in Chapter 2;

- a *conditional inference tree* (CTree), as will be discussed later in this chapter (see Section 3.4).

Figure 3.1 shows the frequency with which each variable was selected as the primary splitter in each scenario. Notice how CART selects m20, the categorical predictor with 20 equiprobable categories, more than 80% of the time, and m10, the categorical predictor with 10 equiprobable categories, roughly 10% of the time! If CART's split variable selection strategy was unbiased, we would expect each feature to be selected roughly $1/7 \approx 14.29\%$ of the time. CTree, on the other hand, which uses an unbiased split variable selection procedure, selects each feature with a roughly equal frequency close to $1/7$. While this doesn't necessarily imply that CART is less accurate, it does make interpretation more difficult (e.g., variable importance plots will be biased).

3.2 Early attempts at unbiased recursive partitioning

The basic idea behind unbiased recursive partitioning is to separate the split search into two sequential steps: 1) selecting the primary split variable, then 2) selecting an optimal split point. Typically, the primary splitter is selected first by comparing appropriate statistical tests (e.g., a chi-square test if both X and Y are nominal, or a correlation-type test if X and Y are both univariate continuous). Once a splitting variable has been identified, the optimal split point can be found using any number of approaches, including further statistical tests, or the impurity measures discussed in Section 2.2.1.

The idea of using statistical tests for split variable selection in recursive partitioning is not new. In fact, the CTree algorithm discussed later in this chapter

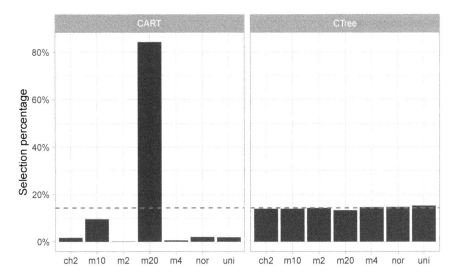

FIGURE 3.1: Split variable selection frequencies for two regression tree procedures (CART and CTree); the y-axis shows the proportion of times each feature—all of which are unrelated to the response—was selected to split the root node in 5,000 Monte Carlo simulations. A horizontal dashed red line is given at $1/7$, the frequency corresponding to unbiased split variable selection.

was inspired by a number of approaches that came before it, like CHAID (Section 1.2.1). While these algorithms help to reduce variable selection bias compared to exhaustive search techniques like CART, most of them only apply to special circumstances. CHAID, for example, requires both the features and the response to be categorical, although, CHAID was eventually extended to handle ordered outcomes. CHAID can be used with ordered features if they're binned into categories; this induces a bias and defeats the purpose of unbiased split variable selection in the first place. CTree, on the other hand, provides a unified framework for unbiased recursive partitioning that's applicable to data measured on arbitrary scales (e.g., continuous, nominal categorical, censored, ordinal, and multivariate data). Before we introduce the details of CTree, it will be helpful to have a basic understanding of the conditional inference framework it relies on.

The following section involves a bit of mathematical detail and notation, mostly around linear algebra, and can be skipped by the uninterested reader.

3.3   A quick digression into conditional inference

This section offers a quick detour into the essentials of a general framework for conditional inference procedures commonly known as *permutation tests*; our discussion follows closely with Hothorn et al. [2006b][a], which is based on the theoretical results derived by Strasser and Weber [1999] for a special class of *linear statistics*. For a more traditional take on (nonparametric) permutation tests, see, for example, Davison and Hinkley [1997, Sec. 4.3].

Suppose $\{(\boldsymbol{X}_i, \boldsymbol{Y}_i)\}_{i=1}^{N}$ are a random sample from some population of interest, where $\boldsymbol{X}$ and $\boldsymbol{Y}$ are from sample spaces $\mathcal{X}$ and $\mathcal{Y}$, respectively, which may be multivariate (hence the **bold** notation) and measured at arbitrary scales (e.g., ordinal, nominal, continuous, etc.). Our primary interest is in testing the null hypothesis of independence between $\boldsymbol{X}$ and $\boldsymbol{Y}$, namely,

$$H_0 : D\left(\boldsymbol{Y}|\boldsymbol{X}\right) = D\left(\boldsymbol{Y}\right),$$

where $D\left(\right)$ is the distribution function for $\boldsymbol{Y}$.

There are a number of ways to carry out such a test. For example, if $\boldsymbol{X}$ and $\boldsymbol{Y}$ are both $N \times 1$ continuous variables, then a simple correlation test based on the Pearson correlation coefficient, or Spearman's ρ, can be used. If $\boldsymbol{X}$ and $\boldsymbol{Y}$ are both $N \times 1$ nominal categorical variables, then a simple chi-square test is appropriate. And so on.

It would be rather convenient to have a standard approach to testing H_0, regardless of the shape or scale of $\boldsymbol{X}$ and $\boldsymbol{Y}$. To that end, Strasser and Weber [1999] suggest using a linear statistic of the general form

$$\boldsymbol{T} = \text{vec}\left(\sum_{i=1}^{N} g\left(\boldsymbol{X}_i\right) h\left(\boldsymbol{Y}_i\right)^{\top}\right) \in \mathbb{R}^{pq \times 1}, \tag{3.1}$$

where vec $\left(\right)$ is the *matrix vectorization operator* that I'll explain in the examples that follow, $g : \mathcal{X} \rightarrow \mathbb{R}^{p \times 1}$ is a transformation function specifying a possible transformation of the $\boldsymbol{X}$ values, and $h : \mathcal{Y} \rightarrow \mathbb{R}^{q \times 1}$ is called the *influence function* and specifies a possible transformation to the $\boldsymbol{Y}$ values. Appropriate choices for $g\left(\right)$ and $h\left(\right)$ allow us to perform general tests of independence, including:

---

[a]An updated version of this paper is freely available in the "A Lego System for Conditional Inference" vignette accompanying the R package **coin** [Hothorn et al., 2021c]; to view, use `utils::vignette("LegoCondInf", package = "coin")` at the R console.

- correlation tests, similar to the Pearson and Spearman's rank correlation tests;

- two-sample tests, similar to the *t*-test and Wilcoxon rank-sum test;

- more general *K*-sample tests, similar to the one-way ANOVA *F*-test and Kruskal-Wallis test;

- independence tests for contingency tables of arbitrary dimension;

- and many more.

Typically, $g\left(\right)$ and $h\left(\right)$ are chosen to be the identity function (e.g., $g\left(\boldsymbol{X}_i\right) = \boldsymbol{X}_i$).

Any test of H_0 based on (3.1) requires knowledge of the sampling distribution of $\boldsymbol{T}$, which is rarely (if ever) known in practice. However, under the null hypothesis (H_0), we can dispose of this dependency by fixing the predictor values, and *conditioning* on all possible permutations S of the response values—whence the term conditional inference. This principle leads to test procedures known as permutation tests.

Let $\mu_h = \mathrm{E}\left(h\left(\boldsymbol{Y}_i\right)|S\right) = \sum_{i=1}^{N} h\left(\boldsymbol{Y}_i\right)/N$ be the conditional expectation for the influence function with corresponding $q \times q$ covariance matrix

$$\Sigma_h = \mathrm{V}\left(h\left(\boldsymbol{Y}_i\right)|S\right)$$

$$= \sum_{i=1}^{N} \left[h\left(\boldsymbol{Y}_i\right) - \mathrm{E}\left(h\left(\boldsymbol{Y}_i\right)|S\right)\right]\left[h\left(\boldsymbol{Y}_i\right) - \mathrm{E}\left(h\left(\boldsymbol{Y}_i\right)|S\right)\right]^{\top}.$$

Strasser and Weber [1999] derived the conditional mean (μ) and variance (σ^2) of $\boldsymbol{T}|S$, which are:

$$\mu = \mathrm{E}\left(\boldsymbol{T}|S\right) = \mathrm{vec}\left(\left(\sum_{i=1}^{N} g\left(\boldsymbol{X}_i\right)\right)\mu_h^{\top}\right) \in \mathbb{R}^{pq \times 1},$$

$$\sigma^2 = \mathrm{V}\left(\boldsymbol{T}|S\right) = \frac{N}{N-1} \mathrm{V}_h \otimes \left(\sum_{i=1}^{N} g\left(\boldsymbol{X}_i\right) \otimes g\left(\boldsymbol{X}_i\right)^{\top}\right) \qquad (3.2)$$

$$- \frac{1}{N-1} \mathrm{V}_h \otimes \left(\sum_{i=1}^{N} g\left(\boldsymbol{X}_i\right)\right) \otimes \left(\sum_{i=1}^{N} g\left(\boldsymbol{X}_i\right)\right)^{\top} \in \mathbb{R}^{pq \times pq},$$

where $\otimes$ denotes the *Kronecker product*. While the equations listed in (3.1)–(3.2) might seem very complex, they simplify quite a bit in many standard situations—specific examples are given in Sections 3.3.1–3.3.2.

The next step is to construct a *test statistic* for testing H_0. In order to do this, we can standardize the (possibly multivariate) linear statistic using μ and Σ. Let $c\,()$ be a function that maps $\boldsymbol{T} \in \mathbb{R}^{pq \times 1}$ to the real line (i.e., a scalar or single number). Hothorn et al. [2006b] suggest using a quadratic form or the maximum of the absolute values of the standardized linear statistic:

$$c_q = c_q\left(\boldsymbol{T}, \mu, \Sigma\right) = \left(\boldsymbol{T} - \mu\right)\Sigma^{+}\left(\boldsymbol{T} - \mu\right)^{\top},$$

$$c_m = c_m\left(\boldsymbol{T}, \mu, \Sigma\right) = \max\left|\frac{\boldsymbol{T} - \mu}{\operatorname{diag}\left(\Sigma\right)^{1/2}}\right|, \tag{3.3}$$

where Σ^{+} denotes the *Moore-Penrose inverse* of Σ.

In order to construct p-values for testing H_0, we need to know the sampling distribution of c_q and c_m. Resampling approaches can be used to approximate the sampling distribution to any desired accuracy by evaluating the test statistic under all possible permutations S. This is often not feasible in practice as there are $N!$ possible permutations to consider. However, approximations based on a random sample of permutations can be used to great effect; see Davison and Hinkley [1997, Sec. 4.3] for details on traditional permutation tests. For certain special cases, the exact sampling distribution of some test statistics is obtainable for small to moderate sample sizes [Hothorn et al., 2006b].

For larger sample sizes, a normal approximation can be used, regardless of the choice for $g\,()$ and $h\,()$. In particular, Strasser and Weber [1999] showed that the conditional distribution of $\boldsymbol{T}$ tends to a multivariate normal distribution with mean μ and covariance Σ as $N \to \infty$. Consequently, the proposed test statistics, c_q and c_m, also have asymptotic approximations. For example, c_q has an asymptotic $\chi^2\,(df)$ distribution with degrees of freedom (df) given by the rank of Σ; if $p = q = 1$, then asymptotically (i.e., as $N \to \infty$), $c_q \sim \chi^2\,(1)$ and $c_m \sim N\,(0,1)$. With the conditional distribution of the test statistic in hand—whether asymptotic, approximate, or exact—we can compute a p-value (p) for testing H_0, where we would reject H_0 whenever $p \leq \alpha$, where α is some prespecified threshold.

Akin to a "one size fits all" approach to statistical tests of independence, this is a powerful idea that opens the door to a wide range of possibilities. For example, if the outcome is censored, then $h\,()$ can be chosen to map the response values to *log-rank scores*, which can be obtained in R using the `logrank_trafo()` function from package **coin**; see, for example, Hothorn et al. [2006b] or the "A Lego System for Conditional Inference" vignette from package **coin** mentioned earlier.

Enough with the mathematical jujitsu; let's illustrate the main ideas with some concrete examples.

3.3.1 Example: X and Y are both univariate continuous

The simplest case occurs when X and Y are both univariate continuous variables; in the univariate case, we drop the bold notation and just write X and Y. In this case, $p = q = 1$ and $T = \sum_{i=1}^{N} g(X_i) h(Y_i)$. If we take $g()$ and $h()$ to be the identity function (e.g., $g(X_i) = X_i$), then

$$T = \sum_{i=1}^{N} X_i Y_i,$$

$$\mu = \left(\sum_{i=1}^{N} X_i \right) \bar{Y},$$

$$\sigma^2 = S_Y^2 \sum_{i=1}^{N} X_i^2 - S_Y^2 \left(\sum_{i=1}^{N} X_i \right)^2 / N,$$

where $\bar{Y} = \sum_{i=1}^{N} Y_i / N$ and $S_Y^2 = \sum_{i=1}^{N} (Y_i - \bar{Y})^2 / (N-1)$ are the sample mean and variance of Y, respectively. Since T is univariate, the standardized test statistics (3.3) are $c_m = \left| (T - \mu) / \sqrt{\Sigma} \right|$ and $c_q = c_m^2$; hence, it makes no difference which test statistic we use in this case, as the results will be identical.

Let's revisit the New York air quality data set (Section 1.4.2) to demonstrate the required computations in R. Let $X = $ Temp and $Y = $ Ozone be the variables of interest. To test the null hypothesis of general independence between X and Y at the $\alpha = 0.05$ level, I'll use the quadratic test statistic (c_q), and compute a p-value for the test using an asymptotic $\chi^2(1)$ approximation. I'll also choose $g()$ and $h()$ to be the identity function. The first line below removes any rows with a missing response value, which I'll be using later.

```
aq <- airquality[!is.na(airquality$Ozone), ]
N <- nrow(aq)      # sample size
gX <- aq$Temp      # g(X)
gY <- aq$Ozone     # h(Y)
Tstat <- sum(gX * gY)    # linear statistic
mu <- sum(gX) * mean(gY)
Sigma <- var(gY) * sum(gX ^ 2) - var(gY) * sum(gX) ^ 2 / N

# Quadratic test statistic (1.3)
(cq <- ((Tstat - mu) / sqrt(Sigma)) ^ 2)

#> [1] 56.1

1 - pchisq(cq, df = 1)   # p-value

#> [1] 6.94e-14
```

Here we would reject the null hypothesis at the $\alpha = 0.05$ level $(p < 0.001)$ and conclude that there is some degree of association between `Temp` and `Ozone`.

For comparison, we can use the `independence_test()` function from package **coin**, which provides a flexible implementation of the conditional inference procedures described in Hothorn et al. [2006b]; see `?coin::independence_test` for details. This is demonstrated in the code snippet below:

```
library(coin)
independence_test(Ozone ~ Temp, data = aq, teststat = "quadratic")

#>
#>   Asymptotic General Independence Test
#>
#> data:  Ozone by Temp
#> chi-squared = 56, df = 1, p-value = 7e-14
```

Happily, we obtain the exact same results. Note that c_m and c_q will only differ when the linear statistic (3.1) is multivariate.

3.3.2   Example: X and Y are both nominal categorical

When X and Y are both categorical, T is essentially the vectorized (i.e., flattened) contingency table between X and Y; in this section, I'll continue to drop the bold notation. Assume X and Y have q and p unique categories, respectively. A contingency table between X and Y is nothing more than a $q \times p$ table containing the observed frequencies of each qp pair of categories from X and Y.[b] Recall the appearance of the matrix vectorization operator, vec (), in formulas (3.1)–(3.2). The vec() operator turns an $m \times n$ matrix into an $mn \times 1$ column vector. So a vectorized $q \times p$ contingency table is just a $qp \times 1$ column vector containing the individual frequencies, where the vectorization happens columnwise. For example, if $A = \left(\begin{smallmatrix} a & b \\ c & d \end{smallmatrix} \right)$ is a 2×2 matrix (or contingency table), then vec $(A) = (a, c, b, d)^\top$, where $\top$ represents the transpose operator.

Let's return to the mushroom edibility example (Section 1.4.4). Let $X =$ `odor` and $Y =$ `Edibility`; both are nominal categorical with nine and two categories, respectively. A contingency table cross-classifying each variable is given below using the `xtabs()` function (see `?stats::xtabs` for details). The vectorized contingency table is also constructed by calling `as.vector()`:

```
mushroom <- treemisc::mushroom
(ctab <- xtabs(~ odor + Edibility, data = mushroom))
```

---

[b]Here we assume that the q categories of X define the rows of the contingency table, but in general, it does not matter.

```
#>              Edibility
#> odor          Edible Poison
#>    almond        400      0
#>    anise         400      0
#>    creosote        0    192
#>    fishy           0    576
#>    foul            0   2160
#>    musty           0     36
#>    none         3408    120
#>    pungent         0    256
#>    spicy           0    576
```

```
(Tstat <- as.vector(ctab))   # multivariate linear statistic
```

```
#>  [1]   400   400     0     0     0     0 3408     0     0     0
#> [11]     0   192   576  2160    36   120   256   576
```

Typically, when X and Y are categorical, the transformation function, $g()$, and influence function, $h()$, map each category to a vector of dummy encoded values. For example, using the order of categories from the above contingency table (ctab),

$$g(X_i) = (0, 0, 0, 0, 1, 0, 0, 0, 0)^{\top}$$

and

$$h(Y_i) = (0, 1)^{\top}.$$

would indicate a foul smelling, poisonous mushroom. In essence, $g(X) = \boldsymbol{X}$ and $h(Y) = \boldsymbol{Y}$ are the associated one-hot encoded model matrices for X and Y, respectively.

In the special case where X and/or Y are binary, the identity transformation will lead to the same results as long as they're encoded as 0/1. For example, if X and Y are both binary 0/1 encoded variables, then the formulas in Section 3.3.1 will lead to identical results when $g()$ and $h()$ are both the identity function.

The Kronecker product, denoted $\otimes$, also appears in (3.2). The Kronecker product between two matrices can be computed in R using the kronecker() function or the %x% operator. The Moore-Penrose inverse, denoted Σ^+ in (3.2), can be computed using the ginv() function from the recommended **MASS** package [Ripley, 2022].

In the next code chunk, we compute the conditional expectation and variance using (3.2). Note that the formulas simplify greatly if we work directly with the one-hot encoded matrices $g(X) = \boldsymbol{X}$ and $h(Y) = \boldsymbol{Y}$; in R, these can be obtained using the `model.matrix()` function.[c] For example, the Kronecker product $\sum_{i=1}^{N} g\left(\boldsymbol{X}_i\right) \otimes g\left(\boldsymbol{X}_i\right)^{\top}$ in (3.2) reduces to $\boldsymbol{X}^{\top}\boldsymbol{X}$—which, in this case, is a $p \times p$ diagonal matrix, whose j-th diagonal entry is equal to the frequency of the j-th category of X (or the sum of the j-th column of $\boldsymbol{X}$). Also, $\sum_{i=1}^{N} g\left(\boldsymbol{X}_i\right)$ in (3.2) is just a $p \times 1$ column vector, whose j-th entry is equal to the frequency of the j-th category of X. Note that we use R's built-in `qr()` function to compute the rank of Σ, which is used for the degrees of freedom of the asymptotic chi-square distribution of c_q; in this example, $df = \operatorname{rank}(\Sigma) = 8$. (Due to rounding, the printed output displays a p-value of zero.)

```
gX <- model.matrix(~ odor - 1, data = mushroom)   # g(X)
hY <- model.matrix(~ Edibility - 1, data = mushroom)   # h(Y)
mu <- as.vector(colSums(gX) %*% t(colMeans(hY)))
Sigma <- var(hY) %x% (t(gX) %*% gX) -
  var(hY) %x% (colSums(gX) %x% t(colSums(gX))) / nrow(hY)

# Quadratic test statistic (1.3)
(cq <- t(Tstat - mu) %*% MASS::ginv(Sigma) %*% (Tstat - mu))

#>        [,1]
#> [1,] 7659

1 - pchisq(cq, df = qr(Sigma)$rank)   # p-value

#>        [,1]
#> [1,]   0
```

Again, we can compare the results with the output from **coin**'s `independence_test()` function. Once again, the results are equivalent.

```
independence_test(Edibility ~ odor, data = mushroom,
                  teststat = "quadratic")

#>
#>  Asymptotic General Independence Test
#>
#> data:  Edibility by
#>   odor (almond, anise, creosote, fishy, foul, musty, none, pungent...
#> chi-squared = 7659, df = 8, p-value <2e-16
```

3.3.3   Which test statistic should you use?

In the previous examples, we used the quadratic form of the test statistic in (3.3) and its asymptotic chi-square distribution, but what about the maxi-

---

[c]Here, I use `model.matrix(~ variable.name - 1)` to suppress the intercept—a column of all ones—and ensure that each category of `variable.name` gets dummy encoded.

mally selected test statistic (c_m) in (3.3)? When $\boldsymbol{X}$ and $\boldsymbol{Y}$ are both univariate continuous, or binary variables encoded as $0/1$, then the choice between c_q and c_m makes no difference, since $c_q = c_m^2$ in this case. However, if $\boldsymbol{X}$ and/or $\boldsymbol{Y}$ are multivariate (e.g., when $\boldsymbol{X}$ and/or $\boldsymbol{Y}$ are multi-level categorical variables), then the two statistics can lead to different, although usually similar results. Some guidance on when one test statistic may be more useful than the other is given in Hothorn et al. [2006b]. For example, if $\boldsymbol{X}$ and $\boldsymbol{Y}$ are both categorical, then working with c_m and the standardized contingency table can be useful in gaining insight into the association structure between $\boldsymbol{X}$ and $\boldsymbol{Y}$. For the general test of independence, it often doesn't matter which form of the test statistic you use. As we'll see in the next few sections, the CTree algorithm often defaults to using the quadratic test statistic (i.e., c_q) from (3.3).

3.4 Conditional inference trees

Conditional inference trees provide a unified framework for unbiased recursive partitioning based on conditional inference, the same idea behind permutation tests, and is general enough to handle data of many different types (e.g., continuous, categorical, ordinal, censored, multivariate, and more). CTree uses a general two-stage process for recursive partitioning based on null significance tests, which is described in Algorithm 3.1 below.

The next two sections dive a bit deeper into steps 1)–2), respectively.

3.4.1 Selecting the splitting variable

Step 1) of Algorithm 3.1 is to decide whether there is any (statistically significant) association between the response and any of the m predictors. This is accomplished via m partial tests of hypothesis $H_0^j : D\left(\boldsymbol{Y} | X_j\right) = D\left(\boldsymbol{Y}\right)$, for $j = 1, 2, \ldots, m$. The test statistic used for assessing the association between each feature and the response depends on the scale on which both are measured (e.g., multivariate, censored, continuous, ordinal, or nominal categorical). For example, if X and Y are both univariate continuous, then a correlation-based test can be carried out. If X is categorical and Y is continuous, then a K-sample test—like an ANOVA F-test or Kruskal-Wallis test—can be used. And so on.

In practice, the predictors will often be measured on different scales; hence, different test statistics need to be employed. Consequently, the test statistics associated with each test cannot be directly compared without biasing the selection of the splitting variable. Fortunately, using p-values provides a

---

Algorithm 3.1 Unbiased recursive partitioning via conditional inference.

---

1) Individually test the null hypothesis of independence between each of the m features $X_1, X_2, \ldots, X_m$ and the response Y using the conditional inference approach outlined in Section 3.3. If none of these hypotheses can be rejected at a prespecified α level, then stop the procedure (i.e., no further splits occur). Otherwise, select the predictor X_j with the "strongest association" to Y, as measured by the corresponding multiplicity adjusted p-values (e.g., *Bonferroni corrected p-values*).

2) Use X_j, the partitioning variable selected in step 1), to partition the data into two disjoint subsets (or child nodes), A_L and A_R. For each possible split S, a standardized test statistic (3.3) is computed, and the partition associated with the largest test statistic is used to partition the data into two child nodes.

3) Repeat steps 1)–2) in a recursive fashion on the resulting child nodes until the global hypothesis in step 1) cannot be rejected at a prespecified α level.

---

standard scale by which to compare the strength of association between each feature and the response, and results in an unbiased method for selecting split variables, regardless of the scale on which each variable is measured [Hothorn et al., 2006c].

In particular, if we have m features, $X_1, X_2, \ldots, X_m$, then we construct m general tests of independence $H_0^j : D\left(Y|X_j\right) = D\left(Y\right)$, for $j = 1, 2, \ldots, m$, using the conditional inference framework briefly discussed in Section 3.3. The features themselves can be measured on different scales; hence, we compare the m tests on the basis of their p-values. Since this involves multiple hypothesis tests, the p-values need to be adjusted to keep the overall *family-wise error rate* $\leq \alpha$.[d] The simplest approach is to use a *Bonferroni adjustment*; that is, multiply each p-value by the total number of tests: $p_j^\star = m \times p_j$, where m is the number of features being considered and p_j is the p-value from the j-th test of independence H_0^j. (R has a built-in function for adjusting p-values using a number of different approaches; see **?stats::p.adjust** for details.) The predictor associated with the test having the smallest adjusted p-value meeting a pre-selected threshold (α) is chosen as the splitter.

---

[d] For an overview of the problems associated with multiple tests of hypothesis, see Shaffer [1995] and Wright [1992]—the latter discusses the use of adjusted p-values.

3.4.1.1 Example: New York air quality measurements

To illustrate, let's write a simple function, called `gi.test()`, that uses the conditional inference procedure described in Section 3.3 to test the null hypothesis of general independence between two variables X and Y.[e] To keep it simple, this function applies only to univariate continuous variables, and computes an approximate p-value assuming an asymptotic $\chi^2(1)$ distribution (see Section 3.3.1 for details)—although, it would not be too difficult to modify `gi.test()` to return approximate p-values using the permutation distribution instead. The arguments g and h allow for suitable transformations of the variables x and y, respectively; for example, if the relationship between X and Y is monotonic, but not necessarily linear, or if we suspect outliers, then we might consider converting X and/or Y to ranks (e.g., g = rank)—converting both X and Y to ranks is similar in spirit to conducting a correlation test based on Spearman's ρ. Both arguments default to R's built-in `identity()` function, which has no effect on the given values.

```
gi.test <- function(x, y, g = identity, h = identity) {
  xy <- na.omit(cbind(x, y))   # only retain complete cases
  gx <- g(xy[, 1L])   # transformation function applied to x
  hy <- h(xy[, 2L])   # influence function applied to y
  lin <- sum(gx * hy)   # linear statistic
  mu <- sum(gx) * mean(hy)   # conditional expectation
  sigma <- var(hy) * sum(gx ^ 2) -   # conditional covariance
    var(hy) * sum(gx) ^ 2 / length(hy)
  c.quad <- ((lin - mu) / sqrt(sigma)) ^ 2   # quadratic test statistic
  pval <- 1 - pchisq(c.quad, df = 1)   # p-value
  c("chisq" = c.quad, "pval" = pval)   # return results
}
```

Continuing with the New York air quality example, let's see which variable, if any, is selected to split the root node. Following convention, I'll use $\alpha = 0.05$ as the set threshold for failing to reject the global null hypothesis in step 1) of Algorithm 3.1.

The following code chunk applies the previously defined `gi.test()` function to test the null hypothesis of general independence between each of the five features and the response—if you skipped Section 3.3, then you can think of this as a simple test of association that defaults to using a test statistic whose asymptotic distribution (i.e., the approximate distribution for sufficiently large N) is $\chi^2(1)$. Note that the `p.adjust()` function mentioned earlier is used to adjust the resulting p-values to account for multiple tests using a simple Bonferroni adjustment:

---

[e]We could also use the much more flexible `independence_test()` function from package **coin**, but writing your own function can help solidify your basic understanding of how the procedure actually works.

```
xnames <- setdiff(names(aq), "Ozone")  # feature names
set.seed(1938)  # for reproducibility
res <- sapply(xnames, FUN = function(x) {  # test each feature
  gi.test(airquality[[x]], y = airquality[["Ozone"]])
})
t(res)  # print transpose of results (nicer printing)
```

```
#>          chisq     pval
#> Solar.R 13.3476 2.59e-04
#> Wind    41.6137 1.11e-10
#> Temp    56.0863 6.94e-14
#> Month    3.1127 7.77e-02
#> Day      0.0201 8.87e-01
```

```
# Bonferroni adjusted p-values (same as 5 * pval in this case)
p.adjust(res["pval", ], method = "bonferroni")
```

```
#>  Solar.R     Wind     Temp    Month      Day
#> 1.29e-03 5.56e-10 3.47e-13 3.88e-01 1.00e+00
```

In this example, the predictor associated with the smallest adjusted p-value is Temp, and since $p^\star_{\text{Temp}} \approx 3.469 \times 10^{-13} < \alpha = 0.05$, Temp is the first variable that will be used to partition the data. The next step is to determine the optimal split point of Temp to use when partitioning the data (step 2) of Algorithm 3.1), which will be discussed in Section 3.4.2.

3.4.1.2   Example: Swiss banknotes

Let's try a binary classification problem as well. If you followed Section 3.3 and paid close attention, then you might have figured out that our gi.test() function should also work for 0/1 encoded binary variables.

Using the Swiss banknote data (Section 1.4.1), let's see which, if any, of the available features can be used to effectively partition the root node—recall that all the features are numeric and that the binary response (y) is already coded as 0 (for genuine banknotes) and 1 (for counterfeit banknotes):

```
bn <- treemisc::banknote  # start with the root node
xnames <- setdiff(names(bn), "y")  # feature names
res <- sapply(xnames, FUN = function(x) {  # test each feature
  gi.test(bn[[x]], y = bn[["y"]])
})
t(res)  # print transpose of results (nicer printing)
```

```
#>          chisq     pval
#> length    7.52 6.11e-03
#> left     48.89 2.71e-12
#> right    68.51 1.11e-16
#> bottom  118.61 0.00e+00
#> top      72.22 0.00e+00
```

```
#> diagonal 160.90 0.00e+00

# Bonferroni adjusted p-values (same as 6 * pval in this case)
p.adjust(res["pval", ], method = "bonferroni")

#>    length      left     right    bottom       top diagonal
#> 3.67e-02 1.62e-11 6.66e-16 0.00e+00 0.00e+00 0.00e+00
```

Using $\alpha = 0.05$, we would select **diagonal** as the primary splitter (since **bottom**, **top**, and **diagonal** are essentially tied in terms of minimum adjusted p-value, we can just select the one with the max χ^2 statistic).

We can double check our computations by comparing the results to those produced by **coin**'s **independence_test()** function, which are given below; spoiler alert, the results are a match. The results are a match, nice! The **independence_test()** function is far more general than my **gi.test()** function, and can handle univariate or multivariate variables measured at arbitrary scales (e.g., censored response value, categorical variables with more than two categories, etc.).

```
res <- sapply(xnames, FUN = function(x) {
  it <- independence_test(bn[["y"]] ~ bn[[x]], teststat = "quadratic")
  c("chisq" = statistic(it), "pval" = pvalue(it))
})
t(res)  # print transpose of results (nicer printing)

#>              chisq      pval
#> length        7.52 6.11e-03
#> left         48.89 2.71e-12
#> right        68.51 1.11e-16
#> bottom      118.61 0.00e+00
#> top          72.22 0.00e+00
#> diagonal    160.90 0.00e+00

# Bonferroni adjusted p-values (same as 6 * pval in this case)
p.adjust(res["pval", ], method = "bonferroni")

#>    length      left     right    bottom       top diagonal
#> 3.67e-02 1.62e-11 6.66e-16 0.00e+00 0.00e+00 0.00e+00
```

Hopefully, by this point, you have a basic understanding of how CTree selects the splitting variable in step 1) of Algorithm 3.1. Let's now turn our attention to finding the optimal split condition for the selected splitter.

3.4.2   Finding the optimal split point

Once a splitting variable has been selected, the next step is to find the optimal split point. CTree uses binary splits like those discussed for CART in Chapter 2; in particular, continuous and ordinal variables produce binary splits of the form $x \leq c$ vs. $x > c$, where c is in the domain of x, and categorical variables produce binary splits of the form $x \in S$ vs. $x \notin S$, where S is a subset of the unique categories of x.

Finding the optimal split point can be done using any number of strategies, including those discussed in Chapter 2 (e.g., maximizing reduction in node impurity). However, the choice of impurity function depends on the scale of the response (e.g., the Gini index for classification and sum of squares for regression). For this reason, CTree uses the same conditional inference framework for selecting the optimal split point as it does for selecting the optimal splitting variable. Instead of using p-values, however, the optimal split point is chosen using the individual test statistics (since we don't have to worry about different scales).

Note that for continuous predictors, CTree chooses a cut point from the observed predictor values. This is in contrast to CART, which uses the midpoints of the observed values; see, for example, Breiman et al. [1984, p. 30]. Other tree algorithms have different methods for selecting the split point values for ordered features (e.g., C4.5, which is discussed in the online complements), but this detail rarely matters in practice.

Every binary partition induces a two-sample test between the response values in each group (e.g., $\{Y_i | X_{ji} \in S\}$ and $\{Y_i | X_{ji} \notin S\}$). The conditional inference framework discussed in Section 3.3 is employed again at this step, and a test statistic (3.3) is computed for each possible split. The split associated with the largest test statistic is used to partition the data, before returning to step 1) of Algorithm 3.1.

3.4.2.1   Example: New York air quality measurements

Continuing with the New York air quality example, let's find the optimal split point for `Temp`, the feature selected previously in step 1) of Algorithm 3.1 (p. 122), to partition the root node. The code chunk below computes the test statistics for testing $H_0 : D(\text{Ozone}|\text{Temp} \leq c) = D(\text{Ozone})$, for each unique value c of `Temp`; the results are plotted in Figure 3.2.

```
set.seed(912)  # for reproducibility
xvals <- sort(unique(aq$Temp))  # potential cut points
splits <- matrix(0, nrow = length(xvals), ncol = 2)
colnames(splits) <- c("cutoff", "chisq")
for (i in seq_along(xvals)) {
  x <- ifelse(aq$Temp <= xvals[i], 0, 1)  # binary indicator
  y <- aq$Ozone
  # Ignore pathological splits or splits that are too small
  if (length(table(x)) < 2 || any(table(x) < 7)) {
    res <- NA
  } else {
    res <- gi.test(x, y)["chisq"]
```

```
  }
  splits[i, ] <- c(xvals[i], res)  # store cutpoint and test statistic
}
splits <- na.omit(splits)
splits[which.max(splits[, "chisq"]), ]

#> cutoff  chisq
#>   82.0   55.3

# Plot the test statistic for each cutoff (Figure 3.2)
plot(splits, type = "b", pch = 19, col = 2, las = 1,
     xlab = "Temperature split value (degrees Fahrenheit)",
     ylab = "Test statistic")
abline(v = 82, lty = "dashed")
```

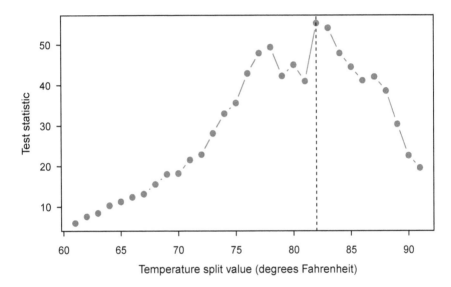

FIGURE 3.2: Test statistics from `gi.test()` comparing the two groups of `Ozone` values for every binary partition using `Temp`. A dashed line shows the optimal split point $c = 82$.

From the results, we see that the maximum of all the test statistics is $c_q = 53.282$ and is associated with the split point $c = 82$, giving us our first split in the tree (i.e., `Temp <= 82`).

Following Algorithm 3.1, we would continue splitting each of the resulting child nodes until the global null hypothesis in step 1) of Algorithm 3.1 cannot be rejected at the specified α level. For example, applying the previous code to the resulting left child node (`Temp` $\leq$ 82) would result in a further partition using `Wind` $\leq$ 6.9. I'll confirm these calculations using specific CTree software in Section 3.5.1.

3.4.3   Pruning

Unlike CART and other exhaustive search procedures, CTree uses statistical stopping criteria (e.g., Bonferroni-adjusted p-values) to determine tree size and does not require pruning—although, pruning can still be beneficial in certain circumstances (see Section 3.4.5). That's not to say that CTree doesn't overfit. As we'll see in Sections 3.4.5 and 3.5, the threshold α has a direct impact on the size and therefore complexity of the tree and is often treated as a tuning parameter.

3.4.4   Missing values

Similar to CART, CTree can use the idea of surrogates to handle missing values, but it is not the default in current implementations of CTree. By default, observations which can't be classified to a child node because of missing values are either 1) randomly assigned according to the node distribution (as in the newer **partykit** package [Hothorn and Zeileis, 2021]), or 2) go with the majority (as in the older **party** package [Hothorn et al., 2021b]).

Observations with missing values in predictor X are simply ignored when computing the associated test statistics during step 1) of Algorithm 3.1 (p. 122). Similarly, missing values associated with the splitting variable are also ignored when computing the test statistics in step 2). Once a split has been found, surrogates can be constructed using an approach similar to the one described in Section 2.7 for CART, in particular, creating a binary decision stump using the binary split in question as the response and trying to find the best splits associated with it using Algorithm 3.1 (p. 122).

3.4.5   Choice of α, $g\,()$, and $h\,()$

From an inferential standpoint, α is the prespecified nominal level of the general independence tests used for feature selection in step 1) of Algorithm 3.1, which controls the probability of type I error. Recall that the type I error (or a *false positive*) occurs when we reject a true null hypothesis—or in our case, when we conclude that X and Y are not independent, when in fact they are (i.e., potentially using an irrelevant splitting variable). Obviously, we want the probability of a type I error to be low, which is why we fix α to a small number (e.g., $\alpha = 0.05$).

Although α controls the probability of falsely rejecting H_0 in each node, we still need to consider the *statistical power* of each test; that is, the probability of rejecting H_0 when it is false. In the context of recursive partitioning, power essentially dictates the chance of selecting a relevant predictor at each node. As noted in Hothorn et al. [2006c], the general tests of independence used in CTree will only have high power for certain directions of deviation from independence and depends on the choice of $g()$ and $h()$. A useful guide for selecting $g()$ and $h()$ can be found in Table 4 of **coin**'s "Implementing a Class of Permutation Tests: The **coin** Package" vignette; to view, use `vignette("Implementation", package = "coin")` at the R console.

In the presence of outliers, the general test of independence discussed in Section 3.3 would be more powerful at a given sample size and α if $g()$ and $h()$ converted $\boldsymbol{X}$ and $\boldsymbol{Y}$ to ranks (because ranks are more robust to outlying observations).

To illustrate, let's run a quick Monte Carlo experiment. Suppose $\boldsymbol{X} = X$ has a standard normal distribution and that $\boldsymbol{Y} = Y$ is equal to X with a tad bit of noise: $Y = X + \epsilon$, where $\epsilon \sim N(0, \sigma = 0.1)$. Figure 3.3 shows a scatterplot for two random samples of size $N = 100$ generated from X and Y. The left panel in Figure 3.3 shows a clear association between X and Y. The right panel shows a scatterplot of the same sample, but with three of the observations replaced by outliers. Even with the outliers, there is still a clear relationship between X and Y.

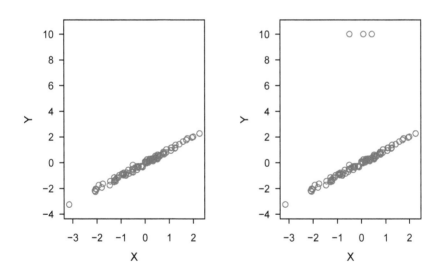

FIGURE 3.3: Scatterplots of two linearly related variables. Left: original sample. Right: original sample with three observations replaced by outliers.

The code chunk below applies the `gi.test()` function, using both the identity and rank transformations, to 100 random samples from X and Y for sample sizes ranging from 10–100; note that each sample includes three outlying Y values. For each sample size, the approximate power of each test is computed as the proportion of times out of 100 that the null hypothesis was rejected at the $\alpha = 0.05$ level. The results are plotted in Figure 3.4. Clearly, using ranks provides a more powerful test across the range of sample sizes in this example.

```
set.seed(2142)   # for reproducibility
N <- seq(from = 5, to = 100, by = 5)   # range of sample sizes
res <- sapply(N, FUN = function(n) {
  pvals <- replicate(1000, expr = {   # simulate 1,000 p-values
    x <- rnorm(n, mean = 0)                    # from each test
    y <- x + rnorm(length(x), mean = 0, sd = 0.1)
    y[1:3] <- 10   # insert outliers
    test1 <- gi.test(x, y)   # no transformations
    test2 <- gi.test(x, y, g = rank, h = rank)   # convert to ranks
    c(test1["pval"], test2["pval"])   # extract p-values
  })
  apply(pvals, MARGIN = 1, FUN = function(x) mean(x < 0.05))
})

# Plot the results (Figure 3.4)
plot(N, res[2L, ], xlab = "Sample size", ylab = "Power", type = "l",
     ylim = c(0, 1), las = 1)
lines(N, res[1L, ], col = 2, lty = 2)
legend("bottomright",
       legend = c("Rank transformation", "No transformation"),
       lty = c(1, 2), col = c(1, 2), inset = 0.01,
       box.col = "transparent")
```

Remember, in recursive partitioning, the sample size decreases as you proceed further down the tree. Consequently, the tests become less powerful the further down the tree they are. Similar to CART-like decision trees, the more accurate splits tend to occur at the top.

Hothorn et al. [2006c] suggest that increasing α can help assure that any type of dependence structure is detected. However, increasing α will result in more splits and therefore a more complex tree. To avoid overfitting in this situation, pruning can be applied in a variety of ways, for example, collapsing terminal nodes that don't meet a second threshold $\alpha' < \alpha$.

Similar to the cost-complexity parameter (*cp*) in CART (Section 2.5.1), you can also think of α as a tuning parameter controlling the overall complexity of the tree (i.e., the number of terminal nodes)—with smaller values of α leading to shallower trees. Hence, α can be optimized using cross-validation or similar techniques, in which case, the former definition related to the type I error rate no longer applies.

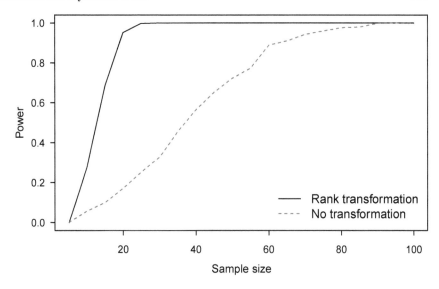

FIGURE 3.4: Power vs. sample size for the general test of independence between two univariate continuous variables, X and Y, using the conditional
inference procedure outlined in Section 3.3. The solid black curve corresponds
to using ranks, whereas the dashed red curve corresponds to the identity (i.e.,
no transformation).

3.4.6 Fitted values and predictions

For univariate regression and classification trees, fitted values and predictions
are obtained in the same manner as they are in CART. For example, the
within-terminal node class proportions can be used for class probability estimates. For regression, the within-terminal node sample mean of the response
values can be used as the fitted values or for predicting new observations.

However, CTree is quite flexible and can handle many other situations beyond simple classification and regression. With censored outcomes, for example, each terminal node can be summarized using the Kaplan-Meier estimator;
see Section 3.5.3. The median survival time can be used for making predictions.

3.4.7 Imbalanced classes

Unlike CART, CTree cannot explicitly take into account prior class probabilities, nor can it account for unequal misclassification costs. However, it can
assign increased weight to specific observations (for brevity, I omitted the case

weights from the formulas in Section 3.3, so see the references listed there for full details). For instance, we can assign higher weights to observations associated with higher misclassification costs. There is a drawback to this approach, however. Increasing the case weights essentially corresponds to increasing the sample size N in the statistical tests used in Algorithm 3.1, which will result in smaller p-values. Consequently, the resulting tree can be much larger since more splits will be significant. Decreasing α and/or employing the tuning strategy discussed in Section 3.4.5 can help, but it can be a difficult balancing act.

3.4.8   Variable importance

In contrast to the impurity-based variable importance measure used in CART (Section 2.8), CTree uses a *permutation*-based framework similar to the procedure outlined in Algorithm 6.1 (p. 205). This is not the same "permutation" as in "permutation test," but rather, in how the individual feature columns are randomly permuted, one at a time, before recording the difference in prediction performance (as compared to the baseline performance when none of the feature columns are permuted).

3.5   Software and examples

Conditional inference trees are available in the **party** package via the `ctree()` function.[f] However, the **partykit** package contains an improved re-implementation of `ctree()` and is recommended for fitting individual conditional inference trees in general. The `ctree()` function in **partykit** is much more modular and written almost entirely in R; this flexibility seems to come at a price, however, as `partykit::ctree()` can be much slower than `party::ctree()` (the latter of which is implemented in C). It's also worth noting that **partykit** is quite extensible and allows you to coerce tree models from different sources into a standard format that **partykit** can work with (e.g., importing trees saved in the *predictive model markup language* (PMML) format [Data Mining Group, 2014]). A good example of this is the R package **C5.0** [Kuhn and Quinlan, 2021], which provides an interface to C5.0 (C5.0, which evolved out of C4.5, is discussed in the online complements).

The R package **boot** [Canty and Ripley, 2021] can be used to carry out general permutation tests, as well as more general bootstrap procedures; for permutation tests, see the example code in Davison and Hinkley [1997, Sec. 4.3].

---

[f]The package name is apparently a play on the words "**part**ition **y**".

The **coin** package implements the conditional inference procedures briefly discussed in Section 3.3; in short, **coin** provides a common framework for general tests of independence, including: two-sample tests, K-sample tests, and correlation-based tests, for continuous, nominal, ordered, and multivariate data.

If you're not an R user, then you may be out of luck, as I'm unaware of any non-R implementations of CTree—all the more reason to be open to more than one opensource language!

Although there are many differences between `partykit::ctree()` and `party::ctree()`, both will produce the same tree almost every time under the default settings. For more information on the differences between the two, see Section 7.4 of **partykit**'s "ctree: Conditional Inference Trees" vignette, which can be viewed in R by calling `utils::vignette("ctree", package = "partykit")`.

The following examples illustrate the basic use of **party** and **partykit** for unbiased recursive partitioning via conditional inference. I'll confirm the results I computed manually in previous sections, as well as construct conditional inference trees for new data sets, including both regression and survival examples.

3.5.1 Example: New York air quality measurements

In earlier sections, we used our own `gi.test()` function to split the root node of the `airquality` data set. Using conditional inference, we found that the first best split occurred with `Temp <= 82`. Now, let's use **partykit** to apply Algorithm 3.1, and recursively split the `airquality` data set until we can no longer reject the null hypothesis of general independence between `Ozone` and any of the five numeric features at the $\alpha = 0.05$ level.

You can use `ctree_control()` to specify a number of parameters governing the CTree algorithm; in the code chunk below, we stick with the defaults; see `?partykit::ctree_control` for a description of all the parameters that can be set. In this chapter, we used the quadratic form of the test statistic c_q for steps 1)–2) of Algorithm 3.1, which is the default in both **party** and **partykit** (`teststat = "quad"`). We also stick with the default Bonferroni adjusted p-values. To specify α, set either the `alpha` or `mincriterion` arguments in `ctree_control()`, where the value of `mincriterion` corresponds to $1 - \alpha$ (only the `mincriterion` argument is available in **party**); both packages use a default significance level of $\alpha = 0.05$. In **party**'s implementation of `ctree()`, the transformation functions $g()$ and $h()$ can be specified via the `xtrafo` and `ytrafo` arguments, respectively; in **partykit**'s implementation, only `ytrafo` is available.

Next, I call `ctree()` to recursively partition the data and plot the resulting tree diagram using **partykit**'s built-in `plot()` method (see Figure 3.5):

```
library(partykit)

# Fit a default CTree using Bonferroni adjusted p-values
aq <- airquality[!is.na(airquality$Ozone), ]
(aq.cit <- ctree(Ozone ~ ., data = aq))

#>
#> Model formula:
#> Ozone ~ Solar.R + Wind + Temp + Month + Day
#>
#> Fitted party:
#> [1] root
#> |   [2] Temp <= 82
#> |   |   [3] Wind <= 6.9: 56 (n = 10, err = 21946)
#> |   |   [4] Wind > 6.9
#> |   |   |   [5] Temp <= 77: 18 (n = 48, err = 3956)
#> |   |   |   [6] Temp > 77: 31 (n = 21, err = 4621)
#> |   [7] Temp > 82
#> |   |   [8] Wind <= 10.3: 82 (n = 30, err = 15119)
#> |   |   [9] Wind > 10.3: 49 (n = 7, err = 1183)
#>
#> Number of inner nodes:    4
#> Number of terminal nodes: 5

plot(aq.cit)   # Figure 3.5
```

Note that I again removed the rows with missing response values. The fitted tree contains four splits (i.e., five terminal nodes) on only two predictors: `Temp` and `Wind`. The `plot()` method for `ctree()` objects is quite flexible, and I encourage you to read the documentation in `?partykit::plot`. By default, the terminal nodes are summarized using an appropriate plot that depends on the scale of the response variable—in this case, boxplots. The p-values from step 1) of Algorithm 3.1 are printed in each node, along with the selected splitting variable and the node number:

In **partykit**, we can print the test statistics and adjusted p-values associated with any node using the `sctest()` function from package **strucchange** [Zeileis et al., 2019], which is illustrated below; the 1 specifies the node of interest, which, according to the printed output and tree diagram, corresponds to the root node.. These correspond to the tests carried out in step 1) of Algorithm 3.1. The results are a match to our earlier computations using `gi.test()` and `p.adjust()`, woot! As far as I'm aware, you cannot currently obtain the test statistics from step 2) in **partykit**, although this is possible in **party**'s implementation of `ctree()`, which I'll demonstrate next.

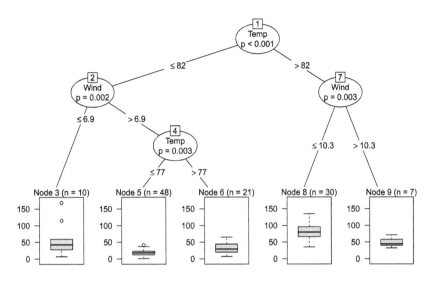

FIGURE 3.5: A default CTree fit to the New York air quality measurements data set.

```
strucchange::sctest(aq.cit, 1)

#>              Solar.R      Wind     Temp Month     Day
#> statistic 13.34761 4.16e+01 5.61e+01 3.113 0.0201
#> p.value    0.00129 5.56e-10 3.47e-13 0.333 1.0000
```

When fitting conditional inference trees with **party**, the `nodes()` function can be used to extract a list of nodes of a tree; the `where` argument specifies the node ID (i.e., the node numbers used to label the nodes in the associated tree diagram). Below, I'll refit the same tree using `party::ctree()`, extract the split associated with the root node, and plot the corresponding test statistics comparing the different cut points of the split variable (in this case, `Temp`). Note that `party::ctree()` only uses the maximally selected statistic (c_m) for step 2) of Algorithm 3.1[8], but recall that in the univariate case, $c_m^2 = c_q$, so I'll square them and compare them to the results I plotted earlier in Figure 3.2 (p. 127). As they should, the results from `party::ctree()`, which are displayed in Figure 3.6, match with what I obtained earlier using my `gi.test()` function.

```
aq.cit2 <- party::ctree(Ozone ~ ., data = aq)  # refit the same tree
root <- party::nodes(aq.cit2, where = 1)[[1L]]  # extract root node
split.stats <- root$psplit$splitstatistic  # split statistics
cutpoints <- aq[[root$psplit$variableName]][split.stats > 0]
cq <- split.stats[split.stats > 0] ^ 2
```

---

[8]In contrast, **partykit** lets you choose which test statistic to use in step 2) of Algorithm 3.1, and defaults to the quadratic form c_q we used earlier in `gi.test()`.

```
# Plot split statistics (Figure 3.6; compare to Figure 3.2)
plot(cutpoints[order(cutpoints)], cq[order(cutpoints)], col = 4,
    pch = 19, type = "b", las = 1,
    xlab = "Temperature split value (degrees Fahrenheit)",
    ylab = "Test statistic")
abline(v = root$psplit$splitpoint, lty = "dashed")
```

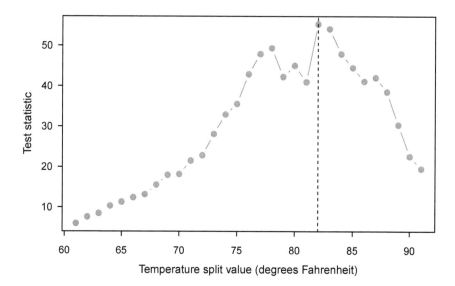

FIGURE 3.6: Test statistics from `party::ctree()` comparing the two groups of `Ozone` values for every binary partition using `Temp`. A dashed line shows the optimal split point $c = 82$. Compare these results to those from Figure 3.2.

You can coerce decision trees produced by various other implementations into `"party"` objects using the `partykit::as.party()` function. This means, for example, we can fit a decision tree using **rpart**, and visualize it using **partykit**'s `plot()` method.

To illustrate, the next code chunk fits an **rpart** tree to the same `aq` data, coerces it to a `"party"` object, and plots the associated tree diagram. Here, I'll set the complexity parameter c_p to zero (i.e., no penalty on the size of the tree) and use the default 10-fold cross-validation along with the 1-SE rule to prune the tree (Section 2.5.2.1). In this example, CART produced a decision stump (i.e., a tree with only a single split).

```
set.seed(1525)  # for reproducibility
aq.cart <- rpart::rpart(Ozone ~ ., data = aq, cp = 0)
aq.cart.pruned <- treemisc::prune_se(aq.cart, se = 1)  # 1-SE rule
plot(partykit::as.party(aq.cart.pruned))
```

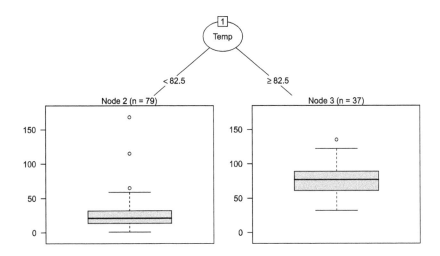

FIGURE 3.7: CART-like decision tree fit to the New York air quality measurements data set. The tree was pruned according to the 1-SE rule discussed in Section 2.5.2.1.

3.5.2 Example: wine quality ratings

Ordinal outcomes are common in scientific research and everyday practice. In medical and epidemiological studies, for example, the ordinal response often represents the levels of a standard measurement scale such as severity of pain (e.g., none < mild < moderate < severe) [Harrell, 2015, p. 311].

In this example, I'll take up the wine quality data introduced in Section 1.4.8. The goal is to model the quality of red wines based solely on physicochemical tests, like acidity, and interpret the results. The target variable, the quality of the wine, is an ordinal variable from 0–10 derived from wine tasting by human experts. Note that in this data set, however, we only have observed ratings in the range 3–9. See `?treemisc::wine` for additional background and column information.

As mentioned in Section 1.4.8, this is a well-known data set and is often used in statistical learning tutorials. However, many individuals treat this as a classification problem by arbitrarily dichotomizing the wine quality score and treating it as a binary classification problem. This is poor statistical practice and results in loss of information. Such "dichotomania" [Senn, 2005] is unfortunately also prevalent in medical and epidemiological research.

With CTree, we can easily model the outcome as an ordinal variable. To start, I'll load the data from **treemisc** and coerce the (integer) response to an *ordered factor*; see ?as.ordered for details:

```
wine <- treemisc::wine
reds <- wine[wine$type == "red", ]   # reds only
rm(wine)  # remove from global environment
reds$type <- NULL  # remove column
reds$quality <- as.ordered(reds$quality)  # coerce to ordinal
head(reds$quality)  # print first few quality scores

#> [1] 5 5 5 6 5 5
#> Levels: 3 < 4 < 5 < 6 < 7 < 8
```

Next, I'll fit a (default) conditional inference tree via **partykit**. Unfortunately, the tree diagram is too large to print neatly on this page, so I'll show a printout of the fitted tree instead:

```
(reds.cit <- ctree(quality ~ ., data = reds))

#>
#> Model formula:
#> quality ~ fixed.acidity + volatile.acidity + citric.acid + residua...
#>       chlorides + free.sulfur.dioxide + total.sulfur.dioxide +
#>       density + pH + sulphates + alcohol
#>
#> Fitted party:
#> [1] root
#> |     [2] alcohol <= 10.5
#> |     |   [3] volatile.acidity <= 0.3
#> |     |   |   [4] sulphates <= 0.7: 5 (n = 27, err = 48%)
#> |     |   |   [5] sulphates > 0.7: 6 (n = 58, err = 41%)
#> |     |   [6] volatile.acidity > 0.3
#> |     |   |   [7] volatile.acidity <= 0.7
#> |     |   |   |   [8] alcohol <= 9.8
#> |     |   |   |   |   [9] total.sulfur.dioxide <= 39: 5 (n = 171, er...
#> |     |   |   |   |   [10] total.sulfur.dioxide > 39
#> |     |   |   |   |   |   [11] pH <= 3.4: 5 (n = 205, err = 22%)
#> |     |   |   |   |   |   [12] pH > 3.4: 5 (n = 53, err = 42%)
#> |     |   |   |   [13] alcohol > 9.8: 6 (n = 228, err = 54%)
#> |     |   |   [14] volatile.acidity > 0.7
#> |     |   |   |   [15] fixed.acidity <= 8.5: 5 (n = 172, err = 26%)
#> |     |   |   |   [16] fixed.acidity > 8.5: 5 (n = 69, err = 35%)
#> |     [17] alcohol > 10.5
#> |     |   [18] volatile.acidity <= 0.9
#> |     |   |   [19] sulphates <= 0.6
#> |     |   |   |   [20] volatile.acidity <= 0.3: 6 (n = 33, err = 45%)
#> |     |   |   |   [21] volatile.acidity > 0.3: 6 (n = 207, err = 45%)
#> |     |   |   [22] sulphates > 0.6
#> |     |   |   |   [23] alcohol <= 11.5
#> |     |   |   |   |   [24] total.sulfur.dioxide <= 49
#> |     |   |   |   |   |   [25] volatile.acidity <= 0.4: 7 (n = 72, e...
```

```
#> |  |  |  |  |  |     [26] volatile.acidity > 0.4: 6 (n = 80, er...
#> |  |  |  |  |     [27] total.sulfur.dioxide > 49: 6 (n = 55, err...
#> |  |  |  |    [28] alcohol > 11.5: 7 (n = 142, err = 48%)
#> |  |    [29] volatile.acidity > 0.9: 5 (n = 27, err = 59%)
#>
#> Number of inner nodes:    14
#> Number of terminal nodes: 15
```

To see how well the model performs (on the learning sample), we can cross-classify the observed quality ratings with the fitted values (i.e., the prediction from the learning sample):

```
p <- predict(reds.cit, newdata = reds)  # fitted values
table(predicted = p, observed = reds$quality)  # contingency table

#>          observed
#> predicted  3   4   5   6   7   8
#>        3   0   0   0   0   0   0
#>        4   0   0   0   0   0   0
#>        5   9  33 483 194   5   0
#>        6   1  20 191 361  85   3
#>        7   0   0   7  83 109  15
#>        8   0   0   0   0   0   0
```

For example, of all the red wines with a rating quality score of 7, 5 were predicted to have a quality rating of 5, 85 were predicted to have a quality rating of 6, and the rest (109) were predicted to have a quality rating of 7.

So which variables seem to be the most predictive of the wine quality rating? At first glance, alcohol by volume (`alcohol`) and volatile acidity (`volatile.acidity`) seem to be important predictors, as they appear at the top of the tree and are used multiple times to partition the data. We can quantify this in CTree using **partykit**'s `varimp()` function. This function computes importance using a permutation-based approach akin to the procedure discussed in Section 6.1.1. For now, just think of the returned importance scores as an estimate of the decrease in performance as a result of removing the effect of the predictor in question. By default, performance is measured by the *negative log-likelihood*[h].

```
set.seed(2023)  # for reproducibility
(vi <- varimp(reds.cit, nperm = 100))  # variable importance scores
dotchart(vi, pch = 19, xlab = "Variable importance")  # Figure 3.8

#>           alcohol     volatile.acidity
#>            0.5465               0.3537
#>          sulphates total.sulfur.dioxide
```

[h] For ordinal outcomes in CTree, the log-likelihood is defined as $\sum_{i=1}^{N} \log(p_i)/N$, where p_i is the proportion of observations in the same node as case i sharing the same class.

```
#>            0.1852              0.0395
#>               pH          fixed.acidity
#>            0.0235              0.0188
```

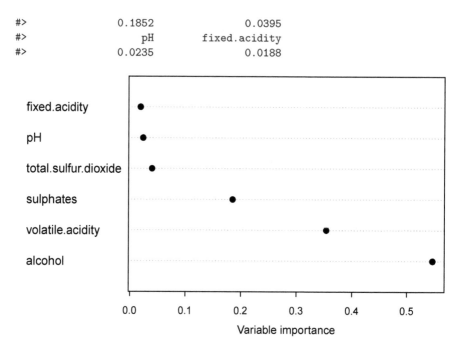

FIGURE 3.8: Variable importance plot for the red wine quality conditional inference tree.

As we suspected from looking at the tree's output, alcohol by volume and volatile acidity, followed by sulphate, are the most important predictors of wine quality rating in this model. The partial dependence (Section 6.2.1) of wine quality rating against each of the top three predictors is given in Figure 3.9[i]; note that the y-axis is interpreted on the same scale as the response. Here we can see the functional effect of each predictor. For example, alcohol has a monotonic increasing relationship with the predicted quality score. This makes sense and is probably why I never buy any red wine that's less than 14% alcohol by volume. Do the effects of the other two predictors make sense to you?

3.5.3   Example: Mayo Clinic liver transplant data

In this example, I'll revisit the PBC data described in Section 1.4.9. A tree-based analysis of the data was briefly discussed in Ahn and Loh [1994]. Below we load the **survival** package and prepare the data:

---

[i]As always, the code to reproduce this plot is available on the book website.

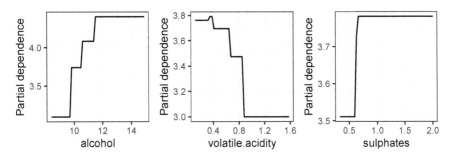

FIGURE 3.9: Partial dependence of wine quality rating on alcohol by volume (left), volatile acidity (middle), and level of potasium sulphate (right).

```
library(survival)

# Prep the data a bit
pbc2 <- pbc[!is.na(pbc$trt), ]  # use randomized subjects
pbc2$id <- NULL  # remove ID column
# Consider transplant patients to be censored at day of transplant
pbc2$status <- ifelse(pbc2$status == 2, 1, 0)
facs <- c("sex", "spiders", "hepato", "ascites", "trt", "edema")
for (fac in facs) {  # coerce to factor
  pbc2[[fac]] <- as.factor(pbc2[[fac]])
}
```

As briefly discussed in Section 1.4.9, survival (or reliability) analysis is concerned with the distribution of lifetimes, typically of humans, animals, or components of a machine. The response variable in survival analysis is *time-to-event*, for example, time until death or readmission (e.g., for subjects in a randomized clinical trial comparing two drugs) or time until failure of some component in a machine (e.g., a running motor).

Kaplan-Meier estimates of the survival function for both the control and treatment (i.e., drug) group were given in Figure 1.11. As with all survival curves, the chance of survival decreases with time. Often we don't care as much about the overall survival curve but rather how it varies between groups. In this example, our goal is to see if the additional covariates can be used to usefully discriminate groups with different survival distributions.

An overview on tree-structured survival models is provided in Loh [2014]. The approach taken in CTree is rather straightforward. The follow-up times (`time`) and event indicator (`status`) are combined into a single numeric response that is treated as univariate continuous; hence, the techniques discussed earlier apply directly. In CTree, right-censored data are converted to log-rank scores using the `logrank_trafo()` function from package **coin**; that is, the influence function $h()$ converts the right-censored survival times to log-rank scores.

To illustrate, let's test the null hypothesis of general independence between the predictor `bili` (serum bilirubin (mg/dl)) and the log-rank scores of the response:

```
library(coin)

independence_test(Surv(time, status) ~ bili, data = pbc2,
                  teststat = "quadratic")

#>
#>   Asymptotic General Independence Test
#>
#> data:  Surv(time, status) by bili
#> chi-squared = 77, df = 1, p-value <2e-16

# Our `gi.test()` function from earlier should also work
lr.scores <- coin::logrank_trafo(Surv(pbc2$time, pbc2$status))
gi.test(pbc2$bili, y = lr.scores)

#> chisq  pval
#>  77.5   0.0
```

Using $\alpha = 0.05$, we would reject the null hypothesis ($p < 0.001$) and conclude that the level of serum bilirubin is associated with survival rate. But this doesn't tell us much beyond that. Do subjects with higher levels of serum bilirubin tend to survive longer? To answer questions like this, we can use CTree to recursively partition the data using conditional inference-based tests of independence between each feature and the log-rank scores:

```
(pbc2.cit <- partykit::ctree(Surv(time, status) ~ ., data = pbc2))

#>
#> Model formula:
#> Surv(time, status) ~ trt + age + sex + ascites + hepato + spiders +
#>      edema + bili + chol + albumin + copper + alk.phos + ast +
#>      trig + platelet + protime + stage
#>
#> Fitted party:
#> [1] root
#> |    [2] bili <= 1.9
#> |    |    [3] edema in 0
#> |    |    |    [4] stage <= 2: Inf (n = 61)
#> |    |    |    [5] stage > 2: 4191 (n = 104)
#> |    |    [6] edema in 0.5, 1: Inf (n = 16)
#> |    [7] bili > 1.9
#> |    |    [8] protime <= 11.2
#> |    |    |    [9] age <= 44.5
#> |    |    |    |    [10] bili <= 5.6: 3839 (n = 29)
#> |    |    |    |    [11] bili > 5.6: 1080 (n = 7)
```

```
#> |   |   |    [12] age > 44.5: 1487 (n = 45)
#> |   |   [13] protime > 11.2
#> |   |   |    [14] albumin <= 3.6: 597 (n = 43)
#> |   |   |    [15] albumin > 3.6: 2540 (n = 7)
#>
#> Number of inner nodes:     7
#> Number of terminal nodes: 8
```

Notice how treatment group (**drug**) was not selected as a splitting variable at any node. This is not surprising since Fleming and Harrington [1991, p. 2] concluded that there was no practically significant difference between the survival times of those taking the placebo and those taking the drug.

We can also display the tree diagram using the **plot()** method; the results are displayed in Figure 3.10. For censored outcomes, the Kaplan-Meier estimate of the survival curve is displayed in each node. The tree diagram in Figure 3.10 makes it clear that subjects with higher serum bilirubin levels tended to have shorter survival times. What other conclusions can you draw from the tree diagram?

```
plot(pbc2.cit)  # Figure 3.10
```

3.6   Final thoughts

CTree is one of the more important developments in recursive partitioning in the last two decades; other important developments are discussed in Chapter 4, as well as the online complements to this book. In summary, compared to CART, CTree:

- uses adjusted statistical tests to separately determine the split variable and split point at each node (CART just uses an exhaustive search);

- provides unbiased split variable selection;

- does not require pruning (or much tuning);

- can naturally take into account the nature of the data—for example, when the variables are of arbitrary type (e.g., multivariate, ordered, right-censored, etc.).

If CTree is competitive with CART in terms of accuracy, doesn't require post pruning, and provides unbiased split variable selection, then why is CART still so popular? As pointed out in the rejoiner to Loh [2014], "This seems to tie in with a third bad effect: Many authors who propose or apply tree algorithms either are not aware of—or choose to ignore—similar work in that area. It happens that even recent papers do not refer to work carried out from 2000 onward, therefore ignoring more than a decade of active development that may be highly relevant." Another important factor is software availability. Many tree algorithms do not have easy to use opensource implementations. For example, of the 99 tree algorithms considered by Rusch and Zeileis (see their discussion at the end of Loh [2014]), roughly one-third had free opensource implementations available (including CART and CTree). CART-like decision trees are also broadly implemented across a variety of opensource platforms (see Section 2.9). CTree, on the other hand, is only available in R—as far as I'm aware.

Should you be concerned about biased variable selection when using CART-like decision trees? Certainly. However, as pointed out in Loh's rejoinder in Loh [2014], "...selection bias may not cause serious harm if a tree model is used for prediction but not interpretation, in some situations." While biased variable selection can lead to more spurious splits on irrelevant features, if the sample size is large and there are not too many such variables, pruning with cross-validation is often effective at removing them.

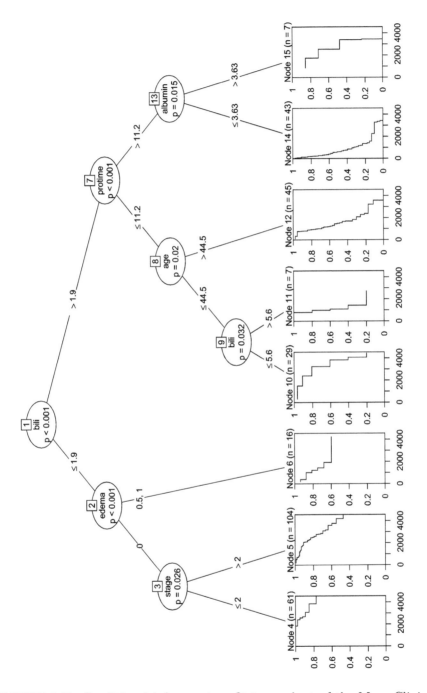

FIGURE 3.10: Conditional inference tree fit to a subset of the Mayo Clinic PBC data. The terminal nodes are summarized using Kaplan-Meier estimates of the survival function. The tree diagram highlights potential risk factors associated with different survival distributions.

4

The hitchhiker's GUIDE to modern decision trees

Of all the trees we could have hit, we had to get one that hits back.

J.K. Rowling

Harry Potter and the Chamber of Secrets

This chapter introduces a powerful, yet seemingly lesser known, decision tree algorithm for *generalized, unbiased, interaction detection, and estimation* (GUIDE), which was introduced in Loh [2002, 2009]. Unfortunately, GUIDE seems much less adopted among practitioners and researchers when compared to other algorithms with easy-to-use open source implementations, like CART (e.g., **rpart**) and CTree (e.g., **partykit**).

Like CTree, GUIDE is based on statistical tests of hypothesis striving to achieve unbiased split variable selection. In particular, GUIDE was specifically designed to solve three problems that can adversely affect the interpretability of decision trees:

1) split variable selection bias;

2) insensitivity to local interactions;

3) overly complicated tree structures (like the Ames housing regression trees from Figure 2.23 on page 97).

4.1   Introduction

GUIDE evolved from earlier tree growing procedures, starting with the *fast and accurate classification tree* (FACT) algorithm [Loh and Vanichsetakul, 1988]. FACT was novel at the time for its use of *linear discriminate analysis* (LDA) to find splits based on linear combinations of two predictors. FACT only applies to classification problems, and any node based on linear splits is partitioned into as many child nodes as there are class labels (i.e., FACT can use multiway splits). Split variable selection for continuous variables is based on comparing ANOVA-based F-statistics; LDA is applied to the variable with the largest F-statistic to find the optimal split point (i.e., $x \leq c$ vs. $x > c$, where c is in the domain of x) to partition the data. Categorical predictors are converted to ordinal variables by using LDA to project their dummy encoded vectors onto the *largest discriminant coordinate* (also called the *canonical variate* in *canonical analysis*); the final splits are expressed back in the form $x \in S$, where S is a subset of the categories of x. Since FACT depends on ANOVA F-tests for split variable selection, it is only unbiased if all the predictors are ordered (i.e., it is biased towards nominal categorical variables) [Loh, 2014].

The *quick, unbiased, and efficient statistical tree* (QUEST) procedure [Loh and Shih, 1997] improves upon the bias in FACT by using chi-square tests for categorical variables (i.e., by forming contingency tables between each categorical variable and the response). Like CART, QUEST only permits binary splits. Let J_t be the number of response categories in any particular node t. Whenever $J_t > 2$, QUEST produces binary splits by merging the J_t classes into two *super classes* before applying the F- and chi-square tests to select a splitting variable. The optimal split point for ordered predictors is found using either an exhaustive search (like in Chapter 2) or *quadratic discriminate analysis* (QDA). For categorical splitting variables, the optimal split point is found in the same way after converting the dummy encoded vectors to the largest discriminate coordinate as in FACT.

Kim and Loh [2001] introduced the *classification rule with unbiased interaction selection and estimation* (CRUISE) algorithm, as a successor to QUEST. In contrast to QUEST, however, CRUISE allows multiway splits depending on the number of response categories in a particular node. Also, while QUEST uses F-tests for ordered variables and chi-square tests for nominal, CRUISE uses chi-square tests for both after discretizing the ordered variables. For split variable selection, CRUISE uses a two-step procedure involving testing both *main effects* and *two-way interactions* at each node.

CRUISE was later succeeded by GUIDE [Loh, 2009][a], which improves upon both QUEST and CRUISE by retaining their strengths and fixing their main weaknesses. One of the drawbacks of CRUISE is that the number of interaction tests greatly outnumbers the number of main effect tests; for example, with k features, there are k main effect tests and $k(k-1)$ pairwise interaction tests. Since most of the p-values come from the interaction tests, CRUISE is biased towards selecting split variables with potentially weak main effects, relative to the other predictors. GUIDE, on the other hand, restricts the number of interaction tests to only those predictors whose main effects are significant based on Bonferroni-adjusted p-values.

The next two sections cover many of the details associated with GUIDE for standard regression and classification, respectively; but note that GUIDE has been extended to handle several other situations as well (e.g., censored outcomes and longitudinal data). In general, GUIDE uses a two-step procedure when selecting the splitting variables. Consequently, GUIDE involves many more steps compared to CART (Chapter 2) and CTree (Chapter 3). The individual steps themselves are not complicated (most of them involve transformations of continuous features and what to do when interaction tests are significant), but for brevity, I'm only going to cover the nitty-gritty details, while pointing to useful references along the way.

Note that the official GUIDE software—which is freely available, but not open source—has evolved quite a lot over the years; the GUIDE program is discussed briefly in Section 4.9. Consequently, some of the fine details on the various GUIDE algorithms may have changed since their original publications. Any important updates are likely to be found in the revision history for the official GUIDE software:

http://pages.stat.wisc.edu/~loh/treeprogs/guide/history.txt.

If you're interested in going deeper on GUIDE for regression and classification, I encourage you to read Loh [2002] (the official reference to GUIDE for regression), Loh [2009] (the official reference to GUIDE for classification), Loh [2011] (an updated overview with some comparisons to other tree algorithms), and Loh [2012] (variable selection and importance). The current GUIDE software manual is also useful and can be obtained from the GUIDE website: http://pages.stat.wisc.edu/~loh/guide.html.

---

[a]GUIDE had already been introduced for regression problems in Loh [2002].

4.2   A GUIDE for regression

Whether building a classification tree or a regression tree, GUIDE uses chi-square tests throughout. This is convenient since chi-square tests are quick to compute, and can detect a large variety of patterns (e.g., curvature and interaction effects). Consequently, the response and any ordered features must be converted to nominal categorical before attempting to split a node[b]. Under the hood, GUIDE treats regression problems like a classification problem. In particular, at any node in the tree, GUIDE fits a model (e.g., a constant or a linear regression model) to the available data and computes the residuals. The sign of the residuals (i.e., positive/negative residuals are mapped to $+1/-1$) is used as a binary response to partition the node. In contrast to CART, GUIDE can fit either constant or non-constant fits in the terminal nodes. Both strategies are discussed in the sections that follow.

4.2.1   Piecewise constant models

Using piecewise constant models is equivalent to a standard regression tree, where a constant (e.g., the average response value) is used to summarize each terminal node. Let N_t be the number of observations in an arbitrary node t, and let $y_{t,1}, y_{t,2}, \ldots, y_{t,N_t}$ be the available response values in t. The residuals for node t are nothing more than the difference between the observed response values and their sample mean: $r_{t,j} = y_{t,j} - \sum_{i=1}^{N_t} y_{t,i}/N_t$. The residuals can then be dichotomized at zero using their sign to create a new binary response variable in node t:

$$y'_{t,j} = \begin{cases} 1 & \text{if } r_{t,j} > 0 \\ -1 & \text{if } r_{t,j} < 0 \end{cases}.$$

Similarly, at any node t, ordered features (e.g., an ordered categorical or continuous feature) are also converted to non-ordered (i.e., nominal) categoricals by discretizing them into either three or four intervals, depending on the sample size in node t (N_t).

Like CTree, GUIDE uses statistical tests (chi-square tests in particular) to select the splitting variable. However, unlike CTree, GUIDE employs a two-stage approach that tests for both main effects (called *curvature tests*) and two-way interaction effects between all pairs of features. The details are quite

---

[b]I mentioned the dangers of "dichotomania" in Section 3.5.2, but keep in mind that discretizing the predictors is only used here for split variable selection, and full predictor information is used in selecting the split point and making predictions.

involved, but the basic steps (skipping the two-way interaction tests) are outlined in Algorithm 4.1 (151); GUIDE's interaction tests are discussed briefly in Section 4.2.2.

Algorithm 4.1 Simplified version of the original GUIDE algorithm for regression. Note that some of the details may have changed as the official software has continued to evolve over the years.

---

1) Start with t being the root node.

2) Obtain the signed residuals from a constant fit to the data (e.g., the mean response).

3) Convert ordered predictors (e.g., continuous features) to categorical by discretizing them into four intervals based on the sample quartiles (or *tertiles* if $N_t < 60$).

4) Using the N_t observations in t, perform a chi-square test of independence between each feature and the signed residuals; the dichotomized residuals form the two rows of the corresponding contingency table. (Call these the main effect, or curvature, tests.) Let $x^\star$ be the feature associated with the smallest Bonferroni-adjusted p-value from the curvature tests.

5) Use an exhaustive search to find the best split on $x^\star$ yielding the greatest reduction in node SSE (see Section 2.3). By default, GUIDE uses univariate splits similar to CART and CTree. In particular, if $x^\star$ is unordered, splits are of the form $x^\star \in S$, where S is a subset of the categories of $x^\star$. If $x^\star$ is ordered, splits have the form $x^\star \leq c$, where c is a midpoint in the observed range of $x^\star$; for speed, GUIDE will optionally use the within node sample median of $x^\star$ for the cutoff c.

6) Recursively apply steps 2)–5) on all the resulting child nodes until all nodes are pure or suitable stopping criteria are met (e.g., the maximum number of allowable splits is reached).

7) Similar to CART, prune the resulting tree using cost-complexity pruning (see Section 2.5 for details).

A few comments regarding step 4) of Algorithm 4.1 are in order. First, any rows or columns with zero margin totals are removed. Second, to avoid difficulties in computing very small p-values and to account for the fact that the degrees of freedom are not fixed across the chi-square tests, GUIDE sometimes uses a modification of the *Wilson-Hilferty transformation* [Wilson and Hilferty, 1931] to ensure all the test statistics approximately correspond to a chi-square distribution with a single degree of freedom.[c] In particular, let x be an observed value from a chi-square distribution with ν degrees of freedom (χ^2_ν). If we define

$$w_1 = \left(\sqrt{2x} - \sqrt{2\nu - 1} + 1\right)^2 / 2,$$

$$w_2 = \max\left\{0, \left(7/9 + \sqrt{\nu}\left[\sqrt[3]{x/\nu} = 1 + 2/9\nu\right]\right)^3\right\},$$

$$w = \begin{cases} w_2 & \text{if } x < \nu + 10\sqrt{2\nu} \\ (w_1 + w_2)/2 & \text{if } x \geq \nu + 10\sqrt{2\nu} \text{ and } w_2 < x, \\ w_1 & \text{otherwise} \end{cases}$$

then it follows that $Pr\left(\chi^2_\nu > x\right) \approx Pr\left(\chi^2_1 > w\right)$; this transformation is implemented in the `wilson_hilferty()` function in package **treemisc** (for details, see `?treemisc::wilson_hilferty`). Finally, for brevity, the tests for two-way interactions that GUIDE carries out by default are omitted; see Section 4.2.2 for details.

Although GUIDE uses chi-square tests throughout (which requires discretizing ordered features into 3–4 groups), Loh [2002] provided a simulation study which gave empirical evidence that GUIDE's split variable selection procedure is indeed unbiased, relative to exhaustive search procedures like CART.

4.2.1.1   Example: New York air quality measurements

To illustrate, let's return to the New York air quality example introduced in Section 1.4.2. Below is a simple function, called `guide.chisq.test()`, for carrying out steps 2)–4) of Algorithm 4.1. For brevity, and since the degrees of freedom are the same for each test, it omits the modified Wilson-Hilferty transformation discussed previously:

```
guide.chisq.test <- function(x, y) {
  y <- as.factor(sign(y - mean(y)))  # discretize response
  if (is.numeric(x)) {  # discretize numeric features
    bins <- quantile(x, probs = c(0.25, 0.5, 0.75), na.rm = TRUE)
    bins <- c(-Inf, bins, Inf)
    x <- as.factor(findInterval(x, vec = bins))  # quartiles
  }
  tab <- table(y, x)  # form contingency table
  if (any(row.sums <- rowSums(tab) == 0)) {  # check rows
    tab <- tab[-which(row.sums == 0), ]  # omit zero margin totals
  }
```

---

[c]CTree (Chapter 3) avoids the small *p*-value problem internally by working with *p*-values on the log scale.

```
  if (any(col.sums <- colSums(tab) == 0)) {  # check columns
    tab <- tab[, -which(col.sums == 0)]  # omit zero margin totals
  }
  chisq.test(tab)$p.value  # p-value from chi-squared test
}
```

Next, I omit any rows with missing response values and compute the Bonferroni-adjusted p-values from step 2) of Algorithm 4.1 for each feature:

```
aq <- airquality[!is.na(airquality$Ozone), ]
pvals <- sapply(setdiff(names(aq), "Ozone"), FUN = function(x) {
  guide.chisq.test(aq[[x]], y = aq[["Ozone"]])
})
p.adjust(pvals, method = "bonferroni")  # Bonferroni adjusted p-values

#>    Solar.R     Wind     Temp    Month      Day
#> 2.23e-03 1.40e-06 2.50e-14 2.83e-06 5.88e-01
```

As we previously found with CART and CTree, `Temp` is selected to split the root node (as it has the smallest adjusted p-value). We previously found `Temp` = `82.5` to be the optimal split point using an exhaustive search in Section 2.3.1.

4.2.2 Interaction tests

Exhaustive search procedures, like CART, can be insensitive to local interactions; according to Loh [2002], splits that are sensitive to two-way interaction effects can produce shorter trees. GUIDE circumvents this issue by explicitly testing for two-way interactions between the response and each pair of features. The basic idea is to partition the feature space between a pair of predictors to form the columns of a new table, then apply the same chi-square test outlined in step 4) of Algorithm 4.1. If there are k predictors in total, then $k(k+1)/2$ chi-square tests are employed each time a split variable is selected to partition the data (when including two-way interaction effects, that is).

To illustrate, suppose we want to test for an interaction between (x_i, x_j) and y. If both x_i and x_j are ordered, we divide the (x_i, x_j) into four quadrants by splitting the range of each feature into two halves using the sample median. From this, a 2×4 contingency table can be formed, where the rows still represent the discretized residuals, and a chi-square test can be applied. If x_i and x_j are both nominal categorical variables, with c_i and c_j unique categories, respectively, then we form the $2 \times c_i c_j$ contingency table, where the columns are based on all the possible pairs of categories between x_i and x_j. Finally, if x_i is ordered and x_j is nominal (with c_j unique categories), then we split the range of x_i into two halves using the sample median and form a $2 \times 2c_j$ contingency table, where the columns correspond to all possible pairs of values between the binned x_i values and x_j. In any of the above cases, rows or columns with zeros in the margin are omitted before applying the chi-square test.

When including interaction tests, we need to modify how the split variable $x^\star$ is selected in step 4) of Algorithm 4.1. If the smallest p-value is from a curvature test, then select the associated predictor to split the node. If the smallest p-value comes from an interaction test, then the choice of splitting variable depends upon whether both features are ordered or not. If x_i and x_j are both ordered, the node is split using the sample mean for each variable (e.g., $x_i \leq \sum_{i=1}^{N_t} x_i / N_t$). For each of the two splits, a constant (e.g., the mean response) is fit to the resulting nodes. The split yielding the greatest reduction in SSE (p. 58) is selected to split the node. On the other hand, if either x_i or x_j is nominal, select the variable with the smallest p-value from the associated curvature tests. For details, see Algorithm 2 in Loh [2002].

Using a split variable selected from an interaction test does not guarantee that the interacting variable will be used to split one of the child nodes. While it may be intuitive to force this behavior to highlight the specific interaction in the tree, Loh [2002] argues that letting variables compete at each individual split can lead to shorter trees.

4.2.3   Non-constant fits

In contrast to CART and CTree, GUIDE is not restricted to fitting a constant in each node. This generality is due to the fact that the residuals are used to select the splitting variable. Hence, any model that produces residuals can be used to construct the tree. The GUIDE software (Section 4.9) allows a wide range of regression models to be used in each node: simple linear regression, Poisson regression, regression for censored outcomes, and more. The benefit to fitting non-constant models in each node is the potential reduction in tree size and increase to predictive accuracy. Of course, the same idea can be applied to exhaustive search procedures like CART, but this can be too computationally expensive. By abandoning a fully-exhaustive search criteria, GUIDE can afford to fit a richer class of models in the nodes, while substantially reducing split variable selection bias—a win-win, so to speak.

Fitting non-constant models in the nodes of the tree means that the predictors can potentially serve more than one role during tree construction. In particular, predictors can compete for splits and/or serve as a regressor in terminal node fits. For simplicity, GUIDE only allows ordered features to serve as regressors, unless the categorical variables are dummy encoded and treated as numeric [Loh, 2002, p. 371].

Borrowing the same terminology in Loh [2002], there are four basic roles a predictor can serve:

- *n*-variable: a numeric feature used to fit regression models and to split nodes;

- *f*-variable: a numeric feature used to fit regression models but not split nodes;

- *s*-variable: a numeric feature used to split nodes but not fit regression models;

- *c*-variable: a categorical feature used to split nodes but not fit regression models.

Therefore, numeric features can fill one of three roles (e.g., an *n*-, *f*-, or *s*-variable), while categorical features can only be used to split nodes. This gives a great deal of flexibility when fitting regression models in the nodes. For example, we can fit a quadratic model in x_j in each of the nodes by specifying x_j^2 as an *f*-variable, so it's not used to split any nodes.

When employing non-constant fits, Algorithm 4.1 requires a few simple modifications, but I'll defer to Loh [2002, Algorithms 3–4] for details.

4.2.3.1 Example: predicting home prices

To illustrate, let's return to the Ames housing data (Section 1.4.7). Recall that I initially split the data into train/test sets using a 70/30 split; since I'm not plotting anything, I did not bother to rescale the response in this example. Using the GUIDE software (Section 4.9), I built a default regression tree with *stepwise linear regression* models in each node.[d] All variables were allowed to compete for splits, and all numeric features were allowed to compete as predictors in the stepwise procedure applied to each node. The tree was pruned using 10-fold cross-validation along with the 1-SE rule. The resulting tree diagram is displayed in Figure 4.1—the inner caption is part of the output from GUIDE and explains the tree diagram.

The 1-SE pruned GUIDE-based tree for the Ames housing data, using non-constant fits, is substantially smaller than the 1-SE pruned CART tree from Figure 2.23 (p. 97); it is also far more accurate, with a test set RMSE of $28,870.78.[e] For further comparison, CTree, using $\alpha = 0.05$, resulted in a tree with 75 terminal nodes and a test RMSE of $35,331.88.

GUIDE will also output a text file containing the variable importance scores (Section 4.7), estimated regression equation for each terminal node, and more.

---

[d]Since linear models are being used to summarize the terminal nodes, it would be wise to consider log-transforming the response first, or use a similar transformation, since it is quite right skewed, but for comparison to tree fits from previous chapters, I elected not to in this example.

[e]While smaller in size, one could argue that the pruned GUIDE tree is no less interpretable, since the terminal nodes are summarized using regression fits in different subsets of the predictors.

For example, the sale price of any home with a garage capacity for three or more cars, excellent basement quality, good basement exposure, and an above ground living area of less than 2,088 sq. ft. would be estimated according to the following equation:

$$\widehat{\texttt{Sale\_Price}} = -149{,}100.00 + 283.30\texttt{First\_Flr\_SF}, \qquad (4.1)$$

where `First_Flr_SF` is the square footage of the first floor. This corresponds to terminal node 24 in Figure 4.1.

The output file from GUIDE also reported that the tree in Figure 4.1 explains roughly 87.89% of the variance in `Sale_Price` on the training data (i.e., $R^2 = 0.8789$).

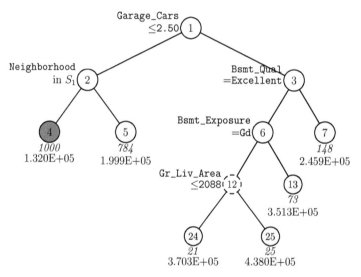

GUIDE v.38.0 0.25-SE piecewise linear least-squares regression tree with stepwise variable selection for predicting `Sale_Price`. Tree constructed with 2051 observations. Maximum number of split levels is 12 and minimum node sample size is 20. At each split, an observation goes to the left branch if and only if the condition is satisfied. Set S_1 = {Briardale, Brookside, Edwards, Iowa_DOT_and_Rail_Road, Landmark, Meadow_Village, Mitchell, North_Ames, Northpark_Villa, Old_Town, Sawyer, South_and_West_of_Iowa_State_University}. Circles with dashed lines are nodes with no significant split variables. Sample size (in *italics*) and mean of `Sale_Price` printed below nodes. Terminal nodes with means above and below value of 1.802E+05 at root node are colored yellow and purple respectively. Second best split variable at root node is `Neighborhood`.

FIGURE 4.1: Example tree diagram produced by GUIDE for the Ames housing example. Stepwise linear regression models were fit in each node. The autogenerated caption produced by the GUIDE software is also included.

4.2.3.2 Bootstrap bias correction

It is difficult to achieve unbiased split variable selection in regression trees that employ regression models in the nodes because the predictors can be used for splitting, referred to as a "split" variable, or as a regressor in the models, where it's referred to as a "fit" variable [Loh, 2014].

When non-constant models are fit to the nodes, the split variable selection procedure in GUIDE is heavily biased towards the *c*- and *s*-variables. This is because the *n*-variables, which are used for both splitting and as regressors, are uncorrelated with the resulting residuals (a property of *ordinary least squares*).

Loh [2002, Algorithm 5] proposed a bootstrap calibration procedure to help correct the split variable selection bias in this situation. I'll omit the details, but the basic idea is to shrink the *p*-values associated with the chi-square tests for the *n*-variables.

4.3 A GUIDE for classification

GUIDE for classification is not that different from its regression counterpart. Instead of residuals, the categorical outcome is used directly in the chi-square tests for split variable selection. Also, once a splitting variable, say $x^\star$, is selected, the optimal split point is found using an exhaustive search, similar to CART's approach based on a weighted sum of Gini impurities (see Section 2.2.1).

4.3.1 Linear/oblique splits

Although orthogonal (or binary) splits are more interpretable, Loh [2009] makes a compelling case for splits based on linear combinations of predictors (which are referred to as either *linear splits* or *oblique splits*, since they are no longer orthogonal to the feature axes). An oblique split on two continuous features, x_i and x_j, takes the form $ax_i + bx_j \leq c$, where a, b, and c are constants determined from the data; see Loh [2009, Sec. 3] for details.

Using orthogonal splits can result in smaller trees and greater predictive accuracy. GUIDE only allows linear splits for classification problems[f] and is restricted to two variables, x_i and x_j (say), only when an interaction test between x_i and x_j is not significant using another Bonferroni correction. The form of the linear split is chosen using LDA; see Loh [2009, Procedure 3.1] for details. In the official GUIDE software, oblique spits can be given higher or lower priority than orthogonal splits (see Section 4.9). Loh [2009] also mentions that while oblique splits are more powerful than orthogonal splits, it is not necessary to apply them to split each node, which he illustrates with an example on classifying fish species.

Even when linear splits are allowed, Loh [2009] showed that the GUIDE procedure for classification is still practically unbiased in terms of split variable selection.

4.3.1.1   Example: classifying the Palmer penguins

To appreciate the benefits of using linear splits, let's look at an example with the Palmer penguins data. The data, which are available in the R package **palmerpenguins** [Horst et al., 2020], contain the size measurements (flipper length, body mass, and bill dimensions) for three species of adult foraging penguins near Palmer Station, Antarctica. For this example, I'll try to classify the three species using just two measurements: bill length in mm (`bill_length_mm`) and bill depth in mm (`bill_depth_mm`); see Figure 4.2.

 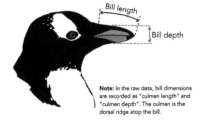

FIGURE 4.2: Artwork by Allison Horst. Source: `https://github.com/allisonhorst/palmerpenguins/`.

Figure 4.3 shows a scatterplot of the bill length vs. bill depth for the three species of penguins. While there does seem to be a good deal of separation between the three species using `bill_depth_mm` and `bill_length_mm`, it will

---

[f]Breiman et al. [1984, p. 248] argue that splits on linear combinations of predictors are less effective in regression problems compared to classification, apparently because linear combination splits tend to produce rectangular-like regions when partitioning the feature space in regression, similar to the more common, but easier to obtain, orthogonal splits.

be challenging for a classification tree that uses splits that are orthogonal to the x- and y-axes (e.g., CART and CTree). If the data come from a multivariate normal distribution with a common covariance matrix across the three species, then LDA would give the optimal linear decision boundary (if the covariance matrices differ between the classes, then QDA would be optimal). If we cannot make those assumptions, then a tree-based approach using oblique splits is a good alternative.

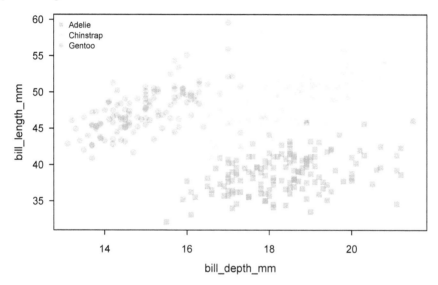

FIGURE 4.3: Scatterplot of bill depth (mm) vs. bill length (mm) for the three species of Palmer penguins.

To illustrate, consider the plots in Figure 4.4, which show the decision boundaries from a GUIDE decision tree with linear splits (top left), LDA (top right), CART (middle left), CTree (middle right), a *random forest* (bottom left), *gradient boosted tree ensemble* (bottom right); the latter two are special types of tree-based ensembles and are discussed in Chapters 7–8. Both GUIDE and CART were pruned using 10-fold cross-validation with the 1-SE rule (for CTree, I used the default $\alpha = 0.05$). Notice the similarity (and simplicity) of the linear decision boundaries produced by GUIDE and LDA; these models are likely to generalize better to new data from the same population. Furthermore, GUIDE only misclassified 8 observations, while LDA, CART, and CTree misclassified 13, 21, and 15 observations, respectively. Compared to CART and CTree, the tree-based ensembles (bottom row) are a bit more flexible and able to adapt to linear decision boundaries, but in this case, they're not as smooth or simple to explain as the LDA or GUIDE decision boundaries.

The associated tree diagram for the fitted GUIDE tree with linear splits is shown in Figure 4.5. The GUIDE tree using linear splits is simpler compared

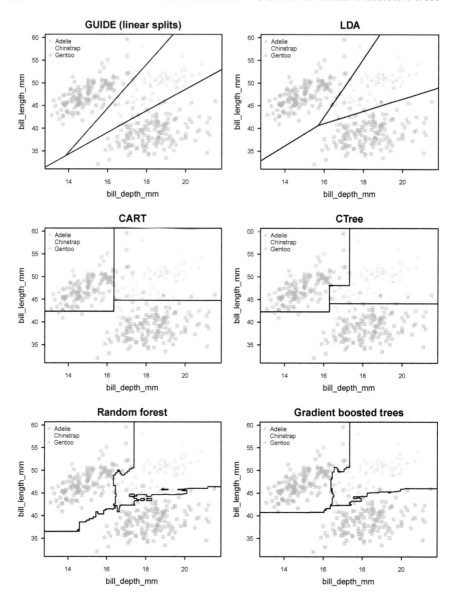

FIGURE 4.4: Decision boundaries from several classification models applied to the Palmer penguins data. Top left: a single GUIDE decision tree with linear splits. Top right: linear discriminant analysis. Middle left: a single CART tree. Middle right: a single CTree. Bottom left: a random forest of 1000 CART-like trees. Bottom right: gradient boosted CART-like trees.

to the associated CART and CTree trees (not shown); the former uses two splits, while the latter two require three and seven splits, respectively.

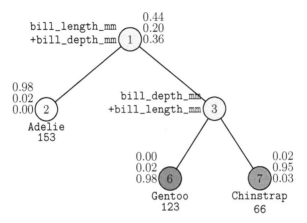

GUIDE v.38.0 1.00-SE classification tree for predicting `species` using linear split priority, estimated priors and unit misclassification costs. Tree constructed with 342 observations. Maximum number of split levels is 10 and minimum node sample size is 3. At each split, an observation goes to the left branch if and only if the condition is satisfied. Intermediate nodes in lightgray indicate linear splits. Predicted classes and sample sizes printed below terminal nodes; class sample proportions for `species` = `Adelie`, `Chinstrap`, and `Gentoo`, respectively, beside nodes.

FIGURE 4.5: Example tree diagram (with description) produced by GUIDE for the Palmer penguins data using linear splits. The autogenerated caption produced by the GUIDE software is also included.

4.3.2   Priors and misclassification costs

Like CART, GUIDE can incorporate different priors and unequal misclassification costs. See Section 2.2.4 and Loh [2009] for details.

4.3.3   Non-constant fits

Similar to GUIDE for regression, we can increase flexibility by fitting non-constant models in the nodes of the classification tree. For increased accuracy, or a greater reduction in the size of the tree, Loh [2009] suggests fitting either a *kernel density estimate* or a k-nearest neighbor model to each node. Owing to the flexibility of these models, however, GUIDE dispenses with linear splits when allowing non-constant fits. When there are weak main effects, but strong two-way interaction effects, classification trees constructed with non-constant fits can achieve substantial gains in accuracy and tree compactness. Although they require more computation than classification trees with constant fits, Loh [2009] empirically shows that their prediction accuracy is often relatively high.

4.3.3.1   Kernel-based and k-nearest neighbor fits

Kernel-based fits essentially select the class with the highest kernel density estimate in each node. For example, if the selected split variable x is ordered, then for each class in node t, a kernel density estimate $\hat{f}(x)$ is computed according to

$$\hat{f}(x) = \frac{1}{N_t h} \sum_{i=1}^{N_t} \phi \left(\frac{x - x_i}{h} \right),$$

where $\phi(\cdot)$ denotes the density function of a standard normal distribution, and

$$h = \begin{cases} 2.5 \min \left(s, 0.7413r \right) N_t^{-1/5} & \text{if } r > 0 \\ 2.5 s N_t^{-1/5} & \text{otherwise} \end{cases},$$

is the *bandwidth*; here, s and r denote the sample standard deviation and *interquartile range* (IQR) of the observed x values. Some motivation for using this bandwidth, which is more than twice as large as the usual bandwidth recommended for density estimation, is given in Ghosh et al. [2006].

A similar idea is used when applying k-nearest neighbor fits to each node. In particular, k is given by

$$k = \max \left(3, \lceil \log \left(N_t \right) \rceil \right),$$

where $\lceil x \rceil$ just means round x up to the nearest integer. The minimum number of neighbors (k) is three to avoid too much trouble in dealing with ties.

4.4   Pruning

The GUIDE algorithm continues to recursively perform splits until some stopping criteria are met (e.g., the minimum number of observations required for splitting has been reached in each node). Like CART, this will often lead to overfitting and an overly complex tree with too many splits. To circumvent the issue, GUIDE adopts the same cost-complexity pruning strategy used in CART (revisit Section 2.5 for the details). Loh [2002] showed via simulation that pruning can reduce the variable selection bias in exhaustive search procedures like CART, provided there are no interaction effects.

4.5 Missing values

GUIDE solves the problem of missing values by assigning missing categorical values to a new "missing" category. The same happens to ordered variables during the construction of the chi-square tests; their missing values are then mapped back to $-\infty$ for split point selection. Hence, the split $x^\star \leq c$ will always send missing values to the left child node. If informative enough, missing values can be isolated completely by choosing $c = -\infty$. See Loh et al. [2020] for more details and comparisons with other approaches to handling missing values.

4.6 Fitted values and predictions

For constant fits, the fitted values and predictions are obtained in the same way as CART and CTree. For example, if y is continuous, then the average within-node response is used. For classification, the within node class proportions can serve as predicted class probabilities. For non-constant fits, the fitted values and predictions for new observations depend on the model fit to each node (e.g., polynomial regression or k-nearest neighbor). The latter tend to produce shorter and more accurate trees, but are more computationally intensive and less interpretable.

4.7 Variable importance

Variable importance scores in GUIDE are based on the sum of the weighted one-*df* chi-square test statistics across the nodes of the tree, where the weights are given by the square root of the sample size of the corresponding nodes. In particular, let $q_t(x)$ be the one-*df* chi-square statistic associated with predictor x at node t. The variable importance score for x is

$$\text{VI}(x) = \sum_t \sqrt{N_t} q_t(x),$$

where N_t is the sample size in node t; see Loh [2012] for details. Later, Loh et al. [2015] suggest using N_t rather than its square root to weight the chi-square statistics, which increases the probability that the feature used to split the root node has the largest importance score. This approach is approximately unbiased unless there is a mix of both ordered and nominal categorical variables. Loh and Zhou [2021] provide an improved version for regression that ensures unbiasedness.

While there are a number of approaches to computing variable importance (especially from decision trees), few include thresholds for identifying the irrelevant (or pure noise) features. In GUIDE, any feature x with a variable importance score less than the 0.95-quantile of the approximate null distribution of $\text{VI}(x)$ is considered unimportant. For example, running GUIDE in variable importance mode (with default settings, which caps the total number of splits at four) on the Ames housing data flagged 64 of the 80 features as being highly important, 4 as being less important, and 12 as being unimportant. The results are displayed in Figure 4.6 (p. 174).

As previously mentioned, the GUIDE software continues to evolve and the details mentioned above may not correspond exactly with what's currently implemented. Fortunately, Loh and Zhou [2021] give a relatively recent account of GUIDE's approach to variable importance, and provide a thorough comparison against other tree-based approaches to computing variable importance, including CART, CTree, and many of the tree-based ensembles discussed later in this book.

4.8   Ensembles

Although I'll defer the discussion of ensembles until Chapter 5, it's worth noting that GUIDE supports two types of GUIDE-based tree ensembles: *bagging* (Section 5.1) and random forest (Chapter 7).

4.9 Software and examples

GUIDE, along with its predecessors, is not open source and exists as a command line program. Compiled binaries for QUEST, CRUISE, and GUIDE are freely available from `http://pages.stat.wisc.edu/~loh/guide.html` and are compatible with most major operating systems. If you're comfortable with the terminal, GUIDE is straightforward to install and use; see the available manual for installation specifics and example usage. Although it's a terminal application, GUIDE will optionally generate R code that can be used for scoring after a tree has been fit. This makes it easy to run simulations and the like after the tree has been built. My only criticism of GUIDE is that it's not currently available via easy-to-use open source code (like R or Python); if it were, it'd probably be much more widely adopted by practitioners.

As discussed in the previous section, GUIDE is a command line program that requires input from the user. For this reason, I'll limit this section to a single example; the GUIDE software manual [Loh, 2020] offers plenty of additional examples. Note that GUIDE can optionally generate R code to reproduce predictions from the fitted tree model, which can be useful for simulations and deployment. Further, I used GUIDE v38.0 for all the examples in this chapter.

4.9.1 Example: credit card default

The credit card default data [Yeh and hui Lien, 2009], available from the UCI Machine Learning Repository at

`https://archive.ics.uci.edu/ml/datasets/default+of+credit+card+clients,`

contains demographic and payment information about credit card customers in Taiwan in the year 2005. The data set contains 30,000 observations on the following 23 variables:

- `default`: A binary indicator of whether or not the customer defaulted on their payment (`yes` or `no`).

- `limit_bal`: Amount of credit (NT dollar) given to the customer.

- `sex`: The customer's gender (`male` or `female`).

- `education`: The customer's level of education (`graduate school`, `university`, `high school`, or `other`).

- `marriage`: The customer's marital status (`married`, `single`, or `other`).

- **age**: The customer's age in years.

- **pay_0**, **pay_2–pay_6**: History of past payment; **pay_0** = the repayment status in September, 2005; **pay_2** = the repayment status in August, 2005; ...; **pay_6** = the repayment status in April, 2005. The measurement scale for the repayment status is: -1 = pay duly; 1 = payment delay for one month; 2 = payment delay for two months; ...; 8 = payment delay for eight months; 9 = payment delay for nine months and above.

- **bill_amt1–bill_amt6**: Amount of bill statement (NT dollar). **bill_amt1** = amount of bill statement in September, 2005; **bill_amt2** = amount of bill statement in August, 2005; ...; **bill_amt6** = amount of bill statement in April, 2005.

- **pay_amt1–pay_amt6**: Amount of previous payment (NT dollar). **pay_amt1** = amount paid in September, 2005; **pay_amt1** = amount paid in August, 2005; ...; **pay_amt6** = amount paid in April, 2005.

The goal is to build a model to predict the probability of a customer defaulting on their credit card payment. For that I'll build a GUIDE classification tree. The R code I used to download and clean the data is shown below. To start, I download the data into a temporary file before reading in the resulting XLS file (i.e., Microsoft Excel spreadsheet) using the **readxl** package [Wickham and Bryan, 2019] and printing a compact summary of the data using **str()**:

```
# Download and read in the credit default data from the UCI ML repo
tf <- tempfile(fileext = ".xls")
url <- paste0("https://archive.ics.uci.edu/ml/",  # sigh, long URLs...
              "machine-learning-databases/",
              "00350/default%20of%20credit%20card%20clients.xls")
download.file(url, destfile = tf)
credit <- as.data.frame(readxl::read_xls(tf, skip = 1))

# Clean up column names a bit
names(credit) <- tolower(names(credit))
names(credit)[names(credit) == "default payment next month"] <-
  "default"

str(credit)  # compactly display structure of the data frame
#> 'data.frame': 30000 obs. of  25 variables:
#>  $ id       : num  1 2 3 4 5 6 7 8 9 10 ...
#>  $ limit_bal: num  20000 120000 90000 50000 50000 500..
#>  $ sex      : num  2 2 2 2 1 1 1 2 2 1 ...
#>  $ education: num  2 2 2 2 2 1 1 2 3 3 ...
#>  $ marriage : num  1 2 2 1 1 2 2 2 1 2 ...
#>  $ age      : num  24 26 34 37 57 37 29 23 28 35 ...
#>  $ pay_0    : num  2 -1 0 0 -1 0 0 0 0 -2 ...
#>  $ pay_2    : num  2 2 0 0 0 0 0 -1 0 -2 ...
#>  $ pay_3    : num  -1 0 0 0 -1 0 0 -1 2 -2 ...
```

```
#>  $ pay_4    : num  -1 0 0 0 0 0 0 0 0 -2 ...
#>  $ pay_5    : num  -2 0 0 0 0 0 0 0 0 -1 ...
#>  $ pay_6    : num  -2 2 0 0 0 0 0 -1 0 -1 ...
#>  $ bill_amt1: num  3913 2682 29239 46990 8617 ...
#>  $ bill_amt2: num  3102 1725 14027 48233 5670 ...
#>  $ bill_amt3: num  689 2682 13559 49291 35835 ...
#>  $ bill_amt4: num  0 3272 14331 28314 20940 ...
#>  $ bill_amt5: num  0 3455 14948 28959 19146 ...
#>  $ bill_amt6: num  0 3261 15549 29547 19131 ...
#>  $ pay_amt1 : num  0 0 1518 2000 2000 ...
#>  $ pay_amt2 : num  689 1000 1500 2019 36681 ...
#>  $ pay_amt3 : num  0 1000 1000 1200 10000 657 38000 0..
#>  $ pay_amt4 : num  0 1000 1000 1100 9000 ...
#>  $ pay_amt5 : num  0 0 1000 1069 689 ...
#>  $ pay_amt6 : num  0 2000 5000 1000 679 ...
#>  $ default  : num  1 1 0 0 0 0 0 0 0 0 ...
```

Note that the categorical variables have been numerically re-encoded. The
next code chunk removes the column ID and cleans up some of the categorical
features by re-encoding them from numeric back to the actual categories based
on the provided column descriptions:

```
# Remove ID column
credit$id <- NULL

# Clean up categorical features
credit$sex <- ifelse(credit$sex == 1, yes = "male", no = "female")
credit$education <- ifelse(
  test = credit$education == 1,
  yes = "graduate school",
  no = ifelse(
    test = credit$education == 2,
    yes = "university",
    no = ifelse(
      test = credit$education == 3,
      yes = "high school",
      no = "other"
    )
  )
)
credit$marriage <- ifelse(
  test = credit$marriage == 1,
  yes = "married",
  no = ifelse (
    test = credit$marriage == 2,
    yes = "single",
    no = "other"
  )
)
```

```
credit$default <- ifelse(credit$default == 1, yes = "yes", no = "no")

# Coerce character columns to factors
for (i in seq_len(ncol(credit))) {
  if (is.character(credit[[i]])) {
    credit[[i]] <- as.factor(credit[[i]])
  }
}
```

Finally, I'll split the data into train/test sets using a 70/30 split, leaving 21,000 observations for training and 9,000 for estimating the generalization performance:

```
set.seed(1342)   # for reproducibility
trn.ids <- sample(nrow(credit), size = 0.7 * nrow(credit),
                  replace = FALSE)
credit.trn <- credit[trn.ids, ]
credit.tst <- credit[-trn.ids, ]
```

The GUIDE program requires two special text files before it can be called:

- the data input file;

- a description file.

See the GUIDE reference manual [Loh, 2020] for full details. The data input file is essentially just a text file containing the training data in a format that can be consumed by GUIDE. The description file provides some basic metadata, like the missing value flag and variable roles. These files can be a pain to generate, especially for data sets with lots of columns, so I included a little helper function in **treemisc** to help generate them; see `?treemisc::guide_setup` for argument details.

Below is the code I used to generate the data input and description files for the credit card default example. By default, numeric columns are used both for splitting the nodes and for fitting the node regression models for non-constant fits, and categorical variables are used for splitting only. In my setup, I have a /guide-v38.0/credit directory containing the GUIDE executable and where the generated files will be written to:

```
treemisc::guide_setup(credit.trn, path = "guide-v38.0/credit",
                      dv = "default", file.name = "credit",
                      verbose = TRUE)

#> Writing data file to guide-v38.0/credit/credit.txt...
#> Writing description file to guide-v38.0/credit/credit_desc.txt...
```

This resulted in the creation of two files in /guide-v38.0/credit (my directory for this example):

- credit.txt (the training data input file in the format required by GUIDE);

- credit_desc.txt (the description file).

Below are the contents of the generated credit_desc.txt file. The first line gives the name of the training data file; if the file is not in the current working directory, its full path must be given with quotes (e.g., "some/path/to/credit.txt"). The second line specifies the missing value code (if it contains non-alphanumeric characters, then it too must be quoted). The remaining lines specify the column number, name, and role for each variable in the data input file. As you can imagine, creating this file for a data set with lots of variables can be tedious, hence the reason for writing a helper function.

```
credit.txt
NA
2
1 limit_bal n
2 sex c
3 education c
4 marriage c
5 age n
6 pay_0 n
7 pay_2 n
8 pay_3 n
9 pay_4 n
10 pay_5 n
11 pay_6 n
12 bill_amt1 n
13 bill_amt2 n
14 bill_amt3 n
15 bill_amt4 n
16 bill_amt5 n
17 bill_amt6 n
18 pay_amt1 n
19 pay_amt2 n
20 pay_amt3 n
21 pay_amt4 n
22 pay_amt5 n
23 pay_amt6 n
24 default d
```

With the credit.txt and credit_desc.txt files in hand (and in the appropriate directories required by GUIDE), we can spin up a terminal and call the GUIDE program. I'll omit the details since it's OS-specific, but the GUIDE reference manual will take you through each step. Once the program is called, GUIDE will ask the user for several inputs (e.g., whether to build a classification or regression tree, whether to use constant or non-constant fits, number of folds to use for cross-validation, etc.). In the end, GUIDE generates a special input text file to be consumed by the software.

Below are the contents of the input file for the credit card default example, called credit_in.txt, highlighting all the options I selected (I basically requested a default classification tree that's been pruned using the 1-SE rule with 10-fold cross-validation, but you can see several of the available options in the output).

```
GUIDE          (do not edit this file unless you know what you are doing)
   38.0        (version of GUIDE that generated this file)
 1             (1=model fitting, 2=importance or DIF scoring, 3=data con...
"credit_out.txt"  (name of output file)
 1             (1=one tree, 2=ensemble)
 1             (1=classification, 2=regression, 3=propensity score group...
 1             (1=simple model, 2=nearest-neighbor, 3=kernel)
 2             (0=linear 1st, 1=univariate 1st, 2=skip linear, 3=skip li...
 1             (0=tree with fixed no. of nodes, 1=prune by CV, 2=by test...
"credit_desc.txt"  (name of data description file)
         10    (number of cross-validations)
 1             (1=mean-based CV tree, 2=median-based CV tree)
     1.000     (SE number for pruning)
 1             (1=estimated priors, 2=equal priors, 3=other priors)
 1             (1=unit misclassification costs, 2=other)
 2             (1=split point from quantiles, 2=use exhaustive search)
 1             (1=default max. number of split levels, 2=specify no. in ...
 1             (1=default min. node size, 2=specify min. value in next l...
 2             (0=no LaTeX code, 1=tree without node numbers, 2=tree wit...
"credit.tex" (latex file name)
 1             (1=color terminal nodes, 2=no colors)
 2             (0=#errors, 1=sample sizes, 2=sample proportions, 3=poste...
 3             (1=no storage, 2=store fit and split variables, 3=store s...
"credit_splits.txt" (split variable file name)
 2             (1=do not save fitted values and node IDs, 2=save in a file)
"credit_fitted.txt" (file name for fitted values and node IDs)
 2             (1=do not write R function, 2=write R function)
"credt_pred.R" (R code file)
 1             (rank of top variable to split root node)
```

Now, all you have to do is feed this input file back into the GUIDE program (again, see the official manual for details). Once the modeling process is complete, you'll end up with several files depending on the options you specified during the initial setup. A portion of the output file produced by GUIDE for my input file is shown below; the corresponding tree diagram is displayed in Figure 4.7.

```
Node 1: Intermediate node
 A case goes into Node 2 if pay_0 <= 1.5000000
 pay_0 mean = -0.15095238E-01
```

```
Class       Number   Posterior
no           16354   0.7788E+00
yes           4646   0.2212E+00
Number of training cases misclassified = 4646
Predicted class is no
----------------------------
Node 2: Terminal node
 Class      Number   Posterior
 no          15686   0.8331E+00
 yes          3143   0.1669E+00
Number of training cases misclassified = 3143
Predicted class is no
----------------------------
Node 3: Terminal node
 Class      Number   Posterior
 no            668   0.3077E+00
 yes          1503   0.6923E+00
Number of training cases misclassified = 668
Predicted class is yes
----------------------------


Classification matrix for training sample:
Predicted      True class
class              no        yes
no              15686       3143
yes               668       1503
Total           16354       4646


Number of cases used for tree construction: 21000
Number misclassified: 3811
Resubstitution estimate of mean misclassification cost: 0.18147619
```

The train and test set accuracies for this tree are 81.85% and 82.21%, respectively (I had to use the R function produced by GUIDE to compute the test accuracy). Despite the reasonably high accuracy, we have a big problem! If you didn't first notice when initially exploring the data on your own, then hopefully you see it now...the model is biased towards predicting yes since the data are imbalanced (and naturally so, since we'd hope that most people are not defaulting on their credit card payments).

The original goal was to build a model to predict the probability of defaulting (default = "yes"), but the train and test accuracy within that specific class are 32.35% and 33.87%, respectively. By default, GUIDE (and many other algorithms) treat the misclassification for both types of error (i.e., predicting a yes as a no and vice versa) as equal. Fortunately, like CART, GUIDE can incorporate a matrix of misclassification costs into the tree construction (see Section 2.2.4).

Suppose, for whatever reason, we considered misclassifying a yes as a no (i.e., predicting that someone will not default on their next payment, when in fact they did) to be five times more costly than predicting a no as a yes. In GUIDE, it's as easy as setting up a loss matrix text file. For this example, I created a file called credit_loss.txt (in the same /guide-v38.0/credit directory as before) with the following two lines:

```
0 5
1 0
```

This corresponds to the following loss (or misclassification cost) matrix

$$L = \begin{array}{c} \\ No \\ Yes \end{array} \begin{array}{c} No \quad Yes \\ \left(\begin{array}{cc} 0 & 5 \\ 1 & 0 \end{array} \right) \end{array},$$

where $L_{i,j}$ denotes the cost of classifying an observation as class i when it really belongs to class j. Note that GUIDE sorts the class values in alphabetical order (i.e., "no" then "yes"). Re-running the previous program, but specifying the cost matrix file when prompted, leads to the much more useful tree structure shown in Figure 4.8. Although the overall test accuracy dropped from 82.21% to 60.00%, the accuracy within the class of interest (the *true positive rate* or *sensitivity*) increased from 7.49% to 17.79%—a significant improvement. I'll leave it to the reader to explore further with linear splits and non-constant fits to see if the results can be improved further.

4.10   Final thoughts

This chapter introduced the GUIDE algorithm for building classification and regression trees. GUIDE was developed to solve three problems often encountered with exhaustive search procedures (like CART):

1. split variable selection bias;

2. insensitivity to local interactions;

3. overly complex tree structures.

Like CTree, GUIDE solves the first problem by decoupling the search for split variables from the split point selection using statistical tests; in contrast to CTree, GUIDE exclusively uses one-*df* chi-square tests throughout. In selecting the splitting variable, GUIDE also looks at two-way interaction effects that can potentially mask the importance of a split when only main effects are considered. Moreover, GUIDE can often produce smaller and more accurate tree structures by allowing splits in linear combinations of (two) predictors, and fitting more complex (i.e., non-constant) models in the nodes (e.g., k-nearest neighbor or polynomial regression). Although this can lead to much shorter tree structures, the trees themselves are not necessarily simpler to interpret (e.g., is the GUIDE-based regression tree for the Ames housing example based on stepwise regression fits actually that simple to interpret?). Fortunately, the post-hoc interpretation procedures outlined in Chapter 6 are model-agnostic, and allow us to easily interpret various aspects of any supervised learning model (including GUIDE-based decision trees with complex terminal node summaries).

GUIDE is not just for simple classification and regression problems, and can be used in all sorts practical situations, including:

- quantile regression;

- Poisson regression (i.e., for modeling count data and rates);

- multivariate outcomes;

- censored outcomes;

- longitudinal data [Loh and Zheng, 2013] (e.g., when multiple subjects are continually measured over time);

- propensity score grouping;

- variable reduction;

- subgroup identification [Loh et al., 2015];

- and more.

For more, visit Wei-Yin Loh's website at `http://pages.stat.wisc.edu/~loh/guide.html`. An example of the effectiveness of *riluzole*, a drug approved for the treatment of ALS (Amyotrophic Lateral Sclerosis) by the US FDA, can be found in Loh and Zhou [2020][g].

---

[g]If you have trouble accessing any of Loh's papers, many of them are freely available on his website at `http://pages.stat.wisc.edu/~loh/guide.html`.

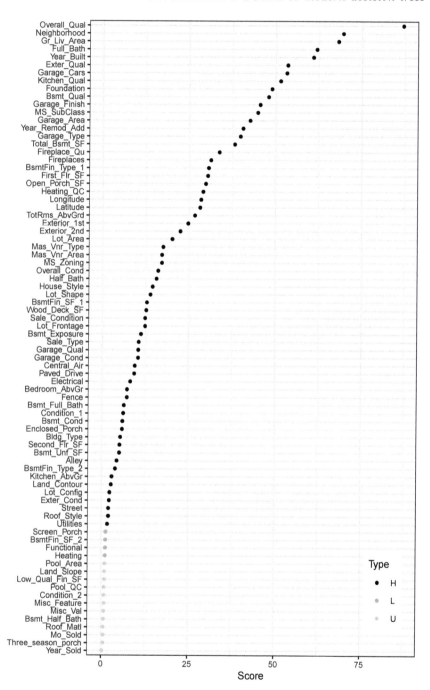

FIGURE 4.6: GUIDE-based variable importance scores for the Ames housing example. GUIDE distinguished between highly important (H), less important (L), and unimportant (U).

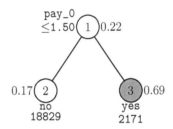

GUIDE v.38.0 1.00-SE classification tree for predicting `default` using estimated priors and unit misclassification costs. Tree constructed with 21000 observations. Maximum number of split levels is 30 and minimum node sample size is 210. At each split, an observation goes to the left branch if and only if the condition is satisfied. Predicted classes and sample sizes printed below terminal nodes; class sample proportion for `default` = `yes` beside nodes. Second best split variable at root node is `pay_2`.

FIGURE 4.7: GUIDE-based classification tree for the credit card default example. The autogenerated caption produced by the GUIDE software is also included.

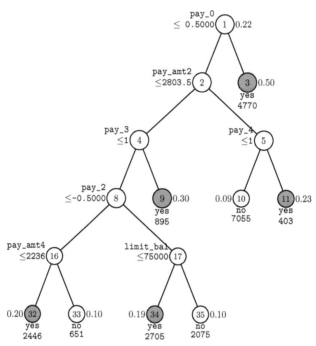

GUIDE v.38.0 1.00-SE classification tree for predicting `default` using estimated priors and specified mis-classification costs. Tree constructed with 21000 observations. Maximum number of split levels is 30 and minimum node sample size is 210. At each split, an observation goes to the left branch if and only if the condition is satisfied. Predicted classes and sample sizes printed below terminal nodes; class sample proportion for `default` = `yes` beside nodes. Second best split variable at root node is `pay_2`.

FIGURE 4.8: GUIDE-based classification tree with unequal misclassification costs for the credit card default example. The autogenerated caption produced by the GUIDE software is also included.

Part II

Tree-based ensembles

5

Ensemble algorithms

You know me, I think there ought to be a big old tree right there. And let's give him a friend. Everybody needs a friend.

Bob Ross

This chapter serves as a basic introduction to *ensembles*; specifically, ensembles of decision trees, although the ensemble methods discussed in this chapter are general algorithms that can also be applied to non-tree-based methods. The idea of ensemble modeling is to combine many models together in an attempt to increase overall prediction accuracy. As we'll see in this chapter, how the individual models are created and combined differs between the various ensembling techniques.

Have you ever watched the show *Who Wants to be a Millionaire?* If not, the game is very simple. A contestant is asked a series of multiple choice questions (each with four choices) of increasing difficulty, with a top prize of $1,000,000. The contestant is allowed to use three "lifelines," a form of assistance to help the contestant with difficult questions. Over the years, a number of different lifelines were made available (e.g., the contestant could choose to randomly eliminate two of the incorrect answers). One of the most useful lifelines (IMO) involved polling the audience. If the contestant chose this lifeline, each audience member was able to use a device to cast their vote as to what they thought was the correct answer. The proportion of votes for each multiple choice answer was displayed to the contestant who could then choose to go with the popular vote or not. This lifeline was notorious for its accuracy; according to some sources, Regis Philbin (one of the hosts) once stated that the audience is right 95% of the time!

So why was polling the audience so accurate? As it turns out, this phenomenon has been observed for centuries and is often referred to as *the wisdom of the crowd*. In particular, it is often the case that the aggregated answers from a large, diverse group of individuals is as accurate, if not more accurate, than the answer from any one individual from the group. For an interesting example, try looking up the phrase "Francis Galton Ox weight guessing" in a search engine. Another neat example is to ask a large number of individuals to guess how many jelly beans are in a jar, after you've eaten a handful, of course. If you look at the individual guesses, you'll likely notice that they vary all over the place. The average guess, however, tends to be closer than most of the individual guesses.

In a way, ensembles use the same idea to help improve the predictions (i.e., guesses) of an individual model and are among the most powerful supervised learning algorithms in existence. While there are many different types of ensembles, they tend to share the same basic structure:

$$f_B(\boldsymbol{x}) = \beta_0 + \sum_{b=1}^{B} \beta_b f_b(\boldsymbol{x}), \tag{5.1}$$

where B is the size of the ensemble, and each member of the ensemble $f_b(\boldsymbol{x})$ (also called a *base learner*) is a different function of the input variables derived from the training data.

In this chapter, our interests lie primarily in using decision trees for the base learners—typically, CART-like decision trees (Chapter 2), but any tree algorithm will work. As discussed in Hastie et al. [2009, Section 10.2], many supervised learning algorithms (not just ensembles) can be seen as some form of additive expansion like (5.1). A single decision tree is one such example of an additive expansion. For a single tree, $f_b(\boldsymbol{x}) = f_b(\boldsymbol{x}; \theta_b)$, where θ_b collectively represents the splits and split points leading to the b-th terminal node region, whose prediction is given by β_b (i.e., the terminal node mean response for ordinary regression trees). Other examples include *single-hidden-layer neural networks* and MARS [Friedman, 1991], among others.

There exist many different flavors of ensembles, and they all differ in the following ways:

- the choice of the base learners $f_b(\boldsymbol{x})$ (although, in this book, the base learners will always be some form of decision tree);

- how the base learners are derived from the training data;

- the method for obtaining the estimated coefficients (or weights) $\{\beta_b\}_{b=1}^{B}$.

The ensemble algorithms discussed in this book fall into two broad categories, to be discussed over the next two sections: *bagging* (Section 5.1), short for bootstrap **agg**regat**ing**), and *boosting* (Section 5.2). First, I'll discuss bagging, one of the simplest approaches to constructing an ensemble.

5.1 Bootstrap aggregating (bagging)

Bagging [Breiman, 1996a] is a *meta-algorithm* based on aggregating the results from multiple bootstrap samples. In the context of machine learning, this means aggregating the predictions from different base learners derived from independent bootstrap samples. When applied to unstable learners that are adaptive to the data, like overgrown/unpruned decision trees, the aggregated predictions can often be more accurate than the individual predictions from a single base learner trained on the original learning sample.

While bagging is a general algorithm, and can be applied to any type of base learner, it is most often successfully applied to decision trees, in particular, trees that have been fully grown to near-maximal depth without any pruning. As we learned in Chapter 2, unpruned decision trees are considered unstable learners and have high variance (i.e., the predictions will vary quite a bit from sample to sample), which often results in overfitting and poor generalization performance. However, through averaging, bagging can often stabilize and reduce variance while maintaining low bias, which can result in improved performance.

We'll illustrate the effect of bagging through a simple example. Consider a training sample of size $N = 500$ generated from the following sine wave with Gaussian noise:

$$Y = \sin(X) + \epsilon,$$

where $X \sim \mathcal{U}(0, 2\pi)$ and $\epsilon \sim \mathcal{N}(0, \sigma = 0.3)$. Figure 5.1 (left) shows the prediction surface from a single (overfit) decision tree grown to near full depth.[a] In contrast, Figure 5.1 (right) shows a bagged ensemble of $B = 1000$ such trees whose predictions have been averaged together; here, each tree was induced from a different bootstrap sample of the original data points. Clearly the individual tree is too complex (i.e., low bias and high variance) and will not generalize well to new samples, but averaging many such trees together resulted in a smoother, more stable prediction. The MSE from an independent test set of 10,000 observations was 0.173 for the single tree and 0.1 for the bagged tree ensemble; the optimal MSE for this example is $\sigma^2 = 0.3^2 = 0.09$.

The general steps for bagging classification and regression trees are outlined in Algorithm 5.1. To help further illustrate, a simple schematic of the process for building a bagged tree ensemble with four trees is given in Figure 5.2. Note that bagged tree ensembles can be extended beyond simple classification and regression trees. For example, it is also possible to bag *survival trees* [Hothorn

---

[a] In this example, each tree was fit using `rpart()` with `minsplit = 2` and `cp = 0`.

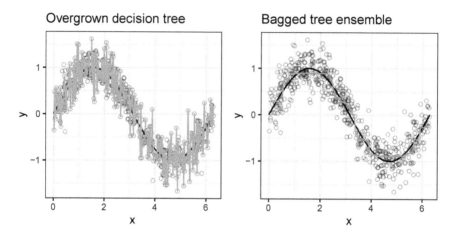

FIGURE 5.1: Simulated sine wave example ($N = 500$). Left: a single (overgrown) regression tree. Right: a bagged ensemble of $B = 1000$ overgrown regression trees whose predictions have been averaged together; here each tree was induced from a different bootstrap sample of the original data points. The individual tree is too complex (i.e., low bias and high variance) but averaging many such trees together results in a more stable prediction and smoother fit.

et al., 2004]. An improved bagging strategy specific to decision trees, called *random forest*, is the topic of Chapter 7.

The (optional) OOB data in step 2) (b) of Algorithm 5.1 will be discussed in Section 7.3, so just ignore it for now. The recommended default value for n_{min} (the minimum size of any terminal node in a tree) depends on the type of response variable:

$$n_{min} = \begin{cases} 1, & \text{for classification} \\ 5, & \text{for regression} \end{cases}.$$

Step 4) of Algorithm 5.1 mentions classification via voting. By voting, I mean that each tree makes a classification (i.e., casts a vote), and the ensemble classification is obtained by a majority vote (or plurality in the multiclass setting), with ties[b] typically handled at random. Class probability estimates for categorical outcomes can also be obtained from bagged tree ensembles and a discussion of different approaches is deferred to Section 7.2.1; often, the class proportions from each tree are just averaged across all the trees in the bagged ensemble (similar to regression).

---

[b]To avoid issues with ties (i.e., an equal number of votes for each class), use an odd number of trees or base learners.

Algorithm 5.1 Bagging for classification and regression trees.

---

1) Start with a training sample, $d_{trn} = \{(x_i, y_i)\}_{i=1}^{N}$, and specify integers n_{min} (the minimum node size of a particular tree), and B (the number of trees in the ensemble).

2) For b in $1, 2, \ldots, B$:

    a) Select a bootstrap sample $d_{trn}^{\star}$ of size N from d_{trn}.

    b) **Optional:** Keep track of which observations from d_{trn} were not selected to be in $d_{trn}^{\star}$; these are called the *out-of-bag* (OOB) observations.

    c) Fit a decision tree T_b to $d_{trn}^{\star}$ by recursively splitting each terminal node until the minimum node size (n_{min}) is reached.

3) Return the ensemble of trees: $\{T_b\}_{b=1}^{B}$.

4) To obtain the bagged prediction for a new case x, denoted $\widehat{f}_B(x)$, pass the observation down each tree—which will result in B separate predictions (one from each tree)—and aggregate as follows:

    • Classification: $\widehat{f}_B(x) = vote\{T_b(x)\}_{b=1}^{B}$, where $T_b(x)$ is the predicted class label for x from the b-th tree in the ensemble (in other words, let each tree vote on the classification for x and take a majority/plurality vote at the end).

    • Regression: $\widehat{f}_B(x) = \frac{1}{B}\sum_{b=1}^{B} T_b(x)$ (in other words, we just average the predictions for case x across all the trees in the ensemble).

Bagging has the same structural form as (5.1) with $\beta_0 = 0$ and $\{\beta_b = 1/B\}_{b=1}^{B}$, and where each tree is induced from an independent bootstrap sample of the original training data and grown to near maximal depth (as specified by n_{min}).

An important aspect of how the trees are constructed in bagging is that they are induced from independent bootstrap samples, which makes the bagging procedure trivial to parallelize. See Boehmke and Greenwell [2020, Sec. 10.4] for details and an example using the Ames housing data (Section 1.4.7) in R using the wonderful **foreach** package [Revolution Analytics and Weston, 2020].

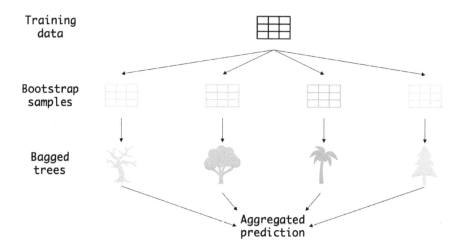

FIGURE 5.2: A simple schematic of the process for building a bagged tree ensemble with four trees.

5.1.1   When does bagging work?

In general, bagging helps to improve the accuracy of unstable procedures that are adaptive, nonlinear functions of the training data [Breiman, 1996a]. Let $\widehat{f}$ represent an individual model (e.g., a single decision tree) trained to the learning sample and $\widehat{f}_B$ represent a bagged ensemble thereof. Specifically, if small changes in d_{trn} lead to small changes in $\widehat{f}$, then $\widehat{f}_B \approx \widehat{f}$ and not much is gained in terms of improvement. On the other hand, if small changes in d_{trn} lead to large changes in $\widehat{f}$, then $\widehat{f}_B$ will often be an improvement over $\widehat{f}$. In the latter case, we often call $\widehat{f}$ an unstable model. Breiman [1996b] noted that algorithms like neural networks, CART, and subset selection in linear regression were unstable, while algorithms like k-nearest neighbor methods were stable; the MARSprocedure [Friedman, 1991], an extension of CART, can also be considered as unstable predictors and therefore potentially benefit from bagging).

Unpruned CART-like decision trees are particularly unstable predictors; that is, the tree structure will often vary heavily from one sample to the next. Hence, bagging is often most successful when applied to unpruned decision trees that have been grown to maximal (or near maximal) depth.

5.1.2   Bagging from scratch: classifying email spam

To illustrate, let's return to the email spam example first introduced in Section 1.4.5. In the code snippet below, we load the data from the **kernlab** package and split the observations into train/test sets using the same 70/30 split as before:

```
data(spam, package = "kernlab")
set.seed(852)  # for reproducibility
id <- sample.int(nrow(spam), size = floor(0.7 * nrow(spam)))
spam.trn <- spam[id, ]  # training data
spam.tst <- spam[-id, ]  # test data
```

Rather than writing our own bagger function, I'll construct a bagged tree ensemble using a basic **for** loop that stores the individual trees in a list called **spam.bag**. Note that I turn off cross-validation (**xval = 0**) when calling **rpart()** to save on computing time. The code is shown below.

```
library(rpart)
```

```
B <- 500  # number of trees in ensemble
ctrl <- rpart.control(minsplit = 2, cp = 0, xval = 0)
N <- nrow(spam.trn)  # number of training observations
spam.bag <- vector("list", length = B)  # to store trees
set.seed(900)  # for reproducibility
for (b in seq_len(B)) {  # fit trees to independent bootstrap samples
  boot.id <- sample.int(N, size = N, replace = TRUE)
  boot.df <- spam.trn[boot.id, ]  # bootstrap sample
  spam.bag[[b]] <- rpart(type ~ ., data = boot.df, control = ctrl)
}
```

Now that we have the individual trees, each of which was fit to a different bootstrap sample from the training data, we can obtain predictions and assess the performance of the ensemble using the test sample. To that end, I'll loop through each tree to obtain predictions on the test set (**spam.tst**), and store the results in an $N \times B$ matrix, one column for each tree in the ensemble. I then compute the test error as a function of B by cumulatively aggregating the predictions from trees 1 through B by means of voting (e.g., if we are computing the bagged prediction using only the first three trees, the final prediction for each observation will simply be the the class with the most votes across the three trees).

To help with the computations, I'll write two small helper functions, **vote()** and **err()**, for carrying out the voting and computing the misclassification error, respectively:

```
vote <- function(x) names(which.max(table(x)))
err <- function(pred, obs) 1 - sum(diag(table(pred, obs))) /
  length(obs)
```

Next, I obtain the $N \times B$ matrix of predictions and cumulatively aggregate them across all the trees in the ensemble to compute the test error as a function of the number of trees:

```
spam.bag.preds <- sapply(spam.bag, FUN = function(tree) {
  predict(tree, newdata = spam.tst, type = "class")
})  # N x B matrix of individual tree predictions

# Compute test error as a function of number of trees
spam.bag.err <- sapply(seq_len(B), FUN = function(b) {
  agg.pred <- apply(spam.bag.preds[, seq_len(b), drop = FALSE],
                    MARGIN = 1, FUN = vote)  # aggregate trees 1:b
  err(agg.pred, obs = spam.tst$type)  # compute test error
})
min(spam.bag.err)  # minimum misclassification error
```

```
#> [1] 0.0485
```

The results are displayed in Figure 5.3. The error stabilizes after around 200 trees and achieves a minimum misclassification error rate of 4.85% (horizontal dashed line). For reference, a single tree (pruned using the 1-SE rule) achieved a test error of 9.99%. Averaging the predictions from several hundred overgrown trees cut the misclassification error by more than half!

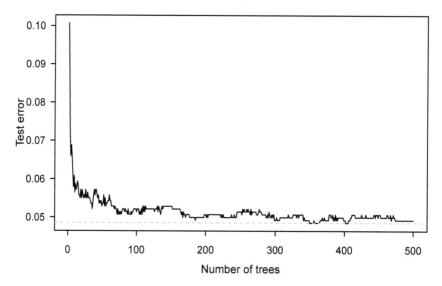

FIGURE 5.3: Test misclassification error for the email spam bagging example. The error stabilizes after around 200 trees and achieves a minimum misclassification error rate of 4.85% (horizontal dashed line).

While bagging was quite successful in the email spam example, sometimes bagging can make things worse. For a good discussion on how bagging can worsen bias and/or variance, see Berk [2008, Sec. 4.5.2–4.5.3].

5.1.3 Sampling without replacement

Inducing trees from independent learning sets that are bootstrap samples from the original training data imitates the process of building trees on independent samples of size N from the true underlying population of interest. While bagging traditionally utilizes bootstrap sampling (i.e., sampling with replacement) for training the individual base learners, it can sometimes be advantageous to use subsampling without replacement; Breiman [1999] referred to this as *pasting*. In particular, if N is "large enough," then bagging using random subsamples of size $N/2$ (i.e., sampling without replacement) can be an effective alternative to bagging based on the bootstrap [Friedman and Hall, 2007]. Strobl et al. [2007b] suggest using a subsample size of 0.632 times the original sample size N—because in bootstrap sampling about 63.2% of the original observations end up in any particular bootstrap sample.

This is quite fortunate since sampling half the data without replacement is much more efficient and can dramatically speed up the bagging process. Applying this to the email spam data from the previous section, which only required modifying one line of code in the previous example, resulted in a minimum test error of 5.21%, quite comparable to the previous results using the bootstrap but much faster to train.

Another reason why subsampling can sometimes improve the performance of bagging is through "de-correlation". Recall that bagging can improve the performance of unstable learners through variance reduction. As discussed in more detail in Section 7.2, correlation limits the variance-reducing effect of averaging. The problem here is that the trees in a bagged ensemble will often be correlated since they are all induced off of bootstrap samples from the same training set (i.e., they will share similar splits and structure, to some degree). Using subsamples of size $N/2$ will help to de-correlate the trees which can further reduce variance, resulting in improved generalization performance. A more effective strategy to de-correlate trees in a bagged ensemble is discussed in Section 7.2.

5.1.4 Hyperparameters and tuning

Bagged tree ensembles are convenient because they don't require much tuning. That's not to say that you can't improve performance by tuning some of the tree parameters (e.g., tree depth). However, in contrast to gradient tree boosting (Chapter 8), increasing the number of trees (B) does not necessarily lead to overfitting (see Figure 5.4 on page 193), and isn't really a tuning parameter—although, computation time increases with B, so it can be advantageous to monitor performance on a validation set to determine when performance has plateaued or reached a point of diminishing return.

5.1.5   Software

Bagging is rather straightforward to implement directly, as is seen in this chapter. Nonetheless, several R packages exist which can be used to implement Algorithm 5.1. The R package **ipred** [Peters and Hothorn, 2021], which stands for **i**mproved **pred**ictors, implements bagged decision trees using the **rpart** package and supports classification, regression, and survival problems. The R package **adabag** [Alfaro et al., 2018] also implements bagging via **rpart**, but only supports classification problems. In Python, bagging is implemented in scikit-learn's **sklearn.ensemble** module and can be applied to any base estimator (i.e., not just decision trees).

In general, I think it's more efficient to use random forest software to implement bagging, especially in R. This is because random forest software typically builds the entire ensemble using highly efficient and compiled code, whereas **ipred**, for example, utilizes existing tree software and builds the ensemble in a way similar to what we did in the email spam example. A regression example using random forest software to implement bagging is given in Section 5.5. Random forest is the topic of Chapter 7.

5.2   Boosting

Recall that bagging reduces variance by averaging several unstable learners together. Boosting, on the other hand, was originally devised for binary classification problems as a way to boost the performance of weak learners—a model that only does slightly better than the *majority vote classifier* (i.e., a model that always predicts the majority class in the learning sample). So how does boosting improve the performance of a weak learner? The basic idea is quite intuitive: fit models in a sequential manner, where each model in the sequence is fit on a resampled version of the training data that gives more weight to the observations that were previously misclassified, hence, "boosting" the performance of the previous models. Each successive model in a boosting sequence effectively homes in on the rows of data where the previous model had the largest errors.

Several different flavors of boosting exist, and the procedure has evolved quite a bit since its initial inception for binary classification. In the following section, I'll discuss one of the earliest and most popular flavors of boosting for binary classification: *AdaBoost.M1* [Freund and Schapire, 1996]. It's important to call out that AdaBoost.M1, along with its many variants, assumes that the base learners (in our case, binary classification trees) can incorporate case weights. A more general (and flexible) boosting strategy will be covered in Chapter 8.

5.2.1 AdaBoost.M1 for binary outcomes

AdaBoost.M1—also referred to as *Discrete AdaBoost* in Friedman et al. [2000] due to the fact that the base learners each return a discrete class label—fits an additive model of the form

$$C\left(\boldsymbol{x}\right) = \text{sign}\left(\sum_{b=1}^{B} \alpha_b C_b\left(\boldsymbol{x}\right)\right),$$

where $\{\alpha_b\}_{b=1}^{B}$ are coefficients that weight the contribution of each respective base learner $C_b\left(\boldsymbol{x}\right)$ and

$$\text{sign}\left(x\right) = \begin{cases} +1 & \text{if } x > 0 \\ -1 & \text{if } x < 0 \\ 0 & \text{otherwise} \end{cases}.$$

In essence, classifiers in the sequence with higher accuracy receive more weight and therefore have more influence on the final classification $C\left(\boldsymbol{x}\right)$.

The details of AdaBoost.M1 are given in Algorithm 5.2. The crux of the idea is this: start with an initial classifier built from the training data using equal case weights $\{w_i = 1/N\}_{i=1}^{N}$, then increase w_i for those cases that have been most frequently misclassified. The process is continued a fixed number of times (B).

Like bagging, boosting is a *meta-algorithm* that can be applied to any type of model, but it's often most successfully applied to shallow decision trees (i.e., decision trees with relatively few splits/terminal nodes). While bagging relies upon aggregating the results from several unstable learners, boosting tends to benefit from sequentially improving the performance of a weak learner (like a simple decision stump). In the next section, I'll code up Algorithm 5.2 and apply it to the email spam data for comparison with the previously obtained bagged tree ensemble.

While AdaBoost.M1 was one of the most accurate classifiers at the time[c], the fact that it only produced a classification was a severe limitation. To that end, Friedman et al. [2000] generalized the AdaBoost.M1 algorithm so that the weak learners return a class probability estimate, as opposed to a discrete class

---

[c] In fact, shortly after its introduction, Leo Breiman referred to AdaBoost as the "...best off-the-shelf classifier in the world."

Algorithm 5.2 Vanilla AdaBoost.M1 algorithm for binary classification.

---

1) Initialize case weights $\{w_i = 1/N\}_{i=1}^{N}$.

2) For $b = 1, 2, \ldots, B$:

    a) Fit a classifier $C_b(x)$ to the training observations using case weights w_i.

    b) Compute the weighted misclassification error

$$err_b = \frac{\sum_{i=1}^{N} w_i I\left(y_i \neq C_b(x)\right)}{\sum_{i=1}^{N}}.$$

    c) Compute $\alpha_b = \log\left(1/err_b - 1\right)$.

    d) Update case weights: $\{w_i \leftarrow \exp\left[\alpha_b I\left(y_i \neq C_b(x)\right)\right]\}_{i=1}^{N}$.

3) Return the weighted majority vote: $C(x) = \text{sign}\left(\sum_{b=1}^{B} \alpha_b C_b(x)\right)$

---

label; the contribution to the final classifier is half the logit-transform of this probability estimate. They refer to this procedure as *Real AdaBoost*. Other generalizations (e.g., to multi-class outcomes) also exist. In Chapter 8, I'll discuss a much more flexible flavor of boosting, called *stochastic gradient tree boosting*, which can naturally handle general outcome types (e.g., continuous, binary, Poisson counts, censored, etc.).

5.2.2   Boosting from scratch: classifying email spam

To illustrate, let's apply AdaBoost.M1 (Algorithm 5.2) to the email spam data and show how it "boosts" the performance of an individual **rpart** tree; I'll continue with the same train/test splits from the previous example in Section 5.1.2. For this example, I'll use $B = 500$ depth-10 decision trees. Since AdaBoost.M1 requires $y \in \{-1, +1\}$, I'll re-code the response (**type**) so that **type = "spam"** corresponds to $y = +1$:

```
spam.trn$type <- ifelse(spam.trn$type == "spam", 1, -1)
spam.tst$type <- ifelse(spam.tst$type == "spam", 1, -1)
spam.xtrn <- subset(spam.trn, select = -type)   # feature columns only
spam.xtst <- subset(spam.tst, select = -type)   # feature columns only
```

Following the previous example on bagging, I'll use a simple **for** loop and **list()** to sequentially construct and and store the fitted trees, respectively.

For AdaBoost.M1, we also have to collect and store the $\{\alpha_b\}_{b=1}^{B}$ coefficients in order to make predictions later. Note that `predict.rpart()` returns a factor—in this case, with factor levels `"-1"` and `"1"`—which needs to be coerced to numeric before further processing; this is the purpose of the `fac2num()` helper function in the code below[d]:

```r
library(rpart)

# Helper function to coerce factors to numeric
fac2num <- function(x) as.numeric(as.character(x))

# Apply AdaBoost.M1 algorithm
B <- 500  # number of trees in ensemble
ctrl <- rpart.control(maxdepth = 10, xval = 0)
N <- nrow(spam.trn)  # number of training observations
w <- rep(1 / N, times = N)  # initialize weights
spam.ada <- vector("list", length = B)  # to store sequence of trees
alpha <- numeric(B)  # to hold coefficients
for (i in seq_len(B)) {  # for b = 1, 2, ..., B
  spam.ada[[i]] <- rpart(type ~ ., data = spam.trn, weights = w,
                         control = ctrl, method = "class")
  # Compute predictions and coerce factor output to +1/-1
  pred <- fac2num(predict(spam.ada[[i]], type = "class"))
  err <- sum(w * (pred != spam.trn$type)) / sum(w)  # weighted error
  if (err == 0 | err == 1) {  # to avoid log(0) and dividing by 0
    err <- (1 - err) * 1e-06 + err * 0.999999
  }
  alpha[i] <- log((1 / err) - 1)  # coefficient from step 2) (c)
  w <- w * exp(alpha[i] * (pred != spam.trn$type))  # update weights
}
```

Next, I'll generate predictions for the test data (`spam.tst`) using the first b trees (where b will be varied over the range $1, 2, \ldots, B$) and compute the misclassification error for each; note that I'm using the same `err()` function defined in the previous example for bagging:

```r
spam.ada.preds <- sapply(seq_len(B), FUN = function(i) {
  class.labels <- predict(spam.ada[[i]], newdata = spam.tst,
                          type = "class")
  alpha[i] * fac2num(class.labels)
})  # (N x B) matrix of un-aggregated predictions

# Compute test error as a function of number of trees
spam.ada.err <- sapply(seq_len(B), FUN = function(b) {
  agg.pred <- apply(spam.ada.preds[, seq_len(b), drop = FALSE],
                    MARGIN = 1, FUN = function(x) sign(sum(x)))
  err(agg.pred, obs = spam.tst$type)
```

---

[d] According to the R FAQ guide (https://cran.r-project.org/doc/FAQ/), a more efficient but harder to remember solution is to use `as.numeric(levels(x))[as.integer(x)]`.

```
})
min(spam.ada.err)   # minimum misclassification error
```

```
#> [1] 0.0406
```

The results are plotted in Figure 5.4, along with those from the previously obtained bagged tree ensembles (i.e., using sampling with/without replacement). The minimum test error from the AdaBoost.M1 ensemble is 0.041. Compare this to the bagged tree ensemble based on sampling with replacement, which achieved a minimum test error of 0.049. In this case, AdaBoost.M1 slightly outperforms bagging.

For comparison, let's see how a single depth-10 decision tree—the base learner for our AdaBoost.M1 ensemble—performs on the same data.

```
spam.tree.10 <- rpart(type ~ ., data = spam.trn,
                      maxdepth = 10, method = "class")
pred <- predict(spam.tree.10, newdata = spam.tst, type = "class")
pred <- as.numeric(as.character(pred))   # coerce to numeric
mean(pred != spam.tst$type)
```

```
#> [1] 0.12
```

Wow, boosting decreased the misclassification error of a single depth-10 tree by roughly 66.27%, nice!

5.2.3 Tuning

In contrast to bagging, the number of base learners is often a critical tuning parameter in boosting algorithms, as they can often overfit for large enough B. While Figure 5.4 doesn't give any indication of overfitting, AdaBoost.M1 (and any boosting algorithm) can certainly overfit; an example of overfitting with AdaBoost is given in Hastie et al. [2009, p. 616; Figure 16.5]. The performance of a boosted tree ensemble can also be sensitive to the tree-specific parameters, such as the tree depth or maximum number of terminal nodes. Further refinements to AdaBoost, like the addition of *shrinkage* and subsampling, introduce other important tuning parameters. These are discussed in more detail in Section 8.3.

5.2.4 Forward stagewise additive modeling and exponential loss

Aside from bagging, additive expansions like (5.1) are often fit by minimizing some *loss function*[e], like *least squares* loss,

---
[e]A loss function measures the error in predicting $f(x)$ instead of y.

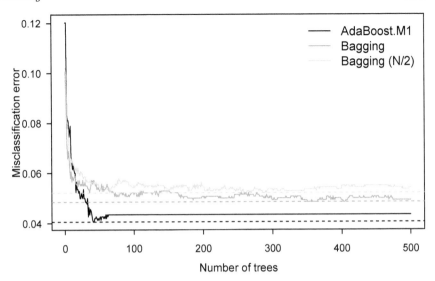

FIGURE 5.4: Misclassification error on the email spam test set from several different tree ensembles: 1) an AdaBoost.M1 classifier with depth-10 classification trees (black curve), 2) a bagged tree ensemble using max depth trees and sampling with replacement (yellow curve), and 3) a bagged tree ensemble using max depth trees and subsampling with replacement (blue curve). The horizontal dashed lines represent the minimum test error obtained by each ensemble.

$$\min_{\{\beta_b, \theta_b\}_{b=1}^B} \sum_{i=1}^N L\left(y_i, \sum_{b=1}^B \beta_b f_b\left(\boldsymbol{x}_i; \theta_b\right)\right).$$

For many combinations of loss functions and base learners, the solution can involve complicated and expensive numerical techniques. Fortunately, a simple approximation can often be used when it is more feasible to solve the optimization problem for a single base learner. This approximate solution is called *stagewise additive modeling*, the details of which are listed in Algorithm 5.3 below.

Friedman et al. [2000] show that AdaBoost.M1 (Algorithm 5.2) is equivalent to *forward stagewise additive modeling* using the exponential loss function

$$L\left(y, f\left(\boldsymbol{x}\right)\right) = \exp\left(-yf\left(\boldsymbol{x}\right)\right). \tag{5.2}$$

The product $yf\left(\boldsymbol{x}\right)$ is referred to as the "margin" and is analogous to the residual, $y - f\left(\boldsymbol{x}\right)$, for continuous outcomes. Hence, AdaBoost.M1 can be

---

Algorithm 5.3 Stagewise additive modeling

---

1) Initialize $f_b(\boldsymbol{x}_i; \theta_b) = 0$ (a constant).

2) For $b = 1, 2, \ldots, B$

    a) Optimize the loss for a single basis function; in particular, solve

$$(\beta_b, \theta_b) = \underset{\beta_b, \theta_b}{\arg\min} \sum_{i=1}^{N} L\left(y_i, \beta f_{b-1}(\boldsymbol{x}_i) + \beta f(\boldsymbol{x}_i; \theta_b)\right).$$

    b) Set $f_b(x) = f_{b-1}(\boldsymbol{x}_i) + \beta_b f(\boldsymbol{x}_i)$.

---

derived in an equivalent way that conforms to exactly the same structure as (5.1).

The boosting procedure discussed in Chapter 8 follows the forwards stagewise fitting approach more explicitly and includes AdaBoost as a special case, so more on this later. The R package **gbm** [Greenwell et al., 2021b] (originally by Greg Ridgeway) implements AdaBoost.M1 via this approach.

5.2.5 Software

Different flavors of AdaBoost (e.g., Discrete AdaBoost, Real AdaBoost, and Gentle AdaBoost) are available in the R package **ada** [Culp et al., 2016]; AdaBoost.M1 and AdaBoost.SAMME are also implemented in the R package **adabag**, as well as scikit-learn's **sklearn.ensemble** module. While R's implementations utilize CART-like decision trees for the base learner (using **rpart**), scikit-learn's implementation allows you to boost any compatible scikit-learn model that supports case weights. As mentioned in the previous section, the R package **gbm** implements AdaBoost.M1 as a forward stagewise additive model; more on **gbm** in Chapter 8.

A boosting strategy similar to AdaBoost is used to boost C5.0 decision trees and is available via the **C5.0** package.

5.3 Bagging or boosting: which should you use?

Among tree-based ensembles, it is generally regarded that boosting outperforms bagging (and its variants, like the random forest procedure discussed in Chapter 7). However, this is not always the case and, as discussed briefly in this chapter and in Chapter 8, boosting tends to require more work up front in terms of tuning in order to see the gains, whereas bagged tree ensembles tend to perform well right out of the box with little to no tuning[f]; this is especially true for random forests. My opinion is summarized by the following relationship:

Gradient boosted trees $\geq$ Random forest $>$ Bagged trees $>$ Single tree.

So, while boosted tree ensembles tend to outperform their bagged counterparts, I don't often find the performance increase to be worth the added complexity and time associated with the additional tuning. It's a trade off that we all must take into consideration for the problem at hand. It should also be noted that sometimes a single decision tree is the right tool for the job, and an ensemble thereof would be overkill; see, for example, Section 7.9.1.

5.4 Variable importance

Recall from Section 2.8 that the relative importance of predictor x is essentially the sum of the squared improvements over all internal nodes of the tree for which x was chosen as the partitioning variable. This idea also extends to ensembles of decision trees, such as bagged and boosted tree ensembles. In ensembles, the improvement score for each predictor is averaged across all the trees in the ensemble. Because of the stabilizing effect of averaging, the aggregated tree-based variable importance score is often more reliable in large ensembles; see Hastie et al. [2009, p. 368], although, as we'll see in Chapter 7, split variable selection bias will also affect the variable importance scores often produced by tree-based ensembles using CART-like decision trees.

---

[f]By "well" here, I mean close to how well they would perform with optimal tuning; the default is usually "in the ballpark."

5.5   Importance sampled learning ensembles

Tree-based ensembles (especially those discussed in Chapters 7–8) often do a good job in building a prediction model, but at the end of the day can involve a lot of trees which can limit their use in production since they can require more memory and take longer to score new data sets.

To help overcome these issues, Friedman and Popescu [2003] introduced the concept of *importance sampled learning ensembles* (ISLEs). Many of the tree-based ensembles discussed in this book—including bagged tree ensembles—are examples of ISLEs. The main point here is that ISLEs can sometimes benefit from post-processing via a technique called the LASSO, which stands for *least absolute shrinkage and selection operator* [Tibshirani, 1996]. Such post-processing can often maintain or, in some cases, improve the accuracy of the original ensemble while dramatically improving computational performance (e.g., lower memory requirements and faster training times). For full details, see Friedman and Popescu [2003], Hastie et al. [2009, Sec. 16.3.1], and Efron and Hastie [2016, 346–347].

The idea is to use the LASSO to select a subset of trees from a fitted ensemble and re-weight them, which can result in an ensemble with far fewer trees and (hopefully) comparable, if not better, accuracy. This is important to consider in real applications since tree ensembles can sometimes require many thousands of decision trees to reach peak performance, often resulting in a large model to maintain in memory and slower scoring times (aspects that are important to consider before deploying a model in a production process).

The LASSO-based post-processing procedure essentially involves fitting an L_1-penalized regression model of the form

$$\min_{\{\beta_b\}_{b=1}^{B}} \sum_{i=1}^{N} L\left[y_i, \sum_{b=1}^{B} \widehat{f}_b\left(\boldsymbol{x}_i\right) \beta_b\right] + \lambda \sum_{b=1}^{B} |\beta_b|,$$

where $\widehat{f}_b\left(\boldsymbol{x}_i\right)$ $(b = 1, 2, \ldots, B)$ is the prediction(s) from the b-th tree for observation i, β_b are fixed, but unknown coefficients to be estimated via the LASSO, and λ is the L_1-penalty to be applied.

The wonderful and efficient **glmnet** package [Friedman et al., 2021] for R can be used to fit the entire LASSO regularization path[g]; that is it efficiently computes the estimated model coefficients for an entire grid of relevant λ values.

---

[g]The **glmnet** package actually implements the entire *elastic net* regularization path for many types of generalized linear models. The LASSO is just a special case of the elastic net, which combines both the LASSO and ridge (i.e., L_2) penalties.

The optimal value of λ can be chosen via cross-validation or an independent test set.

Note that not all ensembles will perform well with post-processing. As discussed in Hastie et al. [2009, section 16.3.1], the individual trees should cover the space of predictors where needed and and be sufficiently different from each other for the post-processor to be effective. Strategies for different tree ensembles are provided in Friedman and Popescu [2003] (e.g., using smaller subsamples when sampling without replacement, like 5–10%, and shallower trees for bagged tree ensembles). The next example shows how to apply this post-processing strategy to a bagged tree ensemble using the Ames housing data (Section 1.4.7).

5.5.1 Example: post-processing a bagged tree ensemble

To illustrate, let's return to the Ames housing example. Below, I'll load the data into R and apply the same 70/30 split from the previous example. Note that I continue to rescale `Sale_Price` by dividing by 1000; this is strictly for plotting purposes.

```
ames <- as.data.frame(AmesHousing::make_ames())
ames$Sale_Price <- ames$Sale_Price / 1000  # rescale response
set.seed(2101)  # for reproducibility
id <- sample.int(nrow(ames), size = floor(0.7 * nrow(ames)))
ames.trn <- ames[id, ]  # training data/learning sample
ames.tst <- ames[-id, ]  # test data
ames.xtst <- subset(ames.tst, select = -Sale_Price)  # features only
```

Next, I'll fit a bagged tree ensemble using the **randomForest** package [Breiman et al., 2018] (computational reasons for doing so are discussed in Section 5.1.5). Random forest, and its open source implementations, are not discussed until Chapter 7. For now, just note that the **randomForest** package, among others, can be used to implement bagged tree ensembles by tweaking a special parameter, often referred to as m_{try} (to be discussed in Section 7.2), and setting this parameter equal to the number of total predictors will result in an ordinary bagged tree ensemble). This will be much more efficient than relying on the **ipred** package and will also allow us to obtain predictions from the individual trees, rather than just the aggregated predictions. Examples of post-processing an RF and boosted tree ensemble are given in Sections 7.9.2 and 8.9.3, respectively.

Here, I'll fit two models, each containing $B = 500$ trees:

- a standard bagged tree ensemble where each tree is fully grown to boot-strap samples of size N (`ames.bag`);

- a bagged tree ensemble consisting of shallow six-node trees, each of which
  is grown using only a 5% random sample of the training data without
  replacement (`ames.bag.6.5`).

Both models are trained in the code chunk below. Note that I recorded the
training time of each fit using **`system.time()`**, which will provide some insight into the potential computational savings offered by this post-processing
method[h]. Although substantially less accurate (see Figure 5.6), notice how
much faster it is to train `ames.bag.6.5`:

```
library(randomForest)
```

```
# Fit a typical bagged tree ensemble
system.time({
  set.seed(942)   # for reproducibility
  ames.bag <-
    randomForest(Sale_Price ~ ., data = ames.trn, mtry = 80,
                 ntree = 500, xtest = ames.xtst,
                 ytest = ames.tst$Sale_Price, keep.forest = TRUE)
})
```

```
#>    user  system elapsed
#> 120.407   0.788 123.070
```

```
# Print results
print(ames.bag)
```

```
#>
#> Call:
#>  randomForest(formula = Sale_Price ~ ., data = ames.trn, mtry = 80...
#>                Type of random forest: regression
#>                      Number of trees: 500
#> No. of variables tried at each split: 80
#>
#>           Mean of squared residuals: 690
#>                     % Var explained: 89
#>                        Test set MSE: 628
#>                     % Var explained: 90.6
```

```
# Fit a bagged tree ensemble using six-node trees on 5% samples
system.time({
  set.seed(1021)
  ames.bag.6.5 <-
    randomForest(Sale_Price ~ ., data = ames.trn, mtry = 80,
                 ntree = 500, maxnodes = 6,
                 sampsize = floor(0.05 * nrow(ames.trn)),
                 replace = FALSE, keep.forest = TRUE,
                 xtest = ames.xtst, ytest = ames.tst$Sale_Price)
})
```

---

[h]Note that there are better ways to benchmark and time expressions in R; see, for
example, the **microbenchmark** package [Mersmann, 2021].

```
#>      user  system elapsed
#>     0.395   0.003   0.402
```

```
# Print results
print(ames.bag.6.5)
```

```
#>
#> Call:
#>   randomForest(formula = Sale_Price ~ ., data = ames.trn, mtry = 80...
#>                   Type of random forest: regression
#>                         Number of trees: 500
#> No. of variables tried at each split: 80
#>
#>             Mean of squared residuals: 1489
#>                       % Var explained: 76.2
#>                         Test set MSE: 1450
#>                       % Var explained: 78.2
```

```
# Test set MSE as a function of the number of trees
mse.bag <- ames.bag$test$mse
mse.bag.6.5 <- ames.bag.6.5$test$mse
```

Next, I'll use **glmnet** to post-process each ensemble using the LASSO. The following steps are conveniently handled by **treemisc**'s isle_post() function, which I'll use to post-process the ames.bag.6.5 ensemble. But first, I think it's prudent to show the individual steps using the ames.bag ensemble.

To start, I'll compute the individual tree predictions for the train and test sets and store them in a matrix

```
preds.trn <- predict(ames.bag, newdata = ames.trn,
                     predict.all = TRUE)$individual
preds.tst <- predict(ames.bag, newdata = ames.tst,
                     predict.all = TRUE)$individual
```

Next, I'll use the glmnet() function to fit the entire regularization path using the training predictions from the $B = 500$ individual trees:

```
library(glmnet)
```

```
# Fit the LASSO regularization path
lasso.ames.bag <- glmnet(
  x = preds.trn,   # individual tree predictions are the predictors
  y = ames.trn$Sale_Price,  # same response variable
  lower.limits = 0,  # coefficients should be strictly positive
  standardize = FALSE,  # no need to standardize
  family = "gaussian"  # least squares regression
)
```

A few things to note about the above code chunk are in order. Since this is a regression problem, I set family = "gaussian" (for least squares) in the call to `glmnet()`. Second, since the individual tree predictions are all on the same scale, there's no need to standardize the inputs (`standardize = FALSE`). Lastly, we could argue that the estimated coefficients (one for each tree) should be non-negative (`lower.limits = 0`).

Figure 5.5 shows the regularization path for the estimated coefficients. In particular, the λ values (on the log scale) are plotted on the x-axis, and the y-axis corresponds to the estimated coefficient value (one curve per coefficient/tree). The top axis highlights the number of non-zero coefficients at each particular value of the penalty parameter λ:

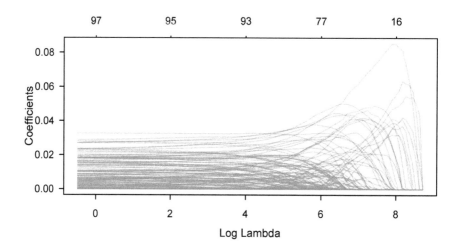

FIGURE 5.5: Profiles of LASSO coefficients for the `ames.bag.6.5` ensemble, as the regularization parameter λ is varied. The top axis indicates the number of non-zero coefficients at a particular value of λ (plotted on the log scale).

From here, we use cross-validation or an independent test set to choose a reasonable value of the penalty parameter λ. This can be done easily using **glmnet**'s `assess.glmnet()` function with the test set predictions using, as shown below (see `?glmnet::assess.glmnet` for details):

```
# Assess performance of fit using an independent test set
perf <- assess.glmnet(
  object = lasso.ames.bag,  # fitted LASSO model
  newx = preds.tst, # test predictions from individual trees
  newy = ames.tst$Sale_Price,  #same response variable (test set)
  family = "gaussian"  # for MSE and MAE metrics
)
perf <- do.call(cbind, args = perf)  # bind results into matrix
```

```
# List of results
ames.bag.post <- as.data.frame(cbind(
  "ntree" = lasso.ames.bag$df, perf,
  "lambda" = lasso.ames.bag$lambda)
)

# Sort in ascending order of number of trees
head(ames.bag.post <- ames.bag.post[order(ames.bag.post$ntree), ])

#>    ntree  mse  mae lambda
#> s0     0 6658 59.6   6164
#> s1     4 5672 55.0   5616
#> s2     5 4851 50.7   5117
#> s3     8 4163 46.8   4663
#> s4     9 3592 43.3   4248
#> s5    10 3114 40.1   3871

# Print results corresponding to smallest test MSE
ames.bag.post[which.min(ames.bag.post$mse), ]

#>     ntree mse  mae lambda
#> s93    97 612 15.9   1.08
```

According to the test MSE, the optimal value of the penalty parameter λ is 1.077, which corresponds to 97 trees or non-zero coefficients in the LASSO model (an appreciable reduction from the original 500).

In the next code chunk, I'll follow the exact same process with the `ames.bag.6.5` ensemble, but using the `isle_post()` function instead:

```
library(treemisc)

# Post-process ames.bag.6.5 ensemble
preds.trn.6.5 <- predict(ames.bag.6.5, newdata = ames.trn,
                         predict.all = TRUE)$individual
preds.tst.6.5 <- predict(ames.bag.6.5, newdata = ames.tst,
                         predict.all = TRUE)$individual
ames.bag.6.5.post <-
  isle_post(preds.trn.6.5, y = ames.trn$Sale_Price,
            family = "gaussian", newX = preds.tst.6.5,
            newy = ames.tst$Sale_Price)
```

The overall results from each ensemble are shown in Figure 5.6. Here, I show the MSE as a function of the number of trees from each model (or non-zero coefficients in the LASSO). In this example, the simpler `ames.bag.6.5` ensemble benefits substantially from post-processing and appears to perform on par with the ordinary bagged tree ensemble (`ames.bag`) in terms of MSE, while requiring only a small fraction of trees and being orders of magnitude faster to train! The original ensemble (`ames.bag`) did not see nearly as much improvement from post-processing.

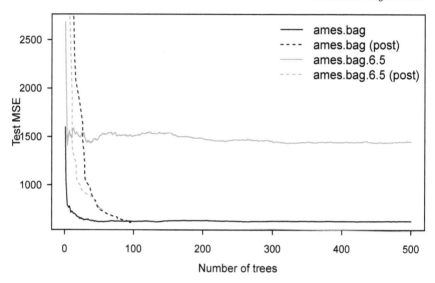

FIGURE 5.6: MSE for the test data from several bagged tree ensembles. The dashed lines correspond to the LASSO-based post-processed versions. Clearly, the `ames.bag.6.5` ensemble benefits the most from post-processing, performing nearly on par with the standard bagged tree ensemble (`ames.bag`).

5.6   Final thoughts

Loh [2014] compared the accuracy of single decision trees to tree ensembles using both real and simulated data sets. He found that, on average, the best single-tree algorithm was about 10% less accurate than that of a tree ensemble. Nonetheless, tree ensembles will not always outperform a simpler individual tree [Loh, 2009]. These points aside, tree ensembles are a powerful class of models that are highly competitive in terms of state-of-the-art prediction accuracy. Chapters 7–8 are devoted to two powerful tree ensemble techniques.

It is also worth pointing out that while tree-based ensembles often out perform carefully tuned individual trees (like CART, CTree, and GUIDE), they are less interpretable compared to a single decision tree; hence, they are often referred to as black box models. Fortunately, post-hoc procedures exist that can help us peek into the black box to understand the relationships uncovered by the model and explain their output to others. This is the topic of Chapter 6.

6

Peeking inside the "black box": post-hoc interpretability

The lack of understanding does not hurt, as much as the lack of effort to understand does!

Wordions

This chapter is dedicated to select topics from the increasingly popular field of *interpretable machine learning* (IML), which easily deserves its own book-length treatment, and it has; see, for example, Molnar [2019] and Biecek and Burzykowski [2021] (both of which are freely available online). The methods covered in this chapter can be categorized into whether they help interpret a black box model at a global or local (e.g., individual row or prediction) level.

To be honest, I don't really like the term "black box," especially when we now have access to a rich ecosystem of interpretability tools. For example, linear regression models are often hailed as interpretable models. Sure, but this is really only true when the model has a simple form. Once you start including transformations and interaction effects—which are often required to boost accuracy and meet assumptions—the coefficients become much less interpretable.

Tree-based ensembles, especially the ones discussed in the next two chapters, can provide state-of-the-art performance, and are quite competitive with other popular supervised learning algorithms, especially on tabular data sets. Even when tree-based ensembles perform as advertised, there's a price to be paid in terms of parsimony, as we lose the ability to summarize the model using a simple tree diagram. Luckily, there exist a number of post-hoc techniques that allow us to tease the same information out of an ensemble of trees that we would ordinarily be able to glean from looking at a simple tree diagram (e.g., which variables seem to be the most important, the effect of each predictor, and potential interactions). Note that the techniques discussed in this chapter are *model-agnostic*, meaning they can be applied to any type of supervised learning algorithm, not just tree-based ensembles. For example, they can also be used to help interpret neural networks or a more complicated tree structure that uses linear splits or non-constant models in the terminal nodes.

The next three sections cover post-hoc methods to help comprehend various aspects of any fitted model:

- feature importance (Section 6.1);
- feature effects (Section 6.2);
- feature contributions (Section 6.3).

6.1 Feature importance

For the purposes of this chapter, we can think of variable importance (VI) as the extent to which a feature has a "meaningful" impact on the predicted outcome. A more formal definition and treatment can be found in van der Laan [2006]. Given that point of view, a natural way to assess the impact of an arbitrary feature x_j is to remove it from the training data and examine the drop in performance that occurs after refitting the model without it. This procedure is referred to as *leave-one-covariate-out* (LOCO) importance; see Hooker et al. [2019] and the references therein.

Obviously, the LOCO importance method is computationally prohibitive for larger data sets and complex fitting procedures because it requires retraining the model once more for each dropped feature. In the next section, I'll discuss an approximate approach based on reassessing performance after randomly permuting each feature (one at a time). This procedure is referred to as *permutation importance*.

6.1.1 Permutation importance

While some algorithms, like tree-based models, have a natural way of quantifying the importance of each predictor, it is useful to have a model-agnostic procedure that can be used for any type of supervised learning algorithm. This also makes it possible to directly compare the importance of features across different types of models. In this section, I'll discuss a popular method for measuring the importance of predictors in any supervised learning model called permutation importance.

Permutation-based VI scores exist in various forms and was made popular in Breiman [2001] for random forests (Chapter 7), before being generalized and extended in Fisher et al. [2018]. A general permutation-based VI procedure is outlined in Algorithm 6.1 below. The idea is that if we randomly permute the values of an important feature in the training data, the training performance would degrade (since permuting the values of a feature effectively destroys any relationship between that feature and the target variable). This of course assumes that the model has been properly tuned (e.g., using cross-validation) and is not overfitting.

The permutation approach uses the difference between some baseline performance measure (e.g., R^2 or RMSE for regression, Brier score or log loss for probability estimation, and AUC for discrimination) and the same performance measure obtained after permuting the values of a particular feature in the training data (Note that the model is NOT refit to the training data after randomly permuting the values of a feature). It is also important to note that this method may not be appropriate when you have, for example, highly correlated features (since permuting one feature at a time may lead to unlikely or unrealistic data instances).

Algorithm 6.1 General steps for constructing permutation-based VI scores for any type of supervised learning model.

Let $x_1, x_2, \ldots, x_p$ be the features of interest and let $\mathcal{M}_{orig}$ be the baseline performance metric for the trained model; for brevity, I'll assume smaller is better (e.g., classification error or RMSE). The permutation-based importance scores can be computed as follows:

1) For $i = 1, 2, \ldots, p$:

   (a) Permute the values of feature x_i in the training data.

   (b) Recompute the performance metric on the permuted data, denoted $\mathcal{M}_{perm}$.

   (c) Record the difference from baseline using $\text{VI}(x_i) = \mathcal{M}_{perm} - \mathcal{M}_{orig}$.

2) Return the VI scores $\text{VI}(x_1), \text{VI}(x_2), \ldots, \text{VI}(x_j)$.

Algorithm 6.1 can be improved or modified in a number of ways. For instance, the process can (and should) be repeated several times and the results averaged together. This helps to provide more stable VI scores, and also the opportunity to measure their variability. Rather than taking the difference in step (c), Molnar [2019, sec. 5.5.4] argues that using the ratio $\mathcal{M}_{perm}/\mathcal{M}_{orig}$ makes the importance scores more comparable across different problems. It's also possible to assign importance scores to groups of features (e.g., by permuting more than one feature at a time); this would be useful if features can be categorized into mutually exclusive groups, for instance, categorical features that have been one-hot encoded.

6.1.2 Software

The permutation approach to variable importance is implemented in several R packages, including: **vip** [Greenwell et al., 2021a], **iml** [Molnar and Schratz, 2020], **ingredients** [Biecek and Baniecki, 2021], and **mmpf** [Jones, 2018]. Further details, some comparisons, and an in-depth explanation of **vip** are provided in Greenwell and Boehmke [2020]. Starting with version 0.22.0, scikit-learn's **inspection** module provides an implementation of permutation importance for any fitted model.

6.1.3 Example: predicting home prices

To illustrate the basic steps, let's compute permutation importance scores for the Ames housing bagged tree ensemble (`ames.bag`) from Section 5.5.1. I'll start by writing a simple function to compute the RMSE, the performance metric of interest, and use it to obtain a baseline value for computing the permutation-based importance scores.

```
rmse <- function(predicted, actual, na.rm = TRUE) {
  sqrt(mean((predicted - actual) ^ 2, na.rm = na.rm))
}
(baseline.rmse <- rmse(predict(ames.bag, newdata = ames.trn),
                       actual = ames.trn$Sale_Price))
```

```
#> [1] 10.6
```

To get more stable VI scores, I'll use 30 independent permutations for each predictor; since the permutations are done independently, Algorithm 6.1 can be trivially parallelized across repetitions or features. This is done using a nested **for** loop in the next code chunk:

```
nperm <- 30  # number of permutation to use per feature
xnames <- names(subset(ames.trn, select = -Sale_Price))
vi <- matrix(nrow = nperm, ncol = length(xnames))
colnames(vi) <- xnames
for (j in colnames(vi)) {
  for (i in seq_len(nrow(vi))) {
    temp <- ames.trn  # temporary copy of training data
    temp[[j]] <- sample(temp[[j]])  # permute feature values
    pred <- predict(ames.bag, newdata = temp)  # score permuted data
    permuted.rmse <- rmse(pred, actual = temp$Sale_Price) ^ 2
    vi[i, j] <- permuted.rmse - baseline.rmse  # smaller is better
```

```
  }
}
```

```
# Average VI scores across all permutations
head(vi.avg <- sort(colMeans(vi), decreasing = TRUE))
```

```
#>  Overall_Qual   Gr_Liv_Area  Neighborhood
#>          2959          1211           725
#> Total_Bsmt_SF  First_Flr_SF      Lot_Area
#>           334           308           167
```

Note that the individual permutation importance scores are computed independently of each other, making it relatively straightforward to parallelize the whole procedure; in fact, many R implementations of Algorithm 6.1, like **vip** and **iml**, have options to do this in parallel using a number of different parallel backends.

A boxplot of the unaggregated permutation scores for the top ten features, as measured by the average across all 30 permutations, is displayed in Figure 6.1. Here, you can see that the overall quality rating of the home and its above grade square footage are two of the most important predictors of sale price, followed by neighborhood. A simple dotchart of the average permutation scores would suffice, but fails to show the variability in the individual VI scores.

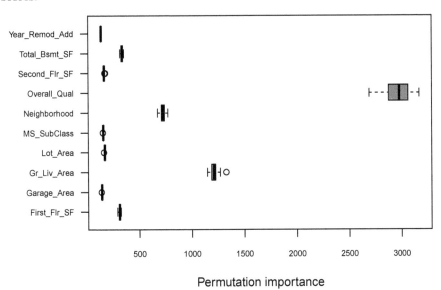

FIGURE 6.1: Permutation-based VI scores for the top ten features in the Ames housing bagged tree ensemble, as measured by the average across all 30 permutations.

Although permutation importance is most naturally computed on the training data, it may also be useful to do the shuffling and measure performance on new data. This is discussed in depth in Molnar [2019, sec. 5.2]. Next, I'll discuss a general technique for interpreting the effect of an individual feature (main effect) or combination of features (interaction effect).

6.2   Feature effects

While determining predictor importance is a crucial task in any supervised learning problem, ranking variables is only part of the story, and once a subset of "important" features is identified it is often necessary to assess the relationship between them (or subset thereof) and the response. This can be done in many ways, but in practice it is often accomplished by constructing plots of *partial dependence* or *individual conditional expectation*; see Friedman [2001] and [Goldstein et al., 2015], respectively, for details.

6.2.1   Partial dependence

Partial dependence (PD) plots (or PDPs) help visualize the relationship between a subset of the features (typically 1–3) and the response while accounting for the average effect of the other predictors in the model. They are particularly effective with black box models like random forests, support vector machines, and neural networks.

Let $x = \{x_1, x_2, \ldots, x_p\}$ represent the predictors in a model whose prediction function is $\widehat{f}(x)$. If we partition x into an interest set, z_s, and its compliment, $z_c = x \setminus z_s$, then the "partial dependence" of the response on z_s is defined as

$$f_s(z_s) = E_{z_c}\left[\widehat{f}(z_s, z_c)\right] = \int \widehat{f}(z_s, z_c)\, p_c(z_c)\, dz_c, \qquad (6.1)$$

where $p_c(z_c)$ is the marginal probability density of z_c: $p_c(z_c) = \int p(x)\, dz_s$. Equation (6.1) can be estimated from a set of training data by

$$\bar{f}_s(z_s) = \frac{1}{n} \sum_{i=1}^{n} \widehat{f}(z_s, z_{i,c}), \qquad (6.2)$$

where $\{z_{i,c}\}_{i=1}^{N}$ are the values of z_c that occur in the training sample; that is, we average out the effects of all the other predictors in the model.

Mathematical gibberish aside, computing partial dependence (6.2) in practice is rather straightforward. To simplify, let $z_s = x_1$ be the predictor variable of interest with unique values $\{x_{1i}\}_{i=1}^k$. The partial dependence of the response on x_1 can be constructed by following the basic steps outlines in 6.2.

Algorithm 6.2 A simple algorithm for constructing the partial dependence of the response on a single predictor x_1.

---

1) For $i \in \{1, 2, \ldots, k\}$:

    (a) Copy the training data and replace the original values of x_1 with the constant x_{1i}.

    (b) Compute the vector of predicted values from the modified copy of the training data.

    (c) Compute the average prediction to obtain $\bar{f}_1(x_{1i})$.

2) Plot the pairs $\{x_{1i}, \bar{f}_1(x_{1i})\}_{i=1}^k$.

Algorithm 6.2 can be quite computationally intensive since it involves k passes over the training records. Fortunately, the algorithm is trivial to parallelize. It can also be easily extended to larger subsets of two or more features as well. See Greenwell [2017] for additional details and examples.

6.2.1.1 Classification problems

Traditionally, for classification problems, partial dependence functions are on a scale similar to the logit; see, for example, Hastie et al. [2009, pp. 369—370]. Suppose the response is categorical with J levels; then for each class we compute

$$f_j(x) = \log\left[p_j(x)\right] - \frac{1}{J}\sum_{j=1}^{J} \log\left[p_j(x)\right], \quad j = 1, 2, \ldots, J, \qquad (6.3)$$

where $p_j(x)$ is the predicted probability for the j-th class. Plotting $f_j(x)$ helps us understand how the log-odds for the j-th class depends on different subsets of the predictor variables. Nonetheless, there's no reason partial dependence can't be displayed on the raw probability scale. The same goes for ICE plots (Section 6.2.3). A multiclass classification example of PD plots on the probability scale is given in Section 6.2.6.

6.2.2   Interaction effects

While partial dependence can be used to help visualize potential interaction effects, it is often desirable to know where to look in the first place. To that end, Friedman and Popescu [2008] proposed a model-agnostic method, called the H-statistic, for identifying predictors that are involved in interactions with other variables, the strength of those interactions, as well as the identities of the other variables with which they interact.

For example, to test for the presence of an interaction effect between predictors x_j and x_k, we can use the statistic

$$H_{jk}^2 = \sum_{i=1}^{N} \left[\bar{f}_{jk}\left(x_{ij}, x_{ik}\right) - \bar{f}_j\left(x_{ij}\right) - \bar{f}_k\left(x_{ik}\right) \right]^2 \Big/ \sum_{i=1}^{N} \bar{f}_{jk}\left(x_{ij}, x_{ik}\right). \qquad (6.4)$$

In essence, (6.4) measures the fraction of variance of $\bar{f}_{jk}\left(x_j, x_k\right)$—the joint partial dependence of y on x_j and x_k—not captured by $\bar{f}_j\left(x_j\right)$ and $\bar{f}_j\left(x_j\right)$ (the individual partial dependence of y on x_j and x_k, respectively) over the training data (or representative sample thereof). Note that $H_{jk}^2 \geq 0$, with zero indicating no interaction between x_j and x_k. To determine whether a single predictor, x_j, say, interacts with any other variables, a similar H-statistic can be computed. Unfortunately, these statistics are not widely implemented; the R **gbm** package [Greenwell et al., 2021b], probably has the most efficient implementation (see `?gbm::interact.gbm` for details), but it's only available for GBMs (Chapter 8).

According to Friedman and Popescu [2008], only predictors with strong main effects (e.g., high relative importance) should be examined for potential interactions; the strongest interactions can then be further explored via two-way PD plots. Be warned, however, that collinearity among predictors can lead to spurious interactions that are not present in the target function.

A major drawback of the H-statistic (6.4) is that it requires computing both the individual and joint partial dependence functions, which can be expensive; the fast recursion method of Section 8.6.1 makes it feasible to compute the H-statistic for binary decision trees (and ensembles of shallow trees). A simpler approach, based on just the joint partial dependence function, is discussed in Greenwell et al. [2018].

6.2.3   Individual conditional expectations

PD plots can be misleading in the presence of strong interaction effects [Goldstein et al., 2015] (akin to interpreting a main effect in a linear model that's also involved in an interaction term). To overcome this issue, Goldstein, Kapelner, Bleich, and Pitkin developed the concept of individual conditional expectation (ICE) plots. ICE plots display the estimated relationship between the response and a predictor of interest for each observation. Consequently, the PD plot for a predictor of interest can be obtained by averaging the corresponding ICE plots across all observations.

As described in [Goldstein et al., 2015], when the individual curves have a wide range of intercepts and consequently overlay each other, heterogeneity in the model can be difficult to discern. For that reason, Goldstein, Kapelner, Bleich, and Pitkin suggest centering the ICE plots to produce a centered ICE plot (or c-ICE plot for short). They also suggest other modifications, like derivative ICE plots (or d-ICE plots), to further explore the presence of interaction effects. Centered ICE plots are obtained after shifting the ICE curves up or down by subtracting off the first value from each curve, effectively pinching them together at the beginning.

6.2.4 Software

PD plots and ICE plots (and many variants thereof) are implemented in several R packages. Historically, PD plots were only implemented in specific tree-based ensemble packages, like **randomForest** [Breiman et al., 2018] and **gbm**. However, they were made generally available in package **pdp**, which was soon followed by **iml** and **ingredients**, among others; these packages also support ICE plots; the R package **ICEbox** [Goldstein et al., 2017] provides the original implementation of ICE plots and several variants thereof, like c-ICE and d-ICE plots. PD plots and ICE plots were also made available in scikit-learn's **inspection** module, starting with versions 0.22.0 and 0.24.0, respectively.

6.2.5 Example: predicting home prices

Using the Ames housing bagged tree ensemble, I'll show how to construct PD plots and ICE curves by hand and using the **pdp** package. To start, let's construct a PD plot for above grade square footage (`Gr_Liv_Area`), one of the top predictors according to permutation-based VI scores from Figure 6.1 (p. 207).

The first step is to create a grid of points over which to construct the plot. For continuous variables, it is sufficient to use a fine enough grid of percentiles, as is done in the example below. Then, I simply loop through each grid point and 1) copy the training data, 2) replace all the values of `Gr_Liv_Area` in the copy with the current grid value, and 3) score the modified copy of the training data and average the predictions together. Lastly, I simply plot the grid points against the averaged predictions obtained from the `for` loop. The results are displayed in Figure 6.2 and show a relatively monotonic increasing relationship between above grade square footage and predicted sale price.

```
x.grid <- quantile(ames.trn$Gr_Liv_Area, prob = 1:30 / 31)
pd <- numeric(length(x.grid))
for (i in seq_along(x.grid)) {
  temp <- ames.trn  # temporary copy of data
  temp[["Gr_Liv_Area"]] <- x.grid[i]
  pd[i] <- mean(predict(ames.bag, newdata = temp))
}

# PD plot for above grade square footage (Figure 6.2)
plot(x.grid, pd, type = "l", xlab = "Above ground square footage",
     ylab = "Partial dependence", las = 1)
rug(x.grid)  # add rug plot to x-axis
```

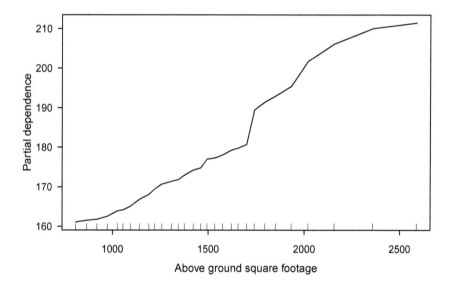

FIGURE 6.2: Partial dependence of sale price on above grade square footage for the bagged tree ensemble.

While looking at the partial dependence of the response on a single feature (i.e., main effect) is informative, it is often useful to look at the dependence on two or three predictors simultaneously[a] (i.e., interaction effects). Fortunately, it is rather straightforward to modify the above `for` loop to accommodate multiple predictors; however, this is a good opportunity to show an alternative

---

[a]Theoretically, we can look at any order of interaction effect of interest; however, the computational complexity usually prohibits going beyond just two- or three-way interactions.

method to computing partial dependence using simple *data wrangling* operations, which can be more efficient if you're working in SQL or Spark.

In essence, we can generate all of the modified training copies in a single
stacked data frame using a *cross-join*, score it once, then aggregate the predictions into the partial dependence values. Below is an example using base
R, but it should be rather straightforward to translate to **data.table** [Dowle
and Srinivasan, 2021], **dplyr** [Wickham et al., 2021b], **SparkR** [Venkatara-
man et al., 2016], **sparklyr** [Luraschi et al., 2021], or any other language that
can perform simple cross-joins—provided you can hold the resulting Cartesian product in memory! (An example of how to accomplish this in Spark is
provided in Section 7.9.5).

To illustrate, I'll construct the partial dependence of sale price on both above
grade and first floor square footage. I still need to construct a grid of points
over which to construct the plot; here, I'll use a Cartesian product between
the percentiles of each feature using `expand.grid()`.

```
x1.grid <- quantile(ames.trn$Gr_Liv_Area, prob = 1:30 / 31)
x2.grid <- quantile(ames.trn$First_Flr_SF, prob = 1:30 / 31)
df1 <- expand.grid("Gr_Liv_Area" = x1.grid,
                   "First_Flr_SF" = x2.grid)   # Cartesian product
```

In the next step, I perform a cross-join between the grid of plotting values (`df1`) and the original training data with the plotting features removed
(`df2`)[b]:

```
df2 <- subset(ames.trn, select = -c(Gr_Liv_Area, First_Flr_SF))

# Perform a cross-join between the two data sets
pd <- merge(df1, df2, all = TRUE)   # Cartesian product
dim(pd)   # print dimensions
#> [1] 1845900      81
```

Then, I simply score the data and aggregate by computing the average prediction within each grid point, as shown in the example below:

```
pd$yhat <- predict(ames.bag, newdata = pd)   # might take a few minutes!
pd <- aggregate(yhat ~ Gr_Liv_Area + First_Flr_SF, data = pd,
                FUN = mean)
```

The code snippet below constructs a *false color level plot* of the data with
contour lines using the built-in **lattice** package; the results are displayed in
Figure 6.3. Here, you can see the joint effect of both features on the predicted
sale price.

---

[b]BE CAREFUL as the resulting data set, which is a Cartesian product, can be quite
large!

```
library(lattice)

# PD plot for above grade and first floor square footage
levelplot(yhat ~ Gr_Liv_Area * First_Flr_SF, data = pd,
          contour = TRUE, col = "white", scales = list(tck = c(1, 0)),
          col.regions = hcl.colors(100, palette = "viridis"))
```

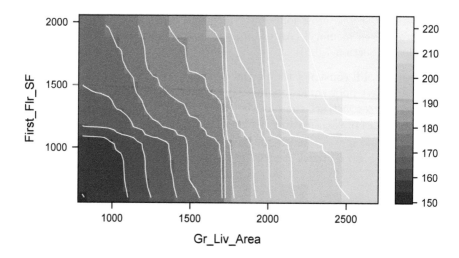

FIGURE 6.3: Partial dependence of sale price on above grade and first floor square footage for the bagged tree ensemble.

It is not wise to draw conclusions from PD plots (and ICE plots) in regions outside the area of the training data. Greenwell [2017] describes two ways to mitigate the risk of extrapolation in PD plots: rug displays, like the one I used in Figure 6.2, and *convex hulls* (which can be used with bivariate displays, like in Figure 6.3).

Constructing ICE curves is just as easy; just skip the aggregation step and plot each of the individual curves. In the example below, I'll use the **pdp** package to construct c-ICE curves showing the partial dependence of above grade square footage on sale price for each observation in the learning sample. There's no need to construct a curve for each sample, especially when you have thousands (or more) data points; here, I'll just plot a random sample of 500 curves. I'll use the same percentiles to construct the plot as I did for the PD plot in Figure 6.2 (p. 212) by invoking the **quantiles** and **probs** arguments in the call to **partial()**; note that **partial()**'s default is to use an evenly spaced grid of points across the range of predictor values.

The results are displayed in Figure 6.4; the red line shows the average c-ICE value at each above grade square footage (i.e., the centered partial dependence). The heterogeneity in the c-ICE curves indicates a potential interaction effect between `Gr_Liv_Area` and at least one other feature. The c-ICE curves also indicate a relatively monotonic increasing relationship for the majority of houses in the training set, but you can see a few of the curves at the bottom deviate from this overall pattern.

```
ice <- partial(ames.bag, pred.var = "Gr_Liv_Area", ice = TRUE,
               center = TRUE, quantiles = TRUE, probs = 1:30 / 31)
set.seed(1123)  # for reproducibility
samp <- sample.int(nrow(ames.trn), size = 500)  # sample 500 homes
autoplot(ice[ice$yhat.id %in% samp, ], alpha = 0.1) +
  ylab("Conditional expectation")
```

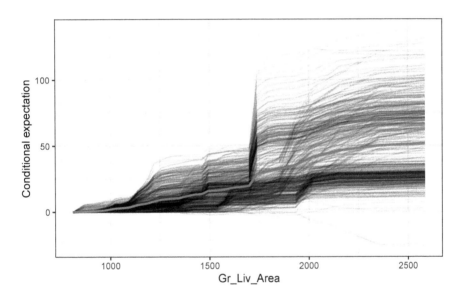

FIGURE 6.4: A random sample of 500 c-ICE curves for above grade square footage using the Ames housing bagged tree ensemble. The curves indicate a relatively monotonic increasing relationship for the majority of houses in the sample. The average of the 500 c-ICE curves is shown in red.

6.2.6   Example: Edgar Anderson's iris data

For a classification example, I'll consider Edgar Anderson's iris data from the **datasets** package in R. The `iris` data frame contains the sepal length, sepal width, petal length, and petal width (in centimeters) for 50 flowers from each of three species of iris: setosa, versicolor, and virginica. Below, I fit a bagged tree ensemble to the data using the **randomForest** package:

```
library(randomForest)

# Fit a bagged tree ensemble
set.seed(1452)  # for reproducibility
(iris.bag <- randomForest(Species ~ ., data = iris, mtry = 4))

#>
#> Call:
#>  randomForest(formula = Species ~ ., data = iris, mtry = 4)
#>                Type of random forest: classification
#>                      Number of trees: 500
#> No. of variables tried at each split: 4
#>
#>          OOB estimate of  error rate: 4.67%
#> Confusion matrix:
#>            setosa versicolor virginica class.error
#> setosa         50          0         0        0.00
#> versicolor      0         47         3        0.06
#> virginica       0          4        46        0.08
```

Next, I plot the partial dependence of `Species` on `Petal.Width` for each
of the three classes using the **pdp** package. The code chunk below exploits
a simple trick to computing partial dependence with `partial()` for several
classes simultaneously. The results are displayed in Figure 6.5. Here, you can
clearly see the average effect petal width has on the probability of belonging
to each species.

```
library(pdp)
library(ggplot2)

# Prediction wrapper that returns average prediction for each class
pfun <- function(object, newdata) {
  colMeans(predict(object, newdata = newdata, type = "prob"))
}

# Partial dependence of probability for each class on petal width
p <- partial(iris.bag, pred.var = "Petal.Width", pred.fun = pfun)
ggplot(p, aes(Petal.Width, yhat, color = as.factor(yhat.id))) +
  geom_line() +
  theme(legend.title = element_blank(),
        legend.position = "top")
```

Note that without the aid of a user-supplied prediction function (via the
`pred.fun` argument), **pdp**'s `partial()` function can only compute partial
dependence in regards to a single class; see Greenwell [2017] for more details
on the use of this package.

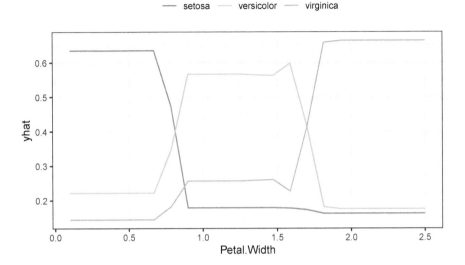

FIGURE 6.5: Partial dependence of species probability on petal width for each of the three iris species using a bagged tree ensemble.

6.3   Feature contributions

In general, a *local feature contribution* is the worth assigned to a feature's value that is proportional to the feature's share in the model's prediction for a particular observation. You can think of feature contributions as a directional variable importance at the individual prediction level. While there are a wide variety of feature contribution methodologies, the next section briefly covers one of the most popular methods in current use: *Shapley values* (or *Shapley explanations*). Shapley values necessarily involve a lot of mathematical notation, but I will try to avoid as much as possible while trying to convey the main concepts. For more details, start with Štrumbelj and Kononenko [2014] and Lundberg and Lee [2017].

6.3.1   Shapley values

The Shapley value [Shapley, 2016] is an idea from coalitional/cooperative game theory. In a coalitional game, assume there are p players that form a grand coalition (S) worth a certain payout (Δ_S). Suppose it is also known how much any smaller coalition ($Q \subseteq S$) (i.e., any subset of p players) is worth (Δ_Q). The goal is to distribute the total payout Δ_S to the individual p players in a "fair" way; that is, so that each player receives their "fair" share. The Shapley value is one such solution and the only one that uniquely satisfies a particular set of "fairness properties."

Let v be a *characteristic function* that assigns a value to each subset of players; in particular, $v : 2^p \to \mathbb{R}$, where $v(S) = \Delta_S$ and $v(\emptyset) = 0$, with $\emptyset$ denoting the empty set (i.e., zero players). Let $\phi_i(v)$ be the contribution (or portion of the total payout) attributed to player i in a particular game with total payout $v(S) = \Delta_S$. The Shapley value satisfies the following properties:

- efficiency: $\sum_{i=1}^{p} \phi_i(v) = \Delta_S$;

- null player: $\forall W \subseteq S \setminus \{i\} : \Delta_W = \Delta_{W \cup \{i\}} \implies \phi_i(v) = 0$;

- symmetry: $\forall W \subseteq S \setminus \{i,j\} : \Delta_{W \cup \{i\}} = \Delta_{W \cup \{j\}} \implies \phi_i(v) = \phi_j(v)$;

- linearity: If v and w are functions describing two coalitional games, then $\phi_i(v + w) = \phi_i(v) + \phi_i(w)$.

The above properties can be interpreted as follows:

- the individual player contributions sum to the total payout, hence, are implicitly normalized;

- if a player does not contribute to the coalition, they receive a payout of zero;

- if two players have the same impact across all coalitions, they receive equal payout;

- the local contributions are additive across different games.

Shapley [2016] showed that the unique solution satisfying the above properties is given by

$$\phi_i(v) = \frac{1}{p!} \sum_{\mathcal{O} \in \pi(p)} \left[v\left(S^{\mathcal{O}} \cup i\right) - v\left(S^{\mathcal{O}}\right)\right], \quad i = 1, 2, \ldots, p, \qquad (6.5)$$

where $\mathcal{O}$ is a specific permutation of the player indices $\{1, 2, \ldots, p\}$, $\pi(p)$ is the set of all such permutations of size p, and $S^{\mathcal{O}}$ is the set of players joining the coalition before player i.

In other words, the Shapley value is the average marginal contribution of a player across all possible coalitions in a game. Another way to interpret (6.5) is as follows. Imagine the coalitions (subsets of players) being formed one player at a time (which can happen in different orders), with the i-th player demanding a fair contribution/payout of $v\left(S^{\mathcal{O}} \cup i\right) - v\left(S^{\mathcal{O}}\right)$. The Shapley value for player i is given by the average of this contribution over all possible permutations in which the coalition can be formed.

A simple example may help clarify the main ideas. Suppose three friends (players)—Alex, Brad, and Brandon—decide to go out for drinks after work (the game). They shared a few pitchers of beer, but nobody paid attention to how much each person drank (collaborated). What's a fair way to split the tab (total payout)?

Suppose we knew the following information, perhaps based on historical happy hours:

- if Alex drank alone, he'd only pay $10;

- if Brad drank alone, he'd only pay $20;

- if Brandon drank alone, he'd only pay $10;

- if Alex and Brad drank together, they'd only pay $25;

- if Alex and Brandon drank together, they'd only pay $15;

- if Brad and Brandon drank together, they'd only pay $13;

- if Alex, Brad, and Brandon drank together, they'd only pay $30.

With only three players, we can enumerate all possible coalitions. In Table 6.1, I list all possible permutations of the three players and list the marginal contribution of each. Take the first row, for example. In this particular permutation, we start with Alex. We know that if Alex drinks alone, he'd spend $10, so his marginal contribution by entering first is $10. Next, we assume Brad enters the coalition. We know that if Alex and Brad drank together, they'd pay a total of $25, leaving $15 left over for Brad's marginal contribution. Similarly, if Brandon joins the party last, his marginal contribution would be only $5 (the difference between $30 and $25). The Shapley value for each player is the average across all six possible permutations (these are the column averages reported in the last row). In this case, Brandon would get away with the smallest payout (i.e., have to pay the smallest portion of the total tab). The next time the bartender asks how you want to split the tab, whip out a pencil, and do the math!

6.3.2 Explaining predictions with Shapley values

From this section forward, let $\{x_i\}_{i=1}^p$ represent the p feature values comprising $x^\star$, the observation whose prediction we want to try to explain. Štrumbelj and Kononenko [2014] suggested using the Shapley value (6.5) to help explain predictions from a supervised learning model. In the context of statistical and machine learning,

- a game is represented by the prediction task for a single observation $x^\star$;

| | Marginal contribution | | |
Permutation/order of players	Alex	Brad	Brandon
Alex, Brad, Brandon	$10	$15	$5
Alex, Brandon, Brad	$10	$15	$5
Brad, Alex, Brandon	$5	$20	$5
Brad, Brandon, Alex	$10	$20	$0
Brandon, Alex, Brad	$5	$15	$10
Brandon, Brad, Alex	$17	$3	$10
Shapley contribution:	$9.50	$14.67	$5.83

TABLE 6.1: Marginal contribution for each permutation of the players/beer drinkers {Alex, Brad, Brandon} (i.e., the order in which they arrive). The Shapley contribution is the average marginal contribution across all permutations. (Notice how each row sums to the total bill of $30.)

- the total payout/worth (Δ_S) for $x^\star$ is the prediction for $x^\star$ minus the average prediction for all training observations (the latter is referred to as the baseline and denoted $\bar{f}$): $\hat{f}(x^\star) - \bar{f}$;

- the players are the individual feature values of $x^\star$ that collaborate to receive the payout Δ_S (i.e., predict a certain value).

The second point, combined with the efficiency property stated in the previous section, implies that the p Shapley explanations (or feature contributions) for an observation of interest $x^\star$, denoted $\{\phi_i(x^\star)\}_{i=1}^p$, are inherently standardized since $\sum_{j=1}^p \phi_i(x^\star) = \hat{f}(x^\star) - \bar{f}$.

6.3.2.1 Tree SHAP

Several methods exist for estimating Shapley values in practice. The most common is arguably Tree SHAP [Lundberg et al., 2020], an efficient implementation of exact Shapley values for decision trees and ensembles thereof.

Tree SHAP is a fast and exact method to estimate Shapley values for tree-based models (including tree ensembles), under several different possible assumptions about feature dependence. The specifics of Tree SHAP are beyond the scope of this book, so I'll defer to [Lundberg et al., 2020] for the details. It's implemented in the Python **shap** module, and embedded in several tree-based modeling packages across several open source languages (like **xgboost** [Chen et al., 2021] and **lightgbm** [Shi et al., 2022]). While the details of Tree SHAP are beyond the scope of this book, we'll see an example of it in action in Section 8.9.4.

In the following section, I'll discuss a general way to estimate Shapley values for any supervised learning model using a simple *Monte Carlo* approach.

6.3.2.2 Monte Carlo-based Shapley explanations

Except in special circumstances, like Tree SHAP, computing the exact Shapley value is computationally infeasible in most applications. To that end, Štrumbelj and Kononenko [2014] suggest a Monte Carlo approximation, which I'll call Sample SHAP for short, that assumes independent features[c]. Their approach is described in Algorithm 6.3 below.

Here, a single estimate of the contribution of feature x_i to $f(\boldsymbol{x}^\star) - \bar{f}$ is nothing more than the difference between two predictions, where each prediction is based on a set of "Frankenstein instances"[d] that are constructed by swapping out values between the instance being explained ($\boldsymbol{x}^\star$) and an instance selected at random from the training data ($\boldsymbol{w}^\star$). To help stabilize the results, the procedure is repeated a large number, say, R, times, and the results averaged together:

Algorithm 6.3 Approximating the i-th feature's contribution to $\hat{f}(\boldsymbol{x}^\star)$ for some instance with predictor values $\boldsymbol{x}^\star = (x_1, x_2, \ldots, x_p)$.

---

1) For $j = 1, 2, \ldots, R$:

    (a) Select a random permutation $\mathcal{O}$ of the sequence $1, 2, \ldots, p$.

    (b) Select a random instance $\boldsymbol{w}$ from the set of training observations $\boldsymbol{X}$.

    (c) Construct two new instances as follows:

- $\boldsymbol{b}_1 = \boldsymbol{x}^\star$, but all the features in $\mathcal{O}$ that appear after feature x_i get their values swapped with the corresponding values in $\boldsymbol{w}$.

- $\boldsymbol{b}_2 = \boldsymbol{x}^\star$, but feature x_i, as well as all the features in $\mathcal{O}$ that appear after x_i, get their values swapped with the corresponding values in $\boldsymbol{w}$.

    (d) $\phi_{ij}(\boldsymbol{x}^\star) = f(\boldsymbol{b}_1) - f(\boldsymbol{b}_2)$.

2) $\phi_i(\boldsymbol{x}^\star) = \frac{1}{R} \sum_{j=1}^{R} \phi_{ij}(\boldsymbol{x}^\star)$.

---

[c]While Sample SHAP, along with many other common Shapley value procedures, assumes independent features, several arguments can be made in favor of this assumption; see, for example, Chen et al. [2020] and the references therein.

[d]The terminology used here takes inspiration from Molnar [2019, p. 231].

A simple R implementation of Algorithm 6.3 is given below. Here, `obj` is a fitted model with scoring function `f` (e.g., `predict()`), `nsim` is the number of Monte Carlo repetitions to perform, `feature` gives the name of the corresponding feature in x to be explained, and X is the training set of features.

```r
sample.shap <- function(f, obj, R, x, feature, X) {
  phi <- numeric(R)   # to store Shapley values
  N <- nrow(X)   # sample size
  p <- ncol(X)   # number of features
  b1 <- b2 <- x
  for (m in seq_len(R)) {
    w <- X[sample(N, size = 1), ]
    ord <- sample(names(w))   # random permutation of features
    swap <- ord[seq_len(which(ord == feature) - 1)]
    b1[swap] <- w[swap]
    b2[c(swap, feature)] <- w[c(swap, feature)]
    phi[m] <- f(obj, newdata = b1) - f(obj, newdata = b2)
  }
  mean(phi)   # return approximate feature contribution
}
```

To illustrate, let's continue with the Ames housing example (`ames.bag`). Below, I use the `sample.shap()` function to estimate the contribution of the value of `Gr_Liv_Area` to the prediction of the first observation in the learning sample (`ames.trn`):

```r
X <- subset(ames.trn, select = -Sale_Price)   # features only
set.seed(2207)   # for reproducibility
sample.shap(predict, obj = ames.bag, R = 100, x = X[1, ],
            feature = "Gr_Liv_Area", X = X)
```

```
#> [1] -6.7
```

So, having `Gr_Liv_Area` = 1474 helped push the predicted sale price down toward the baseline average; in this case, the baseline average is just the average predicted sale price across the entire training set: $\bar{f} = \$181.53$ (don't forget that I rescaled the response in this example).

If there are p features and m instances to be explained, this requires $2 \times R \times p \times m$ predictions (or calls to the scoring function f). In practice, this can be quite computationally demanding, especially since R needs to be large enough to produce good approximations to each $\phi_i(x^\star)$. How large does R need to be to produce accurate explanations? It depends on the variance of each feature in the observed training data, but typically $R \in [30, 100]$ will suffice. The R package **fastshap** [Greenwell, 2021a] provides an optimized implementation of Algorithm 6.3 that only requires $2mp$ calls to f; see the package documentation for details.

Sample SHAP can be computationally prohibitive if you need to explain large data sets (optimized or not). Fortunately, you often only need to explain a handful of predictions, the most extreme ones, for example. However, generating explanations for the entire training set, or a large enough sample thereof, can be useful for generating aggregated global model summaries. For example, Shapley-based dependence plots [Lundberg et al., 2020] show how a feature's value impacts the prediction of every observation in a data set of interest.

6.3.3 Software

Various flavors of Shapley values are starting to become widely available in R. Implementations of Sample SHAP, for example, are provided in **fastshap**, **iml**, and **iBreakDown** [Biecek et al., 2021]. Maksymiuk et al. [2021] discuss several others.

The **shap** module in Python is arguably one of the first and most well known implementations of Shapley values for statistical and machine learning. It offers several different flavors of Shapley explanations, including Tree SHAP, Kernel SHAP [Lundberg and Lee, 2017], Sample SHAP, and many more specific to different applications, like *deep learning*.

6.3.4 Example: predicting home prices

In the example below, I'll use **fastshap** to estimate feature contributions for the record in the test set with the highest predicted sale price using $R = 100$ Monte Carlo repetitions. Note that **fastshap**'s explain() function includes an adjustment argument to ensure the efficiency property; see https://github.com/bgreenwell/fastshap/issues/6 for details.

As with **pdp**, **fastshap** defines its own autoplot() method for automatically producing various **ggplot2**-based Shapley plots; IMO it's far better (and more flexible) to manually produce your own plots from the raw output. In the code chunk below, I use explain() and autoplot() to produce a bar plot of the feature contributions for the training observation with the highest predicted sale price[e]:

```
library(fastshap)
library(ggplot2)

# Find observation with highest predicted sale price
pred <- predict(ames.bag, newdata = ames.tst)
highest <- which.max(pred)
```

---

[e]An alternative way to visualize individual feature contributions using a *waterfall chart* is given in Section 8.9.1.

```
pred[highest]

#> 433
#> 503

# fastshap needs to know how to compute predictions from your model
pfun <- function(object, newdata) predict(object, newdata = newdata)

# Need to supply feature columns only in fastshap::explain()
X <- subset(ames.trn, select = -Sale_Price)  # feature columns only
newx <- ames.tst[highest, names(X)]

# Compute feature contributions for observation with highest prediction
set.seed(1434)  # for reproducibility
ex <- explain(ames.bag, X = X, nsim = 100, newdata = newx,
              pred_wrapper = pfun, adjust = TRUE)
ex[1, 1:5]  # peek at a few

#> # A tibble: 1 x 5
#>   MS_SubClass MS_Zoning Lot_Frontage Lot_Area Street
#>         <dbl>     <dbl>        <dbl>    <dbl>  <dbl>
#> 1       0.930    0.0275        0.472     3.80      0

autoplot(ex, type = "contribution", num_features = 10,
         feature_values = newx)
```

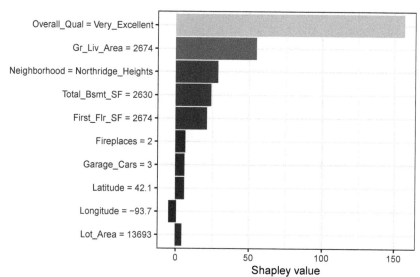

FIGURE 6.6: Top ten (Shapley-based) feature contributions to the highest training prediction from the Ames housing bagged tree ensemble.

Next, I'll construct a Shapley dependence plot for `Gr_Liv_Area` using **fast-shap** with $R = 50$ Monte Carlo repetitions. The results are displayed in Figure 6.2. As with Figures 6.2 and 6.4, the predicted sale price tends to increase with above grade square footage. As with the c-ICE curves in Figure 6.4, the increasing dispersion in the plot indicates a potential interaction with at least one other feature. Coloring the Shapley dependence plot by the values of another feature can help visualize such an interaction, if you know what you're looking for.

```
ex <- explain(ames.bag, feature_names = "Gr_Liv_Area", X = X,
              nsim = 50, pred_wrapper = pfun)

# Shapley dependence plot
autoplot(ex, type = "dependence", X = X, alpha = 0.3)
```

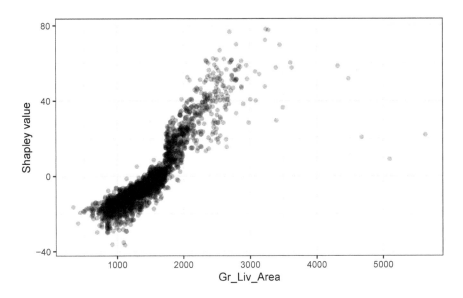

FIGURE 6.7: Shapley dependence of above grade square footage on predicted sale price.

6.4 Drawbacks of existing methods

As discussed in Hooker et al. [2019], *permute-and-predict* methods—like PD plots, ICE plots, and permutation importance—can produce results that are highly misleading.[f] For example, the standard approach to computing permutation-based VI scores involves independently permuting individual features. This implicitly makes the assumption that the observed features are statistically independent. In practice, however, features are often not independent which can lead to nonsensical VI scores. One way to mitigate this issue is to use the conditional approach described in Strobl et al. [2008b]; Hooker et al. [2019] provides additional alternatives, such as *permute-and-relearn importance*. Unfortunately, to the best of my knowledge, this approach is not yet available for general purposes. A similar modification can be applied to PD plots [Parr and Wilson, 2019].

I already mentioned that PD plots can be misleading in the presence of strong interaction effects. As discussed earlier, this can be mitigated by using ICE plots instead. Another alternative would be to use *accumulated local effect* (ALE) plots [Apley and Zhu, 2020]. Compared to PD plots, ALE plots have the advantage of being faster to compute and less affected by strong dependencies among the features. The downside, however, is that ALE plots are more complicated to implement. ALE plots are available in the **ALEPlot** [Apley, 2018] and **iml** packages in R.

Hooker [2007] also argues that feature importance (which concerns only main effects) can be misleading in high dimensional settings, especially when there are strong dependencies and interaction effects among the features, and suggests an approach based on a *generalized functional ANOVA decomposition*—though, to my knowledge, this approach is not widely implemented in open source software.

6.5   Final thoughts

IML is on the rise, and so is IML-related open source software. There are simply too many methods and useful packages to discuss in one chapter, so I only just covered a handful. If you're looking for more, I'd recommend starting with the IML awesome list hosted by Patrick Hall at

```
https://github.com/jphall663/awesome-machine-learning-
                        interpretability.
```

---

[f]It's been argued that approximate Shapley values share the same drawback; however, Janzing et al. [2019] makes a compelling case against those arguments.

A good resource for R users is Maksymiuk et al. [2021]. And of course, Molnar [2019] is a freely available resource, filled with intuitive explanations and links to relevant software in both R and Python. Molnar et al. [2021] is also worth reading, as they discuss a number of pitfalls to watch out for when using model-agnostic interpretation methods.

7

Random forests

In a forest of a hundred thousand trees, no two leaves are alike.
And no two journeys along the same path are alike.

Paulo Coelho

7.1 Introduction

Random forests (RFs) are essentially bagged tree ensembles with an added twist, and they tend to provide similar accuracy to many state-of-the-art supervised learning algorithms on tabular data, while being relatively less difficult to tune. In other words, RFs tend to be competitive right out of the box. But be warned, RFs—like any statistical and machine learning algorithm—enjoy their fair share of disadvantages. As we'll see in this chapter, RFs also include many bells and whistles that data scientists can leverage for non-prediction tasks, like detecting anomalies/outliers, imputing missing values, and so forth.

7.2 The random forest algorithm

Recall that a bagged tree ensemble (Section 5.1) consists of hundreds (sometimes thousands) of independently grown decision trees, where each tree is trained on a different bootstrap sample from the original training data. Each tree is intentionally grown deep (low bias), and variance is reduced by averaging the predictions across all the trees in the ensemble. For classification, a plurality vote among the individual trees is used.

Unfortunately, correlation limits the variance-reducing effect of averaging. Take the following example for illustration. Suppose $\{X_i\}_{i=1}^N \overset{iid}{\sim} (\mu, \sigma^2)$ is a random sample from some distribution with mean μ and variance σ^2. Let $\bar{X} = \sum_{i=1}^N X_i / N$ be the sample mean. If the observations are independent (as is the usual connotation of a random sample), then $\mathrm{E}\left(\bar{X}\right) = \mu$ and $\mathrm{V}\left(\bar{X}\right) = \sigma^2/N$. In other words, the variance of the average is less than the variance of the sample elements. This of course assumes that the X_i are uncorrelated. If the pairwise correlation between any two observations is $\rho = \rho\left(X_i, X_j\right) (i \neq j)$, then

$$\mathrm{V}\left(\bar{X}\right) = \rho\sigma^2 + \frac{1-\rho}{N}\sigma^2,$$

which converges to $\rho\sigma^2$ as $N \to \infty$. In other words, regardless of sample size, correlation limits the variance-reducing effect of averaging. This is illustrated in Figure 7.1, where each boxplot is constructed from 30,000 sample means, each of which is based on a sample of size $N = 30$ from a centered Gaussian distribution with specified pairwise correlation (x-axis); note the increasing variability in the sample means as we go from $\rho = 0$ to $\rho = 1$ (left to right).

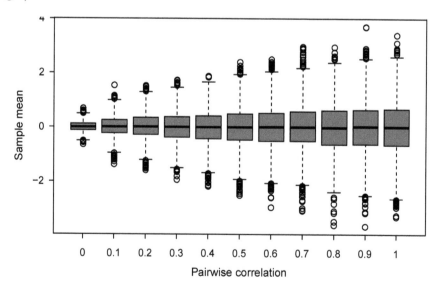

FIGURE 7.1: 100 simulated averages from samples of size $N = 30$ with pairwise correlation increasing from zero to one.

Similarly, bagging a set of correlated predictors (i.e., models producing similar predictions) will only reduce the variance to a certain point. Since each tree is built using an independent bootstrap sample from the original training data, the trees in the bagged ensemble will be somewhat correlated. If we can reduce correlation between the trees, the trees will be more diverse and averaging can further improve the prediction and generalization performance of the ensemble.

Figure 7.2 shows six bagged decision trees applied to the email spam training data; each tree was constrained to a max depth of three to help with the visualization. Since each tree was induced from an independent bootstrap sample from the same training set, the trees are naturally very similar to each other. Notice, for example, that the path to terminal node 15 (highlighted in green) is the same in four of the six trees, albeit the split points are slightly different. The performance of the combined ensemble might improve if we could make the trees more diverse.

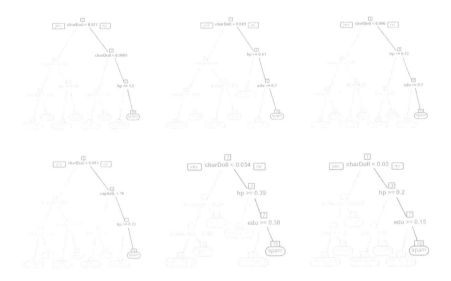

FIGURE 7.2: Six bagged decision trees applied to the email spam training data. The path to terminal node 15 is highlighted in each tree.

Luckily, Leo Breiman and Adele Cutler thought of a clever way to reduce correlation in a bagged tree ensemble; that is, make the trees more diverse. The idea is to limit the potential splitters at each node in a tree to a random subset of the available predictors, which will often result in a much more diverse ensemble of trees. In essence, bagging constructs a diverse tree ensemble by introducing randomness into the rows via sampling with replacement, while an RF further increases tree diversity by also introducing randomness into the columns via subsampling the features.

In other words, an RF is just a bagged tree ensemble with an additional layer of randomness produced by selecting a random subset of candidate splitters prior to partitioning the data at every node in every tree—*extremely randomized trees* (Section 7.8.4), take this randomization a step further in an attempt to reduce variance even more. Let $m_{try} \leq p$ be the number of candidate splitters selected at random from the entire set of p features prior to each split in each tree. Setting $m_{try} << p$ can often dramatically improve performance compared to bagged decision trees; note that setting $m_{try} = p$ will results in an ordinary bagged tree ensemble.

That's it. That's essentially the difference between an ordinary bagged tree ensemble and an RF.

The general steps for constructing a traditional RF are given in Algorithm 7.1 (compare this to Algorithm 5.1).

The recommended default values for m_{try} (the number of features randomly sampled before each split) and n_{min} (the minimum size of any terminal node) depend on the type of outcome:

- For classification, the typical defaults are $m_{try} = \lfloor \sqrt{p} \rfloor$, where p is the number of features, and $n_{min} = 1$.

- For regression, the typical defaults are $m_{try} = \lfloor p/3 \rfloor$ and $n_{min} = 5$.

These default values may vary slightly from one implementation to another. Note that $\lfloor x \rfloor$ is just x rounded down to the nearest integer. As we'll see in Section 7.4, generalization performance is typically most sensitive to the value of m_{try}, but the default is usually in the ballpark, and the tuning space for m_{try} is simple since $m_{try} \in \{1, 2, \ldots, p\}$, where p is the total number of available features.

Traditionally, CART was used for the base learners in an RF, but any decision tree algorithm will work (e.g., GUIDE or CTree); I'll use CTree in Section 7.2.3 to build an RF from scratch. Also, the traditional RF algorithm used the Gini splitting criterion (2.1) for classification and the SSE splitting criterion (2.6) for regression. However, many other splitting criteria can be used to great affect. For example, Ishwaran et al. [2008] proposed several splitting rules more appropriate for right-censored outcomes, including a log-rank splitting rule that splits nodes by maximization of the log-rank test statistic; see also Segal [1988], LeBlanc and Crowley [1992].

7.2.1   Voting and probability estimation

The voting scheme for classification outlined in step 4) of Algorithms 5.1 and 7.1 is called *hard voting*. In hard voting, each base learner casts a direct vote on the class label, and a majority vote (binary outcome) or plurality

---

Algorithm 7.1 Traditional RF algorithm for classification and regression.

---

1) Start with a training sample, d_{trn}, and specify integers, n_{min} (the minimum node size), B (the number of trees in the forest), and $m_{try} \leq p$ (the number of predictors to select at random as candidate splitters prior to splitting the data at each node in each tree).

2) For b in $1, 2, \ldots, B$:

    (a) Select a bootstrap sample $d^{\star}_{trn}$ of size N from the training data d_{trn}.

    (b) **Optional:** Keep track of which observations from the original training data were not selected to be in the bootstrap sample; these are called the out-of-bag (OOB) observations.

    (c) Fit a decision tree $\mathcal{T}_b$ to the bootstrap sample $d^{\star}_{trn}$ according to the following rules:

        (i) Before each attempted split, select a random sample of m_{try} features to use as candidate splitters.

        (ii) Continue recursively splitting each terminal node until the minimum node size n_{min} is reached.

3) Return the "forest" of trees $\{\mathcal{T}_b\}_{b=1}^{B}$.

4) To obtain the RF prediction for a new case x, pass the observation down each tree and aggregate as follows:

    • Classification: $\hat{C}_B^{rf}(x) = vote\left\{\hat{C}_b(x)\right\}_{b=1}^{B}$, where $\hat{C}_b(x)$ is the predicted class label for x from the b-th tree in the forest (in other words, let each tree vote on the classification for x and take the majority/plurality vote).

    • Regression: $\hat{f}_B^{rf}(x) = \frac{1}{B}\sum_{b=1}^{B}\hat{f}_b(x)$ (in other words, we just average the predictions for case x across all the trees in the forest).

---

vote (multiclass outcome) is used to determine the overall classification of an observation.

With categorical outcomes, however, we often care more about the predicted probability of class membership, as opposed to directly predicting a class label. In an RF (or a bagged tree ensemble) there are two ways to obtain predicted probabilities:

1) Take the proportion of votes for each class over the entire forest.

2) Average the class probabilities from each tree in the forest. (In this case, n_{min} should be considered a tuning parameter; see, for example, Malley et al. [2012].)

The first approach can be problematic. For example, suppose the probability that x belongs to class j is $\Pr(Y = j|x) = 0.91$. If each tree correctly predicts class j for x, then $\widehat{\Pr}(x) = 1$, which is incorrect. If $n_{min} = 1$, the two approaches are equivalent and neither will produce consistent estimates of the true class probabilities (see, for example, Malley et al. [2012]). So which approach is better for probability estimation? Hastie et al. [2009, p. 283] argue that the second method tends to provide improved estimates of the class probabilities with lower variance, especially for small B.

Malley et al. [2012] make a similar argument for the binary case, but from a different perspective. In particular, they suggest treating the 0/1 outcome as numeric and fitting a regression forest using the standard MSE splitting criterion (an example of a so-called *probability machine*). It seems strange to use MSE on a 0/1 outcome, right? Not really. Recall from Section 2.2.1 that the Gini index for binary outcomes is equivalent to using the MSE. Malley et al. recommend using a minimum node size equal to 10% of the number of training cases: $n_{min} = \lfloor 0.1 \times N \rfloor$. However, for probability estimation, it seems natural to treat n_{min} as a tuning parameter. Devroye et al. [1997, Chap. 21–22] provide some guidance on the choice of n_{min} for consistent probability estimation in decision trees.

The predicted probabilities can be converted to class predictions (i.e., by comparing each probability to some threshold), which gives us an alternative to hard voting called *soft voting*. In soft voting, we classify x to the class with the largest averaged class probability. This approach to classification in RFs tends to be more accurate since predicted probabilities closer to zero or one are given more weight during the averaging step; hence, soft voting attaches more weight to votes with higher confidence (or smaller standard errors; Section 7.7).

7.2.1.1   Example: Mease model simulation

To illustrate the difference between a classification and regression forest for probability estimation with binary outcomes, I'll expand upon one of the simulation studies in Malley et al. [2012], the Mease example, in particular [Mease et al., 2007]. This is a two-dimensional circle problem with a binary outcome (0/1) and two independent features. The features are independent $\mathcal{U}(0, 50)$ random variables (i.e., the points are generated at random in the square $[0, 50]^2$). The probability function is defined as

$$p\left(\boldsymbol{x}\right) = \Pr\left(Y = 1 | \boldsymbol{x}\right) = \begin{cases} 1, & r\left(\boldsymbol{x}\right) < 8 \\ \frac{28 - r(\boldsymbol{x})}{20}, & 8 \le r\left(\boldsymbol{x}\right) \le 20 \\ 0, & r\left(\boldsymbol{x}\right) \ge 28 \end{cases}, \quad (7.1)$$

where $r\left(\boldsymbol{x}\right)$ is the Euclidean distance from $\boldsymbol{x} = (x_1, x_2)$ to the point $(25, 25)$. A sample of $N = 1000$ observations from the Mease model is displayed in Figure 7.3; note that the observed 0/1 outcomes were generated according to the above probability rule $p\left(\boldsymbol{x}\right)$. (As always, the code to reproduce the simulation is available on the companion website.)

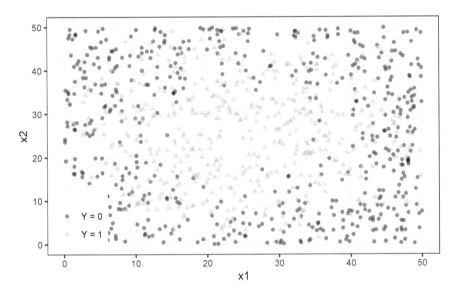

FIGURE 7.3: A sample of $N = 1000$ observation from the Mease model.

Figures 7.4–7.5 display the results of the simulation. In Figure 7.4, the median predicted probability across all 250 simulations was computed and plotted vs. the true probability of class 1 membership; the dashed 45-degree line corresponds to perfect agreement. Here it is clear that the regression forest (i.e., treating the 0/1 outcome as continuous and building trees using the MSE splitting criterion) outperforms the classification forest (except when $n_{min} = 1$, in which case they are equivalent.) This is also evident from Figure 7.5, which shows the distribution of the MSE between the predicted class probabilities and the true probabilities for each case. In essence, for binary outcomes, regression forests produce consistent estimates of the true class probabilities.[a] This goes to show that m_{try} isn't the only important tuning parameter when

---

[a] By consistent, I mean that $\widehat{\Pr}\left(Y = 1 | \boldsymbol{x}\right) \to \Pr\left(Y = 1 | \boldsymbol{x}\right)$ as $N \to \infty$.

it comes to probability estimation, and you should make an effort to tune n_{min} as well.

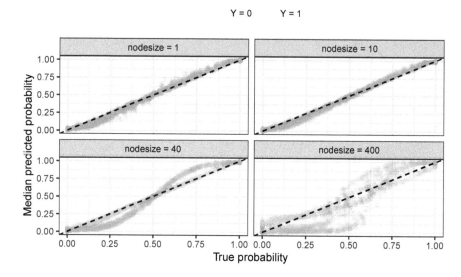

FIGURE 7.4: Class probability estimation using regression forests (yellow) and classification forests (black) in the Mease simulation. Starting from the top-left and moving clockwise, we have: $n_{min} = 1$ (the typical default for classification forests), $n_{min} = 10$ (the current default for probability forests in R's **ranger** package [Wright et al., 2021]), $n_{min} = 40$ (1% of the learning sample), and $n_{min} = 400$ (10% of the learning sample), respectively. The dashed 45-degree line corresponds to perfect agreement.

7.2.2 Subsampling (without replacement)

While a traditional RF (Algorithm 7.1) uses bootstrap sampling (i.e., sample with replacement), it can be useful to subsample the training data without replacement, prior to constructing each tree. This was noted in Section 5.1.3 for bagged tree ensembles, and that discussion equally applies to RFs as well. Furthermore, as I'll discuss in Section 7.5, subsampling with replacement can help eliminate certain bias in computing predictor importance, resulting in variable importance scores that can be used reliably for variable selection even in situations where the potential predictors vary in their scale of measurement or their number of categories. Most RF software includes the option to use subsampling with or without replacement.

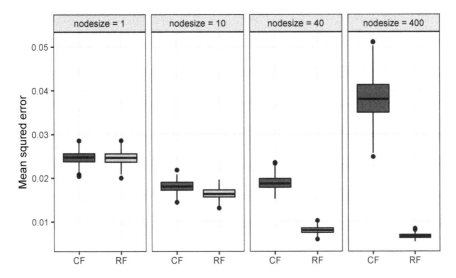

FIGURE 7.5: Mean squared errors (MSEs) from the Mease simulation. Here the MSE between the predicted probabilities and the true probabilities for each simulation are displayed using boxplots. Clearly, in this example, the regression forest (RF) with $n_{min} > 1$ produces more accurate class probability estimates.

7.2.3 Random forest from scratch: predicting home prices

To help solidify the basic concepts of an RF, let's construct one from scratch.[b] To do that, we need a decision tree implementation that will allow us to randomly select a subset of features for consideration at each node in the tree. Such arguments are available in the **sklearn.tree** module in Python, as well as R's **party** and **partykit** packages—unfortunately, this option is not currently available in **rpart**. In this example, I'll go with **party**, since it's `ctree()` function is faster, albeit less flexible, than **partykit**'s implementation.

Below is the definition for a function called `crforest()`, which constructs a *conditional random forest* (CRF) [Hothorn et al., 2006a, Strobl et al., 2008a, 2007b], that is, an RF using conditional inference trees (Chapter 3) for the base learners.[c] The `oob` argument will come into play in Section 7.3, so just

---

[b]The code I'm about to show is for illustration purposes only. It will not be nearly as efficient as actual RF software, which is typically written in a compiled language, like C, C++, or Fortran.

[c]Note that **party** and **partykit** both contain a `cforest()` function for fitting CRFs.

ignore that part of the code for now. Note that the function returns a list of
fitted CTrees that we can aggregate later for the purposes of prediction.

```r
crforest <- function(X, y, mtry = NULL, B = 5, oob = TRUE) {
  min.node.size <- if (is.factor(y)) 1 else 5
  N <- nrow(X)   # number of observations
  p <- ncol(X)   # number of features
  train <- cbind(X, "y" = y)   # training data frame
  fo <- as.formula(paste("y ~ ", paste(names(X), collapse = "+")))
  if (is.null(mtry)) {   # use default definition
    mtry <- if (is.factor(y)) sqrt(p) else p / 3
    mtry <- floor(mtry)   # round down to nearest integer
  }
  # CTree parameters; basically force the tree to have maximum depth
  ctrl <- party::ctree_control(mtry = mtry, minbucket = min.node.size,
                               minsplit = 10, mincriterion = 0)
  forest <- vector("list", length = B)   # to store each tree
  for (b in 1:B) {   # fit trees to bootstrap samples
    boot.samp <- sample(1:N, size = N, replace = TRUE)
    forest[[b]] <- party::ctree(fo, data = train[boot.samp, ],
                                control = ctrl)
    if (isTRUE(oob)) {   # store row indices for OOB data
      attr(forest[[b]], which = "oob") <-
        setdiff(1:N, unique(boot.samp))
    }
  }
  forest   # return the "forest" (i.e., list) of trees
}
```

Let's test out the function on the Ames housing data, using the same 70/30
split from previous examples (Section 1.4.7). Here, I'll fit a default CRF (i.e.,
$m_{try} = \lfloor p/3 \rfloor$ and $n_{min} = 5$) using our new `crforest()` function. (Be warned,
this code may take a few minutes to run; the code on the book website includes
an optional progress bar and the ability to run in parallel using the **foreach**
package [Revolution Analytics and Weston, 2020].)

```r
X <- subset(ames.trn, select = -Sale_Price)   # feature columns
set.seed(1408)   # for reproducibility
ames.crf <- crforest(X, y = ames.trn$Sale_Price, B = 300)
```

To obtain predictions from the fitted model, we can just loop through each
tree, extract the predictions, and then average them together at the end. This
can be done with a simple `for` loop, which is demonstrated in the code chunk
below. Here, I obtain the averaged predictions from `ames.crf` on the test data
and compute the test RMSE.

```r
B <- length(ames.crf)   # number of trees in forest
preds.tst <- matrix(nrow = nrow(ames.tst), ncol = B)
for (b in 1:B) {   # store predictions from each tree in a matrix
  preds.tst[, b] <- predict(ames.crf[[b]], newdata = ames.tst)
```

```
}
pred.tst <- rowMeans(preds.tst)  # average predictions across trees

# Root-mean-square error function
rmse <- function(pred, obs, na.rm = FALSE) {
  sqrt(mean((pred - obs) ^ 2, na.rm = na.rm))
}

# Root mean square error on test data
rmse(pred.tst, obs = ames.tst$Sale_Price)

#> [1] 24.3
```

Rather than reporting the test RMSE for the entire forest, we can compute it for each sub-forest of size $b \le B$ to see how it changes as the forest grows. We can do this using a simple **for** loop, as demonstrated in the code chunk below. (Note that I use **drop = FALSE** here so that the subset matrix of predictions doesn't lose its dimension when **b = 1**.)

```
rmse.tst <- numeric(B)  # to store RMSEs
for (b in 1:B) {
  pred <- rowMeans(preds.tst[, 1:b, drop = FALSE], na.rm = TRUE)
  rmse.tst[b] <- rmse(pred, obs = ames.tst$Sale_Price, na.rm = TRUE)
}
```

The above test RMSEs are displayed in Figure 7.6 (black curve). For comparison, I also included the test error for a single CTree fit (horizontal dashed line). Here, the CRF clearly outperforms the single tree, and the test error stabilizes after about 50 trees. Next, I'll discuss an internal cross-validation strategy based on the OOB data.

7.3 Out-of-bag (OOB) data

One of the most useful by-products of an RF (or bagging in general, for that matter) is the so-called OOB data (see Step 2) (b) in Algorithm 7.1)[d]. Recall that an RF, or any bagged tree ensemble, is constructed by combining predictions from decision trees trained on different bootstrap samples.[e] Since bootstrapping involves sampling with replacement, each tree in the forest only uses a subset of the original learning sample; hence, for each tree in the forest,

---

[d]While the concept of OOB is usually discussed in the context of an RF, it equally applies to bagging and boosting when sampling is involved, regardless if the sampling is done with or without replacement.

[e]This discussion also applies to subsampling without replacement.

a portion of the original learning sample isn't used—these observations are referred to as out-of-bag (or OOB for short). The OOB data associated with a particular tree can be used to obtain an unbiased estimate of prediction error. The OOB errors can then be aggregated across all the trees in the forest to obtain an overall out-of-sample, albeit unstructured, estimate of the overall prediction performance of the forest.

Since bagging/bootstrapping involves sampling with replacement, the probability that a particular case is not selected in a particular bootstrap sample is

$$\Pr\left(\text{case } i \notin \text{ bootstrap sample } b\right) = \left(1 - \frac{1}{N}\right)^N.$$

As $N \to \infty$ it can be shown that $\left(1 - \frac{1}{N}\right)^N \to e^{-1} \approx 0.368$. In other words, on average, each bootstrap sample contains approximately $1 - e^{-1} \approx 0.632$ of the original training records; the remaining $e^{-1} \approx 0.368$ observations are OOB and can be used as an independent validation set for the corresponding tree. This is rather straightforward to observe without a mathematical derivation. The code below computes the proportion of non-OOB observations in $B = 10000$ bootstrap samples of size $N = 100$, and averages the results together:

```
set.seed(1226)  # for reproducibility
N <- 100  # sample size
obs <- 1:N  # original observations
res <- replicate(10000, sample(obs, size = N, replace = TRUE))
inbag <- apply(res, MARGIN = 2, FUN = function(boot.sample) {
  mean(obs %in% boot.sample)  # proportion in bootstrap sample
})
mean(inbag)

#> [1] 0.634
```

Let $w_{b,i} = 1$ if observation i is OOB in the b-th tree and zero otherwise. Further, if we let $B_i = \sum_{i=1}^{B} w_{b,i}$ be the number of trees in the forest for which observation i is OOB, then the OOB prediction for the i-th training observation is given by

$$\hat{y}_i^{OOB} = \frac{1}{B_i} \sum_{b:w_{b,i}=1} \hat{y}_i^b, i = 1, 2, \dots, N. \tag{7.2}$$

The OOB error estimate is just the error computed from these OOB predictions. (See [Hastie et al., 2009, Sec. 7.11] for a more general discussion on using the bootstrap to estimate prediction error and its apparent bias.)

To illustrate, I'm going to compute the OOB RMSE for the CRF I previously fit to the Ames housing data. There are numerous ways in which this can be done programmatically given our setup; I chose the easy route. Recall that each tree in our `rfo` object contains an attribute called `"oob"` which stores the row numbers for the training records that were OOB for that particular tree. From these we can easily construct an $N \times B$ matrix, where the (i, j)-th element is given by

$$\begin{cases} \hat{y}_i^b & \text{if } w_{b,i} = 1 \\ \text{NA} & \text{if } w_{b,i} = 0 \end{cases}.$$

The reason for using `NA`s in place of the predictions for the non-OOB observations will hopefully become apparent soon.

```
preds.oob <- matrix(nrow = nrow(ames.trn), ncol = B)  # OOB predictions
for (b in 1:B) {  # WARNING: Might take a minute or two!
  oob.rows <- attr(ames.crf[[b]], which = "oob")  # OOB row IDs
  preds.oob[oob.rows, b] <-
    predict(ames.crf[[b]], newdata = ames.trn[oob.rows, ])
}
pred.oob <- rowMeans(preds.oob)  # average OOB predictions across trees

# Peek at results
preds.oob[1:3, 1:6]
```

```
#>        [,1] [,2] [,3] [,4] [,5] [,6]
#> [1,]    184   NA   NA   NA   NA  163
#> [2,]     NA  143   NA   NA   NA  154
#> [3,]     NA  136  115   NA  136   NA
```

Peeking at the first few rows and columns you can see that the first training observation (which corresponds to the first row in the above matrix) was OOB in the first and sixth trees (since the rest of the columns are `NA`), whereas the second observation was OOB for trees two and six, so I obtained the corresponding OOB predictions for these. Next, I compute $\hat{y}_i^{OOB}$ as in Equation (7.2) by computing the row means of our matrix `pred.oob`—setting `na.rm = TRUE` in the call to `rowMeans()` ensures that the `NA`s in the matrix aren't counted, so that the average is taken only over the OOB predictions (i.e., the correct denominator B_i will be used). Note that the OOB error is slightly larger than the test error I computed earlier; this is typical in many common settings, as noted in Janitza and Hornung [2018].

```
pred.oob <- rowMeans(preds.oob, na.rm = TRUE)
rmse(pred.oob, obs = ames.trn$Sale_Price, na.rm = TRUE)
```

```
#> [1] 26.6
```

Similar to what I did in the previous section, I can compute the OOB RMSE as a function of the number of trees in the forest. The results are displayed in Figure 7.6, along with the test RMSEs from the same forest (black curve) and test error from a single CTree fit (horizontal blue line). Here, we can see that the OOB error is consistently higher than the test error, but both begin to stabilize at around 50 trees.

```
rmse.oob <- numeric(B)   # to store RMSEs
for (b in 1:B) {
  pred <- rowMeans(preds.oob[, 1:b, drop = FALSE], na.rm = TRUE)
  rmse.oob[b] <- rmse(pred, obs = ames.trn$Sale_Price, na.rm = TRUE)
}
```

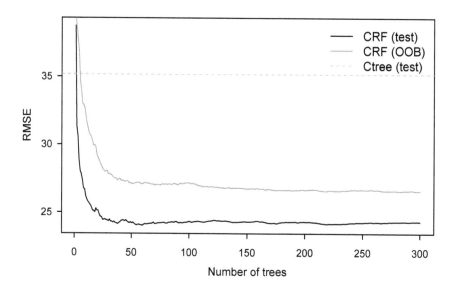

FIGURE 7.6: RMSEs for the Ames housing example: the CRF test RMSEs (black curve), the CRF OOB RMSEs (yellow curve), and the test RMSE from a single CTree fit (horizontal blue line).

As noted in Hastie et al. [2009], the OOB error estimate is almost identical to that obtained by N-fold cross-validation, where N is the number of rows in the learning sample; this is also referred to as *leave-one-out cross-validation* (LOOCV). Hence, algorithms that produce OOB data can be fit in one sequence, with cross-validation being performed along the way. The OOB error can be monitored during fitting and the training can stop once the OOB error has "stabilized". In the Ames housing example (Figure 7.6) it can be seen that the test and OOB errors both stabilize after around 50 trees.

While the OOB error is computationally cheap, Janitza and Hornung [2018] observed that it tends to overestimate the true error in many practical situations, including

- when the class frequencies are reasonably balanced in classification settings;

- when the sample size N is large;

- when there is a large number of predictors;

- when there is correlation between the predictors;

- when the main effects are weak.

The positive bias of OOB error estimates was also noted in Bylander [2002]. In light of this, it seems reasonable to consider the OOB error estimate as an upper bound on the true error. Fortunately, Janitza and Hornung [2018] argue that the OOB error can be used effectively for hyperparameter tuning in RFs—which I'll discuss in the next section—without substantially affecting performance. This is good news since k-fold cross-validation can be computationally expensive, especially when tuning more complex models, like tree-based ensembles.

7.4 Hyperparameters and tuning

RF is one of the most useful *off-the-shelf* statistical learning algorithms you can know. By off-the-shelf, I mean a procedure that can be used effectively without much tweaking or tuning. Don't get me wrong, you can (and should try to) improve performance with a bit of tuning, but relative to other algorithms, the RFs often do reasonably well at their default settings. In contrast, *gradient tree boosting* (Chapter 8) can often outperform RFs, but typically require a lot more tuning.

The most important tuning parameter in an RF is m_{try}. But I'd argue that the typical defaults (i.e., $m_{try} = \lfloor \sqrt{p} \rfloor$ for classification and $m_{try} = \lfloor p/3 \rfloor$ for regression) are quite good. For selecting m_{try}, a simple heuristic is to try the default, half of the default, and twice the default, and pick the best [Liaw and Wiener, 2002]. According to Liaw and Wiener [2002], the results generally do not change dramatically, and even setting $m_{try} = 1$ can give very good performance for some data. Setting $m_{try} < p$ also lessens the computational burden of split variable selection (e.g., CART's exhaustive search for the best split), making RFs more computationally efficient than bagged tree ensembles, especially for larger data sets. On the other hand, if you only suspect a small fraction of the predictors to be "important," then larger values of m_{try} may give better generalization performance.

The number of trees in the forest (B) is arguably not a tuning parameter. You just need to make sure enough trees are aggregated for the error to stabilize (see, for example, Figure 7.6). However, it can be wasteful to fit more trees than necessary, especially when dealing with large data sets. For this reason, some RF implementations have the option to "stop early" if the validation error stops improving; the R package **h2o** [LeDell et al., 2021] includes an RF implementation that supports early stopping with a wide variety of performance metrics. Such early stopping can be based on an independent test set, cross-validation, or the OOB error.

What about using the OOB error estimate or tuning? Although its been argued that the OOB error tends to overestimate the true error in certain cases (see, for example, Mitchell [2011]), Janitza and Hornung [2018] noted that the overestimation seems to have little to no impact on tuning parameter selection, at least in their simulations. If ordinary cross-validation is too expensive, and you don't have access to separate validation and test sets, then using the OOB error is a certainly reasonable thing to do, and in many cases, more efficient.

To illustrate, I carried out a small simulation using the Friedman 1 benchmark problem introduced in Section 1.4.3. For these data, there are 10 possible values for the m_{try} parameter. For each value, I generated 100 separate train and test sets of $N = 1,000$ observations each. For each repetition, I computed the OOB and test MSE and plotted the results; the results are displayed in Figure 7.7. In this example, you can see that the OOB error is quite in line with the test error, and both suggest an optimal value of m_{try} around 5 or 6 (the traditional default here, indicated by a dashed vertical line, is $m_{try} = \lfloor 10/3 \rfloor = 3$).

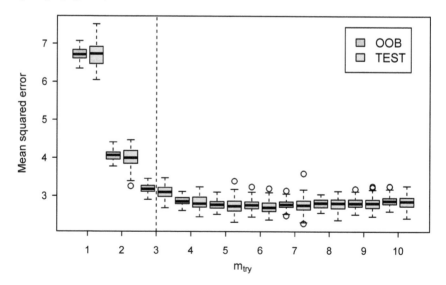

FIGURE 7.7: OOB and test error vs. m_{try} for the Friedman benchmark data using $N = 2000$ with a 50/50 split. The dashed line indicates the standard default for regression; in this case, $m_{try} = 3$.

7.5   Variable importance

Breiman [2002] proposed two measures of variable importance for RFs:

- A measure based on the mean decrease in node impurity (Section 7.5.1), which I'll refer to as the impurity-based measure. Originally, only the Gini splitting criterion (Section 2.2.1) was used. This variable importance measure was discussed for general tree ensembles in Section 5.4.

- A novel permutation-based measure (Section 7.5.2), which I'll refer to as the OOB-based permutation measure. This variable importance measure was discussed for general use in any supervised learning model in Section 6.1.1.

While these variable importance measures were originally introduced for the classification case, they naturally extend to the regression case as well. Note that many other measures have also been defined. An up-to-date and thorough overview of quantifying predictor importance in different tree-based methods is given in Loh and Zhou [2021].

7.5.1   Impurity-based importance

As discussed in Section 5.4, the importance of any predictor can be measured by aggregating the individual variable importance scores across all the trees in the forest. For an arbitrary feature x, we can use the total decrease in node impurities from splitting on x (e.g., as measured by MSE for regression or the Gini index for classification), averaged over all B trees in the forest:

$$\text{VI}(x) = \frac{1}{B} \sum_{b=1}^{B} \text{VI}_{T_b}(x), \qquad (7.3)$$

where $\text{VI}_{T_b}(x)$ is the relative importance of x in tree T_b (Section 2.8). Since averaging helps to stabilize variance, $\text{VI}(x)$ tends to be more reliable than $\text{VI}_{T_b}(x)$ [Hastie et al., 2009, p. 368].

The split variable selection bias inherent in CART-like decision trees also affects the impurity-based importance measure in their ensembles (7.3). The bias tends to result in higher variable importance scores for predictors with more potential split points (e.g., categorical variables with many categories). Several authors have proposed methods for eliminating the bias when the Gini index is used as the splitting criterion; see, for example, Sandri and Zuccolotto [2008] and the references therein. An interesting (and rather simple) approach is provided in Sandri and Zuccolotto [2008], and later modified in Nembrini et al. [2018] for RFs.

The idea in Sandri and Zuccolotto [2008] is to realize that the impurity-based importance measure from a single CART-like decision tree can be expressed as the sum of two components:

$$\mathrm{VI}_{T_b}(x) = \mathrm{VI}_{T_b}^{true}(x) + \mathrm{VI}_{T_b}^{bias}(x),$$

where $\mathrm{VI}_{T_b}^{true}(x_i)$ is the part attributable to informative splits and is related to the "true" importance of x_i, and $\mathrm{VI}_{T_b}^{bias}(x_i)$ is the part attributable to uninformative splits and is a source of bias. The algorithm they propose attempts to eliminate the bias in $\mathrm{VI}_{T_b}(x_i)$ by subtracting off an estimate of $\mathrm{VI}_{T_b}^{bias}(x_i)$. This is done many times and the results averaged together. The basic steps are outlined in Algorithm 7.2 below.

Algorithm 7.2 Bias-corrected Gini importance.

---

1) For $r = 1, 2, \ldots, R$:

   1) Given the original $N \times p$ matrix of predictor values $\boldsymbol{X}$, generate an $N \times p$ matrix of *pseudo predictors* $\boldsymbol{Z}_r$ using one of the following techniques:

      - Randomly permuting each column of the original predictor values $\boldsymbol{X}$ (hence, the j-th column of $\boldsymbol{Z}_r$ can be obtained by randomly shuffling the values in the j-th column of $\boldsymbol{X}$).

      - Randomly permuting the rows of $\boldsymbol{X}$; this has the advantage of maintaining any existing relationships between the original predictors.

   2) Apply the ensemble procedure (e.g., bagging, boosting, or RF) using $\tilde{\boldsymbol{X}}_r = (\boldsymbol{X}, \boldsymbol{Z}_r)$ as the set of available predictors (i.e., use both the original predictors, as well as the randomly generated pseudo predictors).

   3) Use Equation 7.3 to compute both $\mathrm{VI}(x_i)$ and $\mathrm{VI}(z_i)$; that is, compute the usual impurity-based variable importance measure for each predictor x_i and pseudo predictor z_i, for $i = 1, 2, \ldots, p$.

2) Compute the bias-adjusted impurity-based importance measure for each predictor x_i $(i = 1, 2, \ldots, p)$ as $\mathrm{VI}^{\star}(x_i) = R^{-1} \sum_{r=1}^{R} (\mathrm{VI}(x_i) - \mathrm{VI}(z_i))$.

Algorithm 7.2 can be used to correct biased variable importance scores from a single CART-like tree or an ensemble thereof. Also, while the original algorithm was developed for the Gini-based importance measure, Sandri and Zuccolotto [2008] suggest it is also effective at eliminating bias for other impurity measures, like cross-entropy and SSE. One of the drawbacks of Algorithm 7.2., however, is that it effectively doubles the number of predictors to $2p$ and requires multiple (R) iterations. This can be computationally prohibitive for large data sets, especially for tree-based ensembles. Fortunately, Nembrini et al. [2018] proposed a similar technique specific to RFs that only requires a single replication. I'll omit the details, but the procedure is available in the **ranger** package for R (which has also been ported to Python and is available in the **skranger** package [Flynn, 2021]); an example is given in Figure 7.8.

Even though our quick-and-dirty `crforest()` function in Section 7.2.3 used bootstrap sampling, the actual CRF procedure described in Strobl et al. [2007b], and implemented in R packages **party** and **partykit**, defaults to growing trees on random subsamples of the training data without replacement (by default, the size of each sample is given by $\lfloor 0.632N \rfloor$), as opposed to bootstrapping. Strobl et al. [2007b] showed that this effectively removes the bias in CRFs due to the presence of predictor variables that vary in their scale of measurement or their number of categories.

7.5.2 OOB-based permutation importance

RFs offer an additional (and unbiased) variable importance method; the approach is quite similar to the more general permutation approach discussed in Section 6.1.1, but it's based on permuting observations in the OOB data instead. The idea is that if predictor x is important, then the OOB error will go up when x is perturbed in the OOB data. In particular, we start by computing the OOB error for each tree. Then, each predictor is randomly shuffled in the OOB data, and the OOB errors are computed again. The difference in the two errors is recorded for the OOB data, then averaged across all trees in the forest.

As with the more general permutation-based importance measure, these scores can be unreliable in certain situations; for example, when the predictor variables vary in their scale of measurement or their number of categories [Strobl et al., 2007a], or when the predictors are highly correlated [Strobl et al., 2008b]. Additionally, the corrected Gini-based importance discussed in Nembrini et al. [2018] has the advantage of being faster to compute and more memory efficient.

Figure 7.8 shows the results from three difference RF variable importance measures on the simulation example from Section 3.1; the simulation comparing the split variable selection bias between CART and CTree. Here, we can see that the traditional Gini-based variable importance measure is biased towards the categorical variables, while the corrected Gini and permutation-based variable importance scores are relatively unbiased.

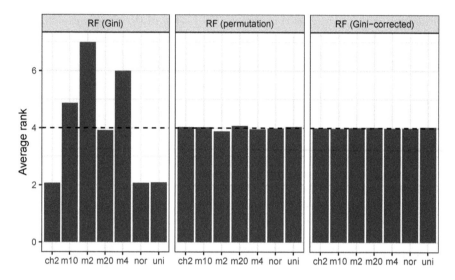

FIGURE 7.8: Average feature importance ranking for three RF-based variable importance measures. Left: the traditional Gini-based measure. Middle: the OOB-based permutation measure. Right: The corrected Gini-based measure.

The next two sections discuss more contemporary permutation schemes for RFs that deserve some consideration.

7.5.2.1 Holdout permutation importance

One drawback to computing variable importance in general is the lack of a natural cutoff that can be used to discriminate between "important" and "non-important" predictors. A number of approaches based on null hypothesis testing and *thresholding* have been developed for addressing this problem; see, for example, Altmann et al. [2010] and Loh and Zhou [2021, Sec. 6]. Janitza et al. [2018] argued that the null distribution of the OOB-based permutation measure is not necessarily symmetric; in particular, for irrelevant features. This makes the OOB-based permutation variable importance scores less suitable for selecting relevant features using a hypothesis-driven approach. Instead, Janitza et al. [2018] propose a method referred to as *holdout variable importance*, which has a symmetric null distribution for both relevant and irrelevant predictors. The idea is to split the data into two halves, grow an RF on one half, and use the leftover half to compute a permutation-based variable importance score. This method is available in the R package **ranger**.

7.5.2.2 Conditional permutation importance

A major drawback of permutation-based importance measures is the inherent assumption of independent features (e.g., the features are uncorrelated). For example, if x_1 and x_2 have a strong dependency, then it doesn't make sense to randomly permute x_1 while holding x_2 constant (or vice versa). To this end, [Strobl et al., 2008b] describe a *conditional permutation importance* measure that adjusts for correlations between predictor variables. In particular, the conditional permutation importance of each variable is computed by permuting within a grid defined by the covariates that are associated with the variable of interest. According to Strobl et al. [2008b], the resulting variable importance scores are conditional in the sense of coefficients in a regression model, but represent the effects of a variables in both main effects and interactions. When missing values are present in the predictors, the procedure described in Hapfelmeier et al. [2014] can be used to measure variable importance. While this idea applies in general to any type of RF, its implementations currently seem limited to the conditional inference trees and forest provided by the **party** and **partykit** packages in R; see, for example, `?partykit::varimp`.

7.6 Casewise proximities

So far in this book, we've mainly discussed tree-based methods for supervised learning problems. However, not every problem is supervised. For example, it is often of interest to understand how the data clusters—that is, whether the rows of the data form any "interesting" groups. (This is an application of unsupervised learning.) Many clustering methods rely on computing the pairwise distances between any two rows in the data, but the challenge becomes choosing the right distance metric. Euclidean distance (i.e., the "ordinary" straight-line, or "as the crow flies" distance between two points), for example, is quite sensitive to the scale of the inputs. It's also rather awkward to compute the Euclidean distance between two rows of data when the features are a mix of both numeric and categorical types. Fortunately, other distance (or distance-like) measures are available which more naturally apply to mixed data types.

Another useful output that can be obtained from an RF, provided it's implemented, are pairwise case *proximities*. RF proximities are distance-like measures of how similar any two observations are, and can be used for

- clustering in supervised and unsupervised (Section 7.6.3) settings;

- detecting outliers/novel cases (Section 7.6.1);

- imputing missing values (Section 7.6.2).

To compute the proximities between all pairs of training observations in an RF, do the following:

1) pass all of the data, both training and OOB, down each tree;

2) every time records i and j cohabitate in the same terminal node of a tree, increase their proximity by one;

3) At the end, normalize the proximities by dividing by the number of trees in the forest.

So how does this measure similarity between cases? Recall that RF's (and bagged decision trees in general) intentionally build deep, overgrown decision trees. In order for two observations to land in the same terminal node, they have to satisfy all of the same conditions leading to it. If two observations occupy the same terminal node across a majority of the trees in the forest, then they are likely very similar to each other in terms of feature values. Note that using all the training data can lead to unrealistic proximities. To circumvent this, proximities can be computed on only the OOB cases. It is also possible to compute proximities for new cases (an example application is given in Section 7.6.4).

The end result is an $N \times N$ proximity matrix, where N is the sample size of the data set proximities are being computed for. As it turns out, this matrix is symmetric (since prox $(i, j) = $ prox (j, i)), positive definite (i.e., has all positive eigenvalues), and bounded above by one, with the diagonal elements equal to one (since prox $(i, i) = 1$). Consequently, for any two cases i and j, we can treat $1 - $ prox (i, j) as a squared distance-like metric, which can be used as input into any distance-based clustering algorithm. For example, Shi et al. [2005] used RF proximities to help identify fundamental subtypes of cancer. A brief example using the Swiss banknote data is provided in Section 7.6.3.1.

The proximities from an RF provide a natural measure of similarity between records when the predictor variables are of mixed types (e.g., numeric and categorical) and measured on different scales; they are invariant to monotone transformations and naturally support categorical variables. The biggest drawback, as with any pairwise distance-like metric, is that it requires storing an $N \times N$ matrix; although, since the casewise proximity matrix is symmetric, you only need to store the upper or lower triangular part (see, for example, `?treemisc::proximity`). Proximities are also not implemented in most open source RF software. However, if you can obtain the $N \times B$ matrix of terminal node assignments (which is available in most open source RF software), then it is rather straightforward to compute the proximities yourself; an example, specific to the R package **ranger** [Wright et al., 2021], can be found at `https://mnwright.github.io/ranger/r/oob-proximity-matrix/`, while a

C++ implementation is available in **treemisc**'s `proximity()` function. The next two sections discuss more specific uses of proximities that are useful in a supervised learning context.

7.6.1 Detecting anomalies and outliers

Outliers (or anomalies) are generally defined as cases that are removed from the main body of the data. In the context of an RF, Leo Breiman defined outliers as cases whose proximities to all other cases in the data are generally small. For classification, he proposed a simple measure of "outlyingness" based on the RF proximity values. Define the average proximity from case m in class j to the rest of the training data in class j as

$$\mathrm{prox}^\star(m) = \sum_{k \,\in\, \mathrm{class}\ j} \mathrm{prox}^2(m, k),$$

where the sum is over all training instances belonging to class j. The outlyingness of case m in class j to all other cases in class j is defined as

$$\mathrm{out}(m, j) = \frac{N}{\mathrm{prox}^\star(m)},$$

where N is the number of training instances. Generally, a value above 10 is reason to suspect the case of being an outlier [Breiman, 2002]. Obviously, this measure is limited to smaller data sets and RF implementations where proximities can be efficiently computed. In Section 7.8.5, I'll look at a specialized RF extension that's more suitable for detecting outliers and anomalies, especially in higher dimensions. An interesting use case for the proximity-based outlyingness measure is presented in the next section.

7.6.1.1 Example: Swiss banknotes

An interesting use case for the proximity-based outlyingness measure is in detecting potentially mislabeled response classes in classification problems. Consider, for example, the Swiss banknote data from Section 1.4.1. Before fitting a default RF, I switched the label for observation 101; this observation is supposedly a counterfeit banknote ($y = 1$), but I switched the class label to genuine ($y = 0$). The proximity-based outlier scores are displayed in Figure 7.9. There are two obvious potential outliers, labeled with their corresponding row number. Here, you can see that the counterfeit banknote I mislabeled as genuine (observation 101) received the largest outlier score. Observation 70 is also interesting and worth investigating; perhaps it was also mislabeled?

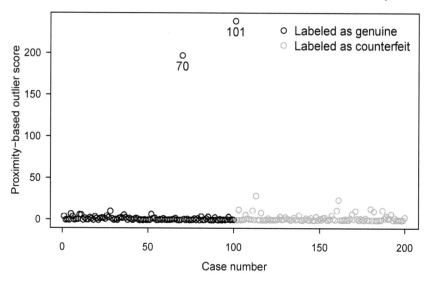

FIGURE 7.9: Proximity-based outlier scores for the Swiss banknote data. The largest outlier score corresponds to observation 101, which was a counterfeit banknote that was mislabeled as genuine.

7.6.2   Missing value imputation

Many decision tree algorithms can can naturally handle missing values; CART and CTree, for example, employ surrogate splits to handle missing values (Section 2.7). Unfortunately, the idea does not carry over to RFs. I suppose that makes sense: searching for surrogates would greatly increase the computation time of the RF algorithm. Although some RF software can handle missing values without casewise deletion (e.g., the **h2o** package in both R and Python), most often they have to be imputed or otherwise dealt with.

Breiman also developed a clever way to use RF proximities for imputing missing values. The idea is to first impute missing values with a simple method (such as using the mean or median for numeric predictors and the most common value for categorical ones). Next, fit an initial RF to the N complete observations and generate the $N \times N$ proximity matrix. For a numeric feature, the initial imputed values can be updated using a weighted mean over the non-missing values where the weights are given by the proximities. For categorical variables, the imputed values are updated using the most frequent non-missing values where frequency is weighted by the proximities. Then just iterate until some convergence criterion is met (typically 4–6 runs). In other words, this method just imputes missing values using a weighted mean/mode with more weight on non-missing cases.

Breiman [2002] noted that the OOB estimate of error in RFs tends to be overly optimistic when fit to training data that has been imputed. As with proximity-based outlier detection, this approach to imputation does not scale well (especially since it requires fitting multiple RFs and computing proximities). Further, this imputation method is often not as accurate as more contemporary techniques, like those implemented in the R package **mice** [van Buuren and Groothuis-Oudshoorn, 2021].

Perhaps the biggest drawback to proximity-based imputation, like many other imputation methods, is that it only generates a single completed data set. As discussed in van Buuren [2018, Chap. 1], our level of confidence in a particular imputed value can be expressed as the variation across a number of completed data sets. In Section 7.9.3, I'll use the CART-based multiple imputation procedure discussed in Section 2.7.1 and show how we can have confidence in the interpretation of the RF output by incorporating the variability associated with multiple imputation runs.

7.6.3 Unsupervised random forests

As it turns out, RFs can be used in unsupervised settings as well (i.e., when there is no defined response variable). In this case, the goal is to cluster the data, that is, see if the rows from the learning sample form any 'interesting" groups.

In an unsupervised RF, the idea is formulate a two-class problem. The first class corresponds to the original data, while the second class corresponds to a synthetic data set generated from the original sample. There are two ways to generate the synthetic data corresponding to the second class [Liaw and Wiener, 2002]:

1) a bootstrap sample is generated from each predictor column of the original data;

2) a random sample is generated uniformly from the range of each predictor column of the original data.

These two data sets are then stacked on top of each other, and an ordinary RF is used to build a binary classifier to try and distinguish between the real and synthetic data. (A necessary drawback here is that the resulting data set is twice as large as the original learning sample.) If the OOB misclassification error rate in the new two-class problem is, say, $\geq 40\%$, then the columns look too much like independent variables in the eyes of the RF; in other words, the dependencies among the columns do not play a large role in discriminating between the two classes. On the other hand, if the OOB misclassification rate is lower, then the dependencies are playing an important role. If there is some discrimination between the two classes, then the resulting proximity matrix can be used as an input into any distance-based clustering algorithm (like k-means or hierarchical clustering).

7.6.3.1   Example: Swiss banknotes

Continuing with the Swiss banknote example, I generated a synthetic version of the data set using the bootstrap approach outlined in the previous section, and then stacked the data together into a two-class problem:

```
bn <- treemisc::banknote
X.original <- subset(bn, select = -y)  # features only
X.synthetic <- X.original
set.seed(1034)
for (i in seq_len(ncol(X.original))) {
  X.synthetic[[i]] <- sample(X.synthetic[[i]], replace = TRUE)
}
X <- rbind(X.original, X.synthetic)

# Add binary indicator (doesn't)
X$y <- rep(c("original", "synthetic"), each = nrow(bn))
```

I then fit an RF of 1000 trees using the newly created binary indicator y and generated proximities for the original (i.e., first 200) observations. So how well did the unsupervised RF cluster the data? Well, we could convert the proximity matrix into a dissimilarity matrix and feed it into any distance-based clustering algorithm. Another approach, which I'll take here, is to visualize the dissimilarities using *multidimensional scaling* (MDS). MDS is one of many methods for displaying (transformed) multidimensional data in a lower-dimensional space; for details, see Johnson and Wichern [2007, Sec. 12.6]. Essentially, MDS takes a set of dissimilarities—one minus the proximities, in this case—and returns a set of points such that the distances between the points are approximately equal to the dissimilarities. Figure 7.10 shows the best-fitting two-dimensional representation. Here you can see a clear separation between the genuine bills (black) and counterfeit bills (yellow).

7.6.4   Case-specific random forests

The *case-specific RF* [Xu et al., 2016] is another interesting application of RF proximities (Section 7.6. The idea is to build a new RF to more accurately predict each individual observation in the test set. The individual RFs give more weight to the training observations that have higher proximity to the observations in the test set.

Let d_{trn} and d_{tst} be the train and test data sets with N and N_{tst} observations, respectively. The general steps for growing a case-specific RF are outlined in Algorithm 7.3 below:

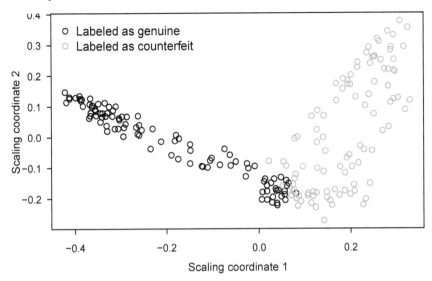

FIGURE 7.10: MDS coordinates of the proximities from an unsupervised RF fit to the Swiss banknote data. As can be seen, there are two noticeable clusters, although not perfectly separated. The genuine banknotes (black circles) generally fall into one cluster, while the counterfeit banknotes (yellow circles) tend to fall in the other.

Algorithm 7.3 Case-specific random forest algorithm.

---

1) Grow an ordinary RF of size B to the training data d_{trn}.

2) For each observation in the test set, say, x_0, do the following:

    a) Compute the proximities between x_0 and each observation in d_{trn} using the initial RF from Step 1); let $\{\text{prox}_i(x_0)\}_{i=1}^{N}$ be the proximities between x_0 and each case from d_{trn}; that is, the fraction of times x_0 cohabitates with each training instance across the B trees in the initial RF.

    b) Define the case weight for training case i relative to x_0 as $w_i^\star = \text{prox}_i(x_0) / \sum_{i=1}^{N} \text{prox}_i(x_0)$.

    c) Predict x_0 with a new RF grown to d_{trn} using case weights $\{w_i^\star\}_{i=1}^{N}$.

In essence, a case-specific RF predicts a new case x_0 using a new RF that gives more weight to the original training observations that have higher proximity (i.e., are more similar) to x_0. Note that most open source RF software provide the option to specify case weights for the training observations, which are used to weight each row when taking bootstrap samples (but many implementations do not provide proximities). While the idea of case-specific RFs makes sense, it has a couple of limitations. First off, it requires fitting $N_{tst} + 1$ RFs, which can be expensive whenever N or N_{tst} are large. Second, it requires computing $N \times N_{tst}$ proximities from an RF, which aren't always available from software.

Case-specific RFs are relatively straightforward to implement with traditional RF software, provided you can compute proximity scores[f]. The R package **ranger** provides an implementation of case-specific RFs. I applied this methodology to the Ames housing example (Section 1.4.7), which actually resulted in a slight increase to the test RMSE when compared to a traditional RF; the code to reproduce the example is available on the companion website for this book.

7.7 Prediction standard errors

Using a similar technique to OOB error estimation, Wager et al. [2014] proposed a method for estimating the variance of an RF prediction using a technique called the *jackknife*. The jackknife procedure is very similar to LOOCV, but specifically used for estimating the variance of a statistic of interest. If we have a statistic, $\hat{\theta}$, estimated from N training records, then the jackknife estimate of the variance of $\hat{\theta}$ is given by:

$$\hat{V}_{jack}\left(\hat{\theta}\right) = \frac{N-1}{N} \sum_{i=1}^{N} \left(\hat{\theta}_{(i)} - \hat{\theta}_{(\cdot)}\right)^2, \tag{7.4}$$

where $\hat{\theta}_{(i)}$ is the statistic of interest using all the N training observations except observation i, and $\hat{\theta}_{(\cdot)} = \sum_{i=1}^{N} \hat{\theta}_{(i)}/N$.

For brevity, let $\hat{f}(x) = \hat{f}_B^{rf}(x)$, for some arbitrary observation x (see Algorithm 7.1). A natural jackknife variance estimate for the RF prediction $\hat{f}(x)$ is given by

---

[f]Even if you don't have access to an implementation of RFs that can compute proximities, they're still obtainable as long as you can compute terminal node assignments for new observations (i.e., compute which terminal node a particular observation falls in for each of the B trees), which is readily available in most RF software. See Section 7.6 for details.

$$\hat{V}_{jack}\left(\hat{f}\left(\boldsymbol{x}\right)\right) = \frac{N-1}{N}\sum_{i=1}^{N}\left(\hat{f}_{(i)}\left(\boldsymbol{x}\right) - \hat{f}\left(\boldsymbol{x}\right)\right)^{2}. \tag{7.5}$$

This is derived under the assumption that $B = \infty$ trees were averaged together in the forest, which, of course, is never the case. Consequently, (7.5) has a positive bias. Fortunately, the same B bootstrap samples used to derive the forest can also be used to provide the bias corrected variance estimate

$$\hat{V}_{jack}^{BC}\left(\hat{f}\left(\boldsymbol{x}\right)\right) = \hat{V}_{jack}\left(\hat{f}\left(\boldsymbol{x}\right)\right) - (e-1)\frac{N}{B}\hat{v}\left(\boldsymbol{x}\right), \tag{7.6}$$

where $e = 2.718...$ is Euler's constant and

$$\hat{v}\left(\boldsymbol{x}\right) = \frac{1}{B}\sum_{b=1}^{B}\left(\hat{f}_{b}\left(X\right) - \hat{f}\left(\boldsymbol{x}\right)\right)^{2} \tag{7.7}$$

is the bootstrap estimate of the variance of a prediction from a single RF tree. Fortunately, all of the required quantities for computing (7.6) are readily available in the output from most open source RF software. This procedure is implemented in the R packages **ranger** and **grf** [Tibshirani et al., 2021]; the latter is a pluggable package for nonparametric statistical estimation and inference based on RFs, also known as *generalized random forests* [Athey et al., 2019]. The **forestci** package [Polimis et al., 2017] provides a Python implementation compatible with scikit-learn RF objects.

Once the estimated standard error of a prediction $\hat{y}$ is obtained, it can be useful to summarize it using a Gaussian-based confidence interval of the form $\hat{y} \pm z_{\alpha}\hat{\sigma}$, where z_{α} is a quantile from a standard normal distribution and $\hat{\sigma}$ is the estimated standard error of $\hat{y}$; see Wager et al. [2014] for details. Zhang et al. [2020] further discuss the use of confidence/prediction intervals for RF predictions using several methods, including *split conformal prediction* [Lei et al., 2018] and *quantile regression forests* (Section 7.8.2).

7.7.1 Example: predicting email spam

Switching back to the email spam data, let's compute jackknife-based standard errors for the test set predicted class probabilities. Following Wager et al. [2014], I fit an RF using $B = 20,000$ trees and three different values for m_{try}: 5, 19 (based on Breiman's default for classification), and 57 (an ordinary bagged tree ensemble).

The predicted class probabilities for `type = "spam"`, based on the test data, from each RF are displayed in Figure 7.11 (x-axis), along with their bias-corrected jackknife estimated standard errors (y-axis). Notice how the misclassified cases (solid black points) tend to correspond to observations where the predicted class probability is closer to 0.5. It also appears that the more constrained RF with $m_{try} = 5$ produced smaller standard errors, while the default RF ($m_{try} = 19$) and bagged tree ensemble ($m_{try} = 57$) produced noticeably larger standard errors, with the bagged tree ensemble performing the worst.

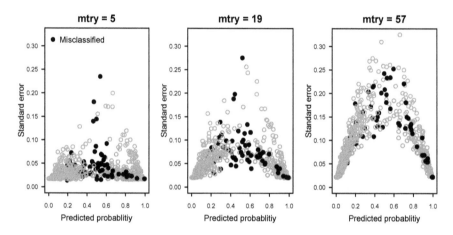

FIGURE 7.11: Bias-corrected jackknife variance estimates for the predicted probabilities using the spam training data. The solid black points correspond to the misclassified cases which appear to be more closely concentrated near the center (i.e., 0.5).

7.8 Random forest extensions

The following subsections highlight some notable extensions to the RF algorithm. Note that this is by no means an exhaustive list, so I tried to choose extensions that have been shown to be practically useful and that also have available (and currently maintained) open source implementations.

7.8.1 Oblique random forests

As briefly mentioned in Section 4.3.1, linear (or oblique) splits—which are splits based on linear combinations of the predictors—can sometimes improve predictive accuracy, even if they do make the tree harder to interpret. Many decision tree algorithms support linear splits (e.g., CART and GUIDE) and Breiman [2001] even proposed a variant of RFs that employed linear splits based on random linear combinations of the predictors. This approach did not gain the same traction as the traditional RF algorithm based on univariate splits. In fact, I'm not aware of any open source RF implementations that support his original approach based on random coefficients.

Menze et al. [2011] proposed a variant, called *oblique random forests* (ORFs), that explicitly learned optimal split directions at internal nodes using linear discriminative models[g], as opposed to random linear combinations. Similar to *random rotation ensembles* (Section 7.8.3), ORFs tend to have a smoother topology; see, for example, Figure 7.16. Menze et al. [2011] even go as far as to recommend the use of ORFs over the traditional RF when applied to mostly numeric features. Nonetheless, the idea of non-axis oriented splits in RFs has still not caught on. The only open source implementation of ORFs that I'm aware of is in the R package **obliqueRF** [Menze and Splitthoff, 2012], which has not been updated since 2012.

A more recent approach, called *projection pursuit random forest* (PPforest) [da Silva et al., 2021a] uses splits based on linear combinations of randomly chosen inputs. Each linear combination is found by optimizing a projection pursuit index [Friedman and Tukey, 1974] to get a projection of the features that best separates the classes; hence, this method is also only suitable for classification. PPforests are implemented in the R package **PPforest** [da Silva et al., 2021b]. Individual *projection pursuit trees* (PPtrees) [Lee et al., 2013], which are used as the base learners in a PPforest, can be fit using the R package **PPtreeViz** [Lee, 2019] (which seems to have superseded the older **PPtree** package).

7.8.2 Quantile regression forests

The goal of many supervised learning algorithms is to infer something about the relationship between the response and a set of predictors. In regression, for example, the goal is often to estimate the conditional mean $\mathrm{E}\,(Y|\boldsymbol{x})$, for some observation $\boldsymbol{x}$. In a typical regression tree, an estimate of the conditional mean is given by the mean response associated with the terminal node observation $\boldsymbol{x}$ falls into. In an RF, the terminal node means are simply averaged across all the trees in the forest.

The conditional mean response, however, provides only a limited summary of the conditional distribution function $F\,(y|\boldsymbol{x}) = \mathrm{Pr}\,(Y \le y|\boldsymbol{x})$. Denote the

---

[g]Similar to the LDA-based approach used in GUIDE (Section 4.3.1). However, GUIDE restricts itself to linear splits in only two features at a time to help with interpretation and reduce the impact of missing values.

α-quantile of $Y|x$ as $Q_\alpha(x)$. In other words, $\Pr(Y \le y | Q_\alpha(x)) = \alpha$. Compared to the conditional mean, the quantiles give a more useful summary of the distribution. For example, $\alpha = 0.5$ corresponds to the median. If the conditional distribution of $Y|x$ were symmetric, then the conditional mean and median would be the same. However, if $Y|x$ is skewed, then, compared to the conditional median, the conditional mean can be a misleading summary of what a typical value of $Y|x$ is. Furthermore, estimating $\Pr(Y \le y | Q_\alpha(x))$ for various values of α can give insight into the variability around a single point estimate, like the conditional median. This is the idea behind *quantile regression*.

The same idea can applied to an RF, and was formerly introduced in Meinshausen [2006] as *quantile regression forests* (QRF). In a QRF, the conditional distribution of $Y|x$ is approximated by the weighted mean

$$\hat{F}(y|x) = \sum_{i=1}^{N} w_i(x) I(Y_i \le y),$$

where $I(expression)$ is the indicator function that evaluates to one whenever *expression* is true, and zero otherwise. The weights $w_i(x)$ are estimated from the terminal node observations across all the tress in the forest and are defined in Meinshausen [2006]. In contrast to a traditional regression forest, this requires storing all the observations in each node, as opposed to just the mean.

As it turns out, there's not much difference between QRFs and RFs, aside from what information gets stored from each tree and how fitted values and predictions are obtained. RFs only need to keep track of the terminal node means. To estimate the full conditional distribution of $Y|x$, QRFs need to retain all observations across all terminal nodes.

7.8.2.1   Example: predicting home prices (with prediction intervals)

To illustrate, I fit a traditional RF and a QRF to the Ames housing data using the same train/test split discussed in Section 1.4.7; similar to before, I used `Sale_Price/1000` as the response. For each house in the test set, the predicted 0.025, 0.5, and 0.975 quantiles were obtained (which corresponds to a predicted median sale price and 95% prediction interval). Following the Boston housing example in Meinshausen [2006], the test set observations were ordered according to the length of the corresponding prediction intervals, and each observation was centered by subtracting the midpoint from the corresponding prediction interval. The observed sale prices are shown in black and the estimated conditional medians are given shown in yellow. The black lines show the corresponding 95% prediction bounds. (Note that the prediction intervals here are pointwise prediction intervals.)

The most expensive house in the test set sold for a (rescaled) sale price of $610. A traditional RF estimated a conditional mean sale price of $472.59, whereas the QRF produced a conditional median sale price of $479.07 with a 0.025 quantile of $255.24 and a 0.975 quantile of $745. Here, the QRF gives a much better sense of the variability in the predicted outcome, as well as a sense of the skewness of its distribution.

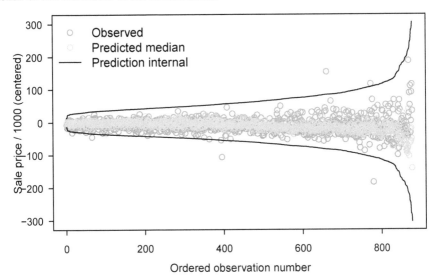

FIGURE 7.12: Rescaled sale prices for each home in the test set, along with the predicted 0.025, 0.5, and 0.975 quantiles from a QRF. To enhance visualization, the observations were ordered according to the length of the corresponding prediction intervals, and the mean of the upper and lower end of the prediction interval is subtracted from all observations and prediction intervals.

7.8.3 Rotation forests and random rotation forests

Before talking about *rotation forests* and random rotation ensembles, it might help to briefly discuss *rotation matrices*. A rotation matrix $\boldsymbol{R}$ of dimension p is a $p \times p$ square transformation matrix that's used to perform a rotation in N-dimensional Euclidean space.

A common application of rotation matrices in statistics is *principal component analysis* (PCA). The details of PCA are beyond the scope of this book, but the interested reader is pointed to Johnson and Wichern [2007, Chap. 8], among others. While PCA has many use cases, it is really just an unsupervised dimension reduction technique that seeks to explain the variance-covariance structure of a set of variables through a few linear combinations of these variables.

To illustrate, consider the $N = 100$ data points shown in Figure 7.13 (left); the axes for x_1 and x_2 are shown using dashed yellow and blue lines, respectively. The data were generated from a simple linear regression defined by

$$X_{2i} = X_{1i} + \epsilon_i, \quad i = 1, 2, \ldots, 100, \tag{7.8}$$

where $X_{1i} \overset{iid}{\sim} \mathcal{U}(0,1)$ and $\epsilon_i \overset{iid}{\sim} N(0,1)$. Further, let $\boldsymbol{X}$ be the 100×2 matrix whose first and second columns are given by X_{1i} and X_{2i}, respectively. As a rotation in two dimensions, PCA finds the rotation of the axes that yields maximum variance. The rotated axes for this example are shown in Figure 7.13 (middle). Notice that the first (i.e., yellow) axis is aligned with the direction of maximum variance in the sample. An alternative would be to rotate the data points themselves (right side of Figure 7.13). In this case, the variable loadings from PCA form a 2×2 rotation matrix, $\boldsymbol{R}$, that can be used to rotate $\boldsymbol{X}$ so that the direction of maximal variance aligns with the first (i.e., yellow) axis; this is shown in the right side of Figure 7.13. The rotated matrix is given by $\boldsymbol{X'} = \boldsymbol{X}\boldsymbol{R}$. Notice how the relative position of the points between x_1 and x_2 is preserved, albeit rotated about the axes.

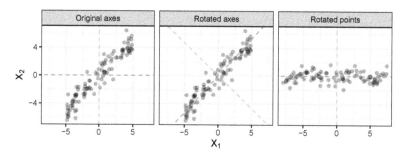

FIGURE 7.13: Data generated from a simple linear regression model. Left: Original data points and axes. Middle: Original data with rotated axes (notice how the first/yellow axis aligns with the direction of maximum variance in the sample). Right: Rotated data points on the original axes (here the data are rotated so that the direction of maximal variance aligns with the first/yellow axis).

So what does any of this have to do with RFs? Recall that the key to accuracy with model averaging is diversity. In an RF, diversity is achieved by choosing a random subset of predictors prior to each split in every tree. A rotation forest [Rodríguez et al., 2006], on the other hand, introduces diversity to a bagged tree ensemble by using PCA to construct a rotated feature space prior to the construction of each tree. Rotating the feature space allows adaptive nonparametric learning algorithms, like decision trees, to learn potentially interesting patterns in the data that might have gone unnoticed in the original feature space. Applying PCA to all the predictors prior to the construction each tree, even when using sampling with replacement, won't be enough to diversify the ensemble. Instead, prior to the construction of each tree, the predictor set is randomly split into K subsets, PCA is run separately on each, and a new set of linearly extracted features is constructed by pooling all the principal components (i.e., the rotated data points). K is treated as a tuning parameter, but the value of K that results in roughly three features per subset seems to be the suggested default [Kuncheva and Rodríguez, 2007]. Rotation forests can be thought of as a bagged tree ensemble with a random feature transformation applied to the predictors prior to constructing each tree. In this sense, PCA can be thought of as a feature extraction method. By performing PCA on random subsets of features prior to fitting each tree, rotation forests can improve the performance of a bagged tree ensemble. In this case, the derived features come from PCA applied to random subsets of the data, and while other feature extraction methods have also been considered, PCA was found to be the most suitable [Kuncheva and Rodríguez, 2007].

Rotation forests have been shown to be competitive with RFs and can achieve better performance on data sets with mostly quantitative variables; although, this seems to be true mostly for smaller ensemble sizes [Rodríguez et al., 2006, Kuncheva and Rodríguez, 2007]. However, most comparative studies I've seen seem to focus on classification accuracy for comparison, which we know is not the most appropriate metric for comparing models in classification settings. Rotation forests are available in the R package **rotationForest** [Ballings and Van den Poel, 2017].

7.8.3.1 Random rotation forests

A similar approach, called random rotation ensembles [Blaser and Fryzlewicz, 2016], apply a random rotation to all the features prior to constructing each tree. Blaser and Fryzlewicz [2016] discuss two algorithms for generating random rotation matrices and provide general R and C++ code for doing so. The **treemisc** function `rrm()` uses the *indirect method* discussed in Blaser and Fryzlewicz [2016] and is shown below. Note that rotations are only applied to numeric features and can be sensitive to both scale and outliers, hence, rescaling the numeric features is often required; see Blaser and Fryzlewicz [2016] for several recommendations. Random rotation forests are not generally available in open source software, but you can find my poor man's implementation in package **treemisc**, which I'll demonstrate in the next section; see `?treemisc::rforest` for details.

Let X_c be the subset of numeric/continuous features from the full feature set X. In a random rotation forest, before fitting the b-th tree, the numeric features are randomly transformed using $X_{c,b} = X_c R_b$ (for $b = 1, 2, \ldots, B$), where R_b is a randomly generated rotation matrix of dimension equal to the number of columns of X_c.

```
treemisc::rrm
```

```
#> function(n) {
#>     QR <- qr(matrix(rnorm(n ^ 2), ncol = n))   # A = QR
#>     M <- qr.Q(QR) %*% diag(sign(diag(qr.R(QR))))
#>     if (det(M) < 0) M[, 1L] <- -M[, 1L]   # det(M) = +1
#>     M
#> }
#> <bytecode: 0x7ff296e1fe90>
#> <environment: namespace:treemisc>
```

To illustrate the effect of applying random rotations to a set of features, let's continue with the simulated data from the previous section. In the code chunk below, I re-generate the same $N = 100$ points from (7.8); the original data are displayed in Figure 7.14 (black points), along with the observations under various random rotations, including PCA (orange points). Such rotations preserve the inter-relationships between predictors, but cast them into a different space resulting in equally accurate, but more diverse trees.

```
set.seed(1038)
X1 <- runif(100, min = -5, max = 5)
X2 <- X1 + rnorm(length(X1))
X <- cbind(X1, X2)
palette("Okabe-Ito")  # colorblind-friendly color palette
plot(X, xlim = c(-8, 8), ylim = c(-8, 8), col = 1, las = 1,
     xlab = expression(x[1]), ylab = expression(x[2]))
pcR <- loadings(princomp(X, cor = FALSE, fix_sign = FALSE))  # PCA
points(X %*% pcR, col = 2)  # plot PCA rotation
abline(0, 1, lty = 2, col = 1)  # original axis
abline(h = 0, lty = 2, col = 2)  # axis after PCA rotation
for (i in 3:5) {  # plot random rotations
  R <- treemisc::rrm(2)  # generate a random 2x2 rotation matrix
  points(X %*% R, col = adjustcolor(i, alpha.f = 0.5))
}
legend("topleft", legend = "Original sample", pch = 1, col = 1,
       inset = 0.01, bty = "n")
palette("default")
```

7.8.3.2   Example: Gaussian mixture data

In this section, I'll use the Gaussian mixture data from Hastie et al. [2009] to compare the results of an RF, rotation forest, and random rotation forest. The data for each class come from a mixture of ten normal distributions,

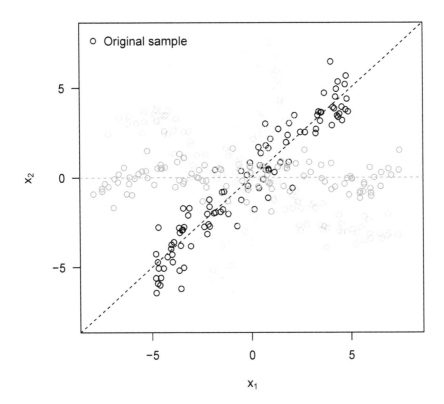

FIGURE 7.14: Scatterplot of x_1 vs. x_2. The original data are shown in black and a dashed black line gives the direction of maximum variance. The rotated points under PCA are shown in dark yellow along with the new axis of maximal variance; notice how in two dimensions this shifts the points to have maximal variance along the x-axis. The rest of the colors display the data points under random rotations.

whose individual means are also normally distributed. A full description of the data generating process can be found in Hastie et al. [2009, Sec. 2.3.3] and an application to RFs is provided in Hastie et al. [2009, Sec. 15.4.3]. The raw data are available at `https://web.stanford.edu/~hastie/ElemStatLearn/data.html`. For convenience, the data are also available in the **treemisc** R package that accompanies this book, and can be read in using:

```
library(treemisc)
```

```
eslmix <- load_eslmix()
class(eslmix)  # should be a list

#> [1] "list"

names(eslmix)  # names of components

#> [1] "x"         "y"         "xnew"      "prob"
#> [5] "marginal"  "px1"       "px2"       "means"
```

Note that this is not a data frame, but rather a list with several components; for a description of each, see `?treemisc::load_eslmix`.

The code chunk below constructs a scatterplot of the training data (i.e., component x) along with the Bayes decision boundary[h]; see Figure 7.15. The Bayes error rate for these data—that is, the theoretically optimal error rate—is 0.210.

```
x <- as.data.frame(eslmix$x)     # training data
xnew <- as.data.frame(eslmix$xnew)   # evenly spaced grid of points
x$y <- as.factor(eslmix$y)   # coerce to factor for plotting
xnew$prob <- eslmix$prob  # Pr(Y = 1 | xnew)

# Colorblind-friendly palette
oi.cols <- unname(palette.colors(8, palette = "Okabe-Ito"))

# Construct scatterplot of training points
p <- ggplot(x, aes(x = x1, y = x2, color = y)) +
  geom_point(alpha = 1, show.legend = FALSE) +
  scale_colour_manual(values = oi.cols) +
  theme_bw()

# Add optimal (i.e., Bayes) decision boundary
p + geom_contour(data = xnew, aes(x = x1, y = x2, z = prob),
                 breaks = 0.5, color = oi.cols[4],
                 inherit.aes = FALSE, linetype = 2)
```

Next I fit three tree-based ensembles: a traditional RF, a rotation forest, and a random rotation forest. The rotation forest was fit using the **rotationForest** package, while the RF and random rotation forest were fit using **treemisc**'s `rforest()` function. Note that this is a poor man's implementation of Breiman's RF algorithm I wrote that optionally rotates the features at random prior to the construction of each tree. It is based on the well-known **randomForest** package [Breiman et al., 2018] and is missing many of the bells and whistles, hence not recommended for general use, but it works. Also, this

---

[h]As noted in Hastie et al. [2009, Chap. 2], since the data generating mechanism is known for each of thew two classes, the theoretically optimal decision boundary can be computed exactly. This makes it useful to compare classifiers visually in terms of their estimated decision boundaries.

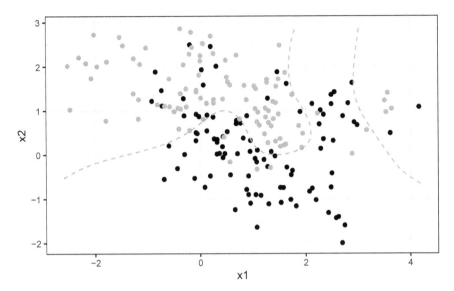

FIGURE 7.15: Simulated mixture data with optimal (i.e., Bayes) decision boundary.

implementation only uses regression trees for the base learners, hence, only regression and binary classification are supported. For the latter, the probability machine approach discussed in Section 7.2.1 is implemented.

The resulting decision boundaries from each forest are in Figure 7.16. Here, you can see that the axis-oriented nature of the individual trees in a traditional RF leads to a decision boundary with an axis-oriented flavor (i.e., the decision boundary is rather "boxy"). The RF also exhibits more signs of overfitting, as suggested by the little islands of decision boundaries. On the other hand, using feature rotation (with PCA or random rotations) prior to building each tree results in a noticeably smoother and non-axis-oriented decision boundary. The test error rates for the RF, rotation forest, and random rotation forest, under this random seed, are 0.235, 0.239, and 0.226, respectively. (As always, the code to reproduce this example is available on the companion website for this book.)

7.8.4 Extremely randomized trees

Just as RFs offer an additional layer of randomization to bagged decision trees, *extremely randomized trees* (or extra-trees) [Geurts et al., 2006] are essentially an RF with an additional randomization step. In particular, the split point for any feature at each node in a tree is essentially selected at random from a uniform distribution. The extra randomization can further

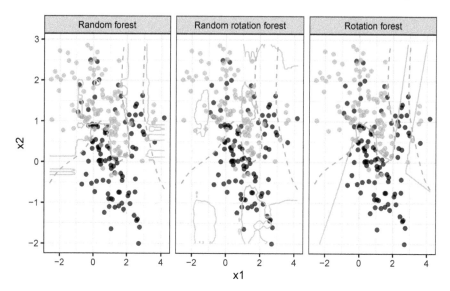

FIGURE 7.16: Traditional RF vs. random rotation forest on the mixture data from Hastie et al. [2009]. The random rotation forest produces a noticeably smoother decision boundary than the axis-oriented decision boundary from a traditional RF.

decrease variance, but sometimes at the cost of additional bias [Geurts et al., 2006]; this is especially true if the data contain irrelevant features. To combat the extra bias, extra-trees utilize the full learning sample to grow each tree, rather than bootstrap replicas (another subtle difference from the bagging and RF algorithms). Note that bootstrap sampling can be used in extra-trees ensembles, but Geurts et al. argue that it can often lead significant drops in accuracy[i].

The primary tuning parameters for an extra-trees ensemble are K and n_{min}, where K is the number of random splits to consider for each candidate splitter, and n_{min} is the minimum node size, which is a common parameter in many tree-based models and can act as a smoothing parameter. A common default for K is

$$K = \begin{cases} \sqrt{p} & \text{for classification} \\ p & \text{for regression} \end{cases},$$

where p is the number of available predictors. The optimal split is chosen from the sample of K splits in the usual way (i.e., the split that results in the

---

[i]Most open source implementations of extra-trees optionally allow for bootstrap sampling.

largest reduction in node impurity). Note that n_{min} has the same defaults as it does in an RF (Section 7.2).

The extra-trees ensemble still makes use of the RF m_{try} parameter, but note that, in the extreme case where $K = 1$, an extra-trees tree is unsupervised in that the response variable is not needed in determining any of the splits. Such a *totally randomized tree* [Geurts et al., 2006] can be useful in detecting potential outliers and anomalies, as will be discussed in Section 7.8.5. Extra-trees can be fit in R via the **ranger** package. In Python, an implementation of the extra-trees algorithm is provided by the **sklearn.ensemble** module.

7.8.5 Anomaly detection with isolation forests

While the proximities of an RF can be used to detect novelties and potential outliers, they're rather computationally expensive to compute and store, especially for large data sets; also, as previously mentioned, many RF implementations do not support proximities. A more general approach to anomaly detection, called an *isolation forest*, was proposed in Liu et al. [2008]. An isolation forest is essentially an ensemble of *isolation trees* (IsoTrees). IsoTrees are similar to extra-trees with $K = 1$ (Section 7.8.4), except that the splitting variables are also chosen at random; hence, IsoTrees are unsupervised in the sense that the tree building process does not make use of any response variable information.

So how does it work? Isolation forests are quite simple actually. The core idea is to "isolate" anomalous observations, rather than creating a profile for "normal" ones—the latter seems to be the more common approach taken by other methods in practice. Isolation forests assume that

1) anomalies are rare in comparison to "normal" observations;

2) anomalies differ from "normal" instances in terms of the values of their features.

In other words, isolation forests assume anomalies are "few and different" [Liu et al., 2008]. If anomalies are "few and different", then they are susceptible to *isolation* (i.e., easy to separate from the rest of the observations).

In an IsoTree, observations are recursively separated until all instances are isolated to their own terminal node; since the split variables and split points are determined completely at random, no response information is needed. Anomalous observations tend to be easier to isolate with fewer random partitions compared to normal instances. That is to say, the relatively few instances of anomalies tend to have shorter path lengths in an IsoTree when compared to normal observations. The path length to each observation can be computed for a forest of independently grown trees and aggregated into a single *anomaly score*. The anomaly score for an arbitrary observation x is given by

$$s\left(\boldsymbol{x}, N\right) = 2^{-\frac{1}{B}\sum_{b=1}^{B} h_b(\boldsymbol{x})/c(N)}, \tag{7.9}$$

where N is the sample size, $h_b\left(\boldsymbol{x}\right)$ is the path length to $\boldsymbol{x}$ in the b-th tree, and $c\left(N\right)$ is the *average path length of unsuccesful searches*; in a binary tree constructed from N observations, $c\left(N\right)$ is given by

$$c\left(N\right) = 2H\left(N-1\right) - 2\left(N-1\right)/N,$$

where $H\left(i\right)$ is the i-th *harmonic number* (`https://mathworld.wolfram.com/HarmonicNumber.html`).

Let $\bar{h}\left(\boldsymbol{x}\right) = \frac{1}{B}\sum_{b=1}^{B} h_b\left(\boldsymbol{x}\right)$ be the average path length for instance $\boldsymbol{x}$ across all trees in the forest, and note that $0 < s\left(\boldsymbol{x}, N\right) \leq 1$ and $0 < \bar{h}\left(\boldsymbol{x}\right) \leq N-1$. A few useful relationships regarding (7.9) are worth noting:

- $s\left(\boldsymbol{x}, N\right) \to 0.5$ as $\bar{h}\left(\boldsymbol{x}\right) \to c\left(N\right)$;
- $s\left(\boldsymbol{x}, N\right) \to 1$ as $\bar{h}\left(\boldsymbol{x}\right) \to 0$;
- $s\left(\boldsymbol{x}, N\right) \to 0$ as $\bar{h}\left(\boldsymbol{x}\right) \to N-1$.

Since the assumption is that anomalies are easier to isolate, they are likely to have shorter path lengths on average. Hence, any instance $\boldsymbol{x}$ with values of $s\left(\boldsymbol{x}, N\right)$ close to one tend to be highly anomalous. If $s\left(\boldsymbol{x}, N\right)$ is much smaller than 0.5, then it is safe to regard $\boldsymbol{x}$ as a "normal" instance. If all the instances return a value close to 0.5, then it is safe to say the sample does not really contain any anomalies. Note that these are just guidelines.

Isolation forests are a top-performing unsupervised method for detecting potential outliers and anomalies [Domingues et al., 2018]. Compared to other outlier detection algorithms, isolation forests are scalable (e.g., they have relatively little computational and memory requirements), fully nonparametric, and do not require a distance-like matrix. The Anti-Abuse AI Team at LinkedIn uses isolation forests to help detect abuse on LinkedIn (e.g., fake accounts, account takeovers, and profile scraping) [Verbus, 2019].

Let's illustrate the overall idea with a simple simulated example. Consider the data in Figure 7.17 (left). Here, the points $\{(x_{1i}, x_{2i})\}_{i=1}^{2,000}$ were independently sampled from a standard normal distribution. However, I changed the first observation to $(x_{11}, x_{21}) = (5.5, 5.5)$ (purple point), which is quite anomalous compared to the rest of the sample.

A single IsoTree from an isolation forest is displayed in Figure 7.17 (right), along with the path taken by $\boldsymbol{x}$ (in purple). Here $\boldsymbol{x}$ has a path length of two and you can see that the path $\boldsymbol{x}$ takes in the tree is relatively shorter than most of the other available paths.

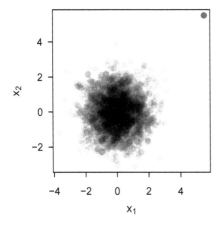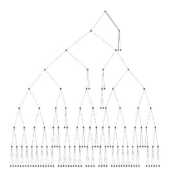

FIGURE 7.17: Isolation tree example. Left: scatterplot of two independent samples from a standard normal distribution; the observation with coordinates (5.5, 5.5) is a clear anomaly. Right: isolation tree diagram; the branch isolating the purple point is also highlighted in purple.

7.8.5.1 Extended isolation forests

An *extended isolation forest* [Hariri et al., 2021] improves the consistency and reliability of the anomaly score produced by a standard isolation forest by using random oblique splits (in this case, hyperplanes with random slopes)—as opposed to axis-oriented splits—which often results in improved anomaly scores. The tree in Figure 7.17 is from an extended isolation forest fit using the **eif** package [Hariri et al., 2021], which is also available in R.

7.8.5.2 Example: detecting credit card fraud

To illustrate the basic use of an isolation forest, I'll use a data set from Kaggle[j] containing anonymized credit card transactions, labeled as fraudulent or genuine, obtained over a 48 hour period in September of 2013; the data can be downloaded from Kaggle at

> `https://www.kaggle.com/mlg-ulb/creditcardfraud`.

Recognizing fraudulent credit card transactions is an important task for credit card companies to ensure that customers are not charged for items that they did not purchase. Since fraudulent transactions are relatively rare (as one

---

[j]Kaggle is an online community of data scientists and machine learning practitioners who can find and publish data sets and enter competitions to solve data science challenges; for more, visit `https://www.kaggle.com/`.

would hope), the data are highly imbalanced, with 492 frauds (0.17%) out of the $N = 284{,}807$ transactions.

For reasons of confidentiality, the original features have been transformed using PCA, resulting in 28 numeric features labeled V1, V2, ..., V28. Two additional variables, Time (the number of seconds that have elapsed between each transaction and the first transaction in the data set) and Amount (the transaction amount), are also available. These are labeled data, with the binary outcome Class taking on values of 0 or 1, where a 1 represents a fraudulent transaction. While this can certainly be framed as a supervised learning problem, I'll only use the class label to measure the performance of our isolation forest-based anomaly detection, which will be unsupervised. I would argue that it is probably more often that you will be dealing with unlabeled data of this nature, as it is rather challenging to accurately label each transaction in a large database.

To start, I'll split the data into train/test samples using only $N = 10{,}000$ observations (3.51%) for training; the remaining 274,807 observations (96.49%) will be used as a test set. However, before doing so, I'm going to shuffle the rows just to make sure they are in random order first. (Assume I've already read the data into a data frame called ccfraud.)

```
# ccfraud <- data.table::fread("some/path/to/ccfraud.csv")

# Randomly permute rows
set.seed(2117)  # for reproducibility
ccfraud <- ccfraud[sample(nrow(ccfraud)), ]

# Split data into train/test sets
set.seed(2013)  # for reproducibility
trn.id <- sample(nrow(ccfraud), size = 10000, replace = FALSE)
ccfraud.trn <- ccfraud[trn.id, ]
ccfraud.tst <- ccfraud[-trn.id, ]

# Check class distribution in each
proportions(table(ccfraud.trn$Class))
proportions(table(ccfraud.tst$Class))

#>
#>       0      1
#> 0.9982 0.0018
#>
#>        0       1
#> 0.99828 0.00172
```

Next, I'll use the **isotree** package [Cortes, 2022] to fit a default isolation forest to the training set and provide anomaly scores for the test set. (Notice how I exclude the true class labels (column 31) when constructing the isolation forest!)

```
library(isotree)

# Fit a default isolation forest
ccfraud.ifo <- isolation.forest(ccfraud.trn[, -31], nthreads = 1,
                                seed = 2223)

# Compute anomaly scores for the test observations
head(scores <- predict(ccfraud.ifo, newdata = ccfraud.tst))

#>     1     2     3     4     5     6
#> 0.320 0.341 0.324 0.325 0.340 0.325
```

Although this isn't necessarily a classification problem, we can treat the anomaly scores as probabilities and construct informative graphics. While a *precision-recall (PR) curve*[k] could be useful here, I think a simple *cumulative lift chart*[l] would be more informative. Below I compute both; see Figure 7.18. Looking at the lift chart, for example, we can see that if we were to audit 5% of the highest scoring transactions in the test set, then we will have found roughly 87% of the fraudulent cases.

The PR curve doesn't look good, as you might expect after looking at the lift chart. For example, even though we can identify roughly 87% of the fraudulent transactions by looking at only 5% of the test sample, that still leaves more than 1300 non-fraudulent transactions that also have to be audited. In this example, it seems that we're not able to detect the majority of frauds without accepting a large number of false positives.

```
#cutoff <- sort(unique(scores))
# Compute precision and recall across various cutoffs
cutoff <- seq(from = min(scores), to = max(scores), length = 999)
cutoff <- c(0, cutoff)
precision <- recall <- numeric(length(cutoff))
for (i in seq_along(cutoff)) {
  yhat <- ifelse(scores >= cutoff[i], 1, 0)
  tp <- sum(yhat == 1 & ccfraud.tst$Class == 1)  # true positives
  tn <- sum(yhat == 0 & ccfraud.tst$Class == 0)  # true negatives
  fp <- sum(yhat == 1 & ccfraud.tst$Class == 0)  # false positives
  fn <- sum(yhat == 0 & ccfraud.tst$Class == 1)  # false negatives
  precision[i] <- tp / (tp + fp)  # precision (or PPV)
  recall[i] <- tp / (tp + fn)  # recall (or sensitivity)
}
precision <- c(precision, 0)
recall <- c(recall, 0)
```

---

[k]Precision (or *positive predictive value*) is directly proportional to the prevalence of the positive outcome. PR curves are not appropriate for case-control studies (e.g., which also includes case-control sampling—like down sampling—with imbalanced data sets) and should only be used when the true class priors are reflected in the data.

[l]The cumulative gains (or lift) chart shows the fraction of the overall number of cases in a given category "gained" by targeting a percentage of the total number of cases.

```
head(cbind(recall, precision))
```

```
#>      recall precision
#> [1,]      1   0.00172
#> [2,]      1   0.00172
#> [3,]      1   0.00172
#> [4,]      1   0.00172
#> [5,]      1   0.00173
#> [6,]      1   0.00173
```

```
# Compute data for lift chart
ord <- order(scores, decreasing = TRUE)
y <- ccfraud.tst$Class[ord]   # order according to sorted scores
prop <- seq_along(y) / length(y)
lift <- cumsum(y) / sum(ccfraud.tst$Class)  # convert to proportion
head(cbind(prop, lift))
```

```
#>           prop    lift
#> [1,] 3.64e-06 0.00000
#> [2,] 7.28e-06 0.00000
#> [3,] 1.09e-05 0.00000
#> [4,] 1.46e-05 0.00000
#> [5,] 1.82e-05 0.00000
#> [6,] 2.18e-05 0.00211
```

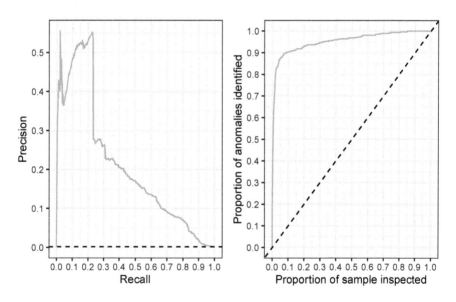

FIGURE 7.18: Precision-recall curve (left) and lift/cumulative gain chart (right) for the isolation forest applied to the credit card fraud detection test set.

We can take the analysis a step further by using Shapley values (Section 6.3.1) to help explain the observations with the highest/lowest anomaly scores, whichever is of more interest. To illustrate, let's estimate the feature contributions for the test observation with the highest anomaly score. Keep in mind that the features in this data set have been anonymized using PCA, so we won't be able to understand much of the output from a contextual perspective, but the idea applies to any application of anomaly detection based on a model that produces anomaly scores, like isolation forests. I'm just treating the scores as ordinary predictions and applying Shapley values in the usual way.

In the code chunk below, I find the observation in the test data that corresponds to the highest anomaly score. Here, we see that `max.x` corresponds to an actual instance of fraud (`Class = 1`) and was assigned an anomaly score of 0.843. The average anomaly score on the training data is 0.336, for a difference of 0.507. The question we want to try and answer is: how did each feature contribute to the difference 0.507? This is precisely the type of question that Shapley values can help with.

```
max.id <- which.max(scores)  # row ID for max anomaly score
(max.x <- ccfraud.tst[max.id, ])

#>       Time     V1    V2    V3   V4    V5   V6  V7    V8
#> 1: 166198 -35.5 -31.9 -48.3 15.3 -114 73.3 121 -27.3
#>       V9 V10  V11   V12  V13   V14   V15  V16   V17
#> 1: -3.87 -12 6.85 -9.19 7.13 -6.8 8.88 17.3 -7.17
#>      V18 V19   V20   V21  V22   V23  V24  V25  V26
#> 1: -1.97 5.5 -54.5 -21.6 5.71 -1.58 4.58 4.55 3.42
#>      V27   V28 Amount Class
#> 1: 31.6 -15.4  25691     0

max(scores)

#> [1] 0.843
```

Next, I'll use **fastshap** to generate Shapley-based feature contributions using the Monte Carlo approach discussed in Section 6.3.2.2 with 1000 repetitions. Note that we have to tell **fastshap** how to generate scores from an **isotree** `isolation.forest()` model by providing a helper prediction function. The estimated contributions are displayed in Figure 7.19. Here you can see that `Amount=25691.16` had the largest (positive) contribution (well above the 99-th percentile for the entire data set) to this observation having a higher than average anomaly score.

```
library(fastshap)

X <- ccfraud.trn[, 1:30]  # feature columns only
max.x <- max.x[, 1:30]    # feature columns only!
pfun <- function(object, newdata) {  # prediction wrapper
  predict(object, newdata = newdata)
}
```

```
# Generate feature contributions
set.seed(1351)  # for reproducibility
ex <- explain(ccfraud.ifo, X = X, newdata = max.x,
              pred_wrapper = pfun, adjust = TRUE,
              nsim = 1000)
sum(ex)  # should sum to f(x) - baseline whenever `adjust = TRUE`

#> [1] 0.507
```

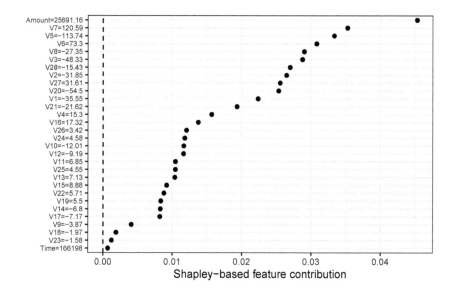

FIGURE 7.19: Estimated feature contributions for the test observation with highest anomaly score. There's a dashed vertical line at zero to differentiate between features with a positive/negative contribution. In this case, all feature values contributed positively to the difference $f(x^\star) - \mathrm{E}\left[\hat{f}(x)\right] = 0.501$.

7.9   Software and examples

RFs are available in numerous software, both open source and proprietary. The R packages **randomForest**, **ranger**, and **randomForestSRC** [Ishwaran and Kogalur, 2022] implement the traditional RF algorithm for classification and regression; the latter two also support survival analysis, as well as several other extensions. It's important to point out that **ranger**'s implementation of the RF algorithm treats categorical variables as ordered by default; for details, see the description of the `respect.unordered.factors` argument in `?ranger::ranger`. The **party** and **partykit** packages offer an implementation using conditional inference trees as the base learners (Section 3.4) for the base learners; our `crforest()` function from Section 7.2.3 follows the same approach. The CRAN task view on "Machine Learning & Statistical Learning" includes a section dedicated to RFs in R, so be sure to check that out as well: `https://cran.r-project.org/view=MachineLearning`.

While **randomForest** is a close port of Breiman's original Fortran code, the **ranger** package is far more scalable and implements a number of modern extensions and improvements discussed in this chapter (e.g., RF as a probability machine, Gini-corrected importance, quantile regression, case-specific RFs, extra-trees, etc.). Another scalable implementation is available from **h2o** [LeDell et al., 2021]. RFs are also part of Spark's MLlib library [Meng et al., 2016], which includes several R and Python interfaces (in particular, **SparkR**, **sparklyr**, and **pyspark**; an example using **SparkR** is provided in Section 7.9.5).

In Python, RFs, extra-trees, and isolation forests are available in the **sklearn.ensemble** module. Julia users can fit RFs via the **DecisionTree.jl** package. The official GUIDE software (Section 4.9) has the option to construct an RF from individual GUIDE trees; see Loh et al. [2019] and Loh [2020] for details.

Let's now work through several problems using random forest software.

7.9.1   Example: mushroom edibility

No! These data are easy and an ensemble would be overkill here. Remember, the original goal of the problem was to come up with an accurate but simple rule for determining the edibility of a mushroom. This was easily accomplished using a single decision tree (e.g., CART with some manual pruning) or a rule-based model like CORELS; see, for example, Figure 2.22.

7.9.2   Example: "deforesting" a random forest

In Section 5.5, I showed how the LASSO can be used to effectively post-process a tree-based ensemble by essentially zeroing out the predictions from some of the trees and reweighting the rest. The idea is that we can often reduce the number of trees quite substantially without sacrificing much in the way of performance. A smaller number of trees means we could, at least in theory, compute predictions faster, which has important implications for model deployment (e.g., when trying to score large data sets on a regular basis). However, unless we have a way to remove the zeroed out trees from the fitted RF object, we can't really reap all the benefits. This is the purpose of the new `deforest()` function in the **ranger** package[m], which I'll demonstrate in this section using the Ames housing example.

Keep in mind that this method of post-processing is not specific to bagged tree ensembles and RFs, and can be fruitfully applied to other types of ensembles as well; see Section 8.9.3 for an example using a *gradient boosted tree ensemble*.

To start, I'll load a few packages, prep the data, and create a helper function for computing the RMSE as a function of the number of trees in a **ranger**-based RF:

```
library(ranger)
library(treemisc)  # for isle_post() function

# Load the Ames housing data and split into train/test sets
ames <- as.data.frame(AmesHousing::make_ames())
ames$Sale_Price <- ames$Sale_Price / 1000  # rescale response
set.seed(2101)  # for reproducibility
trn.id <- sample.int(nrow(ames), size = floor(0.7 * nrow(ames)))
ames.trn <- ames[trn.id, ]  # training data/learning sample
ames.tst <- ames[-trn.id, ]  # test data
xtst <- subset(ames.tst, select = -Sale_Price)  # test features only

# Function to compute RMSE as a function of number of trees
rmse <- function(object, X, y) {  # only works with "ranger" objects
  p <- predict(object, data = X, predict.all = TRUE)$predictions
  sapply(seq_len(ncol(p)), FUN = function(i) {
    pred <- rowMeans(p[, seq_len(i), drop = FALSE])
    sqrt(mean((pred - y) ^ 2))
  })
}
```

Next, I'll fit two different RFs:

RFO a default RF with $B = 1,000$ maximal depth trees;

RFO.4.5 an RF with $B = 1,000$ shallow (depth-4) trees, where each tree is built using only a 5% random sample (with replacement) from the training data.

I'll record the computation time of each fit using `system.time()` (this function will also be used later to measure scoring time), which will provide some insight into the potential computational savings offered by this post-processing method[n]:

---

[m]The `deforest()` function is not available in versions of **ranger** $\leq 0.13.0$.

[n]Note that there are better ways to benchmark and time expressions in R; see, for example, the **microbenchmark** package [Mersmann, 2021].

```
# Fit a default RF with 1,000 maximal depth trees
set.seed(942)   # for reproducibility
system.time({
  rfo <- ranger(Sale_Price ~ ., data = ames.trn, num.trees = 1000)
})

#>    user  system elapsed
#>   5.845   0.112   1.899

# Fit an RF with 1,000 shallow (depth-4) trees on 5% bootstrap samples
set.seed(1021)   # for reproducibility
system.time({
  rfo.4.5 <- ranger(Sale_Price ~ ., data = ames.trn, num.trees = 1000,
                    max.depth = 4, sample.fraction = 0.05)
})

#>    user  system elapsed
#>   0.275   0.009   0.113

# Test set MSE as a function of the number of trees
rmse.rfo <- rmse(rfo, X = xtst, y = ames.tst$Sale_Price)
rmse.rfo.4.5 <- rmse(rfo.4.5, X = xtst, y = ames.tst$Sale_Price)
c("Test RMSE (RFO)" = rmse.rfo[1000],
  "Test RMSE (RFO.4.5)" = rmse.rfo.4.5[1000])

#>    Test RMSE (RFO) Test RMSE (RFO.4.5)
#>             24.8                36.7
```

The test RMSE for the RFO model is comparable to the test RMSE from the conditional RF fit in Section 7.2.3. In comparison, the RFO.4.5 model has a much larger test RMSE, which we might have expected given the shallowness of each tree and the tiny fraction of the learning sample each was built from. Consequently, the RFO.4.5 model finished training in only a fraction of the time it took the RFO model. As we'll see shortly, post-processing will help improve the performance of RFO.4.5 so that it is comparable to RFO in terms of performance, while substantially reducing the number of trees (i.e., comparable performance, faster training time, and fewer trees in the end).

Next, I'll obtain the individual tree predictions from each forest and post-process them using the LASSO via **treemisc**'s isle_post() function. Note that k-fold cross-validation can be used here instead of (or in conjunction with) a test set; see ?treemisc::isle_post for details. For brevity, I'll use a simple prediction wrapper, called treepreds(), to compute and extract the individual tree predictions from each RF model:

```
treepreds <- function(object, newdata) {
  p <- predict(object, data = newdata, predict.all = TRUE)
  p$predictions   # return predictions component
}

# Post-process RFO ensemble using an independent test set
```

```
preds.trn <- treepreds(rfo, newdata = ames.trn)
preds.tst <- treepreds(rfo, newdata = ames.tst)
rfo.post <- treemisc::isle_post(
  X = preds.trn,
  y = ames.trn$Sale_Price,
  newX = preds.tst,
  newy = ames.tst$Sale_Price,
  family = "gaussian"
)

# Post-process RFO.4.5 ensemble using an independent test set
preds.trn.4.5 <- treepreds(rfo.4.5, newdata = ames.trn)
preds.tst.4.5 <- treepreds(rfo.4.5, newdata = ames.tst)
rfo.4.5.post <- treemisc::isle_post(
  X = preds.trn.4.5,
  y = ames.trn$Sale_Price,
  newX = preds.tst.4.5,
  newy = ames.tst$Sale_Price,
  family = "gaussian"
)
```

The results are plotted in Figure 7.20. Here, we can see that both models benefited from post-processing, but the RFO model only experienced a marginal increase in performance compared to RFO.4.5. Is the slightly better performance in the default RFO model enough to justify its larger training time? Maybe in this particular example, but for larger data sets, the difference in training time can be huge, making it extremely worthwhile. For the post-processed RFO.4.5 model, the test RMSE is minimized using only 93 (reweighted) trees.

```
palette("Okabe-Ito")
plot(rmse.rfo, type = "l", ylim = c(20, 50),
     las = 1, xlab = "Number of trees", ylab = "Test RMSE")
lines(rmse.rfo.4.5, col = 2)
lines(sqrt(rfo.post$results$mse), col = 1, lty = 2)
lines(sqrt(rfo.4.5.post$results$mse), col = 2, lty = 2)
legend("topright", col = c(1, 2, 1, 2), lty = c(1, 1, 2, 2),
       legend = c("RFO", "RFO.4.5","RFO (post)", "RFO.4.5 (post)"),
       inset = 0.01, bty = "n")
palette("default")
```

To make this useful in practice, we need a way to remove trees from a fitted RF (i.e., to "deforest" the forest of trees). This could vastly speed up prediction time and reduce the memory footprint of the final model. Fortunately, the **ranger** package includes such a function; see `?ranger::deforest` for details.

In the code snippet below, I "deforest" the RFO.4.5 ensemble by removing trees corresponding to the zeroed-out LASSO coefficients, which requires es-

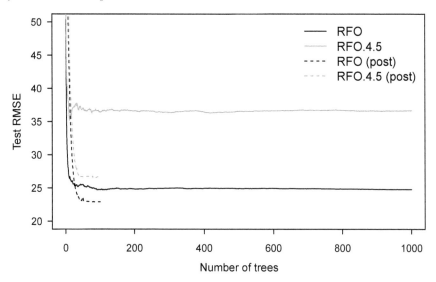

FIGURE 7.20: Test RMSE for the RFO and RFO.4.5 fits. The dashed lines correspond to the post-processed versions of each model. Note how the RFO model only experienced a marginal increase in performance compared to the RFO.4.5 model.

timating the optimal value for the penalty parameter λ (it might be helpful to read the help page for `?glmnet::coef.glmnet`):

```
res <- rfo.4.5.post$results   # post-processing results on test set
lambda <- res[which.min(res$mse), "lambda"]   # optimal penalty parameter
coefs <- coef(rfo.4.5.post$lasso.fit, s = lambda)[, 1L]
int <- coefs[1L]   # intercept
tree.coefs <- coefs[-1L]   # no intercept
trees <- which(tree.coefs == 0)   # trees to remove

# Remove trees corresponding to zeroed-out coefficients
rfo.4.5.def <- deforest(rfo.4.5, which.trees = trees)

# Check size of each object
c(
  "RFO.4.5" = format(object.size(rfo.4.5), units = "MB"),
  "RFO.4.5 (deforested)" = format(object.size(rfo.4.5.def), units = "MB")
)

#>              RFO.4.5 RFO.4.5 (deforested)
#>               "1 Mb"               "0.1 Mb"
```

Notice the impact this had on reducing the overall size of the fitted model. This can often lead to a much more compact model that's easier to save and load when memory requirements are a concern.

We can't just use the "deforested" tree ensemble directly; remember, the estimated LASSO coefficients imply a reweighting of the remaining trees! To obtain the reweighted predictions from the "deforested" model, we need to do a bit more work. Here, I'll create a new prediction function, called `predict.def()`, that will compute the reweighted predictions from the remaining trees using the estimated LASSO coefficients—similar to how predictions in a linear model are computed.

To test it out, I'll stack the learning sample (`ames.trn`) on top of itself 100 times, resulting in $N = 205,100$ observations for scoring. Below, I compare the prediction times for both the original (i.e., non-processed) and "deforested" RFO.4.5 fits:

```
ames.big <-  # stack data on top of itself 100 times
  do.call("rbind", args = replicate(100, ames.trn, simplify = FALSE))

# Compute reweighted predictions from a ``deforested'' ranger object
predict.def <- function(rf.def, weights, newdata, intercept = TRUE) {
  preds <- predict(rf.def, data = newdata,
                   predict.all = TRUE)$predictions
  res <- if (isTRUE(intercept)) {  # returns a one-column matrix
    cbind(1, preds) %*% weights
  } else {
    preds %*% weights
  }
  res[, 1, drop = TRUE]  # coerce to atomic vector
}


# Scoring time for original RFO.4.5 fit
system.time({  # full random forest
  preds <- predict(rfo.4.5, data = ames.big)
})

#>    user  system elapsed
#>   37.15    2.64   13.35

# Scoring time for post-processed RFO.4.5 fit using updated weights
weights <- coefs[coefs != 0]  # LASSO-based weights for remaining trees
system.time({
  preds.post <- predict.def(rfo.4.5.def, weights = weights,
                            newdata = ames.big)
})

#>    user  system elapsed
#>    4.17    0.73    4.47
```

The final model contains only 93 trees and achieved a test RMSE of 26.59, while also being orders of magnitude faster to initially train. The computational advantages are easier to appreciate on even larger data sets.

In summary, I used the LASSO to post-process and "deforest" a large ensemble of shallow trees (which trained relatively fast), producing a much smaller ensemble with fewer trees that scores faster compared to the default RFO. While the default RFO model had a slightly smaller test RMSE of 24.72 compared to the "deforested" RFO.4.5 test RMSE of 111.29, the difference is arguably negligible (especially when you take the differences in both training and scoring time into account).

7.9.3 Example: survival on the Titanic

In this example, I'll walk through a simple RF analysis of the well-known Titanic data set, where the goal is to understand survival probability aboard the ill-fated Titanic. A more thoughtful analysis using logistic regression and spline-based techniques is provided in Harrell [2015, Chap. 12].

Several versions of this data set are publicly available; for example, in the R package **titanic** [Hendricks, 2015]. Here, I'll use a more complete version of the data° which can be loaded using the getHdata() from package **Hmisc** [Harrell, 2021]; the raw data can also be downloaded from https://hbiostat.org/data/. In this example, I'll only consider a handful of the original variables:

```
t3 <- read.csv("https://hbiostat.org/data/repo/titanic3.csv",
               stringsAsFactors = TRUE)
keep <- c("survived", "pclass", "age", "sex", "sibsp", "parch")
t3 <- t3[, keep]  # only retain key variables
```

Note that roughly 20.09% of the values for **age**, the age in years of the passenger, are missing:

```
sapply(t3, FUN = function(x) mean(is.na(x)))
```

```
#> survived   pclass      age      sex    sibsp    parch
#>    0.000    0.000    0.201    0.000    0.000    0.000
```

Following Harrell [2015, Sec. 12.4], I use a decision tree to investigate which kinds of passengers tend to have a missing value for **age**. In the example below, I use the **partykit** package to apply the CTree algorithm (Chapter 3) using a missing value indicator for **age** as the response. From the tree output we can see that third-class passengers had the highest rate of missing **age** values (29.3%), followed by first-class male passengers with no siblings or

---

°A description of the original source of these data is provided in Harrell [2015, p. 291].

spouses aboard (22.8%). This makes sense, since males and third-class passengers supposedly had the least likelihood of survival ("women and children first").

```
library(partykit)

# Fit a conditional inference tree using missingness as response
temp <- t3   # temporary copy
temp$age <- as.factor(ifelse(is.na(temp$age), "y", "n"))
(t3.ctree   <- ctree(age ~ ., data = temp))

#>
#> Model formula:
#> age ~ survived + pclass + sex + sibsp + parch
#>
#> Fitted party:
#> [1] root
#> |     [2] pclass <= 2
#> |     |    [3] sibsp <= 0
#> |     |    |    [4] sex in female: n (n = 135, err = 6%)
#> |     |    |    [5] sex in male
#> |     |    |    |    [6] pclass <= 1: n (n = 123, err = 23%)
#> |     |    |    |    [7] pclass > 1: n (n = 122, err = 11%)
#> |     |    [8] sibsp > 0: n (n = 220, err = 3%)
#> |     [9] pclass > 2: n (n = 709, err = 29%)
#>
#> Number of inner nodes:     4
#> Number of terminal nodes: 5

# plot(t3.ctree)   # plot omitted
```

7.9.3.1   Missing value imputation

Next, I'll use the CART-based multiple imputation procedure outlined in Section 2.7.1 to perform $m = 21$ separate imputations for each missing **age** value. Why did I choose $m = 21$? White et al. [2011] propose setting $m \geq 100f$, where f is the fraction of incomplete cases[p]. Since **age** is the only missing variable, with $f = 0.201$, I chose $m = 21$. Using multiple different imputations will give us an idea of the sensitivity of the results of our (yet to be fit) RF.

```
library(mice)

set.seed(1125)   # for reproducibility
imp <- mice(t3, method = "cart", m = 21, minbucket = 5,
            printFlag = FALSE)

# Display nonparametric densities
densityplot(imp)
```

---

[p]When $f \geq 0.03$, Harrell [2015, p. 57] suggests setting $m = \max(5, 100f)$

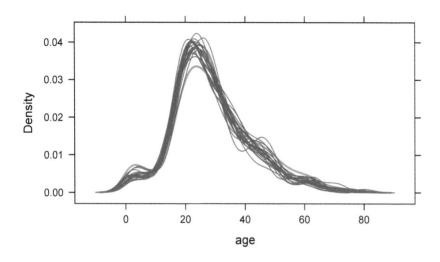

FIGURE 7.21: Nonparametric density estimate of `age` for the complete cases (blue line) and 15 imputed data sets.

Nonparametric densities for passenger age are given in Figure 7.21. There is one density for each of the imputed data sets (red curves) and one density for the original complete case (blue curve). The overall distributions are comparable, but there is certainly some variance across the $m = 21$ imputation runs. Our goal is to run an RF analysis for each of the $m = 21$ completed data sets and inspect the variability of the results. For instance, I might graphically show the $m = 21$ variable importance scores for each feature, along with the mean or median.

Next, I call `complete()` (from package **mice**) to produce a list of the $m = 21$ completed data sets which I can use to carry on with the analysis. The only difference is that I'll perform the same analysis on each for the completed data sets.[q]

```
t3.mice <- complete(
    data = imp,       # "mids" object (multiply imputed data set)
    action = "all",   # return list of all imputed data sets
    include = FALSE   # don't include original data (i.e., data with NAs)
)
length(t3.mice)   # returns a list of completed data sets
#> [1] 21
```

---

[q]This approach is probably not ideal in situations where the analysis is expensive (e.g., because the data are "big" and the model is expensive to tune). In such cases, you may have to settle for a smaller, less optimal value for m.

For comparison, let's look at the results from using the proximity-based RF imputation procedure discussed in Section 7.6.2. The code snippet below uses **rfImpute()** from package **randomForest** to handle the proximity-based imputation. The results are plotted along with those from MICE in Figure 7.22.

```
# Generate completed data set using RF's proximity-based imputation
set.seed(2121)  # for reproducibility
t3.rfimpute <-
    randomForest::rfImpute(as.factor(survived) ~ ., data = t3,
                           iter = 5, ntree = 500)

#> ntree      OOB      1      2
#>   500:  20.70% 12.36% 34.20%
#> ntree      OOB      1      2
#>   500:  19.17% 10.75% 32.80%
#> ntree      OOB      1      2
#>   500:  19.56% 11.62% 32.40%
#> ntree      OOB      1      2
#>   500:  19.71% 11.74% 32.60%
#> ntree      OOB      1      2
#>   500:  19.10% 10.75% 32.60%

# Construct matrix of imputed values
m <- imp$m  # number of MICE-based imputation runs
na.id <- which(is.na(t3$age))
x <- matrix(NA, nrow = length(na.id), ncol = m + 1)
for (i in 1:m) x[, i] <- t3.mice[[i]]$age[na.id]
x[, m + 1] <- t3.rfimpute$age[na.id]

# Plot results
palette("Okabe-Ito")
plot(x[, 1], type = "n", xlim = c(1, length(na.id)), ylim = c(0, 100),
     las = 1, ylab = "Imputed value")
for (i in 1:m) {
  lines(x[, i], col = adjustcolor(1, alpha.f = 0.1))
}
lines(rowMeans(x[, 1:m]), col = 1, lwd = 2)
lines(x[, m + 1], lwd = 2, col = 2)
legend("topright", legend = c("MICE: CART", "RF: proximity"), lty = 1,
       col = 1:2, bty = "n")
palette("default")
```

Here, you can see that the imputed values from both procedures are similar, but that multiple imputations provide a range of plausible values. Also, there are a few instances where there's a bit of a gap between the imputed values for the two procedures. For example, consider observations 956 and 959, whose records are printed below. The first passenger is recorded to be a third-class female with three siblings (or spouses) and one parent (or child) aboard. This individual is likely a child. The proximity-based imputation imputed the age

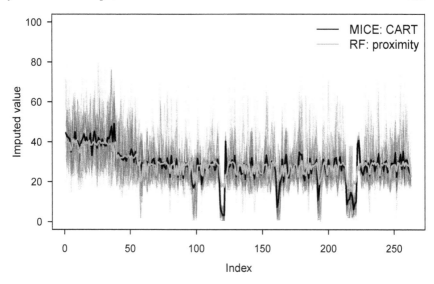

FIGURE 7.22: Imputed values for the 263 missing **age** values. The yellow line corresponds to the proximity-based imputation. The light gray lines correspond to the 15 different imputation runs using MICE, and the black line corresponds to their average.

for this passenger as 17.522 years, whereas MICE gives an average of 4.571 years and a plausible range of 0.75–8.00 years. Similarly, the proximity method imputed the age for case 959—a third-class female with three children—as 23.52 whereas MICE gave an average of 40.238. Which imputations do you think are more plausible?

```
t3[c(956, 959), ]
```

```
#>     survived pclass age    sex sibsp parch
#> 956        0      3  NA female     3     1
#> 959        0      3  NA female     0     4
```

7.9.3.2 Analyzing the imputed data sets

With the $m = 21$ completed data sets in hand, I can proceed with the RF analysis. The idea here is to fit an RF separately to each completed data set. I'll then look at the variance of the results (e.g., OOB error, variable importance scores, etc.) to judge its sensitivity to the different plausible imputations. Below, I use the **ranger** package to fit a (default) RF/probability machine to each of the $m = 21$ completed data sets. In anticipation of looking at the sensitivity of the variable importance scores, I set **importance =**

"permutation" to employ the OOB-based permutation variable importance
procedure discussed in Section 7.5.2.

```
library(ranger)

# Obtain a list of probability forests, one for each imputed data set
set.seed(2147)  # for reproducibility
rfos <- lapply(t3.mice, FUN = function(x) {
  ranger(as.factor(survived) ~ ., data = x, probability = TRUE,
         importance = "permutation")
})

# Check OOB errors (Brier-score, in this case)
sapply(rfos, FUN = function(forest) forest$prediction.error)
#>     1     2     3     4     5     6     7     8     9
#> 0.134 0.133 0.135 0.134 0.134 0.134 0.135 0.134 0.132
#>    10    11    12    13    14    15    16    17    18
#> 0.133 0.132 0.133 0.133 0.135 0.135 0.135 0.133 0.134
#>    19    20    21
#> 0.134 0.134 0.133
```

The OOB errors from each model are comparable; that's a good start! The
average OOB Brier score is 0.134, with a standard deviation of 0.001.

Next, I'll look at variable importance. With multiple imputation I think the
most sensible thing to do is to just plot the variable importance scores from
each run together, so that you can see the variability in the results:

```
# Compute list of VI scores, one for each model. Note: can use
#`FUN = ranger::importance` to be safe
vis <- lapply(rfos, FUN = importance)

# Stack into a data frame
head(vis <- as.data.frame(do.call(rbind, args = vis)))
#>   pclass    age   sex  sibsp  parch
#> 1 0.0531 0.0370 0.122 0.0130 0.0149
#> 2 0.0529 0.0404 0.125 0.0143 0.0145
#> 3 0.0504 0.0323 0.124 0.0126 0.0139
#> 4 0.0498 0.0336 0.125 0.0149 0.0131
#> 5 0.0540 0.0393 0.122 0.0135 0.0161
#> 6 0.0533 0.0407 0.123 0.0137 0.0142

# Display boxplots of results
boxplot(vis, las = 1)
```

Figure 7.23 shows a boxplot of the $m = 21$ variable importance scores for
each feature; these are the OOB-based permutation scores discussed in Section 7.5.2. The results do not vary too much, so we can be somewhat confident in the overall ranking of the features. Clearly sex is the most important

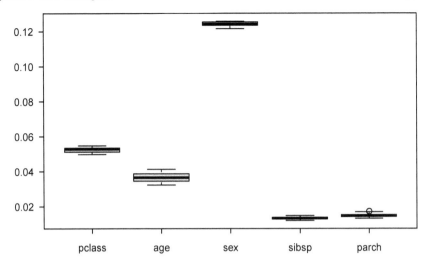

FIGURE 7.23: Boxplot of variable importance scores across all $m = 21$ RF fits.

predictor of survivability in these models, followed by passenger class (**pclass**) and passenger age (**age**). The remaining features are comparatively less important, with no discernible difference between the number of siblings/spouses aboard (codesibsp) and the number of parents/children aboard (**parch**).

It's natural to look at feature effect plots after examining the importance of each variable (i.e., main effects). A common feature effect plot is the partial dependence plot (Section 6.2.1), or PD plot for short. Following a similar strategy, we can compute the partial dependence for each feature across all $m = 21$ fitted RFs, and display the results on the same graph.

The code snippets below rely on the **pdp** package for constructing each partial dependence curve; for details, see Greenwell [2021b]. To start, I define a simple prediction wrapper for extracting predictions from a fitted **ranger** model; see ?ranger::predict.ranger for details on computing and extracting predictions from a "**ranger**" object. Note that PD plots are essentially constructed from averaged predictions, so the function below returns the average predicted probability of survival:

```
pfun <- function(object, newdata) {  # mean(prob(survived=1|x))
  mean(predict(object, data = newdata)$predictions[, "1"])
}
```

Since there are $m = 21$ imputed data sets, I'm essentially computing $m = 21$ 3-way PD plots (i.e., visualizing a 3-way interaction effect), which can be quite cumbersome computationally. Because we only have one numeric predictor (`age`), it will take only a few minutes in this example. For larger problems, or problems with many numeric features, you may want to consider computing PD plots for each feature individually (i.e., main effects), or using parallel computing (or other computational tricks). Below, I instruct **pdp**'s `partial()` function to plot over an evenly spaced grid of 19 percentiles for `age` (from 5-th to 95-th) within each unique combination of `pclass`[r] and `sex`, giving a total of $19 \times 3 \times 2 = 114$ plotting points.

```r
library(pdp)

# Construct PD plots for each model
pdps <- lapply(1:m, FUN = function(i) {
  partial(rfos[[i]], pred.var = c("age", "pclass", "sex"),
          pred.fun = pfun, train = t3.mice[[i]], cats = "pclass",
          quantiles = TRUE, probs = 1:19/20)
})

# Stack into a single data frame for plotting
for (i in seq_along(pdps)) {
  pdps[[i]]$m <- i
}
head(pdps <- do.call(rbind, args = pdps))
#>      age pclass    sex  yhat m
#> 1  5.0      1 female 0.848 1
#> 2 14.5      1 female 0.915 1
#> 3 18.0      1 female 0.935 1
#> 4 19.0      1 female 0.936 1
#> 5 21.0      1 female 0.939 1
#> 6 22.0      1 female 0.934 1
```

Next, I plot the results. There's some R-ninja trickery happening in the code chunk below in order to get the plot I want. Using **ggplot2**, I want to group a set of line plots by two variables, but color by just one of them. We can paste the two grouping variables together into a new column. However, base R's `interaction()` function can accomplish this for us; see `?interaction` for details.

The results are displayed in Figure 7.24; compare this to Figure 12.22 in Harrell [2015, p. 308]. I also included a rug representation (i.e., 1-d plot) in each panel showing the deciles (i.e., the 10-th percentile, 20-th percentile, etc.) of passenger age from the original (incomplete) training set. This helps guide where the plots are potentially extrapolating. Using deciles means that 10% of the observations lie between any two consecutive rug marks; see Greenwell [2017] for some remarks on the importance of avoiding extrapolation when

---

[r]I'm using `cats = "pclass"` here to treat `pclass` as categorical since it's restricted to `pclass` $\in \{1, 2, 3\}$.

interpreting PD plots, as well as some mitigation strategies (e.g., using rug plots and convex hulls).

```
library(ggplot2)

# Plot results
deciles <- quantile(t3$age, prob = 1:9/10, na.rm = TRUE)
ggplot(pdps, aes(age, yhat, color = sex,
                 group = interaction(m, sex))) +
  geom_line(alpha = 0.3) +
  geom_rug(aes(age), data = data.frame("age" = deciles),
           sides = "b", inherit.aes = FALSE) +
  labs(x = "Age (years)", y = "Surival probability") +
  facet_wrap(~ pclass) +
  scale_colour_manual(values = c("black", "orange")) +  # Okabe-Ito
  theme_bw() +
  theme(legend.title = element_blank(),
        legend.position = "top")
```

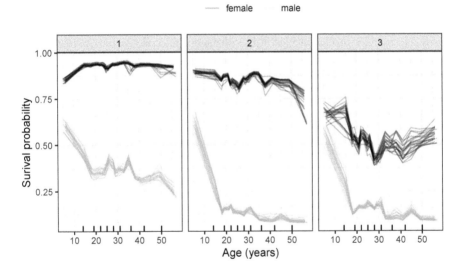

FIGURE 7.24: Partial dependence of the probability of surviving on passenger age, class, and sex. There's one curve for each of the $m = 21$ completed data sets.

While there's some variability in the results, the overall patterns are clear. First-class females had the best chances of surviving, regardless of age or class. Passenger age seems to have a stronger negative effect on passenger survivability for males compared to females, regardless of class. The difference in survivability between males and females is less pronounced for third-class passengers. Do you agree? What other conclusions can you draw from this plot?

Finally, let's use Shapley values (Section 6.3.1) to help us understand individual passenger predictions. To illustrate, let's focus on a single (hypothetical/fictional) passenger. Everyone, especially those who haven't seen the movie, meet Jack[s]:

```
jack.dawson <- data.frame(
  #survived = 0L,  # in case you haven't seen the movie
  pclass = 3L,  # using `3L` instead of `3` to treat as integer
  age = 20.0,
  sex = factor("male", levels = c("female", "male")),
  sibsp = 0L,
  parch = 0L
)
```

Here, I'll use the **fastshap** package for computing Shapley values, but you can use any Shapley value package you like (e.g., R package **iml**). First, we need to set up a prediction wrapper—this is a function that tells **fastshap** how to extract predictions from the fitted **ranger** model, which is the purpose of function pfun() below. Next, I compute approximate feature contributions for Jack's predicted probability of survival using 1000 Monte-Carlo repetitions, which is done for each of the $m = 21$ completed data sets:

```
library(fastshap)

# Prediction wrapper for `fastshap::explain()`; has to return a single
# (atomic) vector of predictions
pfun <- function(object, newdata) {  # compute prob(survived=1|x)
  predict(object, data = newdata)$predictions[, 2]
}

# Estimate feature contributions for each imputed training set
set.seed(754)
ex.jack <- lapply(1:21, FUN = function(i) {
  X <- subset(t3.mice[[i]], select = -survived)
  explain(rfos[[i]], X = X, newdata = jack.dawson, nsim = 1000,
          adjust = TRUE, pred_wrapper = pfun)
})

# Bind together into one data frame
ex.jack <- do.call(rbind, args = ex.jack)

# Add feature values to column names
names(ex.jack) <- paste0(names(ex.jack), "=", t(jack.dawson))
print(ex.jack)
```

---

[s]I guesstimated some of Jack's inputs, based on the movie I saw in seventh grade.

```
#> # A tibble: 21 x 5
#>     `pclass=3`  `age=20`  `sex=male`  `sibsp=0`  `parch=0`
#>         <dbl>      <dbl>       <dbl>      <dbl>      <dbl>
#>  1   -0.0836   -0.0136      -0.141    0.00721   -0.0174
#>  2   -0.0796   -0.0222      -0.144    0.0109    -0.00967
#>  3   -0.0743   -0.000271    -0.144    0.00995   -0.0170
#>  4   -0.0709   -0.0132      -0.139    0.00740   -0.0126
#>  5   -0.0807   -0.0192      -0.134    0.00768   -0.0159
#>  6   -0.0807   -0.0134      -0.136    0.0103    -0.0159
#>  7   -0.0840   -0.00355     -0.145    0.00999   -0.0147
#>  8   -0.0874    0.0110      -0.136    0.0103    -0.0254
#>  9   -0.0754   -0.00982     -0.143    0.00449   -0.0233
#> 10   -0.0663   -0.000338    -0.144    0.00519   -0.0165
#> # ... with 11 more rows
```

Fortunately, again, the results are relatively stable across imputations. A summary of the overall Shapley explanations, along with Jack's predictions, is shown in Figure 7.25. Here, we can see that Jack being a third-class male contributed the most to his poor predicted probability of survival, aside from him not being able to fit on the floating door that Rose was hogging...

```
# Jack's predicted probability of survival across all imputed
# data sets
pred.jack <- data.frame("pred" = sapply(rfos, FUN = function(rfo) {
  pfun(rfo, jack.dawson)
}))

# Plot setup (e.g., side-by-side plots)
par(mfrow = c(1, 2),  mar = c(4, 4, 2, 0.1),
    las = 1, cex.axis = 0.7)

# Construct boxplots of results
boxplot(pred.jack, col = adjustcolor(2, alpha.f = 0.5))
mtext("Predicted probability of surviving", line = 1)
boxplot(ex.jack, col = adjustcolor(3, alpha.f = 0.5), horizontal = TRUE)
mtext("Feature contribution", line = 1)
abline(v = 0, lty = "dashed")
```

We just walked through a simple analysis of the well-known Titanic data set, with a focus on using RFs to understand which passengers were most likely to survive and why. The analysis was complicated by the fact that one variable, **age**, contained many missing values. As a result, I performed multiple imputation, followed by an RF analysis on each of the plausible imputed data sets to gauge the sensitivity of the resulting imputations. In this example, the results seemed relatively stable across imputations, so we can be confident in our conclusions.

Predicted probability of surviving   Feature contribution

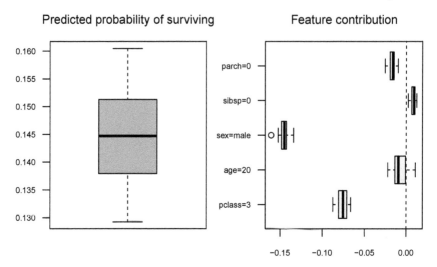

FIGURE 7.25: Predicted probability of survival for Jack across all imputed data sets (left) and their corresponding Shapley-based feature contributions (right).

7.9.4 Example: class imbalance (the good, the bad, and the ugly)

As discussed in Section 7.2.1, we often don't want to predict a class label, but rather the probability of belonging to a particular class—the latter has the benefit of providing some indication of uncertainty (e.g., if we predicted Jack to have a 15% chance of surviving his voyage on the Titanic, then we also estimate that he has an 85% chance of not surviving). Some models, however, can give you poor estimates of the class probabilities. *Calibration* addresses the issue by allowing you to better calibrate the probabilities of a fitted model, or provide estimated probabilities for models that are unable to naturally produce them in the first place (support vector machines, for example).

Calibration refers to the degree of agreement between observed and predicted probabilities, and its utility is well-discussed in the statistical literature (even if not commonly practiced); see, for example, Rufibach [2010], Austin and Steyerberg [2014], and Niculescu-Mizil and Caruana [2005]—the latter discusses calibration in a broader context and includes some discussion on calibrating RFs. A well calibrated (binary) probability model should produce accurate class probabilities such that among those probability estimates close to, say, 0.7, approximately 70% actually belong to the positive class, and so on.

Unless the trees are grown to purity, RFs generally produce consistent and *well-calibrated* probabilities [Malley et al., 2012, Niculescu-Mizil and Caruana, 2005], while boosted trees (Chapter 8) do not[t]. However, we'll see shortly that that's not always the case. Furthermore, *calibration curves* can be useful in comparing the performance of fitted probability models.

Real binary data are often unbalanced. For example, in modeling loan defaults, the target class (default on a loan) is often underrepresented. This is expected since we would hope that most people don't default on their loan over the course of paying it off. However, many practitioners perceive class imbalance as an issue that affects "accuracy." In actuality, the problem is usually that the data are balanced [Matloff, 2017, p. 193].

In this example, I'm going to simulate some unbalanced data. In particular, I'm going to convert the Friedman 1 benchmark regression data (Section 1.4.3) into a binary classification problem using a *latent variable model*. Essentially, I'll treat the observed response values as the linear predictor of a logistic regression model and convert them to probabilities. We can then use a binomial random number generator to simulate the observed class labels. The important thing to remember about this simulation study is that we have access to the true underlying class probabilities!

A simple function to convert the Friedman 1 regression problem into a binary classification problem, as described above, is given below. (Note that the line d$y <- d$y - 23 shifts the intercept term and effectively controls the balance of the generated 0/1 outcomes—here, it was chosen to obtain a 0/1 class balance of roughly 0.95/0.05.)

```
gen_binary <- function(...) {
  d <- treemisc::gen_friedman1(...)  # regression data
  d$y <- d$y - 23  # shift intercept
  d$prob <- plogis(d$y)  # inverse logit to obtain class probabilities
  #d$prob <- exp(d$y) / (1 + exp(d$y))  # same as above
  d$y <- rbinom(nrow(d), size = 1, prob = d$prob)  # 0/1 outcomes
  d
}


# Generate samples
set.seed(1921)  # for reproducibility
trn <- gen_binary(100000)  # training data
tst <- gen_binary(100000)  # test data
```

Let $N = N_0 + N_1$ be the number of observations in the learning sample, where N_0 and N_1 represent the number of observations that belong to class 0 and class 1, respectively. If the learning sample is a random (i.e., representative) sample form the population of interest, then we can estimate the true

---

[t]Surprisingly, in contrast to RFs, bagging decision trees grown to purity produces consistent probability estimates [Malley et al., 2012, Biau et al., 2008].

class priors from the data using $\pi_i = N_i/N$, for $i = 1, 2$; this was discussed for CART-like decision trees in Section 2.2.4. There are three scenarios to consider:

a) the data form a representative sample, and the observed class frequencies reflect the true class priors in the population (**the good**);

b) the class frequencies have been artificially balanced, but the true class frequencies/priors are known (**the bad**);

c) the class frequencies have been artificially balanced, and the true class frequencies/priors are unknown (**the ugly**).

In the code chunk below I use an independent sample of size $N = 10^6$ to estimate π_1 (i.e., the *prevalance* of observations in class 1 in the population):

```
(pi1 <- proportions(table(gen_binary(1000000)$y))["1"])
```

```
#>      1
#> 0.0498
```

Next, I'll define a simple calibration function that can be used for *isotonic calibration*; for a brief overview of different calibration methods, see Niculescu-Mizil and Caruana [2005], Kull et al. [2017]. Note that there are many R and Python libraries for calibration; for example, **val.prob()** from R package **rms** [Harrell, Jr., 2021] and the **sklearn.calibration** module.

```
isocal <- function(prob, y) {  # isotonic calibration function
  ord <- order(prob)
  prob <- prob[ord]   # put probabilities in increasing order
  y <- y[ord]
  prob.cal <- isoreg(prob, y)$yf  # fitted values
  data.frame("original" = prob, "calibrated" = prob.cal)
}
```

To start, let's fit a default RF to the original (i.e., unbalanced) learning sample. Note that I exclude the **prod** column when specifying the model formula.

```
library(ranger)
```

```
# Fit a probability forest (omitting the prob column)
set.seed(1446)   # for reproducibility
(rfo1 <- ranger(y ~ . - prob, data = trn, probability = TRUE,
                verbose = FALSE))
```

```
#> Ranger result
#>
#> Call:
#>  ranger(y ~ . - prob, data = trn, probability = TRUE, verbose = FA...
#>
```

```
#> Type:                            Probability estimation
#> Number of trees:                 500
#> Sample size:                     100000
#> Number of independent variables: 10
#> Mtry:                            3
#> Target node size:                10
#> Variable importance mode:        none
#> Splitrule:                       gini
#> OOB prediction error (Brier s.): 0.0256
```

The OOB prediction error (in this case, the Brier score) is 0.026. The Brier score on the test data can also be computed, but since I have access to the true probabilities, I might as well compare them with the predictions too (for this, I'll compute the MSE between the predicted and true probabilities). In this case, we see that the Brier score on the test set is comparable to the OOB Brier score.

```
prob1 <- predict(rfo1, data = tst)$predictions[, 2]

mean((prob1 - tst$y) ^ 2)  # Brier score

#> [1] 0.0255

mean((prob1 - tst$prob) ^ 2)   # MSE between predicted and true probs

#> [1] 0.00319
```

Looking at a single metric (or metrics) does not paint a full picture, so it can be helpful to look at specific visualizations, like calibration curves, to further assess the accuracy of the model's predicted probabilities (lift charts can also be useful). The leftmost plot in Figure 7.26 shows the actual vs. predicted probabilities for the test set, as well as the isotonic-based calibration curve from the above RF. In this case, the RF seems to be doing a reasonable job in terms of accuracy. The model seems well-calibrated for probabilities below 0.5, but seems to have a slight negative bias for probabilities above 0.5, which makes sense since most of the probability mass is concentrated near zero (as we might have expected given the true class frequencies).

To naively combat the perceived issue of unbalanced class labels, the learning sample is often artificially rebalanced (e.g., using down sampling) so that the class outcomes have roughly the same distribution. In general, THIS IS A BAD IDEA for probability models, and can lead to serious bias in the predicted probabilities—in fact, any algorithm that requires you to remove good data to optimize performance is suspect. Nonetheless, sometimes the data have been artificially rebalanced in a preprocessing step outside of our control, or maybe you decided to down sample the data to reduce computation time (in which case, you should try to preserve the original class frequencies, or at least store them for adjustments later). In any case, let's see what happens to our predictions when we down sample the majority class.

In scenarios b)–c), we cannot estimate π_0 and π_1 from the learning sample; however, in scenario b) we might have estimates of π_0 and π_1, perhaps from historical data. If the data have been artificially balanced, then it's possible to use good estimates of π_0 and π_1 to "correct" (or adjust) the output predicted probabilities. With CART and GUIDE, it's possible to provide the true priors and let the tree algorithm handle the adjustment (we saw how this is handled in CART in Section 2.2.4 and provided an example with **rpart** using the letter image recognition example in Section 2.9.5). What if no priors argument is available in your software? Fortunately, you can apply a simple adjustment to the output predicted probabilities, as discussed in Matloff [2017, pp. 197-198]. A simple function to adjust the predicted probabilities is given below. Here, **p** is a vector of predicted probabilities for the positive class (i.e., $\widehat{\Pr}(Y = 1|\boldsymbol{x})$), **observed.ratio** is the ratio of the observed class frequencies (i.e., N_0/N_1), and **true.ratio** is the ratio of the true class priors (i.e., π_0/π_1).

```
prob.adjust <- function(p, observed.ratio, true.ratio) {
  f.ratio <- (1 / p - 1) * (1 / observed.ratio)
  1 / (1 + true.ratio * f.ratio)
}
```

Let's try this out on an RF fit to a down sampled version of the training data. Below I artificially balance the classes by removing rows corresponding to the dominant class (i.e., **y = 0**):

```
trn.1 <- trn[trn$y == 1, ]
trn.0 <- trn[trn$y == 0, ]
trn.down <- rbind(trn.0[seq_len(nrow(trn.1)), ], trn.1)
table(trn.down$y)

#>
#>    0    1
#> 5018 5018
```

Next, I'll fit another (default) RF, but this time to the down samples training set. I then apply the adjustment formula to the predicted probabilities for the positive class in the test set:

```
set.seed(1146)  # for reproducibility
rfo2 <- ranger(y ~ . - prob, data = trn.down, probability = TRUE)

# Predicted probabilities for the positive class: P(Y=1|x)
prob2 <- predict(rfo2, data = tst)$predictions[, 2]
mean((prob2 - tst$y) ^ 2)  # Brier score

#> [1] 0.0756

mean((prob2 - tst$prob) ^ 2)  # MSE

#> [1] 0.0538
```

```
prob3 <- prob.adjust(prob2, observed.ratio = 1,
                     true.ratio = (1 - pi1) / pi1)
mean((prob3 - tst$y) ^ 2)  # Brier score

#> [1] 0.0285

mean((prob3 - tst$prob) ^ 2)  # MSE between predicted and true probs

#> [1] 0.00609
```

Figure 7.26 shows the predicted vs. true probabilities across three different cases: 1) predicted probabilities (**prob1**) from an RF applied to the original training data (left display), 2) predicted probabilities (**prob2**) from an RF applied to a down-sampled version of the original training data, but adjusted using the original class frequencies (middle display), and 3) predicted probabilities (**prob2**) from an RF applied to a down-sampled version of the original training data with no adjustment (right display).

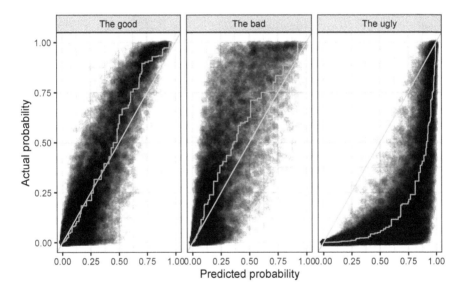

FIGURE 7.26: True vs. predicted probabilities for the test set from an RF fit to the original and down sampled (i.e., artificially balanced) training sets. Left: probabilities from the original RF. Middle: Probabilities from the down sampled RF with post-hoc adjustment. Right: Probabilities from the down sampled RF. The calibration curve is shown in orange, with the blue curve representing perfect calibration.

Compared to the predicted probabilities from an RF fit to the original (i.e., unbalanced) learning sample, down sampling appears to have produced biased and poorly calibrated probabilities, although, the adjustment formula seems to provide some relief but requires access to the true class priors (or good estimates thereof).

Basically, it is ill-advised to choose a model based on a metric that forces a classification based on an arbitrary threshold. Instead, choose a model using a *proper scoring rule* (e.g., or the Brier score) that makes use of the full range of predicted probabilities and is optimized when the true probabilities are recovered. Down sampling, adjusted or not, seems to highly under or over estimate the true class probabilities.

7.9.5   Example: partial dependence with Spark MLlib

In this section, I'll look at a well-known bank marketing data set available from the UC Irvine Machine Learning Repository [Moro et al., 2014]. The data concern the direct marketing campaigns of a Portuguese banking institution, which were based on phone calls. Often, more than one contact to the same client was required, in order to assess if the product, a bank term deposit, would be subscribed or not; hence, this is a binary classification problem with response variable y taking on values yes/no depending on whether or not the client did/did not subscribe to the bank term deposit. For details and a description of the different columns, visit `https://archive.ics.uci.edu/ml/datasets/Bank+Marketing`.

Furthermore, I'll do the analysis via one of the R front ends to Apache Spark and MLlib: **SparkR**[u] [Venkataraman et al., 2016]. Starting with Spark 3.1.1, **SparkR** also provides a distributed data frame implementation that supports common data wrangling operations like selection, filtering, aggregation, etc. (similar to using **data.table** or **dplyr** with R data frames, but on large data sets that don't fit into memory). For instructions on installing Spark, visit `https://spark.apache.org/docs/latest/sparkr.html`.

While the bank marketing data contain 21 columns on 41,188 records, this is by no means "Spark territory." However, you may find yourself in a situation where you need to use Spark MLlib for scalable analytics, so I think it's useful to show how to perform common statistical and machine learning tasks in Spark and MLlib, like fitting RFs, assessing performance, and computing PD plots.

To start, I'll download a zipped file containing a directory with the full bank marketing data set. The code downloads the zipped file into a temporary directory, unzips it, and reads in the CSV file of interest. (If the following code does not work for you, then you may find it easier to just manually download the bank-additional-full.csv and read it into R on your own.)

---

[u]This analysis can easily be translated to any other front end to Spark, including **sparklyr** or **pyspark**.

```
url <- paste0("https://archive.ics.uci.edu/ml/machine-learning",
              "-databases/00222/bank-additional.zip")
temp <- tempfile(fileext = ".zip")  # to store zipped file
download.file(url, destfile = temp)
bank <- read.csv(unz(temp, "bank-additional/bank-additional-full.csv"),
                 sep = ";", stringsAsFactors = FALSE)
unlink(temp)  # delete temporary file
```

Next, I'll clean up the data a bit. First off, I'll replace the dots in the column names with underscores; Spark does not like dots in column names! Second, I'll coerce the response (y) from a factor (no/yes) to a binary indicator (0/1) and treat it as numeric to fit a probability forest/machine. Finally, I'll remove the column called duration. Too often have I seen online analyses of the same data, only for the analyst to be fooled into thinking that duration is a useful indicator of whether or not a client will subscribe to a bank term deposit. If you take care and read the data documentation, you'd notice that the value of duration is not known before a call is made to a client. In other words, the value of duration is not known at prediction time and therefore cannot be used to train a model. This is a textbook example of target leakage. KNOW YOUR DATA! Finally, the data are split into train/test sets using a 50/50 split; I could do this manually, but here I'll use the **caret** package's createDataPartition() function, which uses stratified sampling to ensure that the distribution of classes is similar between the resulting partitions[v]:

```
names(bank) <- gsub("\\.", replacement = "_", x = names(bank))
bank$y <- ifelse(bank$y == "yes", 1, 0)
bank$duration <- NULL  # remove target leakage

# Split data into train/test sets using a 50/50 split
set.seed(1056)
trn.id <- caret::createDataPartition(bank$y, p = 0.5, list = FALSE)
bank.trn <- bank[trn.id, ]  # training data
bank.tst  <- bank[-trn.id, ]  # test data
```

Next, I'll load **ggplot2**, **SparkR**, and initialize a *SparkSession*—the entry point into **SparkR** (see ?SparkR::sparkR.session for details and the various Spark properties that can be set). If you're new to Spark, start with the online documentation: https://spark.apache.org/docs/latest/index.html; you can also find links to **SparkR** here as well. Note that **SparkR** is not on CRAN, but is included with a standard install of Apache Spark. To load **SparkR** in an existing R session[w], say, in RStudio, you need

---

[v]Several other packages could also be used here, like **rsample** [Silge et al., 2021], for example.

[w]You can also start an R session with **sparkR** already available from the terminal by running ./bin/sparkR from your Spark home folder; for details, see https://spark.apache.org/docs/latest/sparkr.html

to tell) the location of the package. (Note that the code snippet below may
need to change for you depending on where you have Spark installed; for me,
it's in `C:\spark\spark-3.0.1-bin-hadoop2.7\R\lib`.)

```
library(SparkR, lib.loc = "C:\\spark")
library(ggplot2)

# Start a local connection to Spark using all available cores
sparkR.session(master = "local[*]")
```

Next, I'll apply MLlib's Spark-enabled RF algorithm by calling
`spark.randomForest()`; here, I'll use $B = 500$ trees with a max depth
of 10. Note that **SparkR** works with Spark DataFrames, not R data
frames, so I have to coerce our train/test sets to Spark DataFrames using
`createDataFrame()` before applying any Spark operations (ideally, I'd read
the original data into a Spark DataFrame directly and process the data using
Spark operations, but I was being lazy):

```
bank.trn.sdf <- createDataFrame(bank.trn)
bank.tst.sdf <- createDataFrame(bank.tst)

# Fit a regression/probability forest
bank.rfo <- spark.randomForest(
  bank.trn.sdf, y ~ ., type = "regression",
  numTrees = 500, maxDepth = 10, seed = 1205
)
```

To assess the performance of the probability forest, I can compute the Brier
score on the test set. A couple of things are worth noting about the code chunk
below. First, the `predict()` method, when applied to a **SparkR** MLlib model,
returns the predictions along with the original columns from the supplied
Spark DataFrame. Second, note that I have to compute the Brier score using
Spark DataFrame operations, like **SparkR**'s `summarize()` function, in this
case).

```
p <- predict(bank.rfo, newData = bank.tst.sdf)  # Pr(Y=yes|x)
head(summarize(p, brier_score = mean((p$prediction - p$y)^2)))

#>    brier_score
#> 1  0.07815544
```

The AUC on the test set for this model, if you care purely about discrimination, is $0.798^{\times}$, which is in line with some of the even more advanced analyses I've seen on these data. Nice! In addition, Figure 7.27 shows an isotonic
regression-based calibration curve (left) and cumulative gains chart, both com-

---

[×]Even if discrimination is the goal, AUC does not take into account the prior class
probabilities and is not necessarily appropriate in situations with severe class imbalance;
in this case the area under the PR curve would be more informative [Davis and Goadrich,
2006].

puted from the test data. The model seems reasonably calibrated (as we would hope from a probability forest). The cumulative gains chart tells us, for example, that we could expect roughly 1,500 subscriptions by contacting the top 20% of clients with the highest predicted probability of subscribing.

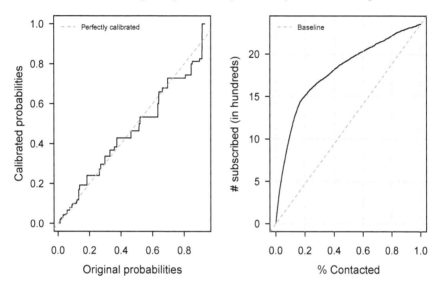

FIGURE 7.27: Graphical assessment of the performance on the test set. Left: isotonic regression-based calibration curve. Right: Cumulative gains chart.

Next, I'll look at the RF-based variable importance scores. Unfortunately, **SparkR** does not return the variable importance scores from tree-based models in a friendly format; it just gives one long nasty string, as can be seen below[y]:

```
rfo.summary <- summary(rfo)  # extract summary information
(vi <- rfo.summary$featureImportances)  # gross...
```

```
#> [1] "(52,[0,1,2,3,4,5,6,7,8,9,10,11,12,13,14,15,16,17,18,19,20,21,...
```

Nonetheless, I can use some *regular expression* (regex) magic to parse the output into a more friendly data frame. (By no means do I claim this to be the best solution; I'm certainly no regexpert—ba dum tsh.)

```
vi <- substr(vi, start = regexpr(",", text = vi)[1] + 1,
             stop = nchar(vi) - 1)
vi <- gsub("\\[", replacement = "c(", x = vi)
```

---

[y] It is not clearly documented which variable importance metric MLlib uses in its implementation of RFs, but I suspect it's the impurity-based metric (Section 7.5) since, as far as I'm aware, MLlib's RF implementation does not support the notion of OOB observations (Section 7.3).

```
vi <- gsub("\\]", replacement = ")", x = vi)
vi <- paste0("cbind(", vi, ")")
vi <- as.data.frame(eval(parse(text = vi)))
names(vi) <- c("feature.id", "importance")
vi$feature.name <- rfo.summary$features[vi[, 1] + 1]
head(vi[order(vi$importance, decreasing = TRUE), ], n = 10)
```

```
#>      feature.id importance       feature.name
#> 52          51     0.1915        nr_employed
#> 51          50     0.1447          euribor3m
#> 44          43     0.1141              pdays
#> 1            0     0.0711                age
#> 50          49     0.0496       cons_conf_idx
#> 48          47     0.0469        emp_var_rate
#> 49          48     0.0341      cons_price_idx
#> 43          42     0.0329           campaign
#> 45          44     0.0192           previous
#> 47          46     0.0171   poutcome_failure
```

The output suggests that the number of employees (`nr_employed`), a quarterly economic indicator, the Euribor 3 month rate (`euribor3m`)[z], a daily economic indicator, and the number of days passed since the client was last contacted from a previous campaign (`pdays`) are important predictors.

To further investigate the effect of these features on the model output, we can look at feature effect plots (such as PD and ICE plots). Here, I'll construct a PD plot for the economic indicator `euribor3m`[aa]. The trick to computing PD plots in Spark, if you can afford the memory, is to generate all the necessary data up front so that you only need one call to a scoring function. Once you have all the predictions, you can just post-process the results into partial dependence values by averaging the predictions within each unique value of the feature of interest; the same idea works for ICE plots as well (Section 6.2.3). We saw how to do this in base R in Section 6.2.5 using the Ames data. Even with large training data sets that don't fit into memory, the aggregated partial dependence values will be small enough to bring into memory as an ordinary R data frame and plotted using your favorite plotting library.

Following the same recipe outlined in Section 6.2.5, I'll start by creating a grid of values we want the PD plot for `euribor3m` to cover. For example, we can use an evenly spaced grid of points that covers the range of the predictor values of interest, or the sample quantiles; the latter has the benefit of potentially excluding outliers/extremes from the resulting plot. Since the data resides in a Spark data frame, we can't just use base R functionality. Luckily, **SparkR** provides the functionality we need via `approxQuantile()`, which we use to

---

[z]The 3 month Euribor rate is the interest rate at which a selection of European banks lend one another funds (denominated in euros) whereby the loans have a 3 month maturity.
[aa]You can find a similar example using **dplyr** with the **sparklyr** front end to Spark here: https://github.com/bgreenwell/pdp/issues/97.

construct a new Spark DataFrame containing only the plotting values for `euribor3m`. Then, we just need to create a Cartesian product with the original training data (excluding the variable `euribor3m`), or representative sample thereof. This is accomplished in the next code chunk.

A word of caution is in order. Even though Spark is designed to work with large data sets in a distributed fashion, Cartesian products can still be costly! Hence, if your learning sample is quite large (e.g., in the millions), which is probably the case if you're using MLlib, then keep in mind that you don't necessarily need to utilize the entire training sample for computing partial dependence and the like. If you have 50 million training records, for example, then consider only using a small fraction, say, 10,000, for constructing feature effect plots.

```
euribor3m.grid <- as.DataFrame(unique(  # DataFrame of unique quantiles
   approxQuantile(bank.trn.sdf, cols = "euribor3m",
                  probabilities = 1:29 / 30, relativeError = 0)
))
names(euribor3m.grid) <- "euribor3m"

# Training data without euribor3m
trn.wo.euribor3m <- bank.trn.sdf  # copy of training data
trn.wo.euribor3m$euribor3m <- NULL  # remove euribor3m

# Create a Cartesian product
pd <- crossJoin(euribor3m.grid, trn.wo.euribor3m)
dim(pd)  # nrow(euribor3m.grid) * nrow(trn.wo.euribor3m)

#> [1] 514850      20
```

Finally, we can compute the partial dependence values by aggregating the predictions using a simple grouping operator combined with a summary function (for PD plots, we just average the predictions). The results are displayed in Figure 7.28. Here you can see that the relative frequency of exclamation marks is positively associated with spam (note that the y-axis is on the probability scale).

```
ggplot(pd, aes(x = euribor3m, y = yhat)) +
  geom_line() +
  geom_rug(data = as.data.frame(euribor3m.grid),
           aes(x = euribor3m), inherit.aes = FALSE) +
  xlab("Euribor 3 month rate") +
  ylab("Partial dependence") +
  theme_bw()

sparkR.stop()  # stop the Spark session
```

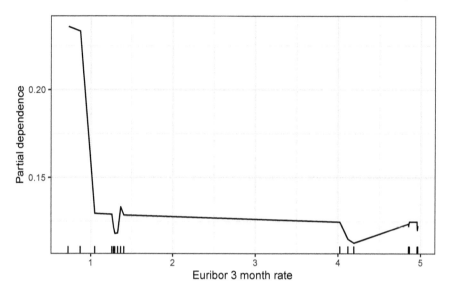

FIGURE 7.28: Partial dependence of subscription probability on Euribor 3 month rate from the bank marketing probability forest. The rug display along the *x*-axis summarizes the distribution of `euribor3m` via a grid of 29 evenly spaced quantiles.

7.10   Final thoughts

Yikes, that was a long chapter, but necessarily so. While RFs were originally introduced in Breiman [2001], many of the ideas have been seen before. For example, the term "random forest" was actually coined by Ho [1995], who used the *random subspace* method to combine trees grown in random subspaces of the original features. Breiman [2001] references several other attempts to further improve bagging by introducing more diversity among the trees in a "forest."

Leo Breiman was a phenomenal statistician (and theoretical probabilist) who had a profound impact on the field of statistical and machine learning. If you're interested in more of his work, especially on the development of RF, and the many collaborations it involved, see Cutler [2010]. Adele Cutler, a close collaborator with Breiman on RFs, still maintains their original RF website at `https://www.stat.berkeley.edu/~breiman/RandomForests/`. This website is still one of the best references to understanding Breiman's original RF and includes links to several relevant papers and the original Fortran source code.

To end this chapter, I'll leave you with a quote listed under the philosophy section of Breiman and Cutler's RF website, which applies more generally than just to RFs:

Random forest is an example of a tool that is useful in doing analyses of scientific data. But the cleverest algorithms are no substitute for human intelligence and knowledge of the data in the problem. Take the output of random forests not as absolute truth, but as smart computer generated guesses that may be helpful in leading to a deeper understanding of the problem.

Leo Breiman and Adele Cutler

`https://www.stat.berkeley.edu/~breiman/RandomForests/`

8

Gradient boosting machines

One step at a time is all it takes to get you there.

Emily Dickinson

I like RFs because they're powerful and flexible, yet conceptually simple: average a bunch of "de-correlated" trees together in the hopes of producing an accurate prediction. However, RF is not always the most efficient or most accurate tree ensemble to use. Gradient tree boosting provides another rich and flexible class of tree-based ensembles that, at a high level, I think is also conceptually simple. However, with gradient tree boosting, the devil is in the details.

In Section 5.2.1, we were introduced to AdaBoost.M1, a particularly simple boosting algorithm for binary classification problems. Boosting initially started off as a way to improve the performance of weak binary classifiers. Over time, boosting has evolved into an incredibly flexible procedure that, like RFs, can handle a wide array of supervised learning problems.

In this chapter, I'll walk through the basics of the most currently popular flavor of boosting: *stochastic gradient boosting*, also known as a *gradient boosting machine* (or GBM for short)[a]. Although GBMs are meant to be more general, in this book, GBM generally refers to stochastic gradient boosted decision trees.

---

[a]This flavor of boosting goes by several names in the literature. For example, the R package **gbm** [Greenwell et al., 2021b] fits this class of models, but stands for *generalized boosted models*.

8.1   Steepest descent (a brief overview)

In parametric modeling (i.e., where the form of the prediction function is known in advance), we are often concerned with estimating the parameters of a prediction function $f(\boldsymbol{x}; \boldsymbol{\theta})$, where $\boldsymbol{\theta} \in \mathbb{R}^p$ is a p-dimensional vector of fixed but unknown parameters (e.g., the coefficients in a linear regression model). We do this by minimizing the "loss" in using $f(\boldsymbol{x}; \boldsymbol{\theta})$ to predict y on a set of training data, where the loss function is defined as

$$L(\boldsymbol{\theta}) = \sum_{i=1}^{N} L\left[y_i, f(\boldsymbol{x}_i; \boldsymbol{\theta})\right].$$

The goal is to minimize $L(\boldsymbol{\theta})$ with respect to $\boldsymbol{\theta}$. A simple example is least squares regression, where loss is defined as the sum of squared residuals (which I'll refer to henceforth as LS loss) and is minimized as a function of the coefficients $\boldsymbol{\theta}$. LS loss is analytically tractable and easy to solve for linear models. A more general approach for any differentiable loss function is to use numerical optimization methods to solve

$$\boldsymbol{\theta}^\star = \arg\min_{\boldsymbol{\theta}} L(\boldsymbol{\theta}). \tag{8.1}$$

A popular method for solving (8.1) is the method of *steepest descent*, a special case of the more general method of *gradient descent*. Gradient descent is a general class of iterative optimization algorithms that express the solution to (8.1) as a sum of components:

$$\boldsymbol{\theta}^\star = \sum_{b=0}^{B} \boldsymbol{\theta}_b,$$

where $\boldsymbol{\theta}_0$ is some initial guess and $\{\boldsymbol{\theta}_b\}_{b=1}^{B}$ are successive "steps" or "boosts" towards the optimal solution $\boldsymbol{\theta}^\star$, and are found through the update equation

$$\boldsymbol{\theta}_b = \boldsymbol{\theta}_{b-1} + \gamma \Delta \boldsymbol{\theta}_{b-1},$$

where $\Delta \boldsymbol{\theta}_{b-1} = -\partial L(\boldsymbol{\theta})/\partial \boldsymbol{\theta}$ is the *negative gradient* of $L(\boldsymbol{\theta})$ with respect to $\boldsymbol{\theta}$ and represents the direction of "steepest descent" of $L(\boldsymbol{\theta})$, and $\gamma > 0$ is the *step size* taken in that direction. Steepest descent methods differ in how γ is determined.

Here's a pretty common analogy for explaining gradient descent without any math or fancy notation. Imagine trying to reach the bottom of a large hill (i.e., trying to find the global minimum) blindfolded. Without being able to see, and assuming you're not playing Marco Polo, you'll have to rely on what you can feel on the ground around you (i.e., local information) to find your way to the bottom. Ideally you'd feel around the ground at your current location $(\boldsymbol{\theta}_{b-1})$ to get a sense of the direction of steepest descent $(\Delta\boldsymbol{\theta}_{b-1})$—the fastest way down—and proceed in that direction while periodically reassessing which direction to go using the current local information available. How far you go in each direction is determined by your step size γ_b.

This simple analogy glosses over a number of details, like getting stuck in a hole (i.e., finding a local minimum), but hopefully the basic idea of gradient descent is relatively clear: to find the global minimum of $L(\boldsymbol{\theta})$, we take incremental steps in the direction of steepest descent, provided by the negative gradient of $L(\boldsymbol{\theta})$ evaluated at the current point. The step size to take at each iteration can be fixed or estimated by solving another minimization problem:

$$\gamma_b = \arg\min_{\gamma} L\left(\boldsymbol{\theta}_{b-1} - \gamma\Delta\boldsymbol{\theta}_b\right). \tag{8.2}$$

In the latter case, finding γ_b by minimizing (8.2) is referred to as the "line search" along the direction of $\Delta\boldsymbol{\theta}_b$, and the overall procedure is referred to as the method of steepest descent. It's worth noting that oftentimes the solution to (8.2) is found via a simple approximation (e.g., using a single Newton-Rhapson step).

8.2   Gradient tree boosting

Now, what does steepest descent have to do with boosting decision trees? Imagine trying to find some generic prediction function f such that

$$f^\star = \arg\min_{f} L(f),$$

where $L(f) = \sum_{i=1}^{N} L[y_i, f(\boldsymbol{x}_i)]$ is a loss function evaluated over the learning sample and encourages f to fit the data well [James et al., 2021, p. 302].

In contrast to (8.1), the parameters to be optimized here are the N fitted values $\boldsymbol{f} \in \mathbb{R}^N$ from $f(\boldsymbol{x}_i)$ found at each iteration evaluated at the training data $\boldsymbol{x}_i$:

$$\boldsymbol{f} = \{f(\boldsymbol{x}_i)\}_{i=1}^{N}.$$

Steepest descent in "function space" can be used to find the optimal $\widehat{\boldsymbol{f}}$ as the sum of B N-dimensional component vectors:

$$\boldsymbol{f}_B = \sum_{b=0}^{B} \boldsymbol{f}_b, \quad \boldsymbol{f}_b \in \mathbb{R}^N,$$

where, similar to before, $\boldsymbol{f}_0$ is an initial guess, and each component $\boldsymbol{f}_b$ depends on the current estimate $\boldsymbol{f}_{b-1}$, and so forth. Steepest descent finds the next update using

$$\boldsymbol{f}_b = \boldsymbol{f}_{b-1} - \gamma \boldsymbol{g}_b,$$

where

$$\boldsymbol{g}_b = \left\{ \left[\frac{\partial L(f)}{\partial f} \right]_{f=f_{b-1}(\boldsymbol{x}_i)} \right\}_{i=1}^{N} \tag{8.3}$$

is a length N column vector whose j-th component (g_{jb}) is the gradient of $L(\boldsymbol{f})$ evaluated at $f = f_{b-1}(\boldsymbol{x}_j)$.

One drawback is that the gradient components $\boldsymbol{g}_b$ in (8.3) are only defined at the observed training observations $\boldsymbol{x}_i$ $(i = 1, 2, \ldots, N)$, whereas we want the final prediction function $f_M(\boldsymbol{x})$ to be defined at new data points; otherwise, how would we make new predictions? Further, the procedure does not take into account the fact that observations with similar feature values are likely to have similar predictions [Ridgeway, 1999]. To this end, Friedman [2001] proposed using a class of functions that make use of the predictor information to approximate the gradient at each step. In gradient tree boosting (the focus of this chapter), for example, a regression tree is used to approximate the gradient at each step. In particular, at each step, we fit a J-terminal node regression tree[b], which has the form

$$f(\boldsymbol{x}_i; \boldsymbol{\theta}, \boldsymbol{R}) = \sum_{j=1}^{J} \theta_j I(\boldsymbol{x}_i \in R_j),$$

where $\boldsymbol{\theta} = \{\theta_j\}_{j=1}^{J}$ represents the terminal node estimates (i.e., the mean response in each terminal node), $\boldsymbol{R} = \{R_j\}_{j=1}^{J}$ represents the disjoint regions

---

[b]Typically, a CART-like tree, but any regression tree would, in theory, work here.

that form the J terminal nodes, and $I(\cdot)$ is the usual indicator function that evaluates to one whenever its argument is true (and zero otherwise). By fitting a model—a regression tree, in this case—to the observed negative gradient means we can define it at new data points. Using regression trees, the update becomes

$$f_b(\boldsymbol{x}) = f_{b-1}(\boldsymbol{x}) + \gamma_b \sum_{j=1}^{J} \theta_{jb} I(\boldsymbol{x} \in R_{jb})$$

$$= f_{b-1}(\boldsymbol{x}) + \sum_{j=1}^{J} \gamma_{jb} I(\boldsymbol{x} \in R_{jb}).$$

Consequently, the line search for choosing the step size γ_b is equivalent to updating the terminal node estimates using the specified loss function:

$$\{\gamma_{jb}\}_{j=1}^{J} = \underset{\{\gamma_j\}_{j=1}^{J}}{\arg\min} \sum_{i=1}^{N} L\left[y_i, f_{b-1}(\boldsymbol{x}_i) + \sum_{j=1}^{J} \gamma_j I(\boldsymbol{x} \in R_{jb})\right]. \qquad (8.4)$$

Following Friedman [2001], since the J terminal node regions are disjoint, we can rewrite (8.4) as

$$\gamma_{jb} = \underset{\gamma}{\arg\min} \sum_{\boldsymbol{x}_i \in R_{jb}} L[y_i, f_{b-1}(\boldsymbol{x}_i) + \gamma], \qquad (8.5)$$

which is the optimal constant update for each terminal node region, $\{R_{jb}\}_{j=1}^{J}$, based on the specified loss function L and the current iteration $f_{b-1}(\boldsymbol{x}_i)$. Solving (8.5) is equivalent to fitting a *generalized linear model* with an *offset*[c] [Efron and Hastie, 2016, p. 349]. This step is quite important since, for some loss functions, the original terminal node estimates will not be accurate enough. For example, with *least absolute deviation* (LAD) loss (see Section 8.2.0.1), the observed negative gradient, $\boldsymbol{g}_b$, only takes on integer values in $\{-1, 1\}$; hence, the fitted values are not likely to be very accurate. In summary, the "line search" step (8.5) modifies the terminal node estimates of the current fit to minimize loss.

---

[c]Roughly speaking, an offset is an adjustment term (in this case, a fixed constant) to be added to the predictions in a model; this is more common in generalized linear models where it's added to the *linear predictor* with a fixed coefficient of one (rather than an estimated coefficient).

For LS loss, (8.5) becomes

$$
\begin{aligned}
\gamma_{jb} &= \arg\min_{\gamma} \sum_{\boldsymbol{x}_i \in R_{jb}} \left([y_i - f_{b-1}(\boldsymbol{x}_i)] - \gamma \right)^2 \\
&= \arg\min_{\gamma} \sum_{\boldsymbol{x}_i \in R_{jb}} \left(r_{i,b-1} - \gamma \right)^2 \\
&= \frac{1}{N_{jb}} \sum_{\boldsymbol{x}_i \in R_{jb}} r_{i,b-1}
\end{aligned}
\qquad ,
$$

where $r_{i,b-1}$ and N_{jb} are the i-th residual and number of observations in the j-th terminal node for the b-th iteration, respectively. This results in the mean of the residuals in each terminal node at the b-th iteration, which is precisely what the original regression tree induced at iteration b uses for prediction (i.e., the terminal node summaries). In other words, for the special case of LS loss, the original terminal node estimates at the b-th iteration are already optimal, and so no update (i.e., line search) is needed.

For LAD loss, $L(f) = |y - f(\boldsymbol{x})|$, and (8.5) results in the median of the current residuals in the j-th terminal node at the b-th iteration. Solving (8.5) can be difficult for general loss functions, like those often used in binary or multinomial classification settings, and fast approximations are often employed (see Section 8.2.0.1).

The full gradient tree boosting algorithm is presented in Algorithm 8.1. Note that this is the original gradient tree boosting algorithm proposed in Friedman [2001]. Several variations have been proposed in the literature, each with their own enhancements, and I'll discuss some of these modifications in the sections that follow.

8.2.0.1   Loss functions

Various boosting algorithms can be defined by specifying different (surrogate) loss functions $L(y, f)$ in Algorithm 8.1. While not a tuning parameter, it is important to use an appropriate loss function for the problem at hand—LS loss, for example, is not appropriate for every regression problem. There are several common loss functions often used in GBMs and their implementations, and a handful of these are described in Table 8.1.

For regression, LS and LAD loss are common choices. LAD loss has the benefit of being robust in the presence of long-tailed error distributions and response outliers. Regardless of loss, gradient tree boosting is already robust to long-tailed distributions or outliers in the feature space due to the robustness of the individual base learners (in this case, regression trees). Recall that trees are invariant to strictly monotone transformations of the predictors (e.g., using x_j, e^{x_j}, or $\log(x_j)$ for the j-th predictor all produce the same results).

Algorithm 8.1 Vanilla gradient tree boosting algorithm.

1) Initialize $f_0(x) = \arg\min_\theta \sum_{i=1}^N L(y_i, \theta)$ (a constant).

2) For b in $1, \dots, B$ do the following:

    a) Compute the negative gradient, evaluated at the training data, to be used as the current working response

$$y_{ib}^\star = - \left[\frac{\partial L(y_i, f(x_i))}{\partial f(x_i)} \right]_{f(x_i)=f_{b-1}(x_i)}, \quad i = 1, 2, \dots, N.$$

    b) Fit a regression tree with J_b terminal node regions using CART's level-wise tree growing strategy: R_{jb}, $j = 1, 2, \dots, J_b$.

    c) Update the terminal node predictions using

$$\gamma_{jb} = \arg\min_\gamma \sum_{x_i \in R_{jb}} L(y_i, f_{b-1}(x_i) + \gamma), \quad j = 1, 2, \dots, J_b.$$

    d) Update $f_b(x)$ as

$$f_b(x) \leftarrow f_{b-1}(x) + \sum_{j=1}^{J_b} \gamma_{jb} I(x \in R_{jb}).$$

3) Return $\widehat{f}(x) = f_B(x)$.

Consequently, there is little need to be concerned with transformations of the features in most tree-based ensembles. For outcomes with normally distributed errors (or at least approximately so), LAD loss will be less efficient than LS loss and generalization performance will suffer. A happy compromise is provided by the Huber loss function for *Huber M-regression* described in Friedman [2001] and Hastie et al. [2009, p. 360]. The Huber loss function provides resistance to outliers and long-tailed error distributions while maintaining high efficiency in cases where errors are more normally distributed.

Compared to exponential loss (Section 5.2.4), using *binomial deviance* (or log loss) for binary outcomes provides some robustness to mislabeled examples [Hastie et al., 2009, Section 10.6]. For the binary case with $y \in \{0, 1\}$, the binomial deviance can be written as

$$L(y, f) = - [y \log(p) + (1 - y) \log(1 - p)]$$
$$= \log[1 + \exp(-2\tilde{y}f)],$$

where $\tilde{y} = 2y - 1 \in \{-1, +1\}^d$ and f refers to half the log odds for $y = +1$. With binomial deviance, there is no closed-form solution to the line search (8.5) in Algorithm 8.1, and approximations are often used instead. For example, a single Newton-Rhapson step yields

$$\gamma_{jb} = \sum_{x_i \in R_{jb}} \tilde{y}_i / \sum_{x_i \in R_{jb}} |\tilde{y}_i| (2 - |\tilde{y}_i|).$$

The final approximation $\hat{f}(x)$, which is half the logit for $y = +1$, can be inverted to produce a predicted probability. The binomial deviance can also be generalized to the case of multiclass classification [Friedman, 2001].

More specialized loss functions also exist when dealing with other types of outcome variables. For example, Poisson loss (Table 8.1) can be used when modeling counts (which are always positive integers). Ridgeway [1999] showed how gradient boosting is extendable to the exponential family[e], via likelihood-based loss functions, as well as Cox proportional hazards regression models for censored outcomes. Greg Ridgeway is also the original creator of the R package **gbm**, which is arguably the first open source implementations of gradient boosted decision trees.

TABLE 8.1: Common loss functions for gradient tree boosting. The top and bottom sections list common loss functions used for ordered and binary outcomes, respectively.

Loss name	Loss function	Negative gradient
Least squares	$\frac{1}{2}[y_i - f(x_i)]^2$	$y_i - f(x_i)$
Least absolute deviation	$|y_i - f(x_i)|$	$\text{sign}[y_i - f(x_i)]$
Poisson deviance[f]	$y_i f(x_i) - e^{f(x_i)}$	$y_i - e^{f(x_i)}$
Exponential	$\exp[-y_i f(x_i)]$	$y_i \exp[-y_i f(x_i)]$
Binomial deviance[g]	$\log[1 + \exp(-2\tilde{y}_i f(x_i))]$	$2\tilde{y}_i / (1 + \exp[2\tilde{y}_i f(x_i)])$

[d] This re-encoding is done for computational efficiency and also results in the same population minimizer as exponential loss (i.e., Adaboost.M1 from Section 5.2.4); see Bühlmann and Hothorn [2007] for details.

[e] The exponential family includes many common loss functions as a special case; for example, the Gaussian family is equivalent to using LS loss, the Laplace distribution is equivalent to using LAD loss, and the Bernoulli/binomial family is equivalent to using binomial deviance (or log loss).

[f] Here, $y_i \in \{0, 1, 2, ...\}$ is a non-negative integer (e.g., the number of people killed by mule or horse kicks in the Prussian army per year, or the number of calls to a customer support center on a particular day).

[g] Here, $\tilde{y}_i \in \{-1, 1\}$ and $f(x_i)$ refers to half the log odds for $y = +1$.

8.2.0.2   Always a regression tree?

Another subtle difference from bagging or RFs is that GBMs always use regression trees for the base learner, even for classification problems! It makes sense if you think about it: gradient tree boosting involves fitting a tree to the negative gradient of the loss function (i.e., pseudo residuals) at each iteration, which is always ordered and treated as continuous.

8.2.0.3   Priors and missclassification cost

In Section 2.2.4, I discussed how CART can naturally incorporate specific class priors and unequal misclassification costs (through a loss/cost matrix) when used for classification. Unfortunately, these concepts do not carry over to GBMs since the base learners are always regression trees; hence, there is no concept of false positives or negatives. One workaround is to use loss functions that incorporate case weights, which would allow us to give more weight to different subsets of the data (e.g., the underrepresented class). For example, we can incorporate case weights into the LS loss function by using

$$\frac{1}{\sum_{i=1}^{N} w_i} \sum_{i=1}^{N} w_i \left[y_i - f\left(\boldsymbol{x}_i\right) \right]^2,$$

where $\{w_i\}_{i=1}^{N}$ are positive case weights that affect the terminal node estimates. For example, a terminal node in a regression tree with values 2, 3, and 8 would be given estimate of $(2+3+8)/3 = 4.333$ under equal case weights (the default). However, if the corresponding case weights were 1, 5, and 1, then the terminal node estimate would become $[(1 \times 2) + (5 \times 3) + (1 \times 8)]/(1 + 5 + 1) = 3.571$. See Kriegler and Berk [2010] for an example on boosted quantile regression with case weights for small area estimation. The discussion in Berk [2008, Sec. 6.5] is also worth reading.

8.3   Hyperparameters and tuning

Gradient tree boosting provides a powerful and flexible approach to predictive modeling. The flexibility in choosing a loss function allows you to fit a rich class of models to all kinds of response outcomes. The flexibility and state-of-the-art performance come at the price of a relatively large number of tuning parameters, which fall into two categories: boosting-specific and tree-specific hyperparameters. These will be discussed in the next two sections. Several modern implementations of GBMs include even more tunable parameters, especially around regularization. These will be discussed briefly in Section 8.8.

8.3.1   Boosting-specific hyperparameters

The two boosting-specific parameters associated with gradient tree boosting are the number of trees in the ensemble (B) and a *shrinkage* parameter ν which Friedman also discussed in Friedman [2001]. These are arguably the two most impact tuning parameters associated with GBMs, and will be discussed first, starting with B.

8.3.1.1   The number of trees in the ensemble: B

If it's not using early stopping, it's crap.

Pafka [2020]

Unlike bagging and RFs, GBMs can overfit as the number of trees in the ensemble (B) increases, all else held constant. This is evident from Figure 8.1, which shows the training error (black curve) and 5-fold CV error (yellow curve) for a sequence of boosted regression stumps fit to the Ames housing data (Section 1.4.7) using LS loss. While the training error will continue to decrease to zero as the number of trees (B) increases, at some point the validation error will start to increase. Consequently, it is important to tune the number of boosting iterations B.

So how many boosting iterations should you try? In part, it depends on a number of things, including the values set for other hyperparameters. In essence, you want B large enough to adequately minimize loss and small enough to avoid overfitting. For smaller data sets, like the PBC data (Section 1.4.9), it's easy enough to fix B to an arbitrarily large value $(B = 2000$, say) and use a method like cross-validation to select an optimal value, assuming B was large enough to begin with. In the baseball hitters example, the optimal number of trees found by 5-fold CV is 52 (dashed blue line). For larger data sets, this approach can be wasteful, which is where *early stopping* comes in.

The idea of early stopping is rather simple. At each iteration, we keep track of the overall performance using some form of cross-validations or a separate validation set. When the model stops improving by a prespecified amount, the

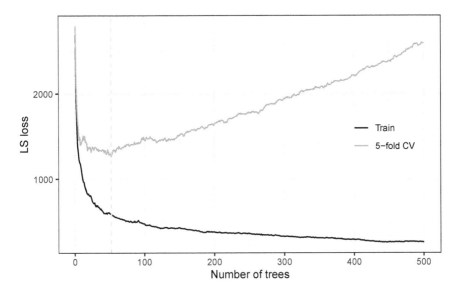

FIGURE 8.1: Gradient boosted decision stumps for the Ames housing example. The training error (black curve) continues to decrease as the number of trees increases, while the error based on 5-fold CV eventually starts to increase after $B = 83$ trees (dashed blue vertical line), indicating a problem with overfitting for larger values of B.

process "stops early" and the model is considered to have converged. Early stopping is really a mechanic of the implementation, not the GBM algorithm itself, and so it is not necessarily supported in all implementations of gradient tree boosting. Note that the concept of early stopping can be applied to other iterative or ensemble methods as well, like RFs (Chapter 7). Implementations of GBMs that support early stopping are discussed in Section 8.9. The next section deals with a tuning parameter that's intimately connected to the number of boosted trees, the *learning rate*.

8.3.1.2 Regularization and shrinkage

Regularization methods attempt to prevent overfitting by constraining the fitting procedure. There are two main approaches to regularization in GBMs: 1) controlling the number of terms/base learners B, which we discussed in Section 8.3.1.1, and 2) an explicit shrinkage parameter; the latter was introduced by Friedman [2001] to help prevent overfitting. It effectively reduces the influence of each individual tree leaving more room for future trees to improve the model. In particular, step 2d) of Algorithm 8.1 is replaced with

$$f_b(\boldsymbol{x}) \leftarrow f_{b-1}(\boldsymbol{x}) + \nu \sum_{j=1}^{J_b} \gamma_{jb} I(\boldsymbol{x} \in R_{jb}),$$

where $\nu \in (0,1]$ is a shrinkage, or regularization parameter, sometimes also referred to as the learning rate. The two parameters B and ν are not independent. Each one can control the degree of fit and thus affect the best value of the other. Decreasing ν increases the best value for B but can also result in an appreciable increase in generalization performance. All else held constant, by decreasing the learning rate, you have to fit more trees to reach optimal performance, which results in a smoother performance curve. It is generally the case that for small shrinkage parameters, say, $\nu = 0.001$, there is a fairly long plateau in which predictive performance is at its best, making it harder to overfit compared to using a relatively larger learning rate.

8.3.1.3 Example: predicting ALS progression

Here, I'll look at a brief example using the ALS data from Efron and Hastie [2016, p. 349]. A description of the data, along with the original source and download instructions, can be found at

https://web.stanford.edu/~hastie/CASI/.

The data concern $N = 1,822$ observations on *amyotrophic lateral sclerosis* (ALS or Lou Gehrig's disease) patients. The goal is to predict ALS progression over time, as measured by the slope (or derivative) of a functional rating score (dFRS), using 369 available predictors obtained from patient visits. The data were originally part of the DREAM-Phil Bowen ALS Predictions Prize4Life challenge. The winning solution [Küffner et al., 2015] used a tree-based ensemble quite similar to an RF, while Efron and Hastie [2016, Chap. 17] analyzed the data using GBMs (as I'll do in this chapter). I'll show a fuller analysis of these data in Sections 8.9.2–8.9.3.

Figure 8.2 shows the performance of a (very) basic implementation of gradient tree boosting with LS loss using **treemisc**'s lsboost() function (see Section 8.5) applied to the ALS data. Here, we can see the test MSE as a function of the number of trees using two different learning rates: 0.02 (black curve) and 0.50 (yellow curve) (following Efron and Hastie [2016, p. 339], these are boosted regression trees of depth three). Using $\nu = 0.50$ results in overfitting much quicker. The performance curve for $\nu = 0.50$ is also less smooth than for $\nu = 0.02$. While not spectacularly different, using $\nu = 0.02$ results in a slightly more accurate model (in terms of MSE on the test set), but requires far more trees. For comparison (and as a sanity check against **treemisc**'s overly simplistic lsboost() function), I also included the results from a popular open source implementation of gradient boosting called

XGBoost (to be discussed further in Section 8.8.1). For comparison, a default RF using 250 trees produced a test MSE of 0.261.

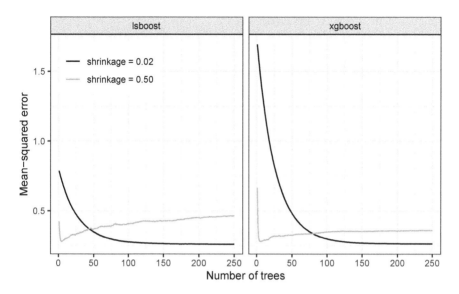

FIGURE 8.2: Gradient boosted depth-three regression trees for the ALS data using two different learning rates: 0.02 (black curve) and 0.50 (yellow curve). Left: results from our own `lsboost()` function. Right: results from XGBoost.

8.3.2 Tree-specific hyperparameters

The number of terminal nodes J controls the size (i.e., complexity) of each tree in gradient tree boosting and plays the role of an important tuning parameter for capturing interaction effects. Alternatively, you can control tree size by specifying the maximum depth. In general, a tree with maximum depth d can capture interactions up to order d. Note that a tree of depth d will have at most 2^d terminal nodes and $2^d - 1$ splits. A binary tree with J terminal nodes contains $J - 1$ splits and can capture interactions up to order $J - 1$. (A $J - 1$-th order interaction is known as a J-way interaction effect; hence, $J = 1$ corresponds to an additive model with no interaction effects). The documentation for scikit-learn's implementation of GBMs[h] notes that controlling tree size with J seems to give comparable results to using $d = J - 1$ "...but is significantly faster to train at the expense of a slightly higher training error."

---

[h] See the "Controlling the tree size" section of the scikit-learn documentation at `https://scikit-learn.org/stable/modules/ensemble`.

Hastie et al. [2009, p. 363] suggest that $4 \leq J \leq 8$ (or $2 \leq d \leq 3$) works well in general (with results not being too sensitive to different values in this range) and that $J > 10$ is rarely necessary. In many cases, a simpler additive model (i.e., $d = 1$ or $J = 2$) is sufficient.

8.3.3   A simple tuning strategy

I don't think that grid searches are all that useful for GBMs, and tend to be too costly for large data sets, especially if early stopping is not available. A simple and effective tuning strategy for GBMs is to leave the tree-specific hyperparameters at their defaults (discussed in the previous sections) and tune the boosting parameters. A rule of thumb proposed by Greg Ridgeway[i] is to set shrinkage as small as possible while still being able to fit the model in a reasonable amount of time and storage. For example, aim for 3,000–10,000 iterations with shrinkage rates between 0.01–0.001; use early stopping, if it's available. More elaborate tuning strategies for GBMs are discussed in Boehmke and Greenwell [2020, Chap. 12].

8.4   Stochastic gradient boosting

Friedman [2002] proposed a minor modification to the original GBM algorithm in Friedman [2001], where he showed that both generalization performance and execution speed of GBMs can often be improved dramatically by incorporating additional randomization into the procedure, similar to how bagging can improve the performance of a single tree. The idea is to forgo using the entire training sample to fit each subsequent tree in the ensemble and instead use a subsample of the training data drawn at random without replacement; typically a 50% random subsample is used to induce each tree (i.e., roughly half the original training data), although, for larger data sets, a smaller fraction can be used. Due to the extra randomization step, the full procedure is referred to as stochastic gradient boosting and is the most common flavor of gradient boosting seen in practice today. Friedman [2002] suggests that $0.5 \leq f \leq 0.8$ generally leads to an improvement for small to moderate sized data sets, where f is the fraction of the original training data sampled at random before building each tree.

As with bootstrap sampling in bagging and RFs, a happy by-product of subsampling in GBMs is the ability to produce an OOB estimate of the generalization error (Section 7.3). OOB estimates of error are similar to that obtained

---

[i]See the **gbm** package vignette: `vignette("gbm", package = "gbm")`.

using N-fold (or leave-one-out) cross-validation and computed at virtually no extra cost to the fitting algorithm. However, as stated in Section 7.3, the OOB approach can provide overly pessimistic estimates of the true error, but can still be used for hyperparameter tuning [Janitza and Hornung, 2018]. Since GBMs have lots of tuning parameters, the OOB approach provides a computationally feasible solution to selecting a reasonable learning rate, number of trees, etc.

It's important to note that Janitza and Hornung [2018] refer specifically to OOB-based error estimates for RFs, not GBMs. To this day, I have yet to see an extensive study on the usefulness of OOB-based error estimates in GBMs compared to more traditional cross-validation approaches.

8.4.1   Column subsampling

Column subsampling is another technique that can be used to improve model performance and speed up fitting. Similar to column subsampling in an RF, a subsample of columns can be used for building each individual tree[j]. Apparently subsampling the columns prior to building each tree, can reduce the chances of overfitting even more than traditional row subsampling [Chen and Guestrin, 2016].

To illustrate, consider the test MSE curves for the ALS data displayed in Figure 8.3. In this example, subsampling the columns appears to outperform subsampling the rows (here, I arbitrarily chose a subsampling rate of 0.3). In practice these parameters need to be tuned, but it's probably safe and more computationally efficient in practice to just deal with one of these two hyperparameters. If you're dealing with a really wide data set, it may be more efficient to consider column subsampling, or both column subsampling and row subsampling if you have many rows as well.

8.5   Gradient tree boosting from scratch

Let's implement a quick-and-dirty gradient tree boosting function based on LS loss. The function, called `lsboost()`, is available in package **treemisc** (see `?treemisc::lsboost` for details and a description of the arguments), but the code is relatively straightforward and reproduced in the code chunk below. It's

---

[j]While similar, an RF chooses a random subsample of features prior to each split of every tree.

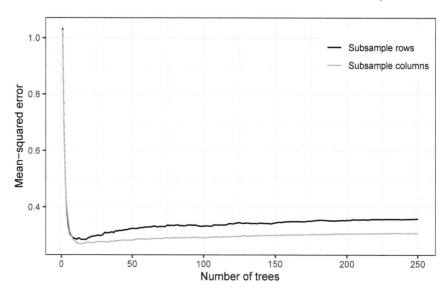

FIGURE 8.3: Effect of subsampling in GBMs on the ALS data. In this case, randomly subsampling the columns (yellow curve) slightly outperforms randomly subsampling the rows (black curve).

worth noting a few things about `lsboost()` and `predict.lsboost()`, before we continue:

- the code uses R's built-in S3 *object-oriented (OO) programming* system [Wickham, 2019, Chap. 13], which allows us to extend R's built-in `predict()` generic via the `predict.lsboost()` function (e.g., so we can compute predictions with, say, `predict(my.lsboost.model, newdata = some.data)`);

- `lsboost()` uses **rpart** to fit the individual regression trees, but other implementations could be used instead (e.g., CTrees via **partykit**);

- `lsboost()` returns an object of class `"lsboost"`, which is essentially a list of **rpart** trees that the `predict()` function knows how to combine;

- these functions are for illustration and not meant for serious use—they are not optimized in any sense.

If you cut out the fluff, gradient tree boosting, at least with LS loss, can be implemented in as little as 10 lines of code (probably less):

```
lsboost <- function(X, y, ntree = 100, shrinkage = 0.1, depth = 6,
                    subsample = 0.5, init = mean(y)) {
  yhat <- rep(init, times = nrow(X))  # initialize fit; f_0(x)
  trees <- vector("list", length = ntree)  # to store each tree
  ctrl <-  # control tree-specific parameters
```

```
      rpart::rpart.control(cp = 0, maxdepth = depth, minbucket = 10)
   for (tree in seq_len(ntree)) {  # Step 2) of Algorithm 8.1
     id <- sample.int(nrow(X), size = floor(subsample * nrow(X)))
     samp <- X[id, ]  # random subsample
     samp$pr <- y[id] - yhat[id]  # pseudo residual
     trees[[tree]] <-  # fit tree to current pseudo residual
       rpart::rpart(pr ~ ., data = samp, control = ctrl)
     yhat <- yhat + shrinkage * predict(trees[[tree]], newdata = X)
   }
   res <- list("trees" = trees, "shrinkage" = shrinkage,
               "depth" = depth, "subsample" = subsample, "init" = init)
   class(res) <- "lsboost"
   res
 }

 # Extend R's generic predict() function to work with "lsboost" objects
 predict.lsboost <- function(object, newdata, ntree = NULL,
                             individual = FALSE, ...) {
   if (is.null(ntree)) {
     ntree <- length(object[["trees"]])  # use all trees
   }
   shrinkage <- object[["shrinkage"]]  # extract learning rate
   trees <- object[["trees"]][seq_len(ntree)]
   pmat <- sapply(trees, FUN = function(tree) {  # all predictions
     shrinkage * predict(tree, newdata = newdata)
   })  # compute matrix of (shrunken) predictions; one for each tree
   if (isTRUE(individual)) {
     pmat  # return matrix of (shrunken) predictions
   } else {
     rowSums(pmat) + object$init  # return boosted predictions
   }
 }
```

Gradient tree boosting with LS loss is simpler to implement because there's no need to perform the line search step in Algorithm 8.1 (i.e., the terminal node estimates are already optimal). A slightly more complicated function that also implements gradient tree boosting with **rpart**, but using LAD loss, is shown below; this function is also part of **treemisc** (see `?treemisc::ladboost` for details). Here, care needs to be taken to update the terminal node summaries accordingly (see the commented section starting with `# Line search`). For LAD loss, we simply use the terminal node sample medians, as discussed in Section 8.2; here, I update the **frame** component of the **rpart** tree, but **partykit** could also be used, as illustrated in the commented out section. Also, note that the initial fit (`init`) defaults to the median response as well.

```
ladboost <- function(X, y, ntree = 100, shrinkage = 0.1, depth = 6,
                     subsample = 0.5, init = median(y)) {
  yhat <- rep(init, times = nrow(X))  # initialize fit
  trees <- vector("list", length = ntree)  # to store each tree
```

```r
  ctrl <-  # control tree-specific parameters
    rpart::rpart.control(cp = 0, maxdepth = depth, minbucket = 10)
  for (tree in seq_len(ntree)) {
    id <- sample.int(nrow(X), size = floor(subsample * nrow(X)))
    samp <- X[id, ]
    samp$pr <- sign(y[id] - yhat[id])  # use signed residual
    trees[[tree]] <-
      rpart::rpart(pr ~ ., data = samp, control = ctrl)
    #-------------------------------------------------------------------
    # Line search; update terminal node estimates using median
    #-------------------------------------------------------------------
    where <- trees[[tree]]$where  # terminal node assignments
    map <- tapply(samp$pr, INDEX = where, FUN = median)
    trees[[tree]]$frame$yval[where] <- map[as.character(where)]
    #
    # Could use partykit instead:
    #
    # trees[[tree]] <- partykit::as.party(trees[[tree]])
    # med <- function(y, w) median(y)  # see ?partykit::predict.party
    # yhat <- yhat +
    #   shrinkage * partykit::predict.party(trees[[tree]],
    #                                       newdata = X, FUN = med)
    #-------------------------------------------------------------------
    yhat <- yhat + shrinkage * predict(trees[[tree]], newdata = X)
  }
  res <- list("trees" = trees, "shrinkage" = shrinkage,
              "depth" = depth, "subsample" = subsample, "init" = init)
  class(res) <- "ladboost"
  res
}
```

8.5.1   Example: predicting home prices

Let's apply the lsboost() function to the Ames housing data. Below, I use the same train/test split for the Ames housing data we've been using throughout this book, then call lsboost() to fit a GBM to the training set; here, I'll use a shrinkage factor of $\nu = 0.1$:

```r
library(treemisc)
```

```r
# Split Ames data into train/test sets using a 70/30 split
ames <- as.data.frame(AmesHousing::make_ames())
ames$Sale_Price <- ames$Sale_Price / 1000  # rescale response
set.seed(4919)  # for reproducibility
id <- sample.int(nrow(ames), size = floor(0.7 * nrow(ames)))
ames.trn <- ames[id, ]
ames.tst <- ames[-id, ]
```

```
# Fit a gradient tree boosted ensemble with 500 trees
set.seed(1110)  # for reproducibility
ames.bst <-
  lsboost(subset(ames.trn, select = -Sale_Price),  # features only
          y = ames.trn$Sale_Price, ntree = 500, depth = 4,
          shrinkage = 0.1)
```

The test RMSE as a function of the number of trees in the ensemble is computed below using the previously defined **predict()** method; the results are shown in Figure 8.4 (black curve). For brevity, the code uses **sapply()** to essentially iterate cumulatively through the $B = 500$ trees and computes the test RMSE for the first tree, first two trees, etc. For comparison, the test RMSEs from a default RF are also computed and displayed in Figure 8.4 (yellow curve). In this example, the GBM slightly outperforms the RF.

```
set.seed(1128)  # for reproducibility
ames.rfo <-  # fit a default RF for comparison
  randomForest(subset(ames.trn, select = -Sale_Price),
               y = ames.trn$Sale_Price, ntree = 500,
               # Monitor test set performance (MSE, in this case)
               xtest = subset(ames.tst, select = -Sale_Price),
               ytest = ames.tst$Sale_Price)

# Helper function for computing RMSE
rmse <- function(pred, obs, na.rm = FALSE) {
  sqrt(mean((pred - obs)^2, na.rm = na.rm))
}

# Compute RMSEs from both models on the test set as a function of the
# number of trees in each ensemble (i.e., B = 1, 2, ..., 500)
rmses <- matrix(nrow = 500, ncol = 2)  # to store results
colnames(rmses) <- c("GBM", "RF")
rmses[, "GBM"] <- sapply(seq_along(ames.bst$trees), FUN = function(B) {
  pred <- predict(ames.bst, newdata = ames.tst, ntree = B)
  rmse(pred, obs = ames.tst$Sale_Price)
})  # add GBM results
rmses[, "RF"] <- sqrt(ames.rfo$test$mse)  # add RF results
```

8.6 Interpretability

Interpreting GBMs is no different from any other nonparametric model. For example, Section 5.4 discussed how the individual tree-based importance scores (Section 2.8) can be aggregated across all the trees in an ensemble

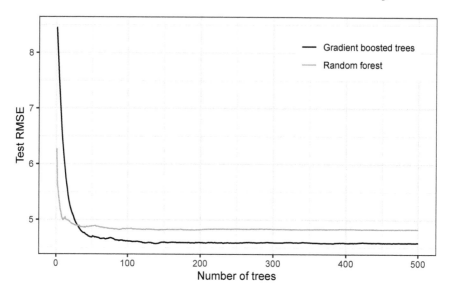

FIGURE 8.4: Root mean-squared error for the Ames housing test set as a function of B, the number of trees in the ensemble. Here, I show both a GBM (black curve) and a default RF (yellow curve). In this case, gradient tree boosting with LS loss, a shrinkage of $\lambda = 0.1$, and a maximum tree depth of $d = 4$ (black curve) slightly outperforms a default RF (yellow curve).

to form a more stable measure of predictor importance; however, as with CART and RFs, this measure is also biased for GBMs, although, the permutation importance method (Section 6.1.1) applies equally well to GBMs, or any supervised learning model, for that matter. PDPs and ICE plots can be used to visualize the global and local effect that subsets of features have on the model's predictions, respectively. Shapley values, among other techniques, can be used to infer the contribution each feature value has on the difference between its associated prediction and the model's baseline (or average training) prediction, which can also be used to generate global measures of both feature importance and feature effects. The next two sections discuss specialized interpretability techniques often associated with GBMs.

8.6.1 Faster partial dependence with the recursion method

For regression trees based on single-variable splits, Friedman [2001] described a fast procedure for computing the partial dependence of $\widehat{f}(\boldsymbol{x})$ on a subset of features using a weight traversal of each tree (henceforth referred to as the *recursion method*). In particular[k], if a split node involves an input feature from the interest set (z_c), the corresponding left or right branch is followed; otherwise both branches are followed, each branch being weighted by the fraction of training observations that entered that branch. Finally, the partial dependence is given by a weighted average of all the visited terminal node values.

The idea is not specific to gradient tree boosting (e.g., it could also be applied to a RF), but as far as I'm aware, it's only implemented in a couple of open source packages: R's **gbm** package (which **pdp** takes full advantage of) and the **sklearn.inspection** module, which supports the recursion method only for certain tree-based estimators. Most other implementations rely on the brute force method described in Section 6.2.1.

8.6.1.1 Example: predicting email spam

Let's illustrate with the email spam example (Section 1.4.5). Here, I used the R package **gbm** to fit a GBM using log loss, $B = 4,043$ depth-2 regression trees (found using 5-fold cross-validation), a shrinkage factor of $\nu = 0.01$.

To gain an appreciation for the computational speed-up of the recursion method (which is implemented in **gbm**), I computed Friedman's H-statistic for all 1,596 pairwise interactions, which took roughly five minutes! The largest pairwise interaction occurred between `address` and `receive`. The partial dependence of the log-odds of spam on the joint frequencies of `address` and `receive` is displayed in Figure 8.5. Using the fast recursion method, this took roughly a quarter of a second to compute, compared to the brute force method, which took almost 500 seconds.

8.6.2 Monotonic constraints

Increasing the interpretability of a model without sacrificing too much in the way of accuracy is useful in many real-world applications. For example, prior knowledge may be available (e.g., from subject matter experts or historical data) indicating that a given feature should in general have a positive (or negative) effect on the expected outcome. With GBMs, we can often increase interpretability by enforcing such *monotonic constraints*.

In gradient tree boosting, monotonic constraints enforce a specific splitting strategy in each of the constituent regression trees, where binary splits of a variable in one direction either always increase (monotone increasing) or decrease (monotone decreasing) the mean response in the resulting child node. For example, in a model with just two features, x_1 and x_2, if we specified a

---

[k]Deets taken from the partial dependence documentation on scikit-learn's website: https://scikit-learn.org/stable/modules/partial_dependence.html.

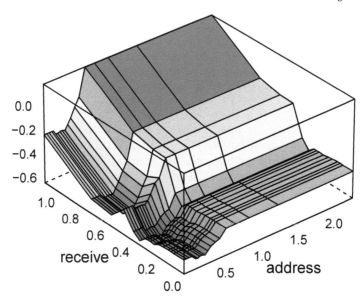

FIGURE 8.5: Partial dependence of log-odds of spam on joint frequency of `address` and `receive`.

monotonic increasing constraint on x_1, then for all $x_1 \leq x'_1$ we would have $f(x_1, x_2) \leq f(x'_1, x_2)$. Such constraints are quite easy to visualize using partial dependence or ICE plots (Section 6.2), as briefly illustrated in the next section.

Enforcing monotonicity, where it makes sense, can make predictions more interpretable. For example, credit score is often used in determining whether to reject a loan or credit card application. If all the relevant features between two applicants are the same, aside from their current credit score, it might make sense to force the model to predict a lower probability of default for the applicant with the higher credit score. Such constraints are suitable for use in more regulated applications; for example, the likelihood of a loan approval is often higher with a better credit score. Gill et al. [2020] propose a mortgage lending workflow based on GBMs with monotonicity constraints, *explainable neural networks*, and Shapley values (Section 6.3.1), which gives careful consideration to US adverse action notice and anti-discrimination requirements.

8.6.2.1 Example: bank marketing data

Returning to the bank marketing example from Section 7.9.5, I fit a GBM with and without a decreasing monotonic constraint on `euribor3m`, the Euribor 3 month rate[1]. In both cases, I used 5-fold cross-validation to fit a GBM with a maximum of 3,000 trees using a shrinkage rate of $\nu = 0.01$ and a maximum depth of $d_{max} = 3$. The partial dependence of the probability of subscribing on `euribor3m` from each model is displayed in Figure 8.6. Both figures tell the same story: the predicted probability of subscribing tends to decrease as the euribor 3 month rate increases. However, it may make sense here to assume the relationship to be monotonic decreasing, as in the left side of Figure 8.6. This can help increase interpretation and understanding by incorporating domain expertise, for example, by removing some of the noise like the little spike in the right side of Figure 8.6 near `euribor3m = 1 = 1`. Compare these to the RF-based PDP from Figure 7.28.

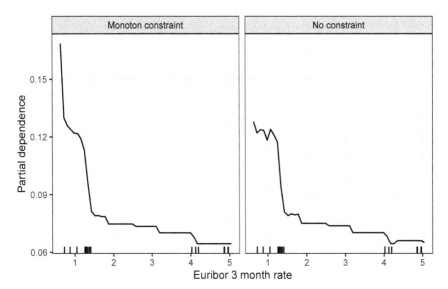

FIGURE 8.6: Partial dependence of subscription probability on euribor 3 month rate from the bank marketing probability forest. Left: with monotonic constraint on `euribor3m`. Right: without constraint. The rug display along the x-axis summarizes the distribution of `euribor3m` via a grid of 29 evenly spaced quantiles.

---

[1]Recall that the 3 month Euribor rate is the interest rate at which a selection of European banks lend each other funds (denominated in euros) whereby the loans have a 3 month maturity.

8.7   Specialized topics

8.7.1   Level-wise vs. leaf-wise tree induction

There two strategies to consider when growing an individual decision tree that we have yet to discuss:

- *Level-wise* (also referred to as *depth-wise* or *depth first*) tree induction is used by many common decision tree algorithms (e.g., CART and C4.5/C5.0, but this probably depends on the implementation) and grows a tree level by level in a fixed order; that is, each node splits the data by prioritizing the nodes closer to the root node.

- *Leaf-wise* tree induction (also referred to as *best-first* splitting), on the other hand, grows a tree by splitting the node whose split leads to the largest reduction of impurity.

When grown to maximum depth, both strategies result in the same tree structure; the difference occurs when trees are restricted to a maximum depth or number of terminal nodes. Leaf-wise tree induction, while not specific to boosting, has primarily only been evaluated in that context; see, for example, Friedman [2001] and Shi [2007].

Figure 8.7 gives an example of a tree grown level-wise (left) and leaf-wise (right). Notice how the overall tree structures are the same, but the order in which the splits are made (i.e., S_1–S_4) is different. In general, level-wise growth tends to work better for smaller data sets, whereas leaf-wise tends to overfit. Leaf-wise growth tends to excel in larger data sets where it is considerably faster than level-wise growth. This is why some modern GBM implementations—like LightGBM (Section 8.8)—default to growing trees leaf-wise.

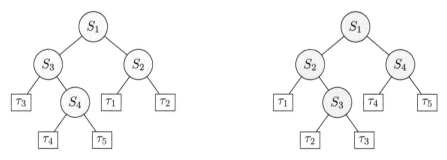

FIGURE 8.7: Hypothetical decision tree grown level-wise/depth-first (left) and leaf-wise/best-first (right).

8.7.2   Histogram binning

Finding the optimal split for a numeric feature in a decision tree can be slow when dealing with many unique values; the more unique values a numeric predictor has, the more split points the tree algorithm has to search through. A much faster alternative is to bucket the numeric features into bins using histograms.

The idea is to first bin the input features into integer-valued bins (255–256 bins seems to be the default across many implementations) which can tremendously reduce the number of split points to search through. Histogram binning is implemented in a number of popular GBM implementations, including XGBoost (Section 8.8.1), LightGBM (Section 8.8.2), and the **sklearn.ensemble** module. LightGBM's online documentation lists several earlier references to this approach; visit `https://lightgbm.readthedocs.io/en/latest/Features.html` for details.

8.7.3   Explainable boosting machines

Explainable boosting machine (EBM) is an interpretable model developed at Microsoft Research [Nori et al., 2019]. In particular, an EBM is a tree-based, cyclic gradient boosting *generalized additive model* (GAM) with automatic interaction detection. The authors claim that EBMs are often as accurate as current state-of-the-art algorithms (like RFs and GBMs) while remaining completely interpretable. And while EBMs are often slower to train than other modern algorithms, they are extremely compact and fast at prediction time, which makes them attractive for deploying in a production process.

In essence, an EMB fits a GAM of the form:

$$g\left(\mathrm{E}\left[y|\boldsymbol{x}\right]\right) = \beta_0 + \sum_j f_j\left(x_j\right), \tag{8.6}$$

where g is a *link function* that connect the random and systematic component (e.g., adapts the GAM to different settings such as classification, regression, or Poisson regression), and f_j is a function of predictor x_j. Compared to a traditional GAM, an EBM:

- estimates each feature function $f_j\left(x_j\right)$ using tree-based ensembles, like gradient tree boosting or bagging;

- can automatically detect and include pairwise interaction terms of the form $f_{ij}\left(x_i, x_j\right)$.

The overall boosting procedure is restricted to train on one feature at a time in a "round-robin" fashion using a small learning rate to ensure that feature order does not matter, which helps limit the effect of collinearity or strong dependencies among the features.

EBMs are considered "glass box" or highly interpretable models because the contribution of each feature (or pairwise interaction) to a final prediction can be visualized and understood by plotting $f_j(x_j)$, similar to a PD plot (Section 6.2.1). And since EBMs are additive models, each feature contributes to predictions in a modular way that makes it easy to reason about the contribution of each feature to the prediction [Nori et al., 2019]. The simple additive structure of an EBM comes at the cost of longer training times. However, at the end of model fitting, the individual trees can be dropped and only the $f_j(x_j)$ and $f_{ij}(x_i, x_j)$ need to be retained, which makes EBMs faster at execution time. EBMs are available in the **interpret** package for Python. For more info, check out the associated GitHub repository at `https://github.com/interpretml/interpret`.

8.7.4 Probabilistic regression via natural gradient boosting

Many classification tasks are inherently probabilistic. For example, probability forests (Section 7.2.1) can be used to obtain consistent probability estimates for the different class outcomes (i.e., $\Pr(y = j|\boldsymbol{x})$). Regression tasks, on the other hand, are typically not probabilistic and the predictions correspond to some location estimate of $y|\boldsymbol{x}$; that is, the distribution of y conditional on a set of predictor values $\boldsymbol{x}$. For instance, the terminal nodes in a regression tree—which are used to compute fitted values and predictions—provide an estimate of the conditional mean $\mathrm{E}(y|\boldsymbol{x})$. Often, it is of scientific interest to know about the probability of specific events conditional on a set of features, rather than a single point estimate like $\mathrm{E}(y|\boldsymbol{x})$. In the ALS example, rather than using an estimate of the conditional mean $\hat{f}(\boldsymbol{x}) = \widehat{\mathrm{E}}(\mathtt{dFRS}|\boldsymbol{x})$ to predict ALS progression for a new patient, it might be more useful to estimate $\Pr(\mathtt{dFRS} < c|\boldsymbol{x})$, for some constant c. This is where probabilistic regression/forecasting comes in.

Probabilistic regression models provide estimates of the entire probability distribution of the response conditional on a set of predictors, denoted $\mathcal{D}_{\boldsymbol{\theta}}(y|\boldsymbol{x})$, where $\boldsymbol{\theta}$ represents the parameters of the conditional distribution. For example, the normal distribution has $\boldsymbol{\theta} = (\mu, \sigma)$; examples include *generalized additive models for shape, scale, and location* (GAMLSS) [Rigby and Stasinopoulos, 2005], *Bayesian additive regression trees* (BART) [Chipman et al., 2010], and Bayesian deep learning. While several approaches to probabilistic regression exist, many of them are inflexible (e.g., GAMSLSS), computationally expensive (e.g., BART), or inaccessible to non-experts (e.g., Bayesian deep learning) [Duan et al., 2020]. *Natural gradient boosting* (NGBoost) extends the simple ideas of gradient boosting to probabilistic regression by treating the parameters $\boldsymbol{\theta}$ as targets for a multiparameter boosting algorithm similar to gradient boosting (Algorithm 8.1). We say "multiparameter" because NGBoost fits a separate model for each parameter at every iteration.

The "natural" in "natural gradient boosting" refers to the fact that NGBoost uses something called the *natural gradient*, as opposed to the ordinary gradient. The natural gradient provides the direction of steepest descent in *Riemannian space*; this is necessary since gradient descent in the parameter space is not gradient descent in the distribution space because distances don't correspond. The important thing to remember is that NGBoost approximates the gradient of a proper scoring rule—similar to a loss function, but for predicted probabilities and probability distributions of the observed data—as a function of $\boldsymbol{\theta}$. Compared to alternative probabilistic regression methods, NGBoost is fast, flexible, scalable, and easy to use. An example, albeit in Python, is given in Section 8.9.2. NGBoost is available in the **ngboost** package for Python. For more info, check out the NGBoost GitHub repository at `https://github.com/stanfordmlgroup/ngboost`.

8.8   Specialized implementations

In this section, I'll take a look at three specialized implementations of GBMs that are quite popular for supervised learning tasks, probably due to their availability across platforms and ability to scale to incredibly large data sets, even on a single machine where the data cannot fit into memory.

8.8.1   eXtreme Gradient Boosting: XGBoost

XGBoost [Chen and Guestrin, 2016] is one of the most popular and scalable implementations of GBMs. While XGBoost follows the same principles as the standard GBM algorithm, there are some important differences, a few of which are listed below:

- more stringent regularization to help prevent overfitting;
- a novel sparsity-aware split finding algorithm;
- weighted quantile sketch for fast and approximate tree learning;
- parallel tree building (across nodes within a tree);
- exploits out-of-core processing for maximum scalability on a single machine;

- employs the deep-learning concept of *dropout* to mitigate the problem of *overspecialization* [Vinayak and Gilad-Bachrach, 2015].

With these differences, XGBoost can scale to billions of rows in distributed or memory-limited settings (e.g., a single machine), hence the "extreme" in extreme gradient boosting. It has also been shown to be more accurate than the more traditional implementation of GBM (e.g., the R package **gbm**. There are a number of interfaces to XGBoost, including R, Python (with a scikit-learn interface), Julia, Scala, Java, Ruby, and C++. There's also a command-line interface. General tuning strategies for XGBoost (with examples in R) are given in Boehmke and Greenwell [2020, Section 12.5.2]. For more details, including installation, visit the XGBoost GitHub repository at `https://github.com/dmlc/xgboost`.

In contrast to ordinary GBMs, which use CART-like regression trees for the base learners (i.e., splits are determined using the sum of squares criteria described in Section 2.3), XGBoost uses a regularized second-order approximation to the loss function in its split search algorithm. Let g_{jb} (defined previously in Section 8.2) and $h_{jb} = \partial^2 L\left(y_j, f_{b-1}\right)/\partial f_{b-1}^2$ be the gradient and *hessian* (i.e., second-order gradient) values of the loss function L at the b-th iteration evaluated at the j-th observation (what a mouthful!). Further, let I_L and I_R be the set of observations in the left and right daughter nodes resulting from split s. The set of instances in the parent node is denoted by $I = I_L \cup I_R$. In XGBoost, splits are selected that result in the largest gain (or reduction in loss) which can be written as

$$\mathcal{L}_{split} = \frac{1}{2}\left[\frac{\left(\sum_{i \in I_L} g_i\right)^2}{\sum_{i \in I_L} h_i + \lambda} + \frac{\left(\sum_{i \in I_R} g_i\right)^2}{\sum_{i \in I_R} h_i + \lambda} - \frac{\left(\sum_{i \in I_I} g_i\right)^2}{\sum_{i \in I_I} h_i + \lambda}\right] - \gamma. \quad (8.7)$$

For brevity, I dropped the b subscript in g_i and h_i. Here, $\gamma \in [0, \infty)$ is the minimum gain required to make a further split in the tree, with larger values resulting in a more conservative algorithm (i.e., fewer splits). Similarly, $\lambda \in [0, \infty)$ can be viewed as an L_2 penalty (similar to the one used in *ridge regression*), with larger values resulting in a more conservative algorithm. Both parameters provide a form regularization that constrains the model's complexity and guards against overfitting. Eventually, XGBoost included an additional regularization parameter, $\alpha \in [0, 1)$, which acts as an L_1 penalty (similar to the one used in the LASSO). Consequently, these can be viewed as additional hyperparameters to be tuned.

Aside from the exhaustive greedy search employed by CART, XGBoost provides an optional approximate search algorithm that is more efficient in distributed settings or when the data cannot easily fit into memory. In essence, the continuous feature values are mapped to buckets formed by the percentiles of each feature's distribution in the training data. The best split for each feature is found by searching through this reduced set of candidate split values. For details, see Chen and Guestrin [2016]. A modification for weighted data, called a *weighted quantile sketch*, is also discussed in Chen and Guestrin [2016].

Around early 2017, XGBoost introduced *fast histogram binning* (Section 8.7.2) to even further push the boundaries of scale and computation speed. In contrast to the original approximate tree learning strategy, which generates a new set of bins for each iteration, the histogram method re-uses the bins over multiple iterations, and therefore is far better suited for large data sets. XGBoost also introduced the option to grow trees leaf-wise, as opposed to just level-wise (the default), which can also speed up fitting, albeit, at the risk of potentially overfitting the training data (Section 8.7.1).

Sparse data are common in many situations, including the presence of missing values and one-hot encoding. In such cases, efficiency can be obtained by making the algorithm aware of any sparsity patterns. XGBoost handles sparsity by learning an optimal "default" direction at each split in a tree. When an observation is missing for one of the split variables, for example, it is simply passed down the default branch. For details, see Chen and Guestrin [2016].

One drawback of XGBoost is that it does not currently handle categorical variables—they have to be re-encoded numerically (e.g., using one-hot encoding). However, at the time of writing this book, XGBoost has experimental support for categorical variables, although it's currently quite limited. An example of using XGBoost is given in Section 8.9.4. Note that XGBoost can also be used to fit RFs in a distributed fashion; see the XGBoost documentation for details.

8.8.2 Light Gradient Boosting Machine: LightGBM

LightGBM [Ke et al., 2017] offers many of the same advantages as XGBoost, including sparse optimization, parallel tree building, a plethora of loss functions, enhanced regularization, bagging, histogram binning, and early stopping. A major difference between the two is that LightGBM defaults to building trees leaf-wise (or best-first). Unlike XGBoost, LightGBM can more naturally handle categorical features in a way similar to what's described in Section 2.4. In addition, the LightGBM algorithm utilizes two novel techniques, *gradient-based one-side sampling* (GOSS) and *exclusive feature bundling* (EFB).

GOSS reduces the number of observations by excluding rows with small gradients, while the remaining instances are used to estimate the information gain for each split; the idea is that observations with larger gradients play a more important role in split selection. EFB, on the other hand, reduces the number of predictors by bundling mutually exclusive features together (i.e., they rarely take nonzero values simultaneously). Both of these features allow LightGBM to speed up the training process, while maintaining accuracy.

LightGBM is available in both C, R, and Python. For details, visit the Light-GBM GitHub repository at `https://github.com/microsoft/LightGBM`. Without access to GPUs, XGBoost and LightGBM are among the most efficient GBM implementations. If you have access to GPUs, XGBoost currently seems to have a slight edge. Like XGBoost, LightGBM can also be run in random forest mode; consult the LighGBM documentation for details.

8.8.3 CatBoost

While XGBoost and LightGBM seem to be the most popular implementations of GBMs, they didn't initially[m] handle categorical variables as well as another GBM variant called CatBoost [Dorogush et al., 2018, Prokhorenkova et al., 2017]. One of the main selling points of CatBoost is the ability to handle categorical variables without the need for numerically encoding them. From the CatBoost website:

> Improve your training results with CatBoost that allows you to use non-numeric factors, instead of having to pre-process your data or spend time and effort turning it to numbers.
>
> `https://catboost.ai/`

They also claim that CatBoost works reasonably well out of the box with less time needed to be spent on hyperparameter tuning.

In CatBoost, a process called *quantization* is applied to numeric features, whereby values are divided into disjoint ranges or buckets—this is similar to the approximate tree growing algorithm in XGBoost whereby numeric features are binned. Before each split, categorical variables are converted to numeric using a strategy similar to mean target encoding, called *ordered target statistics*, which avoids the problem of target leakage and reduces overfitting. CatBoost is currently available as a command line application in C++, but R and Python interfaces are also available. For further details and resources, visit the CatBoost website at `https://catboost.ai/`.

---

[m]Both XGBoost and LightGBM now support categorical features.

8.9 Software and examples

GBMs are implemented in a number of open source software packages. The original implementation of GBMs was called MART(tm), for *multiple additive regression trees*[n]. The R package **gbm** was probably one of the first available open source implementations of Friedman's original GBM algorithm, with extensions to boosting likelihood-based models, like exponential family and proportional hazards regression models. Another general R package for boosting is **mboost** [Hothorn et al., 2021a], which implements boosting for optimizing general risk functions utilizing component-wise (penalized) least squares estimates as base-learners for fitting various kinds of generalized linear and generalized additive models to potentially high-dimensional data (i.e., with **mboost** you can specify different base-learners for individual predictors).

The **sklearn.ensemble** module implements Friedman's original GBM algorithm with some additional optimizations, like early stopping and the option to use histogram binning.

XGBoost, LightGBM, and CatBoost are available in both R and Python, and all three support early stopping; the R packages **xgboost** and **lightgbm** are both available on CRAN (at least at the time of writing this book). H2O's implementation of GBM also supports early stopping. Several benchmarks [Pafka, 2019, 2021] comparing these as well as several other GBM implementations, are available from Szilard Pafka at `https://github.com/szilard/GBM-perf`. Check out his other repositories on GitHub for fantastic talks on GBMs and other machine learning algorithms and their implementations.

8.9.1 Example: Mayo Clinic liver transplant data

In this example, I'll return to the PBC data introduced in Section 1.4.9, where the goal is to model survival in patients with the autoimmune disease PBC. In Section 3.5.3, I fit a CTree model to the randomized subjects using log-rank scores. Here, I'll use the GBM framework to boost a *Cox proportional hazards* (Cox PH) model; see Ridgeway [1999] for details.

The Cox PH model is one of the most widely used models for the analysis of survival data. It is a semi-parametric model in the sense that it makes a parametric assumption regarding the effect of the predictors on the hazard

---

[n]MART, which evolved into a product called TreeNet(tm), is proprietary software available from Salford Systems, which is currently owned by MiniTab; visit `https://www.minitab.com/en-us/predictive-analytics/treenet/` for details.

function (or *hazard rate*) at time t, often denoted $\lambda(t)$, but makes no assumption regarding the shape of $\lambda(t)$; since little is often known about $\lambda(t)$ in practice, the Cox PH model is quite useful. The hazard rate—also referred to as the *force of mortality* or *instantaneous failure rate*—is related to the probability that the event (e.g., death or failure) will occur in the next instant of time, given the event has not yet occurred [Harrell, 2015, Sec. 17.3]; it's not a true probability since $\lambda(t)$ can exceed one. Studying the hazard rate helps understand the nature of risk of time.

Extending Algorithm 8.1 to maximize Cox's log-partial likelihood (which is akin to minimizing an appropriate loss function) allows us to relax the linearity assumption, which assumes that the (possibly transformed) predictors are linearly related to the log hazard, and fit a richer class of models based on regression trees. Below, I load the **survival** package and recreate the same pbc2 data frame I used in Section 3.5.3:

```
library(survival)
```

```
# Prep the data a bit
pbc2 <- pbc[!is.na(pbc$trt), ]  # use randomized subjects
pbc2$id <- NULL  # remove ID column
# Consider transplant patients to be censored at day of transplant
pbc2$status <- ifelse(pbc2$status == 2, 1, 0)
facs <- c("sex", "spiders", "hepato", "ascites", "trt", "edema")
for (fac in facs) {  # coerce to factor
  pbc2[[fac]] <- as.factor(pbc2[[fac]])
}
```

Next, I'll fit a boosted PH regression model using the **gbm** package; details on the deviance/loss function used in Algorithm 8.1 can be found in the **gbm** package vignette: `vignette("gbm", package = "gbm")`. Here, I use $B = 3000$, a maximum tree depth of three, and a learning rate of $\nu = 0.001$. The optimal number of trees (`best.iter`) is determined using 5-fold cross-validation.

```
library(gbm)
```

```
set.seed(1551)  # for reproducibility
pbc2.gbm <- gbm(Surv(time, status) ~ ., data = pbc2,
                distribution = "coxph", n.trees = 3000,
                interaction.depth = 3, shrinkage = 0.001,
                cv.folds = 5)
(best.iter <- gbm.perf(pbc2.gbm, method = "cv", plot.it = FALSE))
```

```
#> [1] 1934
```

Below, I construct a Cleveland dot plot of the overall variable importance scores (Section 5.4) using **gbm**'s summary method; the results are displayed in Figure 8.8:

```
vi <- summary(pbc2.gbm, n.trees = best.iter, plotit = FALSE)
dotchart(vi$rel.inf, labels = vi$var, xlab = "Variable importance")
```

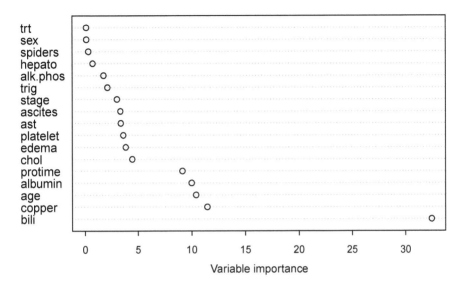

FIGURE 8.8: Variable importance plot from the boosted Cox PH model applied to the PBC data.

It looks as though serum bilirubin (mg/dl) (`bili`) is the most influential feature on the fitted model. We can easily investigate a handful of important features by constructing PDPs or ICE plots. In the next code chunk, I construct c-ICE plots (Section 6.2.3) for the top four features. The results are displayed in Figure 8.9; the average curve (i.e., partial dependence, albeit centered) is shown in red. Note that while **gbm** has built-in support for partial dependence using the recursion method (Section 8.6.1), it does not support ICE plots; hence, I'm using the brute force approach (`recursive = FALSE`) via the **pdp** package. The code essentially creates a list of plots, which is displayed in a 2-by-2 grid using the **gridExtra** package [Auguie, 2017]:

```
library(ggplot2)
library(pdp)

# Create list of c-ICE/PD plots for top 4 predictors
top4 <- c("bili", "copper", "age", "albumin")
pdps.top4 <- lapply(top4, FUN = function(x) {
  partial(pbc2.gbm, pred.var = x, check.class = FALSE,
          recursive = FALSE, n.trees = best.iter, ice = TRUE,
          center = TRUE, plot = TRUE, plot.engine = "ggplot2",
          rug = TRUE, alpha = 0.1) +
    ylab("Log hazard")  # change default y-axis label
```

```
})
```

```
# Display list of plots in a grid
gridExtra::grid.arrange(grobs = pdps.top4, nrow = 2)
```

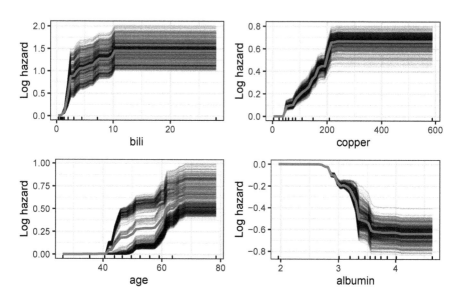

FIGURE 8.9: Main effect plots from a gradient boosted Cox PH model on the PBC data.

Each plot shows a relatively nonlinear monotonic effect on the predicted log hazard rate. For example, the predicted log hazard tends to increase with increasing serum bilirubin (mg/dl). The heterogeneity in the individual c-ICE curves for some of the predictors (e.g., **age**) suggests the presence of potential interaction effects. We can use Friedman's H-statistic (Section 6.2.2) to quantify the strength of potential pairwise interaction effects. The **gbm** function **interact.gbm()** can be used to obtain these statistics; see **?gbm::interact.gbm** for details. Here, I create a simple wrapper function, called **gbm.2way()**, which uses **interact.gbm()** with R's built-in **combn()** function to measure the interaction strength between all possible pairs of predictors:

```
gbm.2way <- function(object, data, var.names = object$var.names,
                     n.trees = object$n.trees) {
  var.pairs <- combn(var.names, m = 2, simplify = TRUE)
  h <- combn(var.names, m = 2, simplify = TRUE, FUN = function(x) {
    interact.gbm(object, data = data, i.var = x, n.trees = n.trees)
  })
  res <- as.data.frame(t(var.pairs))
  res$h <- h
```

```
    names(res) <- c("var1", "var2", "h")
    res[order(h, decreasing = TRUE), ]
}

# Compute H-statistics for all pairs of predictors
pbc2.h <- gbm.2way(pbc2.gbm, data = pbc2, n.trees = best.iter)
head(pbc2.h, n = 5)   # look at top 5

#>            var1      var2       h
#> 22          age      bili 0.1351
#> 29          age  platelet 0.0749
#> 99         bili   protime 0.0703
#> 114     albumin   protime 0.0701
#> 135    platelet     stage 0.0642
```

According to the H-statistic, the strongest interaction effects appear to occur between `age` and `bili`. This should not be surprising since the CTree fit to the same data in Section 3.5.3 showed `bili` as the first splitter, but `age` split below that, with `bili` again splitting below `age`, which suggest a potential interaction effect between the two. We can visualize this effect using a two-dimensional PDP, which I accomplish below using the **pdp** package. The result is displayed in Figure 8.10. Here, you can see that the effect of increasing serum bilirunbin (mg/dl) on the predicted log hazard is stronger for older individuals.

```
pd <- partial(pbc2.gbm, pred.var = c("bili", "age"), chull = TRUE,
              check.class = FALSE, n.trees = best.iter)
autoplot(pd, legend.title = "PD") +
  xlab("Serum bilirunbin (mg/dl)") +
  ylab("Age (years)")
```

Now suppose we wanted to understand why the model predicted a relatively high log hazard rate for a particular patient. One way to accomplish this, as we've already seen, is through Shapley-based feature contributions. To illustrate, I'll use the **fastshap** package to help understand which feature values contributed the most (and how) to the subject in the learning sample with the largest predicted log hazard.

To start, I'll grab the fitted values from the model (i.e., the predictions from the learning sample) and determine which subject had the highest predicted log hazard, before using **fastshap**'s `explain()` function to estimate the Shapley-based feature contributions (recall that we need to define and pass a prediction wrapper to `explain()` so it knows how to compute new predictions from the fitted model):

```
library(fastshap)

p <- predict(pbc2.gbm, newdata = pbc2, n.trees = best.iter)
max.id <- which.max(p)   # row ID highest predicted log hazard
```

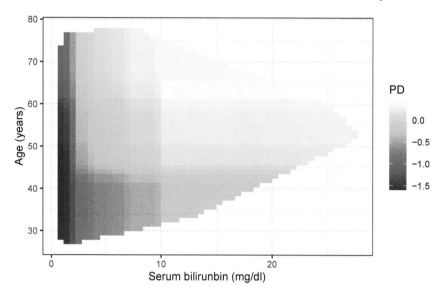

FIGURE 8.10: Partial dependence of log hazard on the joint value of `bili` and `age`.

```
# Define prediction wrapper for explain
pfun <- function(object, newdata) {
  predict(object, newdata = newdata, n.trees = best.iter)
}

# Estimate feature contributions for newx using 1,000 Monte Carlo reps
X <- pbc2[, pbc2.gbm$var.names]  # feature columns only
newx <- pbc2[max.id, pbc2.gbm$var.names]
set.seed(1408)  # for reproducibility
(ex <- explain(pbc2.gbm, X = X, nsim = 1000, pred_wrapper = pfun,
               newdata = newx))

#> # A tibble: 1 x 17
#>        trt     age      sex  ascites  hepato  spiders
#>      <dbl>  <dbl>    <dbl>    <dbl>   <dbl>    <dbl>
#> 1 -0.00168  0.313 -0.000342   0.282 -0.0175 -0.00492
#> # ... with 11 more variables: edema <dbl>, bili <dbl>,
#> #   chol <dbl>, albumin <dbl>, copper <dbl>,
#> #   alk.phos <dbl>, ast <dbl>, trig <dbl>,
#> #   platelet <dbl>, protime <dbl>, stage <dbl>
```

A great way to visualize such contributions is through a *waterfall chart*. I do so below using the R package **waterfall** [Howard, II, 2016]; the results are displayed in Figure 8.11. Here, we can see a bit more clearly the magnitude and direction of each feature value's contribution to the difference between the current predicted log hazard and the overall average baseline—in essence, it shows how this subject went from the average baseline log hazard of -1 to their much higher prediction of 2.701. Note that the `waterfallchart()` function produces a **lattice** graphic, which behaves differently than base R graphics; hence, I use the `ladd()` function from package **mosaic** [Pruim et al., 2021] to add specific details to the plot (e.g., text labels and additional reference lines).

```r
library(waterfall)

# Reshape Shapley values for plotting and include feature values
res <- data.frame("feature" = paste0(names(newx), "=", t(newx)),
                  "shapley.value" = t(ex))

# Waterfall chart of feature contributions
palette("Okabe-Ito")
waterfallchart(feature ~ shapley.value, data = res, origin = mean(p),
               summaryname = "f(x) - baseline", col = 2:3,
               xlab = "Log hazard")
mosaic::ladd(panel.abline(v = max(p), lty = 2, col = 1))
mosaic::ladd(panel.abline(v = mean(p), lty = 2, col = 1))
mosaic::ladd(panel.text(2.5, 8, labels = "f(x)", col = 1))
mosaic::ladd(panel.text(-0.55, 8, labels = "baseline", col = 1))
palette("default")
```

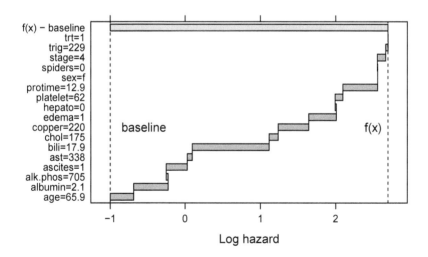

FIGURE 8.11: Waterfall chart showing the individual feature contributions to the subject with the highest predicted log hazard.

8.9.2 Example: probabilistic predictions with NGBoost (in Python)

In addition to **ngboost**, I'll require some additional packages: **pandas** [The Pandas Development Team, 2020], for reading and manipulating data; **numpy** [Harris et al., 2020], for basic numeric computations on arrays; and **scipy**, for basic statistical functions (like computing probabilities from distributions). In the Python chunk below, I load all the functionality I need before reading in the ALS data and splitting it into the same train/test sets used in the previous examples (note that the test set indicator used for splitting the data is included as a column named **testset**).

```python
import numpy as np
import pandas as pd
import scipy.stats
from ngboost import NGBRegressor
from ngboost.distns import Normal
```

```python
# Read in ALS data and split into train/test sets
url = "https://web.stanford.edu/~hastie/CASI_files/DATA/ALS.txt"
als = pd.read_csv(url, sep =" ")
als_trn = als[als["testset"] == False]
als_tst = als[als["testset"] == True]
X_trn = als_trn.drop(["testset", "dFRS"], axis=1)   # features only
X_tst = als_tst.drop(["testset", "dFRS"], axis=1)   # features only
```

Next, I initialize a `NGBRegressor` object, called **ngb**, with several parameters. In this example, I'll assume the distribution of **dFRS** conditional on a set of predictor values is normally distributed with some (unknown) mean and standard deviation, but note that **ngboost** supports several other distributions for regression and classification (yes, **ngboost** can also be used for classification tasks as well). As with ordinary GBMs, I also specify the number of base estimators and the learning rate. (The default base learner in **ngboost** is a depth-3 regression tree.)

```python
ngb = NGBRegressor(Dist=Normal, n_estimators=2000, learning_rate=0.01,
                   verbose_eval=0, random_state=1601)
```

Now I can fit the actual model by calling the `fit()` on our **ngb** object. Here, I provide the test set to the validation parameters and use early stopping to determine at which iteration (or tree) the procedure should stop. Here, the procedure will stop if there has been no improvement in the negative log-likelihood (the default scoring rule in **ngboost**) in five consecutive rounds.

```python
_ = ngb.fit(X_trn, Y=als_trn["dFRS"], X_val=X_tst,
            Y_val=als_tst["dFRS"], early_stopping_rounds=5)
```

```
#> == Early stopping achieved.
```

```
#> == Best iteration / VAL145 (val_loss=0.7495)
```

There are two prediction methods for `NGBRegressor` objects: `predict()`, which returns point predictions as one would expect from a standard regression model, and `pred_dist()`, which returns a distribution object representing the conditional distribution of $Y|x_i$ for each observation x_i in the scoring data set. First, let's compute point predictions for the entire data set and the corresponding MSE for comparison with our earlier GBM models on these data. Here, the results are rather similar in terms of accuracy.

```
pred = ngb.predict(X_tst)
np.mean(np.square(als_tst["dFRS"].values - pred))
```

```
#> 0.26773677346012037
```

Of more interest in probabilistic regression is an estimate of the conditional distribution itself, rather than some point estimate like the mean. In this case, I need estimates of the mean and standard deviation (since those parameters uniquely define a normal distribution). These are easily obtainable from the `params` component of the output from `pred_dist()`.

Below, I estimate these parameters for the first observation in the test set (x_1) and use them to estimate $\Pr(\text{dFRS} < 0|x_1)$. The estimated mean and standard deviation (i.e., location and scale parameters of the normal distribution) for each observation of interest can also be used to obtain prediction intervals.

```
dist = ngb.pred_dist(X_tst.head(1)).params
dist
```

```
#> {'loc': array([-0.4649206]), 'scale': array([0.43861847])}
```

```
scipy.stats.norm(dist["loc"][0], scale=dist["scale"][0]).cdf(0)
```

```
#> 0.8554199300698616
```

8.9.3 Example: post-processing GBMs with the LASSO

Recall from Section 5.5 that it is possible to reduce the number of base learners in a fitted ensemble via post-processing with the LASSO (hopefully without sacrificing accuracy). We saw this using the Ames housing data with a bagged tree ensemble and RF in Sections 5.5.1 and 7.9.2, respectively.

It may seem redundant to include another ISLE post-processing example, but there's a subtle difference that can be overlooked with GBMs: the initial fit $f_0(x)$ in Step 1) of Algorithm 8.1 essentially represents an offset.

To illustrate, I'll continue with the ALS example from Section 8.9.2°. Below, I read in the ALS data and split it into train/test sets using the provided **testset** indicator column; the -1 keeps the indicator column out of the train/test sets.

```
# Read in the ALS data
url <- "https://web.stanford.edu/~hastie/CASI_files/DATA/ALS.txt"
als <- read.table(url, header = TRUE)

# Split into train/test sets
trn <- als[!als$testset, -1]   # training data w/o testset column
tst <- als[als$testset, -1]    # test data w/o testset column
X.trn <- subset(trn, select = -dFRS)
X.tst <- subset(tst, select = -dFRS)
y.trn <- trn$dFRS
y.tst <- tst$dFRS
```

Next, I call on our `lsboost()` function to fit a GBM using $B = 1000$ depth-2 trees with a shrinkage factor of $\nu = 0.01$ and a subsampling rate of 50%. I also compute test predictions from each individual tree and compute the cumulative MSE as a function of the number of trees. (Warning, the model fitting here will take a few minutes.)

```
library(treemisc)

set.seed(1122)  # for reproducibility
lsb.fit <- lsboost(X.trn, y = y.trn, shrinkage = 0.01, ntree = 1000,
                   depth = 2, subsample = 0.5)

# Mean squared error function
mse <- function(y, yhat, na.rm = FALSE) {
  mean((y - yhat) ^ 2, na.rm = na.rm)
}

# Compute test MSE as a function of the number of trees
preds.tst <- predict(lsb.fit, newdata = X.tst, individual = TRUE)
mse.boost <- sapply(seq_len(ncol(preds.tst)), FUN = function(ntree) {
  # Only aggregate predictions from first B/ntree trees
  pred.ntree <- rowSums(preds.tst[, seq_len(ntree), drop = FALSE]) +
    lsb.fit$init  # don't forget to add on the initial fit/mean response
  mse(y.tst, yhat = pred.ntree)
})
```

Next, I'll call upon the `isle_post()` function from package **treemisc** to post-process our fitted GBM using the LASSO. There's one important difference between this example and the previous ones applied to bagging and RF: with

---

°A special thanks to Trevor Hastie for clarification and sharing code from Efron and Hastie [2016, pp. 346–347], which greatly helped in producing this analysis (which is a detailed recreation of their example) and building out **treemisc**'s `isle_post()` function.

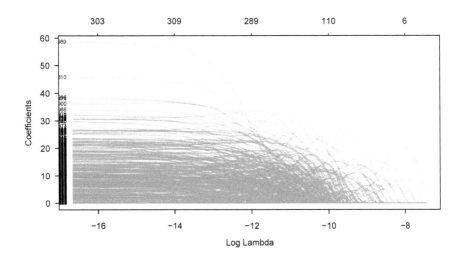

FIGURE 8.12: Coefficient path as a function of the L_1-penalty as λ varies. The top axis indicates the number of nonzero coefficients (i.e., number of trees) at the current value of λ. Here, we have one coefficient per tree in the full ensemble (the intercept is forced to be zero here).

boosting, we need to make sure we include the initial fit $f_0(\boldsymbol{x})$, which is stored in the "init" component of the output from `lsboost()`. Recall that for LS loss $f_0(\boldsymbol{x}) = \bar{y}$, the mean response in the training data. This can be done in one of two ways:

1) arbitrarily add it to the predictions from the first tree;

2) include it as an offset when fitting the LASSO and generating predictions.

In this example, I'll include the initial fit as an offset in the call to `isle_post()`.

The results are displayed in Figure 8.12 which contains the coefficient paths as a function of the L_1-penalty as λ varies. The top axis indicates the number of nonzero coefficients (i.e., number of trees) at the current value of λ. Here, the smallest test error for the LASSO-based post-processed GBM is 25.9% and corresponds to 84 trees; see Figure 8.13. The post-processing has significantly reduced the number of trees in this example resulting in a substantially more parsimonious model while maintaining accuracy. Sweet!

```
library(treemisc)

# Fit a LASSO model to the individual training predictions
preds.trn <- predict(lsb.fit, newdata = X.trn, individual = TRUE)
```

```
als.boost.post <- isle_post(preds.trn, y = y.trn, offset = lsb.fit$init,
                            newX = preds.tst, newy = y.tst,
                            family = "gaussian")

# Plot the coefficient paths from the LASSO model
plot(als.boost.post$lasso.fit, xvar = "lambda", las = 1, label = TRUE,
     col = adjustcolor("forestgreen", alpha.f = 0.3),
     cex.axis = 0.8, cex.lab = 0.8)

# Plot regularization path
palette("Okabe-Ito")
plot(mse.boost, type = "l", las = 1,
     ylim = range(c(mse.boost, als.boost.post$results$mse)),
     xlab = "Number of trees", ylab = "Test MSE")
lines(als.boost.post$results, col = 2)
abline(h = min(mse.boost), lty = 2)
abline(h = min(als.boost.post$results$mse), col = 2, lty = 2)
palette("default")
```

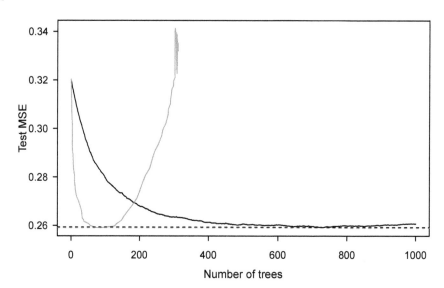

FIGURE 8.13: Test MSE as a function of the number of trees from the full GBM (black curve) and LASSO-based post-processed results (yellow curve). Here, we can see that by re-weighting the trees using an L_1 penalty, which enables some the trees to be dropped entirely, we end up with a smaller more parsimonious model without degrading performance on the test set.

8.9.4 Example: direct marketing campaigns with XGBoost

In this example, I'll revisit the bank marketing data I analyzed back in Section 7.9.5 using an RF in Spark. Here, I'll fit a GBM using the scalable XGBoost library and show the benefits to early stopping. For brevity, I'll omit the necessary code already shown in Section 7.9.5. To that end, assume we already have the full data set loaded into a data frame called `bank`.

Next, similar to the RF-based analysis, I'll clean up some of the columns and column names. Since XGBoost requires all the data to be numeric[p], we have to re-encode the categorical features. The binary variables I'll convert to 0/1, while categorical variables with higher cardinality will be transformed using one-hot encoding (OHE)[q]. First, I'll deal with the binary variables and column names:

```
names(bank) <- gsub("\\.", replacement = "_", x = names(bank))
bank$y <- ifelse(bank$y == "yes", 1, 0)
bank$contact <- ifelse(bank$contact == "telephone", 1, 0)
bank$duration <- NULL  # remove target leakage
```

Next, I'll deal with the one-hot encoding. There are several packages that can help with this (e.g., **caret**'s `dummyVars()` function); I'll do the transformation using pure **data.table**. The code below identifies the remaining categorical variables (`cats`) and uses **data.table**'s `melt()` and `dcast()` functions to handle the heavy lifting; the left hand side of the generated formula (`fo`) tells `dcast()` which variables to not OHE (i.e., the binary and non-categorical features):

```
bank$id <- seq_len(nrow(bank))  # need a unique row identifier
cats <- names(which(sapply(bank, FUN = class) == "character"))
lhs <- paste(setdiff(names(bank), cats), collapse = "+")
fo <- as.formula(paste(lhs, "~ variable + value"))
bank <- as.data.table(bank)  # coerce to data.table
bank.ohe <- dcast(melt(bank, id.vars = setdiff(names(bank), cats)),
                  formula = fo, fun = length)
bank$id <- bank.ohe$id <- NULL
```

Now that we have the data encoded properly for XGBoost, we can split the data into train/test sets using the same 50/50 split as before:

```
set.seed(1056)  # for reproducibility
trn.id <- caret::createDataPartition(bank.ohe$y, p = 0.5, list = FALSE)
bank.trn <- data.matrix(bank.ohe[trn.id, ])   # training data
bank.tst  <- data.matrix(bank.ohe[-trn.id, ])  # test data
```

---

[p]While XGBoost has limited (and experimental) support for categorical features, this does not seem to be accessible via the R interface, at least at the time of writing this book.

[q]Several of the categorical features are technically ordinal (e.g., `day_of_week`) and should probably be converted to integers.

XGBoost does not work with R data frames. The `xgb.train()` function, in particular, only accepts data as an `"xgb.DMatrix"` object. An XGBoost DMatrix is an internal data structure used by XGBoost, which is optimized for both memory efficiency and training speed; see `?xgboost::xgb.DMatrix` for details. We can create such an object using the **xgboost** function `xgb.DMatrix()` (note that I separate the predictors and response in the calls to `xgb.DMatrix()`):

```
library(xgboost)

xnames <- setdiff(names(bank.ohe), "y")
dm.trn <- xgb.DMatrix(bank.trn[, xnames], label = bank.trn[, "y"])
dm.tst <- xgb.DMatrix(bank.tst[, xnames], label = bank.tst[, "y"])
```

XGBoost has a lot of parameters, so it can be helpful to construct a list of such for use when calling the fitting function `xgb.train()`. Below, I create a list containing several boosting and tree-specific parameters:

```
params <- list(
  eta = 0.01,  # shrinkage/learning rate
  max_depth = 3,
  subsample = 0.5,
  objective = "binary:logistic",  # for predicted probabilities
  eval_metric = "rmse",  # square root of Brier score
  nthread = 8
)
```

Finally, I can fit an XGBoost model. I'll fit two in total, one without early stopping and one with, starting with the no early stopping version below. But first, I'll define a "watch list," which is just a named list of data sets to use for evaluating model performance after each iteration that we can use to determine the optimal number of trees (k-fold cross-validation could also be used via **xgboost**'s `xgb.cv()` function):

```
watch.list <- list(train = dm.trn, eval = dm.tst)
```

```
# Train an XGBoost model without early stopping
set.seed(1100)  # for reproducibility
bank.xgb.1 <-
  xgb.train(params, data = dm.trn, nrounds = 3000, verbose = 0,
            watchlist = watch.list)
(best.iter <- which.min(bank.xgb.1$evaluation_log$eval_rmse))
```

```
#> [1] 1296
```

Out of 3,000 total iterations, we really only needed to build 1,296 trees, which can be expensive for large data sets (regardless of which fancy scalable implementation you use). While XGBoost is incredibly efficient, it is still wasteful to fit more trees than potentially necessary. To that end, I can turn on early stopping (Section 8.3.1.1) to halt performance once it detects the potential for overfitting. In XGBoost, early stopping will halt training if model performance has not improved for a specified number of iterations (`early_stopping_rounds`).

In the code chunk below, I fit the same model (random seed and all), but tell XGBoost to stop the training process if the performance on the test set (as specified in the watch list) has not improved for 150 consecutive iterations (5% of the total number of requested iterations)[r]:

```r
set.seed(1100)  # for reproducibility
(bank.xgb.2 <-
   xgb.train(params, data = dm.trn, nrounds = 3000, verbose = 0,
             watchlist = watch.list, early_stopping_rounds = 150))

#> ##### xgb.Booster
#> raw: 1.7 Mb
#> xgb.attributes:
#>   best_iteration, best_msg, best_ntreelimit, best_score, niter
#> niter: 1445
#> best_ntreelimit : 1296
#> best_iteration : 1296
#> best_score : 0.274
#> best_msg : [1296] train-rmse:0.274552 eval-rmse:0.273857
```

In this case, using early stopping resulted in the same optimal number of trees (e.g., 1,296), but only required 1,446 boosting iterations (or trees) in total, a decent savings in terms of both computation time and storage space (1.7 Mb for early stopping compared to 3.6 Mb for the full model)! The overall training results are displayed in Figure 8.14 below.

```r
palette("Okabe-Ito")
plot(bank.xgb.1$evaluation_log[, c(1, 2)], type = "l",
     xlab = "Number of trees",
     ylab = "RMSE (square root of Brier score)")
lines(bank.xgb.1$evaluation_log[, c(1, 3)], type = "l", col = 2)
abline(v = best.iter, col = 2, lty = 2)
abline(v = bank.xgb.2$niter, col = 3, lty = 2)
legend("topright", legend = c("Train", "Test"), inset = 0.01, bty = "n",
       lty = 1, col = 1:2)
palette("default")
```

Unlike R's **gbm** package or Python's **sklearn.inspection** module, XGBoost does not support the recursion method for fast PD and ICE plots. However, XGBoost does support Tree SHAP (Section 6.3.2.1), which we can use to construct Shapley-based variable importance and dependence plots. In R, these can be obtained at prediction time by specifying `predcontrib = TRUE` in the call to `predict()`.

---

[r] I've seen several online blog posts suggest a value of `early_stopping_rounds` equal to 10% of the total number of requested iterations, but no evidence or citations as to why.

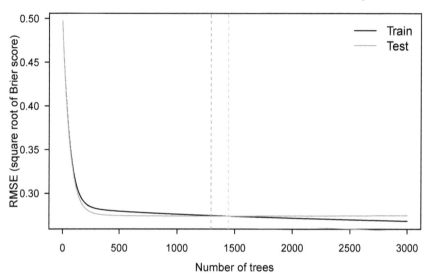

FIGURE 8.14: RMSE (essentially, the square root of the Brier score) from
an XGBoost model fit to the bank marketing data. According to the independent test set (yellow curve), the optimal number of trees is 1,296 (vertical dashed yellow line). Early stopping, which reached the same conclusion,
would've stopped training at 1,446 trees (vertical dashed blue line), which
roughly halves the training time in this case.

Below, I compute Tree SHAP values for the entire training set and use that
to form Shapley-based variable importance scores; here, I'll follow Lundberg
et al. [2020] and the **shap** module's approach by computing the mean absolute Shapley value for each column. (Note that it is not necessary to use the
entire learning sample for doing this, and a large enough subsample should
often suffice, especially when dealing with hundreds of thousands or millions
of records.) A dot plot of the top 10 Shapley-based importance scores is displayed in Figure 8.15. Note that I need to specify the optimal number of trees
(`ntreelimit = best.iter`) when calling `predict()`:

```
shap.trn <- predict(bank.xgb, newdata = dm.trn, ntreelimit = best.iter,
                    predcontrib = TRUE, approxcontrib = FALSE)
shap.trn <- shap.trn[, -which(colnames(shap.trn) == "BIAS")]

# Shapley-based variable importance
shap.vi <- colMeans(abs(shap.trn))
shap.vi <- shap.vi[order(shap.vi, decreasing = TRUE)]
dotchart(shap.vi[1:10], xlab = "mean(|SHAP value|)", pch = 19)
```

For comparison, I computed the usual aggregated tree-based importance
scores from the boosted model; this is what's given by the **Gain** column in the

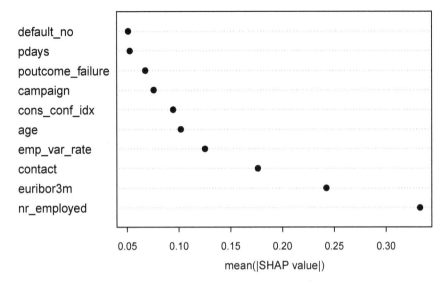

FIGURE 8.15: Shapley-based variable importance scores from an XGBoost model fit to the bank marketing data.

output below (see **?xgboost::xgb.importance** for details on the other two measures **Cover** and **Frequency**). Below, I display the top six features (note that tree indexing in XGBoost, whether using R or not, is zero-based):

```
head(xgb.importance(model = bank.xgb, trees = 0:(best.iter - 1)))
```

```
#>                 Feature    Gain   Cover Frequency
#> 1:           nr_employed 0.3689 0.1012    0.0395
#> 2:            euribor3m 0.1531 0.1693    0.1801
#> 3:         cons_conf_idx 0.0704 0.0960    0.0651
#> 4:                  age 0.0665 0.1219    0.1471
#> 5:                pdays 0.0400 0.0420    0.0363
#> 6: poutcome_success 0.0334 0.0178    0.0137
```

To get a sense of a predictor's effect on the model output, in terms of Shapley values, we can construct a Shapley dependence plot; this is nothing more than a plot of a feature's Shapley values against the raw feature values in a particular sample. Below, I construct a Shapley dependence plot for **age** (Figure 8.16); a nonparametic smooth is also displayed (yellow curve). Here, you can see that individuals in the age range 30–50 (roughly) generally correspond to negative Shapley values, meaning they tend to drive the predicted probability of subscribing towards the average baseline. Whereas younger and older individuals tend to have **age** contribute a positive effect. The non-constant variance in the scatter plot suggests potential interaction effects, which could be explored further by using another feature (or features) to help color the plot.

```
shap.age <- data.frame("age" = bank.trn[, "age"],
                       "shap" = shap.trn[, "age"])

# Shapley dependence plot for age
cols <- palette.colors(3, palette = "Okabe-Ito")
ggplot(shap.age, aes(age, shap)) +
  geom_point(alpha = 0.1) +
  geom_smooth(se = FALSE, color = cols[2]) +
  geom_hline(yintercept = 0, linetype = "dashed", color = cols[3]) +
  xlab("Age (years)") + ylab("Shapley value")
```

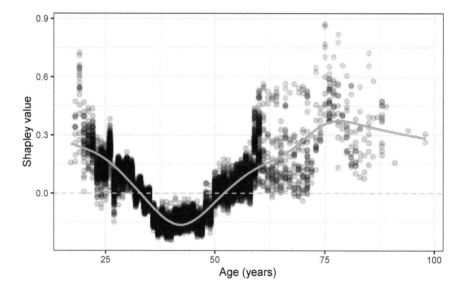

FIGURE 8.16: Shapley dependence plot for `age` from the XGBoost model fit to the bank marketing data; a nonparametric smooth is also shown (yellow curve). Any point below the horizontal dashed blue line corresponds to a negative contribution to the predicted outcome).

8.10   Final thoughts

GBMs are a powerful class of machine learning algorithms that can achieve state-of-the-art performance, provided you train them properly. Due to the existence of efficient libraries (like XGBoost and Microsoft's LightGBM) and the shallower nature of the individual trees, GBMs can also scale incredibly well; see, for example, Pafka [2021]. For these reasons, GBMs are quite popular in applied practice and are often used in the winning entries for many supervised learning competitions with tabular data sets. Just keep in mind that, unlike RFs, GBMs are quite sensitive to several tuning parameters (e.g., the learning rate and number of boosting iterations), and these models should be carefully tuned (ideally, with some form of early stopping, especially if you're working with a large learning sample, using a fairly small learning rate with a large number of boosting iterations, and/or tuning lots of parameters).

Bibliography

Hongshik Ahn and Wei-Yin Loh. Tree-structured proportional hazards regression modeling. *Biometrics*, 50(2):471–485, 1994.

Alfaro, Esteban; Gamez, Matias, Garcia, and Noelia; with contributions from Li Guo. *adabag: Applies Multiclass AdaBoost.M1, SAMME and Bagging*, 2018. URL `https://CRAN.R-project.org/package=adabag`. R package version 4.2.

André Altmann, Laura Toloşi, Oliver Sander, and Thomas Lengauer. Permutation importance: a corrected feature importance measure. *Bioinformatics*, 26(10):1340–1347, 2010. ISSN 1367-4803. doi: 10.1093/bioinformatics/btq134. URL `https://doi.org/10.1093/bioinformatics/btq134`.

Elaine Angelino, Nicholas Larus-Stone, Daniel Alabi, Margo Seltzer, and Cynthia Rudin. Learning certifiably optimal rule lists for categorical data. *Journal of Machine Learning Research*, 18(234):1–78, 2018. URL `http://jmlr.org/papers/v18/17-716.html`.

Dan Apley. *ALEPlot: Accumulated Local Effects (ALE) Plots and Partial Dependence (PD) Plots*, 2018. URL `https://CRAN.R-project.org/package=ALEPlot`. R package version 1.1.

Dan Apley and Jingyu Zhu. Visualizing the effects of predictor variables in black box supervised learning models. *Journal of the Royal Statistical Society Series B*, 82(4):1059–1086, September 2020. doi: 10.1111/rssb.12377. URL `https://ideas.repec.org/a/bla/jorssb/v82y2020i4p1059-1086.html`.

Susan Athey, Julie Tibshirani, and Stefan Wager. Generalized random forests. *The Annals of Statistics*, 47(2):1148–1178, 2019. doi: 10.1214/18-AOS1709. URL `https://doi.org/10.1214/18-AOS1709`.

Baptiste Auguie. *gridExtra: Miscellaneous Functions for "Grid" Graphics*, 2017. URL `https://CRAN.R-project.org/package=gridExtra`. R package version 2.3.

Peter C. Austin and Ewout W. Steyerberg. Graphical assessment of internal and external calibration of logistic regression models by using loess smoothers. *Statistics in Medicine*, 33(3):517–535, 2014. doi: 10.1002/sim.5941. URL `https://doi.org/10.1002/sim.5941`.

Michel Ballings and Dirk Van den Poel. *rotationForest: Fit and Deploy Rotation Forest Models*, 2017. URL https://CRAN.R-project.org/package=rotationForest. R package version 0.1.3.

Richard A. Berk. *Statistical Learning from a Regression Perspective*. Springer Series in Statistics. Springer New York, 2008. ISBN 9780387775012.

Gérard Biau, Luc Devroye, and Gábor Lugosi. Consistency of random forests and other averaging classifiers. *Journal of Machine Learning Research*, 9 (66):2015–2033, 2008. URL http://jmlr.org/papers/v9/biau08a.html.

Przemyslaw Biecek and Hubert Baniecki. *ingredients: Effects and Importances of Model Ingredients*, 2021. URL https://CRAN.R-project.org/package=ingredients. R package version 2.2.0.

Przemyslaw Biecek and Tomasz Burzykowski. *Explanatory Model Analysis*. Chapman and Hall/CRC, New York, 2021. ISBN 9780367135591. URL https://pbiecek.github.io/ema/.

Przemyslaw Biecek, Alicja Gosiewska, Hubert Baniecki, and Adam Izdebski. *iBreakDown: Model Agnostic Instance Level Variable Attributions*, 2021. URL https://CRAN.R-project.org/package=iBreakDown. R package version 2.0.1.

Rico Blaser and Piotr Fryzlewicz. Random rotation ensembles. *Journal of Machine Learning Research*, 17(4):1–26, 2016. URL http://jmlr.org/papers/v17/blaser16a.html.

Bradley Boehmke and Brandon Greenwell. *Hands-On Machine Learning with R*. Chapman & Hall/CRC the R series. CRC Press, 2020. ISBN 9781138495685.

Leo Breiman. Bagging predictors. *Machine Learning*, 24(2):123–140, 1996a. URL https://doi.org/10.1007/BF00058655.

Leo Breiman. Heuristics of instability and stabilization in model selection. *The Annals of Statistics*, 24(6):2350–2383, 1996b.

Leo Breiman. Technical note: Some properties of splitting criteria. *Machine Learning*, 24:41–47, 1996c.

Leo Breiman. Pasting small votes for classification in large databases and online. *Machine Learning*, 36(1):85–103, 1999. doi: 10.1023/A:1007563306331. URL https://doi.org/10.1023/A:1007563306331.

Leo Breiman. Random forests. *Machine Learning*, 45(1):5–32, 2001. URL https://doi.org/10.1023/A:1010933404324.

Leo Breiman. Manual on setting up, using, and understanding random forests v3.1. Technical report, 2002. URL https://www.stat.berkeley.edu/~breiman/Using_random_forests_V3.1.pdf.

Leo Breiman, Jerome H. Friedman, Charles J. Stone, and Richard A. Olshen. *Classification and Regression Trees*. Taylor & Francis, 1984. ISBN 9780412048418.

Leo Breiman, Adele Cutler, Andy Liaw, and Matthew Wiener. *randomForest: Breiman and Cutler's Random Forests for Classification and Regression*, 2018. URL `https://www.stat.berkeley.edu/~breiman/RandomForests/`. R package version 4.6-14.

Peter Bühlmann and Torsten Hothorn. Boosting Algorithms: Regularization, Prediction and Model Fitting. *Statistical Science*, 22(4):477–505, 2007. doi: 10.1214/07-STS242. URL `https://doi.org/10.1214/07-STS242`.

Tom Bylander. Estimating generalization error on two-class datasets using out-of-bag estimates. *Machine Learning*, 48(1):287–297, 2002. doi: 10.1023/A:1013964023376. URL `https://doi.org/10.1023/A:1013964023376`.

Angelo Canty and Brian Ripley. *boot: Bootstrap Functions (Originally by Angelo Canty for S)*, 2021. URL `https://CRAN.R-project.org/package=boot`. R package version 1.3-28.

M. Carlisle. Racist data destruction?, May 2019. URL `https://medium.com/@docintangible/racist-data-destruction-113e3eff54a8`.

Hugh Chen, Joseph D. Janizek, Scott Lundberg, and Su-In Lee. True to the model or true to the data?, 2020. URL `https://arxiv.org/abs/2006.16234`.

Tianqi Chen and Carlos Guestrin. Xgboost: A scalable tree boosting system. In *Proceedings of the 22nd ACM SIGKDD International Conference on Knowledge Discovery and Data Mining*, KDD '16, pages 785–794, New York, NY, USA, 2016. Association for Computing Machinery. ISBN 9781450342322. doi: 10.1145/2939672.2939785. URL `https://doi.org/10.1145/2939672.2939785`.

Tianqi Chen, Tong He, Michael Benesty, Vadim Khotilovich, Yuan Tang, Hyunsu Cho, Kailong Chen, Rory Mitchell, Ignacio Cano, Tianyi Zhou, Mu Li, Junyuan Xie, Min Lin, Yifeng Geng, and Yutian Li. *xgboost: Extreme Gradient Boosting*, 2021. URL `https://github.com/dmlc/xgboost`. R package version 1.5.0.2.

Hugh A. Chipman, Edward I. George, and Robert E. McCulloch. BART: Bayesian additive regression trees. *The Annals of Applied Statistics*, 4(1): 266–298, 2010. doi: 10.1214/09-AOAS285. URL `https://doi.org/10.1214/09-AOAS285`.

Philip A. Chou. Optimal partitioning for classification and regression trees. *IEEE Transactions on Pattern Analysis and Machine Intelligence*, 13(04): 340–354, 1991. ISSN 1939-3539. doi: 10.1109/34.88569.

William S. Cleveland. Robust locally weighted regression and smoothing scatterplots. *Journal of the American Statistical Association*, 74(368):829–836, 1979. doi: 10.1080/01621459.1979.10481038. URL https://doi.org/10.1080/01621459.1979.10481038.

David Cortes. *isotree: Isolation-Based Outlier Detection*, 2022. URL https://github.com/david-cortes/isotree. R package version 0.5.5.

Paulo Cortez, António Cerdeira, Fernando Almeida, Telmo Matos, and José Reis. Modeling wine preferences by data mining from physicochemical properties. *Decision Support Systems*, 47(4):547–553, 2009. ISSN 0167-9236. doi: https://doi.org/10.1016/j.dss.2009.05.016. URL http://www.sciencedirect.com/science/article/pii/S0167923609001377.

Mark Culp, Kjell Johnson, and George Michailidis. *ada: The R Package Ada for Stochastic Boosting*, 2016. URL https://CRAN.R-project.org/package=ada. R package version 2.0-5.

Adele Cutler. Remembering Leo Breiman. *The Annals of Applied Statistics*, 4(4):1621–1633, 2010. doi: 10.1214/10-AOAS427. URL https://doi.org/10.1214/10-AOAS427.

Natalia da Silva, Dianne Cook, and Eun-Kyung Lee. A projection pursuit forest algorithm for supervised classification. 0(0):1–13, 2021a. doi: 10.1080/10618600.2020.1870480. URL https://doi.org/10.1080/10618600.2020.1870480.

Natalia da Silva, Eun-Kyung Lee, and Di Cook. *PPforest: Projection Pursuit Classification Forest*, 2021b. URL https://github.com/natydasilva/PPforest. R package version 0.1.2.

Data Mining Group. Predictive model markup language, 2014. URL http://www.dmg.org/. Version 4.2.

Jesse Davis and Mark Goadrich. The relationship between precision-recall and roc curves. In *Proceedings of the 23rd International Conference on Machine Learning*, ICML '06, pages 233–240, New York, NY, USA, 2006. Association for Computing Machinery. ISBN 1595933832. doi: 10.1145/1143844.1143874. URL https://doi.org/10.1145/1143844.1143874.

Anthony C. Davison and David V. Hinkley. *Bootstrap Methods and Their Application*. Cambridge Series in Statistical and Probabilistic Mathematics. Cambridge University Press, 1997. ISBN 9780521574716.

Dean De Cock. Ames, Iowa: Alternative to the Boston housing data as an end of semester regression project. *Journal of Statistics Education*, 19(3): null, 2011. doi: 10.1080/10691898.2011.11889627. URL https://doi.org/10.1080/10691898.2011.11889627.

Luc Devroye, László Györfi, and Gábor Lugosi. *A Probabilistic Theory of Pattern Recognition*. Stochastic Modelling and Applied Probability. Springer New York, 1997. ISBN 9780387946184.

Stephan Dlugosz. *rpart.LAD: Least Absolute Deviation Regression Trees*, 2020. URL `https://CRAN.R-project.org/package=rpart.LAD`. R package version 0.1.2.

Rémi Domingues, Maurizio Filippone, Pietro Michiardi, and Jihane Zouaoui. A comparative evaluation of outlier detection algorithms: Experiments and analyses. *Pattern Recognition*, 74:406–421, 2018. ISSN 0031-3203. doi: https://doi.org/10.1016/j.patcog.2017.09.037. URL `https://doi.org/10.1016/j.patcog.2017.09.037`.

Lisa Doove, Stef Van Buuren, and Elise Dusseldorp. Recursive partitioning for missing data imputation in the presence of interaction effects. *Computational Statistics & Data Analysis*, 72:92–104, 2014. ISSN 0167-9473. doi: 10.1016/j.csda.2013.10.025. URL `https://doi.org/10.1016/j.csda.2013.10.025`.

Anna Veronika Dorogush, Vasily Ershov, and Andrey Gulin. Catboost: gradient boosting with categorical features support, 2018. URL `https://arxiv.org/abs/1810.11363`.

Matt Dowle and Arun Srinivasan. *data.table: Extension of 'data.frame'*, 2021. URL `https://CRAN.R-project.org/package=data.table`. R package version 1.14.2.

Tony Duan, Anand Avati, Daisy Yi Ding, Khanh K. Thai, Sanjay Basu, Andrew Y. Ng, and Alejandro Schuler. Ngboost: Natural gradient boosting for probabilistic prediction, 2020. URL `https://arxiv.org/abs/1910.03225`.

Bradley Efron and Trevor Hastie. *Computer Age Statistical Inference: Algorithms, Evidence, and Data Science*. Institute of Mathematical Statistics Monographs. Cambridge University Press, 2016. doi: 10.1017/CBO9781316576533.

Ad Feelders. Handling missing data in trees: Surrogate splits or statistical imputation? In *Principles of Data Mining and Knowledge Discovery*, pages 329–334, 03 2000. ISBN 978-3-540-66490-1. doi: 10.1007/978-3-540-48247-5_38. URL `https://doi.org/10.1007/978-3-540-48247-5_38`.

Aaron Fisher, Cynthia Rudin, and Francesca Dominici. All models are wrong, but many are useful: Learning a variable's importance by studying an entire class of prediction models simultaneously. 2018. doi: 10.48550/ARXIV.1801.01489. URL `https://arxiv.org/abs/1801.01489`.

Thomas R. Fleming and David P. Harrington. *Counting Processes and Survival Analysis*. Wiley Series in Probability and Statistics. Wiley, 1991. ISBN 9780471522188.

Bernhard Flury and Hans Riedwyl. *Multivariate Statistics: A Practical Approach*. Statistics texts. Springer Netherlands, 1988. ISBN 9780412300301.

Christopher Flynn. *Python bindings for C++ ranger random forests*, 2021. URL https://github.com/crflynn/skranger. Python package version 0.3.2.

Yoav Freund and Robert E. Schapire. Experiments with a new boosting algorithm. In *Proceedings of the Thirteenth International Conference on Machine Learning*, ICML'96, pages 148–156, San Francisco, CA, USA, 1996. Morgan Kaufmann Publishers Inc. ISBN 1558604197.

Peter W. Frey and David J. Slate. Letter recognition using Holland-style adaptive classifiers. *Machine Learning*, 6(2):161–182, 1991. URL https://doi.org/10.1007/BF00114162.

Jerome Friedman, Trevor Hastie, Rob Tibshirani, Balasubramanian Narasimhan, Kenneth Tay, Noah Simon, and James Yang. *glmnet: Lasso and Elastic-Net Regularized Generalized Linear Models*, 2021. URL https://CRAN.R-project.org/package=glmnet. R package version 4.1-3.

Jerome H. Friedman. Multivariate adaptive regression splines. *The Annals of Statistics*, 19(1):1–67, 03 1991. doi: 10.1214/aos/1176347963. URL https://doi.org/10.1214/aos/1176347963.

Jerome H. Friedman. Greedy function approximation: A gradient boosting machine. *The Annals of Statistics*, 29(5):1189–1232, 2001. URL https://doi.org/10.1214/aos/1013203451.

Jerome H. Friedman. Stochastic gradient boosting. *Computational Statistics & Data Analysis*, 38(4):367–378, 2002. ISSN 0167-9473. doi: https://doi.org/10.1016/S0167-9473(01)00065-2. URL https://doi.org/10.1016/S0167-9473(01)00065-2.

Jerome H. Friedman and Peter Hall. On bagging and nonlinear estimation. *Journal of Statistical Planning and Inference*, 137(3):669–683, 2007. ISSN 0378-3758. doi: https://doi.org/10.1016/j.jspi.2006.06.002. URL http://www.sciencedirect.com/science/article/pii/S0378375806001339. Special Issue on Nonparametric Statistics and Related Topics: In honor of M.L. Puri.

Jerome H. Friedman and Bogdan E. Popescu. Importance sampled learning ensembles. Technical report, Stanford University, Department of Statistics, 2003. URL https://statweb.stanford.edu/~jhf/ftp/isle.pdf.

Jerome H. Friedman and Bogdan E. Popescu. Predictive learning via rule ensembles. *The Annals of Applied Statistics*, 2(3):916–954, 2008. ISSN 19326157. URL https://doi.org/10.2307/30245114.

Jerome H. Friedman and John W. Tukey. A projection pursuit algorithm for exploratory data analysis. *IEEE Transactions on Computers*, C-23(9): 881–890, 1974. doi: 10.1109/T-C.1974.224051. URL https://doi.org/10.1109/T-C.1974.224051.

Jerome H. Friedman, Trevor Hastie, and Robert Tibshirani. Additive logistic regression: a statistical view of boosting (With discussion and a rejoinder by the authors). *The Annals of Statistics*, 28(2):337–407, 2000. doi: 10.1214/aos/1016218223. URL https://doi.org/10.1214/aos/1016218223.

Giuliano Galimberti, Gabriele Soffritti, and Matteo Di Maso. *rpartScore: Classification trees for ordinal responses*, 2012. URL https://CRAN.R-project.org/package=rpartScore. R package version 1.0-1.

Aurélien Géron. *Hands-On Machine Learning with Scikit-Learn, Keras & TensorFlow: Concepts, Tools, and Techniques to Build Intelligent Systems*. O'Reilly Media, Inc., 2nd edition, 2019. ISBN 9781492032649.

Pierre Geurts, Damien Ernst, and Louis Wehenkel. Extremely randomized trees. *Machine Learning*, 63(1):3–42, 2006. doi: 10.1007/s10994-006-6226-1. URL https://doi.org/10.1007/s10994-006-6226-1.

Anil K. Ghosh, Probal Chaudhuri, and Debasis Sengupta. Classification using kernel density estimates. *Technometrics*, 48(1):120–132, 2006. URL https://doi.org/10.1198/004017005000000391.

Navdeep Gill, Patrick Hall, Kim Montgomery, and Nicholas Schmidt. A responsible machine learning workflow with focus on interpretable models, post-hoc explanation, and discrimination testing. *Information*, 11 (3), 2020. ISSN 2078-2489. doi: 10.3390/info11030137. URL https://doi.org/10.3390/info11030137.

Alex Goldstein, Adam Kapelner, Justin Bleich, and Emil Pitkin. Peeking inside the black box: Visualizing statistical learning with plots of individual conditional expectation. *Journal of Computational and Graphical Statistics*, 24(1):44–65, 2015. URL https://doi.org/10.1080/10618600.2014.907095.

Alex Goldstein, Adam Kapelner, and Justin Bleich. *ICEbox: Individual Conditional Expectation Plot Toolbox*, 2017. URL https://CRAN.R-project.org/package=ICEbox. R package version 1.1.2.

Brandon M. Greenwell. pdp: An R package for constructing partial dependence plots. *The R Journal*, 9(1):421–436, 2017. URL https://journal.r-project.org/archive/2017/RJ-2017-016/index.html.

Brandon M. Greenwell. *fastshap: Fast Approximate Shapley Values*, 2021a. URL https://github.com/bgreenwell/fastshap. R package version 0.0.7.

Brandon M. Greenwell. *pdp: Partial Dependence Plots*, 2021b. URL `https://github.com/bgreenwell/pdp`. R package version 0.7.0.9000.

Brandon M. Greenwell. *treemisc: Data Sets and Functions to Accompany "Tree-Based Methods for Statistical Learning in R"*, 2021c. R package version 0.0.1.

Brandon M. Greenwell and Bradley C. Boehmke. Variable importance plots— an introduction to the vip package. *The R Journal*, 12(1):343–366, 2020. URL `https://doi.org/10.32614/RJ-2020-013`.

Brandon M. Greenwell, Bradley C. Boehmke, and Andrew J. McCarthy. A simple and effective model-based variable importance measure, 2018. URL `https://arxiv.org/abs/1805.04755`.

Brandon M. Greenwell, Brad Boehmke, and Bernie Gray. *vip: Variable Importance Plots*, 2021a. URL `https://github.com/koalaverse/vip/`. R package version 0.3.3.

Brandon M. Greenwell, Bradley Boehmke, Jay Cunningham, and GBM Developers. *gbm: Generalized Boosted Regression Models*, 2021b. URL `https://github.com/gbm-developers/gbm`. R package version 2.1.5.

Alexander Hapfelmeier, Torsten Hothorn, Kurt Ulm, and Carolin Strobl. A new variable importance measure for random forests with missing data. *Statistics and Computing*, 24(1):21–34, 2014.

Sahand Hariri, Matias Carrasco Kind, and Robert J. Brunner. Extended isolation forest. *IEEE Transactions on Knowledge and Data Engineering*, 33(4):1479–1489, 2021. doi: 10.1109/TKDE.2019.2947676. URL `https://doi.org/10.1109/TKDE.2019.2947676`.

Frank Harrell. *Regression Modeling Strategies*. Springer Series in Statistics. Springer International Publishing, 2nd edition, 2015. ISBN 978-3-319-19424-0.

Frank E Harrell, Jr. *Hmisc: Harrell Miscellaneous*, 2021. URL `https://hbiostat.org/R/Hmisc/`. R package version 4.6-0.

Frank E. Harrell, Jr. *rms: Regression Modeling Strategies*, 2021. URL `https://CRAN.R-project.org/package=rms`. R package version 6.2-0.

Charles R. Harris, K. Jarrod Millman, Stéfan J. van der Walt, Ralf Gommers, Pauli Virtanen, David Cournapeau, Eric Wieser, Julian Taylor, Sebastian Berg, Nathaniel J. Smith, Robert Kern, Matti Picus, Stephan Hoyer, Marten H. van Kerkwijk, Matthew Brett, Allan Haldane, Jaime Fernández del Río, Mark Wiebe, Pearu Peterson, Pierre Gérard-Marchant, Kevin Sheppard, Tyler Reddy, Warren Weckesser, Hameer Abbasi, Christoph Gohlke, and Travis E. Oliphant. Array programming with NumPy. *Nature*, 585

(7825):357–362, September 2020. doi: 10.1038/s41586-020-2649-2. URL `https://doi.org/10.1038/s41586-020-2649-2`.

David Harrison and Daniel L. Rubinfeld. Hedonic housing prices and the demand for clean air. *Journal of Environmental Economics and Management*, 5(1):81–102, 1978. URL `https://doi.org/10.1016/0095-0696(78)90006-2`.

Trevor Hastie, Robert Tibshirani, and Jerome H. Friedman. *The Elements of Statistical Learning: Data Mining, Inference, and Prediction, Second Edition*. Springer Series in Statistics. Springer-Verlag, 2009. URL `https://web.stanford.edu/~hastie/ElemStatLearn/`.

Paul Hendricks. *titanic: Titanic Passenger Survival Data Set*, 2015. URL `https://github.com/paulhendricks/titanic`. R package version 0.1.0.

Lionel Henry and Hadley Wickham. *purrr: Functional Programming Tools*, 2020. URL `https://CRAN.R-project.org/package=purrr`. R package version 0.3.4.

Tin Kam Ho. Random decision forests. In *Proceedings of 3rd International Conference on Document Analysis and Recognition*, volume 1, pages 278–282, 1995. doi: 10.1109/ICDAR.1995.598994. URL `https://doi.org/10.1007/s10115-013-0679-x`.

Giles Hooker. Generalized functional anova diagnostics for high-dimensional functions of dependent variables. *Journal of Computational and Graphical Statistics*, 16(3):709–732, 2007. URL `https://doi.org/10.1198/106186007X237892`.

Giles Hooker, Lucas Mentch, and Siyu Zhou. There is no free variable importance, 2019. URL `https://arxiv.org/abs/1905.03151`.

Kurt Hornik. *RWeka: R/Weka Interface*, 2021. URL `https://CRAN.R-project.org/package=RWeka`. R package version 0.4-44.

Allison Horst, Alison Hill, and Kristen Gorman. *palmerpenguins: Palmer Archipelago (Antarctica) Penguin Data*, 2020. URL `https://CRAN.R-project.org/package=palmerpenguins`. R package version 0.1.0.

Torsten Hothorn and Achim Zeileis. *partykit: A Toolkit for Recursive Partytioning*, 2021. URL `http://partykit.r-forge.r-project.org/partykit/`. R package version 1.2-15.

Torsten Hothorn, Berthold Lausen, Axel Benner, and Martin Radespiel-Tröger. Bagging survival trees. *Statistics in Medicine*, 23(1):77–91, 2004. doi: 10.1002/sim.1593. URL `https://doi.org/10.1002/sim.1593`.

Torsten Hothorn, Peter Buehlmann, Sandrine Dudoit, Annette Molinaro, and Mark Van Der Laan. Survival ensembles. *Biostatistics*, 7(3):355–373, 2006a.

Torsten Hothorn, Kurt Hornik, Mark A. van de Wiel, and Achim Zeileis. A lego system for conditional inference. *The American Statistician*, 60(3): 257–263, 2006b. URL https://doi.org/10.1198/000313006X118430.

Torsten Hothorn, Kurt Hornik, and Achim Zeileis. Unbiased recursive partitioning: A conditional inference framework. *Journal of Computational and Graphical Statistics*, 15(3):651–674, 2006c.

Torsten Hothorn, Peter Buehlmann, Thomas Kneib, Matthias Schmid, and Benjamin Hofner. *mboost: Model-Based Boosting*, 2021a. URL https://github.com/boost-R/mboost. R package version 2.9-5.

Torsten Hothorn, Kurt Hornik, Carolin Strobl, and Achim Zeileis. *party: A Laboratory for Recursive Partytioning*, 2021b. URL http://party.R-forge.R-project.org. R package version 1.3-9.

Torsten Hothorn, Henric Winell, Kurt Hornik, Mark A. van de Wiel, and Achim Zeileis. *coin: Conditional Inference Procedures in a Permutation Test Framework*, 2021c. URL http://coin.r-forge.r-project.org. R package version 1.4-2.

James P. Howard, II. *waterfall: Waterfall Charts*, 2016. URL https://CRAN.R-project.org/package=waterfall. R package version 1.0.2.

Hemant Ishwaran and Udaya B. Kogalur. *randomForestSRC: Fast Unified Random Forests for Survival, Regression, and Classification (RF-SRC)*, 2022. URL https://luminwin.github.io/randomForestSRC/https://github.com/kogalur/randomForestSRC/https://web.ccs.miami.edu/~hishwaran/. R package version 3.0.0.

Hemant Ishwaran, Udaya B. Kogalur, Eugene H. Blackstone, and Michael S. Lauer. Random survival forests. *Ann. Appl. Stat.*, 2(3):841–860, 09 2008. doi: 10.1214/08-AOAS169. URL https://doi.org/10.1214/08-AOAS169.

Gareth James, Daniella Witten, Trevor Hastie, and Robert Tibshirani. *An Introduction to Statistical Learning: With Applications in R*. Springer Texts in Statistics. Springer New York, 2nd edition, 2021. ISBN 9781071614174. URL https://www.statlearning.com/.

Silke Janitza and Roman Hornung. On the overestimation of random forest's out-of-bag error. *PLOS ONE*, 13(8):1–31, 08 2018. doi: 10.1371/journal.pone.0201904. URL https://doi.org/10.1371/journal.pone.0201904.

Silke Janitza, Ender Celik, and Anne-Laure Boulesteix. A computationally fast variable importance test for random forests for high-dimensional data. *Advances in Data Analysis and Classification*, 12(4):885—-915, 2018. ISSN 1862-5347. doi: 10.1007/s11634-016-0276-4. URL https://doi.org/10.1007/s11634-016-0276-4.

Dominik Janzing, Lenon Minorics, and Patrick Blöbaum. Feature relevance quantification in explainable ai: A causal problem, 2019. URL https://arxiv.org/abs/1910.13413.

Richard A. Johnson and Dean W. Wichern. *Applied Multivariate Statistical Analysis*. Applied Multivariate Statistical Analysis. Pearson Prentice Hall, 2007. ISBN 9780131877153.

Zachary Jones. *mmpf: Monte-Carlo Methods for Prediction Functions*, 2018. URL https://CRAN.R-project.org/package=mmpf. R package version 0.0.5.

Alexandros Karatzoglou, Alex Smola, and Kurt Hornik. *kernlab: Kernel-Based Machine Learning Lab*, 2019. URL https://CRAN.R-project.org/package=kernlab. R package version 0.9-29.

Gordon V. Kass. An exploratory technique for investigating large quantities of categorical data. *Journal of the Royal Statistical Society. Series C (Applied Statistics)*, 29(2):119–127, 1980.

Guolin Ke, Qi Meng, Thomas Finley, Taifeng Wang, Wei Chen, Weidong Ma, Qiwei Ye, and Tie-Yan Liu. Lightgbm: A highly efficient gradient boosting decision tree. In I. Guyon, U. V. Luxburg, S. Bengio, H. Wallach, R. Fergus, S. Vishwanathan, and R. Garnett, editors, *Advances in Neural Information Processing Systems*, volume 30. Curran Associates, Inc., 2017. URL https://proceedings.neurips.cc/paper/2017/file/6449f44a102fde848669bdd9eb6b76fa-Paper.pdf.

Hyunjoong Kim and Wei-Yin Loh. Classification trees with unbiased multiway splits. *Journal of the American Statistical Association*, 96(454):589–604, 2001. ISSN 01621459. URL https://doi.org/10.2307/2670299.

John P. Klein and Melvin L. Moeschberger. *Survival Analysis: Techniques for Censored and Truncated Data*. Statistics for Biology and Health. Springer New York, 2003. ISBN 9780387953991.

Brian Kriegler and Richard Berk. Small area estimation of the homeless in Los Angeles: An application of cost-sensitive stochastic gradient boosting. *The Annals of Applied Statistics*, 4(3):1234–1255, 2010. doi: 10.1214/10-AOAS328. URL https://doi.org/10.1214/10-AOAS328.

Robert Küffner, Neta Zach, Raquel Norel, Johann Hawe, David Schoenfeld, Liuxia Wang, Guang Li, Lilly Fang, Lester Mackey, Orla Hardiman, Merit Cudkowicz, Alexander Sherman, Gokhan Ertaylan, Moritz Grosse-Wentrup, Torsten Hothorn, Jules van Ligtenberg, Jakob H. Macke, Timm Meyer, Bernhard Schölkopf, Linh Tran, Rubio Vaughan, Gustavo Stolovitzky, and Melanie L. Leitner. Crowdsourced analysis of clinical trial data to predict amyotrophic lateral sclerosis progression. *Nature Biotechnology*, 33(1):51–57, 2015. doi: 10.1038/nbt.3051. URL https://doi.org/10.1038/nbt.3051.

Max Kuhn. *AmesHousing: The Ames Iowa Housing Data*, 2020. URL https://github.com/topepo/AmesHousing. R package version 0.0.4.

Max Kuhn. *modeldata: Data Sets Useful for Modeling Packages*, 2021. URL https://CRAN.R-project.org/package=modeldata. R package version 0.1.1.

Max Kuhn and Kjell Johnson. *Applied Predictive Modeling*. Springer-Verlag, 2013. URL https://books.google.com/books?id=xYRDAAAAQBAJ. ISBN 978-1-4614-6848-6.

Max Kuhn and Ross Quinlan. *C50: C5.0 Decision Trees and Rule-Based Models*, 2021. URL https://topepo.github.io/C5.0/. R package version 0.1.5.

Meelis Kull, Telmo M. Silva Filho, and Peter Flach. Beyond sigmoids: How to obtain well-calibrated probabilities from binary classifiers with beta calibration. *Electronic Journal of Statistics*, 11(2):5052–5080, 2017. doi: 10.1214/17-EJS1338SI. URL https://doi.org/10.1214/17-EJS1338SI.

Ludmila I. Kuncheva and Juan J. Rodríguez. An experimental study on rotation forest ensembles. In Michal Haindl, Josef Kittler, and Fabio Roli, editors, *Multiple Classifier Systems*, pages 459–468. Springer Berlin, Heidelberg, 2007. ISBN 978-3-540-72523-7.

Michael LeBlanc and John Crowley. Relative risk trees for censored survival data. *Biometrics*, 48(2):411–425, 1992.

Michael Leblanc and John Crowley. Survival trees by goodness of split. *Journal of the American Statistical Association*, 88(422):457–467, 1993. doi: 10.1080/01621459.1993.10476296. URL https://doi.org/10.1080/01621459.1993.10476296.

Erin LeDell, Navdeep Gill, Spencer Aiello, Anqi Fu, Arno Candel, Cliff Click, Tom Kraljevic, Tomas Nykodym, Patrick Aboyoun, Michal Kurka, and Michal Malohlava. *h2o: R Interface for the H2O Scalable Machine Learning Platform*, 2021. URL https://github.com/h2oai/h2o-3. R package version 3.34.0.3.

Eun-Kyung Lee. *PPtreeViz: Projection Pursuit Classification Tree Visualization*, 2019. URL https://CRAN.R-project.org/package=PPtreeViz. R package version 2.0.4.

Yoon Dong Lee, Dianne Cook, Ji won Park, and Eun-Kyung Lee. Pptree: Projection pursuit classification tree. *Electronic Journal of Statistics*, 7:

1369–1386, 2013. doi: 10.1214/13-EJS810. URL `https://doi.org/10.1214/13-EJS810`.

Jing Lei, Max G'Sell, Alessandro Rinaldo, Ryan J. Tibshirani, and Larry Wasserman. Distribution-free predictive inference for regression. *Journal of the American Statistical Association*, 113(523):1094–1111, 2018. doi: 10.1080/01621459.2017.1307116. URL `https://doi.org/10.1080/01621459.2017.1307116`.

Friedrich Leisch and Evgenia Dimitriadou. *mlbench: Machine Learning Benchmark Problems*, 2021. URL `https://CRAN.R-project.org/package=mlbench`. R package version 2.1-3.

Andy Liaw and Matthew Wiener. Classification and regression by randomforest. *R News*, 2(3):18–22, 2002. URL `https://CRAN.R-project.org/doc/Rnews/`.

Fei Tony Liu, Kai Ming Ting, and Zhi-Hua Zhou. Isolation forest. In *2008 Eighth IEEE International Conference on Data Mining*, pages 413–422, 2008. doi: 10.1109/ICDM.2008.17. URL `https://doi.org/10.1109/ICDM.2008.17`.

Wei-Yin Loh. Regression trees with unbiased variable selection and interaction detection. *Statistica Sinica*, 12(1):361–386, 2002. URL `https://doi.org/10.1214/09-AOAS260`.

Wei-Yin Loh. Regression trees with unbiased variable selection and interaction detection. *Annals of Applied Statistics*, 3(4):1710–1737, 2009.

Wei-Yin Loh. Classification and regression trees. *WIREs Data Mining and Knowledge Discovery*, 1(1):14–23, 2011. URL `https://doi.org/10.1002/widm.8`.

Wei-Yin Loh. *Variable Selection for Classification and Regression in Large p, Small n Problems*, volume 205, pages 135–159. 12 2012. ISBN 978-1-4614-1965-5. doi: 10.1007/978-1-4614-1966-2_10.

Wei-Yin Loh. Fifty years of classification and regression trees. *International Statistical Review / Revue Internationale de Statistique*, 82(3):329–348, 2014.

Wei-Yin Loh. User manual for guide ver. 34.0, 2020. URL `http://pages.stat.wisc.edu/~loh/treeprogs/guide/guideman.pdf`.

Wei-Yin Loh and Yu-Shan Shih. Split selection methods for classification trees. *Statistica Sinica*, 7(4):815–840, 1997.

Wei-Yin Loh and Nunta Vanichsetakul. Tree structured classification via generalized discriminant analysis. *Journal of the American Statistical Association*, 83:715–728, 1988.

Wei-Yin Loh and Wei Zheng. Regression trees for longitudinal and multire-sponse data. *The Annals of Applied Statistics*, 7(1):495–522, 2013. doi: 10.1214/12-AOAS596. URL https://doi.org/10.1214/12-AOAS596.

Wei-Yin Loh and Peigen Zhou. *The GUIDE Approach to Subgroup Identi-fication*, pages 147–165. Springer International Publishing, Cham, 2020. ISBN 978-3-030-40105-4. doi: 10.1007/978-3-030-40105-4_6. URL https://doi.org/10.1007/978-3-030-40105-4_6.

Wei-Yin Loh and Peigen Zhou. Variable importance scores. *Journal of Data Science*, 19(4):569–592, 2021. ISSN 1680-743X. doi: 10.6339/21-JDS1023. URL https://doi.org/10.6339/21-JDS1023.

Wei-Yin Loh, Xu He, and Michael Man. A regression tree approach to identi-fying subgroups with differential treatment effects. *Statistics in Medicine*, 34(11):1818–1833, 2015. URL https://doi.org/10.1002/sim.6454.

Wei-Yin Loh, John Eltinge, Moon Jung Cho, and Yuanzhi Li. Classification and regression trees and forests for incomplete data from sample surveys. *Statistica Sinica*, 29(1):431–453, 2019. ISSN 10170405, 19968507. doi: 10.5705/ss.202017.0225. URL https://doi.org/10.5705/ss.202017.0225.

Wei-Yin Loh, Qiong Zhang, Wenwen Zhang, and Peigen Zhou. Imissing data, imputation and regression trees. *Statistica Sinica*, 30:1697–1722, 2020.

Scott M. Lundberg and Su-In Lee. A unified approach to interpreting model predictions. In I. Guyon, U. V. Luxburg, S. Bengio, H. Wallach, R. Fer-gus, S. Vishwanathan, and R. Garnett, editors, *Advances in Neural Infor-mation Processing Systems 30*, pages 4765–4774. Curran Associates, Inc., 2017. URL http://papers.nips.cc/paper/7062-a-unified-approach-to-interpreting-model-predictions.pdf.

Scott M. Lundberg, Gabriel Erion, Hugh Chen, Alex DeGrave, Jordan M. Prutkin, Bala Nair, Ronit Katz, Jonathan Himmelfarb, Nisha Bansal, and Su-In Lee. From local explanations to global understanding with explainable ai for trees. *Nature Machine Intelligence*, 2(1):2522–5839, 2020.

Javier Luraschi, Kevin Kuo, Kevin Ushey, J. J. Allaire, Hossein Falaki, Lu Wang, Andy Zhang, Yitao Li, and The Apache Software Founda-tion. *sparklyr: R Interface to Apache Spark*, 2021. URL https://spark.rstudio.com/. R package version 1.7.3.

Szymon Maksymiuk, Alicja Gosiewska, and Przemyslaw Biecek. Landscape of r packages for explainable artificial intelligence, 2021. URL https://arxiv.org/abs/2009.13248.

James D. Malley, Jochen Kruppa, Abhijit Dasgupta, Karen Godlove Malley, and Andreas Ziegler. Probability machines: consistent probability estima-tion using nonparametric learning machines. *Methods of Information in*

Medicine, 51(1):74–81, 2012. URL https://doi.org/10.3414/ME00-01-0052.

Christopher D. Manning, Prabhakar Raghavan, and Hinrich Schütze. *Intro-duction to Information Retrieval*. Cambridge University Press, USA, 2008. ISBN 0521865719.

Norm Matloff. *The Art of R Programming: A Tour of Statistical Software Design*. No Starch Press, 2011. ISBN 9781593273842.

Norm Matloff. *regtools: Regression and Classification Tools*, 2019. URL https://github.com/matloff/regtools. R package version 1.1.0.

Norman Matloff. *Statistical Regression and Classification: From Linear Models to Machine Learning*. Chapman & Hall/CRC Texts in Statistical Science. CRC Press, 2017. ISBN 9781351645898.

David Mease, Abraham J. Wyner, and A. Buja. Boosted classification trees and class probability/quantile estimation. *Journal of Machine Learning Research*, 8:409–439, 2007.

Nicolai Meinshausen. Quantile regression forests. *Journal of Machine Learning Research*, 7(35):983–999, 2006. URL http://jmlr.org/papers/v7/meinshausen06a.html.

Xiangrui Meng, Joseph Bradley, Burak Yavuz, Evan Sparks, Shivaram Venkataraman, Davies Liu, Jeremy Freeman, D. B. Tsai, Manish Amde, Sean Owen, Doris Xin, Reynold Xin, Michael J. Franklin, Reza Zadeh, Matei Zaharia, and Ameet Talwalkar. Mllib: Machine learning in apache spark. *Journal of Machine Learning Research*, 17(34):1–7, 2016. URL http://jmlr.org/papers/v17/15-237.html.

Bjoern Menze and Nico Splitthoff. *obliqueRF: Oblique Random Forests from Recursive Linear Model Splits*, 2012. URL https://CRAN.R-project.org/package=obliqueRF. R package version 0.3.

Bjoern H. Menze, B. Michael Kelm, Daniel N. Splitthoff, Ullrich Koethe, and Fred A. Hamprecht. On oblique random forests. In *Machine Learning and Knowledge Discovery in Databases*, pages 453–469. Springer Berlin, Heidelberg, 2011. ISBN 978-3-642-23783-6.

Olaf Mersmann. *microbenchmark: Accurate Timing Functions*, 2021. URL https://github.com/joshuaulrich/microbenchmark/. R package version 1.4.9.

Robert Messenger and Lewis Mandell. A modal search technique for predictibe nominal scale multivariate analys. *Journal of the American Statistical Association*, 67(340):768–772, 1972.

Daniele Micci-Barreca. A preprocessing scheme for high-cardinality categorical attributes in classification and prediction problems. *SIGKDD Explor. Newsl.*, 3(1):27–32, July 2001. ISSN 1931-0145. doi: 10.1145/507533.507538. URL https://doi.org/10.1145/507533.507538.

Stephen Milborrow. *earth: Multivariate Adaptive Regression Splines*, 2021a. URL http://www.milbo.users.sonic.net/earth/. R package version 5.3.1.

Stephen Milborrow. *rpart.plot: Plot rpart Models: An Enhanced Version of plot.rpart*, 2021b. URL http://www.milbo.org/rpart-plot/index.html. R package version 3.1.0.

Matthew W. Mitchell. Bias of the random forest out-of-bag (oob) error for certain input parameters. *Open Journal of Statistics*, 1(3):205–211, 10 2011. doi: 10.4236/ojs.2011.13024. URL https://doi.org/10.4236/ojs.2011.13024.

Christoph Molnar. *Interpretable Machine Learning*. 2019. URL https://christophm.github.io/interpretable-ml-book/.

Christoph Molnar and Patrick Schratz. *iml: Interpretable Machine Learning*, 2020. URL https://CRAN.R-project.org/package=iml. R package version 0.10.1.

Christoph Molnar, Gunnar König, Julia Herbinger, Timo Freiesleben, Susanne Dandl, Christian A. Scholbeck, Giuseppe Casalicchio, Moritz Grosse-Wentrup, and Bernd Bischl. General pitfalls of model-agnostic interpretation methods for machine learning models, 2021.

James N. Morgan and John A. Sonquist. Problems in the analysis of survey data, and a proposal. *Journal of the American Statistical Association*, 58 (302):415–434, 1963.

Sérgio Moro, Paulo Cortez, and Paulo Rita. A data-driven approach to predict the success of bank telemarketing. *Decision Support Systems*, 62: 22–31, 2014. ISSN 0167-9236. doi: https://doi.org/10.1016/j.dss.2014.03.001. URL https://www.sciencedirect.com/science/article/pii/S016792361400061X.

Stefano Nembrini, Inke R König, and Marvin N. Wright. The revival of the gini importance? *Bioinformatics*, 34(21):3711–3718, 05 2018. ISSN 1367-4803. doi: 10.1093/bioinformatics/bty373. URL https://doi.org/10.1093/bioinformatics/bty373.

Anna Neufeld. *treevalues: Selective Inference for Regression Trees*, 2022. URL https://github.com/anna-neufeld/treevalues. R package version 0.1.0.

Anna C. Neufeld, Lucy L. Gao, and Daniela M. Witten. Tree-values: selective inference for regression trees, 2021.

Alexandru Niculescu-Mizil and Rich Caruana. Predicting good probabilities with supervised learning. In *Proceedings of the 22nd International Conference on Machine Learning*, ICML '05, pages 625–632, New York, NY, USA, 2005. Association for Computing Machinery. ISBN 1595931805. doi: 10.1145/1102351.1102430. https://doi.org/10.1145/1102351.1102430.

Deborah Ann Nolan and Duncan Temple Lang. *Data Science in R: a Case Studies Approach to Computational Reasoning and Problem Solving*. Chapman & Hall/CRC the R series. CRC Press, Boca Raton, 2015. ISBN 9781482234817.

Harsha Nori, Samuel Jenkins, Paul Koch, and Rich Caruana. Interpretml: A unified framework for machine learning interpretability. *arXiv preprint arXiv:1909.09223*, 2019.

Szilard Pafka. benchm-ml. `https://github.com/szilard/benchm-ml`, 2019. URL `https://github.com/szilard/benchm-ml`.

Szilard Pafka. Gradient boosting machines (gbm): From zero to hero (with r and python code). `https://docs.google.com/presentation/d/1WdQajKNeJR5gJs437XUuLksBJPm4rowdzH3i1vEWTHA/edit#slide=id.g58411bbf6a_0_15`, 2 2020. URL `https://docs.google.com/presentation/d/1WdQajKNeJR5gJs437XUuLksBJPm4rowdzH3i1vEWTHA/edit#slide=id.g58411bbf6a_0_15`.

Szilard Pafka. Gbm-perf. `https://github.com/szilard/GBM-perf`, 2021. URL `https://github.com/szilard/GBM-perf`.

Terence Parr and James D. Wilson. Technical report: Partial dependence through stratification, 2019. URL `https://arxiv.org/abs/1907.06698`.

Andrea Peters and Torsten Hothorn. *ipred: Improved Predictors*, 2021. URL `https://CRAN.R-project.org/package=ipred`. R package version 0.9-12.

Kivan Polimis, Ariel Rokem, and Bryna Hazelton. Confidence intervals for random forests in python. *Journal of Open Source Software*, 2(1), 2017.

Liudmila Prokhorenkova, Gleb Gusev, Aleksandr Vorobev, Anna Veronika Dorogush, and Andrey Gulin. Catboost: unbiased boosting with categorical features, 2017. URL `https://arxiv.org/abs/1706.09516`.

Randall Pruim, Daniel T. Kaplan, and Nicholas J. Horton. *mosaic: Project MOSAIC Statistics and Mathematics Teaching Utilities*, 2021. URL `https://CRAN.R-project.org/package=mosaic`. R package version 1.8.3.

Ross J. Quinlan. *C4.5: Programs for Machine Learning*. Morgan Kaufmann series in machine learning. Elsevier Science, 1993. ISBN 9781558602380.

Revolution Analytics and Steve Weston. *foreach: Provides Foreach Loop-ing Construct*, 2020. URL `https://github.com/RevolutionAnalytics/foreach`. R package version 1.5.1.

Greg Ridgeway. The state of boosting. *Computing Science and Statistics 31:172-181s*, 31:172–181, 1999.

Robert A. Rigby and Mikis D. Stasinopoulos. Generalized additive models for location, scale and shape. *Journal of the Royal Statistical Society: Series C (Applied Statistics)*, 54(3):507–554, 2005. doi: https://doi.org/10.1111/j.1467-9876.2005.00510.x. URL `https://rss.onlinelibrary.wiley.com/doi/abs/10.1111/j.1467-9876.2005.00510.x`.

Brian Ripley. *tree: Classification and Regression Trees*, 2021. URL `https://CRAN.R-project.org/package=tree`. R package version 1.0-41.

Brian Ripley. *MASS: Support Functions and Datasets for Venables and Rip-ley's MASS*, 2022. URL `http://www.stats.ox.ac.uk/pub/MASS4/`. R package version 7.3-55.

Brian D. Ripley. *Pattern Recognition and Neural Networks*. Cambridge University Press, 1996. ISBN 9780521717700.

Juan J. Rodríguez, Ludmila I. Kuncheva, and Carlos J. Alonso. Rotation forest: A new classifier ensemble method. *IEEE Transactions on Pattern Analysis and Machine Intelligence*, 28(10):1619–1630, 2006. doi: 10.1109/TPAMI.2006.211.

Kaspar Rufibach. Use of brier score to assess binary predictions. *Journal of Clinical Epidemiology*, 63(8):938–939, Aug 2010.

Marco Sandri and Paola Zuccolotto. A bias correction algorithm for the gini variable importance measure in classification trees. *Journal of Computational and Graphical Statistics*, 17(3):611–628, 2008. doi: 10.1198/106186008X344522. URL `https://doi.org/10.1198/106186008X344522`.

Deepayan Sarkar. *lattice: Trellis Graphics for R*, 2021. URL `http://lattice.r-forge.r-project.org/`. R package version 0.20-45.

Mark Robert Segal. Regression trees for censored data. *Biometrics*, 44(1):35–47, 1988.

Stephen J. Senn. Dichotomania: An obsessive compulsive disorder that is badly affecting the quality of analysis of pharmaceutical trials. In *Proceedings of the International Statistical Institute, 55th Session*, Sydney, 2005.

Juliet P. Shaffer. Multiple hypothesis testing. *Annual Review of Psychology*, 46(1):561–584, 1995. URL `https://doi.org/10.1146/annurev.ps.46.020195.003021`.

Lloyd S. Shapley. *17. A Value for n-Person Games*, pages 307–318. Princeton University Press, 2016. URL https://doi.org/10.1515/9781400881970-018.

Haijian Shi. *Best-first Decision Tree Learning*. PhD thesis, Hamilton, New Zealand, 2007. URL https://hdl.handle.net/10289/2317. Masters.

Tao Shi, David Seligson, Arie Belldegrun, Aarno Palotie, and Steve Horvath. Tumor classification by tissue microarray profiling: random forest clustering applied to renal cell carcinoma. *Modern Pathologyc*, 18:547–57, 05 2005. doi: 10.1038/modpathol.3800322. URL https://doi.org/10.1038/modpathol.3800322.

Yu Shi, Guolin Ke, Damien Soukhavong, James Lamb, Qi Meng, Thomas Finley, Taifeng Wang, Wei Chen, Weidong Ma, Qiwei Ye, Tie-Yan Liu, and Nikita Titov. *lightgbm: Light Gradient Boosting Machine*, 2022. URL https://github.com/Microsoft/LightGBM. R package version 3.3.2.

Julia Silge, Fanny Chow, Max Kuhn, and Hadley Wickham. *rsample: General Resampling Infrastructure*, 2021. URL https://CRAN.R-project.org/package=rsample. R package version 0.1.1.

Nora Sleumer. *Hyperplane arrangements: construction visualization and applications*. PhD thesis, Swiss Federal Institute of Technology, 1969. PhD dissertation.

Helmut Strasser and Christian Weber. On the asymptotic theory of permutation statistics. *Mathematical Methods of Statistics*, 2(27), 1999.

Carolin Strobl, Anne-Laure Boulesteix, Achim Zeileis, and Torsten Hothorn. Bias in random forest variable importance measures: Illustrations, sources and a solution. *BMC Bioinformatics*, 8(25), 2007a. doi: 10.1186/1471-2105-8-25.

Carolin Strobl, Anne-Laure Boulesteix, Achim Zeileis, and Torsten Hothorn. Bias in random forest variable importance measures: Illustrations, sources and a solution. *BMC Bioinformatics*, 8(25), 2007b. URL https://doi.org/10.1186/1471-2105-8-25.

Carolin Strobl, Anne-Laure Boulesteix, Thomas Kneib, Thomas Augustin, and Achim Zeileis. Conditional variable importance for random forests. *BMC Bioinformatics*, 9(307), 2008a. doi: 10.1186/1471-2105-9-307.

Carolin Strobl, Anne-Laure Boulesteix, Thomas Kneib, Thomas Augustin, and Achim Zeileis. Conditional variable importance for random forests. *BMC Bioinformatics*, 9(307), 2008b. URL https://doi.org/10.1186/1471-2105-9-307.

The Pandas Development Team. pandas-dev/pandas: Pandas, February 2020. URL https://doi.org/10.5281/zenodo.3509134.

Terry Therneau and Beth Atkinson. *rpart: Recursive Partitioning and Regression Trees*, 2019. URL https://CRAN.R-project.org/package=rpart. R package version 4.1-15.

Terry M. Therneau. *survival: Survival Analysis*, 2021. URL https://github.com/therneau/survival. R package version 3.2-13.

Julie Tibshirani, Susan Athey, Erik Sverdrup, and Stefan Wager. *grf: Generalized Random Forests*, 2021. URL https://github.com/grf-labs/grf. R package version 2.0.2.

Robert Tibshirani. Regression shrinkage and selection via the lasso. *Journal of the Royal Statistical Society. Series B (Methodological)*, 58(1):267–288, 1996. ISSN 00359246. URL http://www.jstor.org/stable/2346178.

Stef van Buuren. *Flexible Imputation of Missing Data*. Chapman & Hall/CRC Interdisciplinary Statistics. CRC Press, Taylor & Francis Group, 2018. ISBN 9781138588318. URL https://books.google.com/books?id=bLmItgEACAAJ.

Stef van Buuren and Karin Groothuis-Oudshoorn. *mice: Multivariate Imputation by Chained Equations*, 2021. URL https://CRAN.R-project.org/package=mice. R package version 3.14.0.

Mark J. van der Laan. Statistical inference for variable importance. *The International Journal of Biostatistics*, 2(1), 2006. URL https://doi.org/10.2202/1557-4679.1008.

Shivaram Venkataraman, Zongheng Yang, Davies Liu, Eric Liang, Hossein Falaki, Xiangrui Meng, Reynold Xin, Ali Ghodsi, Michael Franklin, Ion Stoica, and Matei Zaharia. Sparkr: Scaling r programs with spark. In *Proceedings of the 2016 International Conference on Management of Data*, SIGMOD '16, pages 1099–1104, New York, NY, USA, 2016. Association for Computing Machinery. ISBN 9781450335317. doi: 10.1145/2882903.2903740. URL https://doi.org/10.1145/2882903.2903740.

James Verbus. Detecting and preventing abuse on linkedin using isolation forests, Aug. 2019. URL https://engineering.linkedin.com/blog/2019/isolation-forest.

Rashmi Korlakai Vinayak and Ran Gilad-Bachrach. DART: Dropouts meet Multiple Additive Regression Trees. In Guy Lebanon and S. V. N. Vishwanathan, editors, *Proceedings of the Eighteenth International Conference on Artificial Intelligence and Statistics*, volume 38 of *Proceedings of Machine Learning Research*, pages 489–497, San Diego, California, USA, 09–12 May 2015. PMLR. URL http://proceedings.mlr.press/v38/korlakaivinayak15.html.

Erik Štrumbelj and Igor Kononenko. Explaining prediction models and individual predictions with feature contributions. *Knowledge and Information Systems*, 31(3):647–665, 2014. URL https://doi.org/10.1007/s10115-013-0679-x.

Stefan Wager, Trevor Hastie, and Bradley Efron. Confidence intervals for random forests: The jackknife and the infinitesimal jackknife. *Journal of Machine Learning Research*, 15(48):1625–1651, 2014. URL http://jmlr.org/papers/v15/wager14a.html.

Ian R. White, Patrick Royston, and Angela M. Wood. Multiple imputation using chained equations: Issues and guidance for practice. *Statistics in Medicine*, 30(4):377–399, 2011. doi: 10.1002/sim.4067. URL https://doi.org/10.1002/sim.4067.

Hadley Wickham. *Advanced R, Second Edition*. Chapman & Hall/CRC The R Series. CRC Press, 2019. ISBN 9781351201308. URL https://adv-r.hadley.nz/.

Hadley Wickham and Jennifer Bryan. *readxl: Read Excel Files*, 2019. URL https://CRAN.R-project.org/package=readxl. R package version 1.3.1.

Hadley Wickham, Winston Chang, Lionel Henry, Thomas Lin Pedersen, Kohske Takahashi, Claus Wilke, Kara Woo, Hiroaki Yutani, and Dewey Dunnington. *ggplot2: Create Elegant Data Visualisations Using the Grammar of Graphics*, 2021a. URL https://CRAN.R-project.org/package=ggplot2. R package version 3.3.5.

Hadley Wickham, Romain François, Lionel Henry, and Kirill Müller. *dplyr: A Grammar of Data Manipulation*, 2021b. URL https://CRAN.R-project.org/package=dplyr. R package version 1.0.7.

Edwin B. Wilson and Margaret M. Hilferty. The distribution of chi-square. *Proceedings of the National Academy of Sciences of the United States of America*, 17(12):684–688, 1931.

Marvin N. Wright, Stefan Wager, and Philipp Probst. *ranger: A Fast Implementation of Random Forests*, 2021. URL https://github.com/imbs-hl/ranger. R package version 0.13.1.

Paul S. Wright. Adjusted p-values for simultaneous inference. *Biometrics*, 48(4):1005–1013, 1992. doi: 10.2307/2532694. URL https://doi.org/10.2307/2532694.

Yihui Xie. *knitr: A General-Purpose Package for Dynamic Report Generation in R*, 2021. URL https://yihui.org/knitr/. R package version 1.36.

Ruo Xu, Dan Nettleton, and Daniel J. Nordman. Case-specific random forests. *Journal of Computational and Graphical Statistics*, 25(1):49–65,

2016. doi: 10.1080/10618600.2014.983641. URL https://doi.org/10.
1080/10618600.2014.983641.

I-Cheng Yeh and Che hui Lien. The comparisons of data mining techniques
for the predictive accuracy of probability of default of credit card clients.
Expert Systems with Applications, 36(2, Part 1):2473–2480, 2009. doi: https:
//doi.org/10.1016/j.eswa.2007.12.020. URL https://doi.org/10.1016/
j.eswa.2007.12.020.

Achim Zeileis, Friedrich Leisch, Kurt Hornik, and Christian Kleiber. *strucchange: Testing, Monitoring, and Dating Structural Changes*, 2019. URL
https://CRAN.R-project.org/package=strucchange. R package version
1.5-2.

Haozhe Zhang, Joshua Zimmerman, Dan Nettleton, and Daniel J. Nordman.
Random forest prediction intervals. *The American Statistician*, 74(4):392–
406, 2020. doi: 10.1080/00031305.2019.1585288. URL https://doi.org/
10.1080/00031305.2019.1585288.

Heping Zhang and Burton H. Singer. *Recursive Partitioning and Applications*.
Springer New York, New York, NY, 2010. ISBN 978-1-4419-6824-1. URL
https://doi.org/10.1007/978-1-4419-6824-1_3.

Huan Zhang, Si Si, and Cho-Jui Hsieh. Gpu-acceleration for large-scale tree
boosting, 2017. URL https://arxiv.org/abs/1706.08359.

Index